Integrating Third-Party Logistics with SAP S/4HANA®

SAP PRESS is a joint initiative of SAP and Rheinwerk Publishing. The know-how offered by SAP specialists combined with the expertise of Rheinwerk Publishing offers the reader expert books in the field. SAP PRESS features first-hand information and expert advice, and provides useful skills for professional decision-making.

SAP PRESS offers a variety of books on technical and business-related topics for the SAP user. For further information, please visit our website: *www.sap-press.com*.

Gautam Bhattacharya, Mehfuze Ali Molla

Integrating Third-Party Logistics with SAP S/4HANA®

Editor Rachel Gibson
Acquisitions Editor Emily Nicholls
Copyeditor Julie McNamee
Cover Design Graham Geary
Photo Credit Shutterstock.com: 32214169/© Worldpics
Layout Design Vera Brauner
Production Kyrsten Coleman
Typesetting SatzPro, Germany
Printed and bound in Canada, on paper from sustainable sources

ISBN 978-1-4932-2446-3

1st edition 2024

Library of Congress Cataloging-in-Publication Control Data
Names: Bhattacharya, Gautam, author. | Molla, Mehfuze Ali, author.
Title: Integrating third-party logistics with SAP S/4HANA / by Gautam Bhattacharya and Mehfuze Ali Molla.
Description: Boston, MA : Rheinwerk Publishing, [2024] | Includes index.
Identifiers: LCCN 2023047199 | ISBN 9781493224463 (hardcover) | ISBN 9781493224470 (ebook)
Subjects: LCSH: Business logistics. | SAP HANA (Electronic resource)
Classification: LCC HD38.5 .B4746 2024 | DDC 658.5--dc23/eng/20231027
LC record available at https://lccn.loc.gov/2023047199

Contents at a Glance

Contents

Preface

The main consideration in writing this book was to present a comprehensive view of the third-party logistics (3PL) integration projects, starting from evaluation and selection of logistics service providers, design approach, and SAP functional and technical settings. The idea of writing on this topic came from our experience and struggle while working on similar projects. There is there is no one-size-fits-all solution to 3PL design, and there is always something unique about each of these implementations. The topics covered in this book are essential from the perspective of business and technical requirements for such integration, however, the integration design may vary from project to project.

We also realized that working on such a topic requires knowledge of multiple modules along with basic know-how of SAP technical areas. Project team members embarking on an integration project may also be new to or unfamiliar with 3PL. Keeping these points in mind, our effort has been to gather all relevant topics at one place, sequencing them appropriately and presenting them in a manner that covers all relevant touchpoints and modules.

3PL integration requires the functional and technical teams to work in tandem to understand client requirements and work on design to arrive at the best solution for specific client needs and collaborate with cross-functional teams to implement that solution.

Objective of This Book

This book is an integration guide for connecting SAP S/4HANA with third-party systems for logistics operation, inventory and warehouse management, and transportation. The integrations covered here include delivery and shipment logistics processes, handling and traceability of shipment packages, inventory and warehouse management, and transportation. As the 3PL-relevant processes are embedded within individual processes within logistics modules, we explain not only the process setup itself—including relevant organization entities and master data—but all its touchpoints with cross-functional areas as well as technical aspects. This gives logistics consultants at and technical integration specialists a step-by-step approach to design elements, as also configuration and technical instructions for connecting external 3PL systems to SAP S/4HANA.

The majority of our descriptions and screenshots refer to the SAP S/4HANA 2021/2022 release, however, some of the screenshots and discussions also refer to the earlier version, 1909.

Target Audience

The book is aimed at all SAP logistics consultants, SAP integration specialists, and SAP project managers who are interested to know how SAP S/4HANA supports 3PL integration. You will learn about nuances of design and implementation as well as various configuration and technical touchpoints for integration with third-party logistics service providers. It is assumed that you are fairly conversant with techno-functional topics and are familiar with one or more of SAP's logistics submodules.

Structure of the Book

The is book is structured in such a way so that you don't have to read each chapter in order. It is recommended, of course, to start with this preface, to understand the construct of this book. As this book focuses on integration, it is essential for you to have some basic knowledge of SAP technical areas, as the topics in this book are not entirely functional and cover many technical set-ups and touchpoints.

Chapter 1 and Chapter 2 review basic concepts about 3PL scenarios, business process coverage, and some of the essential considerations for deciding on 3PL integration. Chapter 3 covers various SAP master data elements relevant for 3PL integration and their setup.

Topics covered in Chapter 4, Chapter 5, and Chapter 6 are all logistics- and logistics execution-specific business process areas impacted by 3PL integration. These chapters cover the functional configuration setups needed for compliance with 3PL integration.

Chapter 7 and Chapter 8 cover considerations for when an organization runs its own warehousing and transportation tools or outsources them to logistics service providers. These chapters do not cover full configuration of embedded extended warehouse management (EWM) or SAP Transportation Management (SAP TM) in SAP S/4HANA, rather, they touch upon the relevant integration points for 3PL setup. It is recommended that these topics are read as a group or by readers with cross-functional knowledge of logistics modules.

Chapter 9 and Chapter 10 cover technical aspects and require you to have some technical knowledge of SAP as well as exposure to new-dimensional features like cloud-based solutions and application programming interfaces (APIs). It also covers electronic data interchange (EDI) and its features. This chapter can be referenced for a variety of 3PL setups using a variety of technologies.

Chapter 11 covers Business Rule Framework plus (BRFplus), which comes built into SAP S/4HANA. Chapter 12 provides a glimpse of SAP S/4HANA's reporting features and shows how you can monitor messages to and from 3PL providers.

The following list provides a brief summary of each chapter:

- **Chapter 1**
  This chapter describes how partnerships between supply chain stakeholders and operators can deliver value to end users. It presents an overview of the 3PL model, its integration to supply chain practices, and what makes 3PL an accelerator in end-to-end digital supply chains.
- **Chapter 2**
  This chapter elaborates on the steps involved in a typical 3PL implementation, from boardroom consensus to practice. Before embarking on a 3PL integration journey, management should carefully consider the business needs, determine key performance indicators, and weigh any possible limitations.
- **Chapter 3**
  Organization elements and master data in SAP S/4HANA play an important part in logistics-relevant transactions. This chapter contains a basic introduction to SAP S/4HANA's organizational elements and key master data and describes how they influence supply chain transactions.
- **Chapter 4**
  Batch management functionality has extensive usage in industries, mainly for end-to-end traceability of material from sourcing to manufacturing, storage, sales, and after-sales services. Tracking a material with a unique batch number involves several functions in SAP S/4HANA; this chapter elaborates on all relevant settings that are needed for inbound and outbound logistics. It also explains configuration settings for handling unit management is SAP. A handling unit is the physical arrangement of packing material (e.g., pallet, carton, shrink-wrap, container, or truck) items to be transported, stored, or used. This chapter makes it easier to understand the overall concept of packaging of finished goods prior to their shipment to the end customer.
- **Chapter 5**
  This is one of the main chapters that covers key configuration elements required for the integration of inbound and outbound logistics with 3PL providers. The chapter focuses on inbound and outbound SAP deliveries and all relevant settings controlling the types of deliveries, route determination, criterion for picking, and generation of output messages for transmission to external entities.
- **Chapter 6**
  Inventory management plays a crucial part in 3PL integration, as it necessitates that inventory at the logistics service provider and at the customer are always in sync. This chapter describes goods-movement specific settings in SAP S/4HANA that form the basis of such integration. It also covers output settings for communicating any specific movement across systems.

- **Chapter 7**
  Third-party logistics service providers let you integrate their warehouse management module with your in-house enterprise resource planning (ERP) software. This enables customers to virtually oversee stock availability, inventory movement, shipment status, and document verification processes. Customers then save on the costs required for holding their stock in a certain location, as well as operations costs incurred in storing materials in specific conditions and other overhead costs. SAP Warehouse management and SAP Extended Warehouse Management can be implemented to map in-house and 3PL operations in the SAP S/4HANA system, as we cover in this chapter.
- **Chapter 8**
  This chapter provides an overview of how to use SAP TM, which is embedded in SAP S/4HANA, in transportation planning, freight management, and charge settlement with respect to 3PL integration. However, the idea shared here is not to elaborate on entire SAP TM setup; rather, we focus on the integration points that are critical to such scenarios.
- **Chapter 9**
  This chapter focuses on the widely used EDI IDoc method that allows you to seamlessly integrate with any SAP or non-SAP systems. EDI is a system or method for exchanging business documents with trading partners and uses a highly secure, machine-readable, standardized format.
- **Chapter 10**
  This chapter explains the direct connectivity and middleware integration features available in SAP S/4HANA and SAP S/4HANA Cloud, including recent techniques like Cloud Integration that seamlessly integrate with any SAP or non-SAP systems. This provides an option for 3PL providers to operate some of the functions themselves on behalf of customers.
- **Chapter 11**
  Business Rule Framework plus (BRFplus) is a comprehensive API and user interface for modelling flexible business rules without the need for any coding. These rules can then be integrated easily and quickly with multiple applications in SAP. SAP S/4HANA has provided new features for doing output management using BRFplus. This chapter elaborates on the features of BRFplus, its use in SAP S/4HANA output management, and how BRFplus can be intuitively deployed for 3PL integration.
- **Chapter 12**
  This chapter is dedicated to the information systems and monitoring options available in SAP S/4HANA. It covers some of the advanced features and business intelligence (BI) reporting options that can be utilized to closely monitor the communication flow across systems. Additionally, it shows how you can build custom inventory-specific reports.

Acknowledgments

We would like to express our special thanks to Emily Nicholls, Rachel Gibson, and others at Rheinwerk Publishing. for their expert support and guidance in completing this book project.

Our special thanks to SAP SE for the documents and technical information provided in this book.

We are thankful to our seniors and colleagues at our workplace for providing us with ample time and opportunity to dedicate to writing this book and for their helpful suggestions and tips on SAP topics.

Last but not least, it's really heartwarming to know that we have amazing families that are always looking to support such endeavors, even if it means sacrificing many weekends. We appreciate all you do for us.

Gautam Bhattacharya
Mehfuze Ali Molla

ACKNOWLEDGMENTS

Chapter 1
Introduction to Third-Party Logistics

With the ever-increasing demand for specialized warehouse management and logistics solutions, we've witnessed a rapid growth in third-party logistics used across industries. The idea is straightforward: concentrate your efforts on developing your strengths in product development, sales, and marketing while allowing a specialized business partner to handle your warehousing, inventory control, and logistics needs. Thus, a fully or partly outsourced warehouse and logistics environment for your company is often achieved by integration with third-party logistics service providers that accept stock, ship, and invoice your inventory goods on your behalf.

Customer satisfaction is the key to any company's success, no matter how big or little, and fostering customer growth requires an effective logistics and supply chain infrastructure. While a reliable delivery and fulfillment process may be at the heart of many corporate operations, maintaining the supply chain can frequently become challenging and expensive, harming the company's reputation. This is where third-party logistics services (sometimes shortened to 3PL) come into play, as businesses contract with a third-party logistics service provider to handle their fulfillment and logistical needs.

Third-party logistics service providers provide specialized warehousing, logistics, and transport solutions catering to the vital link between manufacturing operations and end customers. In addition, they also provide value-added services such as packaging, bagging or kitting and shrink-wrapping, and so on. Large third-party logistics organizations with global operations have their own country-specific setups that may provide specialized knowledge of specific country legal and custom requirements, local know-how regarding routes, any special handling or tracking requirements, and last-mile connectivity requirements.

In this chapter, we'll look at how third-party logistics integration plays a major part in streamlining the supply chain processes and acts as an integrator to many interconnected functions cutting across multiple process-specific areas within the logistics value chain. There also will be some discussion on value drivers for third-party logistics integration, its attributes, and various pros and cons that make it an apt business case for manufacturers and distributors.

1.1 Supply Chain Integration

A third-party logistics service provider oversees all or a portion of a client's logistical needs, which may include delivery and inventory management, transportation, order fulfillment, warehousing, or a combination of these and other tasks. In addition, third-party logistics service providers bring in their expertise in the following areas to improve the overall efficiency of supply chain management:

- Strategic capacity management
- Network analysis
- Transportation mode and load optimization
- Cost-containment initiatives
- Vendor compliance
- Business intelligence
- Sharing of best practices
- Intellectual property rights
- Understanding the risk profile of business

The following sections describe the various functional and operational leverages that a typical third-party logistics integration may provide, key operational and financial benefits of such integration, and process coverage from first-party logistics to fifth-party logistics integration.

1.1.1 What Is Third-Party Logistics Integration?

Third-party refers to outsourcing logistics and order fulfillment operations. Some of the activities carried out by the third-party logistics service provider include the following:

- **Inventory management**
  These providers ensure safe storage of the goods and resources to store temperature-controlled goods, high-value products, perishable items, and other types of products.
- **Warehousing**
  Upon receipt of inventory, items are stored in warehouses or fulfillment centers. Each SKU has a separate dedicated storage location, either on a shelf, in a bin, or on a pallet. It's important to consider the storage space of the third-party logistics service provider as it should be more than enough for your current inventory, as well as scalable as your product line and order volume grow.

- **Transportation/freight forwarding**
  A third-party logistics service provider may have its own fleet of carriers or may act as a freight forwarder. Planning and arranging for transportation of goods from one place to another falls under this service. A third-party logistics service provider may choose one of these distribution methods: cross-docking, FCL (full container load), LCL (less than container load), TL (truckload), and intermodal transport. It also covers fleet optimization for the complete utilization of every vehicle's capacity. Execution and real-time tracking of goods also come under transportation services.
- **Order fulfillment**
  This includes order management, fulfillment, and distribution. Example of services offered by third-party logistics service providers are receiving orders, updating inventory, picking and packaging products, and distributing and shipping goods to customers' location.
- **Customs clearance/declaration**
  Logistics service providers may act as authorized clearing and forwarding agents to ensure that all necessary government declarations and security requirements are accurately completed for your freight shipments. They possess a thorough understanding of customs regulations, laws, tariff schedules, and other local regulatory requirements.
- **Material procurement**
  The provider procures products or goods from the vendor or the manufacturer and maintains inventory on behalf of the customer. If your third-party logistics service provider offers these two services, they can identify the stock shortages and raise procurement orders on your behalf. In addition, they facilitate the transportation of goods from the supplier or manufacturer to the third-party logistics service provider's warehouse.
- **Value-added services**
  Some examples of value-added services include the following:
  - **Packaging:** Packing the goods into cartons or containers per specific requirements of clients.
  - **Assembly:** Assembling parts. Sometimes products are shipped as unassembled parts to make the transportation more convenient and cheaper.
  - **Bagging or kitting:** Combining a number of separate but similar items and putting them into a single easy-to-pick package and easy-to-track SKU.
  - **Blister pack/shrink-wrapping:** Wrapping multiple or individual products and of varying types and qualities. This is often used for fragile, perishable, or high-value items.
  - **Labeling:** Perform specialized labeling or relabeling to meet regulatory or legal requirements, as well as country-specific standards.

- **Reverse logistics:** Carrying out minor repackaging or reworking to make a damaged product reusable, or determining if the damaged item needs to be repaired by the manufacturer and, if so, arranging for picking products from the customer's place and delivering them back to the warehouse or manufacturer.
- **Disposal or product destruction:** Getting rid of damaged or obsolete products. Third-party logistics service providers can add value by breaking down those products, recycling them, or disposing of them safely and cost-effectively.

Most commonly, a third-party logistics service provider follows a systematic approach for order fulfillment and customer satisfaction. They usually take care of operations such as receiving orders, warehousing, picking and packing, shipping, delivery, and returns.

1.1.2 Key Benefits and Features

Some of the key benefits and features of third-party logistics include the following:

- **Investment in your core expertise**
  By allowing third-party logistics/logistics service providers, who are specialists in logistics, to manage your shipping and fulfillment, you'll be able to invest more valuable time and effort in core functional areas.
- **Facilitating business expansion**
  Whether it's the addition of more products in their portfolio or expanding to newer geographies, businesses need not worry about additional capital expenses such as setting up warehouses or distribution centers. Third-party logistics with their existing infrastructure and setup can provide immediate help to store and move products.
- **Scalability**
  Seasonal fluctuations in demands, peaks and troughs in product demand can be easily handled without any additional effort. With third-party logistics, businesses can scale logistic operations up or down as and when needed.
- **Save time and cost**
  When lead times are shortened, inventory turnover rates also quicken. This results in freeing up time as well as a reduction in inventory carrying costs.
- **Mitigation of unforeseen risks**
  Unexpected events frequently cause delays in logistics. However, if you've contracted with a third-party logistics service provider to handle your logistics, it's their responsibility to find alternatives and solutions. Additionally, third-party logistics services are covered by insurance in case your products are damaged or lost.
- **Efficient and streamlined movement of goods**
  Third-party logistics service providers have vast experience in storing, moving, and delivering products. They have knowledge of government regulations, country-

specific local requirements, and tie-ups with other delivery partners for the last mile connectivity. Moreover, they have the technology to support such operations. All of these helps avoid inaccuracies, inefficiencies, and other critical mistakes that can lead to loss of money as well as customer's trust.

- **Efficient reverse logistics**
  Reverse logistics, where third-party logistics offer customer support and a smooth return process, is another area of advantage to businesses. This not only improves customer service but also enables you to return defective or damaged products back into the inventory for rework or disposal.

1.1.3 First-Party Logistics to Fifth-Party Logistics

It's imperative to understand the terms used for logistics services and operations, in context of third-party logistics/logistics service providers. The following list describes various scenarios involving the use of logistics service providers:

- **First-party logistics**
  First-party logistics (sometimes shortened to 1PL) is when a person or business stores their own goods and transports them to the final customer. Most small production enterprises with local distribution use first-party logistics. Thus, the logistical services provided by these businesses are entirely self-sufficient.
- **Second-party logistics**
  Second-party logistics companies (sometimes shortened to 2PL) handle the transportation for a supply chain. These businesses rent or lease their transportation equipment, vehicles, or services to other businesses in the supply chain. Second-party logistics firms are businesses that provide transportation for only a section of a supply chain as opposed to the full supply chain. Second-party logistics usually is an asset-based carrier that takes care of transportation. Examples of second-party logistics are shipping lines, airlines, and hauling companies that operate vehicles.
- **Third-party logistics**
  Third-party logistics (sometimes shortened to 3PL) refers to the outsourcing of logistics services to a third party. Third-party logistics, also known as logistics service provider (LSP), manages all (or part) of a company's logistics operations.

  Third-party logistics companies are businesses that can handle a variety of logistics-related tasks, including forecasting, inventory management, order management, and storage and warehouse management. Although they provide a range of services, they don't usually expect clients to use them all. Customers can simply choose the services they require for their company.
- **Fourth-party logistics**
  Fourth-party logistics (sometimes shortened to 4PL) companies can manage, plan, and strategize your logistics, business/supply chain, projects, and supplier and

buyer networks, in addition to executing third-party logistics services. Even consultancy services are provided by some fourth-party logistics suppliers. Its focus is on providing strategic insight and management.

- **Fifth-party logistics**

  In a fifth-party logistics (sometimes shortened to 5PL) system, companies (clients) outsource every aspect of their supply chains to a logistics service provider. The fifth-party logistics company is involved in the organization, planning, and execution of the client's logistics plans.

  Fifth-party logistics controls all the supply chain networks while simultaneously using the services of third-party logistics and fourth-party logistics via the various supply chain activities. Additionally, fifth-party logistics businesses set prices and, occasionally, renegotiate the previously agreed-upon price.

Figure 1.1 provides a comparative and comprehensive view of activities that each of the preceding scenarios entails.

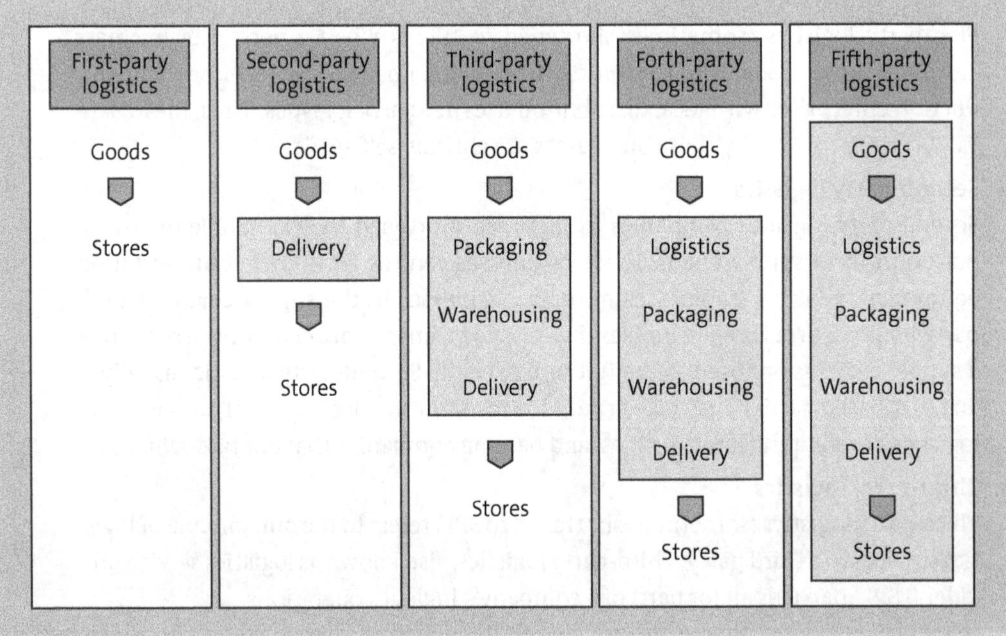

Figure 1.1 Comparison of Logistics Services from First-Party Logistics to Fifth-Party Logistics

The successful integration of IT and computer systems is essential to achieving this because it ensures real-time visibility and management of the entire supply chain, regardless of how many different suppliers are involved.

Fifth-party logistics is a more futuristic option, so let's look at some of the differences between third-party logistics and fourth-party logistics services in Table 1.1.

Third-Party Logistics Service Provider	Fourth-Party Logistics Provider
▪ Provides physical logistics support such as warehousing and freight capabilities (optional) ▪ Supports international logistics services ▪ Flexible collaboration with short- and medium-term partnership ▪ Acts as service provider	▪ Uses third-party logistics for physical logistics execution ▪ Strives for optimum use of the supply chain and maximize cost reduction ▪ Long-term partnership ▪ Treated as strategic partner

Table 1.1 Third-Party Logistics Services versus Fourth-Party Logistics Services

1.2 Business Case for Third-Party Logistics Integration

If you operate with a third-party logistics supplier, you're aware of the significance of assessing their efficacy and worth. However, how can you gauge their value offer and market advantage? This section will cover some pointers to assist you in evaluating the benefits.

1.2.1 Value Drivers

Following are the major reasons that encourage companies to outsource "logistics" businesses to third-party logistics:

- Concentrate on core activities and processes.
- Enhance customer service levels.
- Create a more efficient and integrated supply chain with a global reach.
- Form a responsive supply chain to reduce risk, uncertainty, and fluctuation.
- Cater to a global customer base.
- Avoid excessive capital expenses and reduce overheads.
- Enhance productivity with better utilization of available resources.
- Improve awareness, market knowledge, and data access.
- Create competitive advantage.

The key priorities for outsourcing are depicted in Figure 1.2.

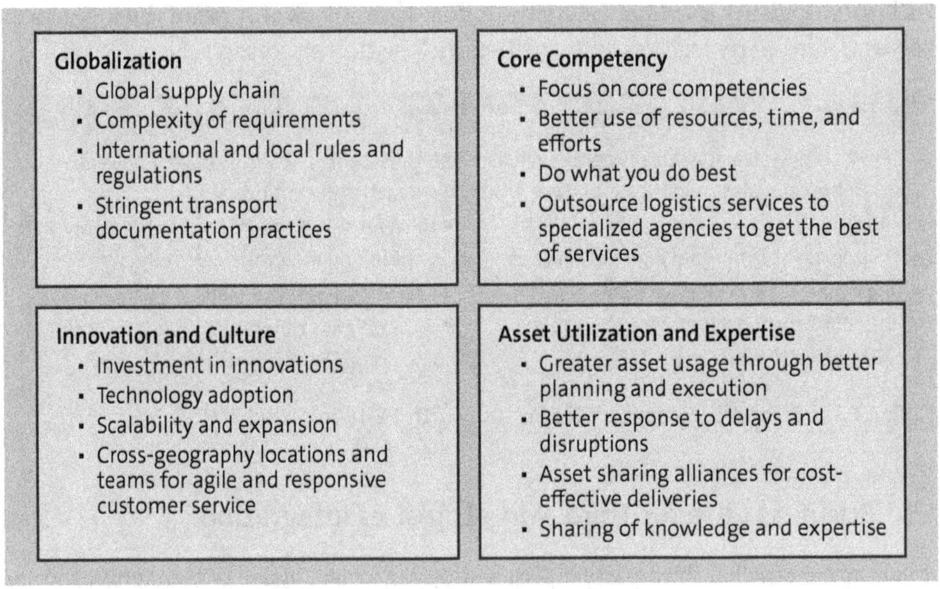

Globalization	Core Competency
▪ Global supply chain ▪ Complexity of requirements ▪ International and local rules and regulations ▪ Stringent transport documentation practices	▪ Focus on core competencies ▪ Better use of resources, time, and efforts ▪ Do what you do best ▪ Outsource logistics services to specialized agencies to get the best of services
Innovation and Culture	**Asset Utilization and Expertise**
▪ Investment in innovations ▪ Technology adoption ▪ Scalability and expansion ▪ Cross-geography locations and teams for agile and responsive customer service	▪ Greater asset usage through better planning and execution ▪ Better response to delays and disruptions ▪ Asset sharing alliances for cost-effective deliveries ▪ Sharing of knowledge and expertise

Figure 1.2 Key Drivers for Outsourcing to Third-Party Logistics

1.2.2 Pros and Cons of Third-Party Logistics

Some of the pros of third-party logistics include the following:

- **Cost savings**
  Businesses can avoid making big infrastructure investments by working with a third-party logistics supply chain management company, as it can provide transportation, storage space, people, and tracking and tracing technology, among other things.
- **Sharing expertise and experience**
  Third-party logistics service providers offer expertise and understanding in a variety of areas, including international compliance, economic regulations, import and export laws, and transport documentation. Businesses wanting to expand into foreign markets can profit from the support and logistics expertise that their partner can offer, thereby reducing expensive delays, shortening the cycle time, and facilitating entry into a new region.
- **Concentrating on core competency**
  By outsourcing logistics, companies are able to concentrate on their core strengths rather than having to manage non-core, yet crucial, operations. In addition, businesses can benefit from the logistical experience and expertise of the partnering agency.

- **Scalability and flexibility**
  Third-party logistics in supply chain management also gives businesses the adaptability and scalability to use supply and distribution resources in accordance with the market demands. As a result, businesses are able to match peaks and troughs in demand with optimum usage of resources and investments.
- **Enable business growth and expansion**
  Third-party logistics partners can provide easy access to new markets and regions where businesses don't have an established presence. With this access and know-how about the newer market and regulations, third-party logistics agencies can help businesses grow quickly without much capital or resource investments.
- **Better customer service**
  Responsive and resilient supply chain practices can result in improved customer services and response time, timely deliveries, and greater reliability.

Some of the cons of third-party logistics are as follows:

- **Reduced control**
  Sharing certain proprietary information with a third party (sourcing, order information, etc.) can lead to businesses becoming vulnerable in the event of a data breach. In addition, businesses have to trust the third-party logistics partner by sharing control of logistics operations.
- **Cost**
  There is an up-front cost of setting up integration with third-party logistics and a recurring subscription cost to maintain a minimum service level. In addition, as the business grows and consumes more resources, it may be difficult to control the cost as it may increase exponentially. It's usually difficult to establish a cost-effective partnership with third-party logistics as other external and market factors may have influence over the cost.
- **Business knowledge**
  Highly regulated industries (e.g., pharma and life sciences) may have special delivery needs (temperature controlled, cold storage, etc.) along with protection against proprietary information and adherence to local market regulations. These specialized services require stringent criteria for selection of third-party logistics partners.

1.3 Integrated Business Processes

The end-to-end logistics operations during order fulfillment involves multiple process areas and requires seamless integration between different SAP S/4HANA modules. Figure 1.3 illustrates the process flow with third-party logistics integration.

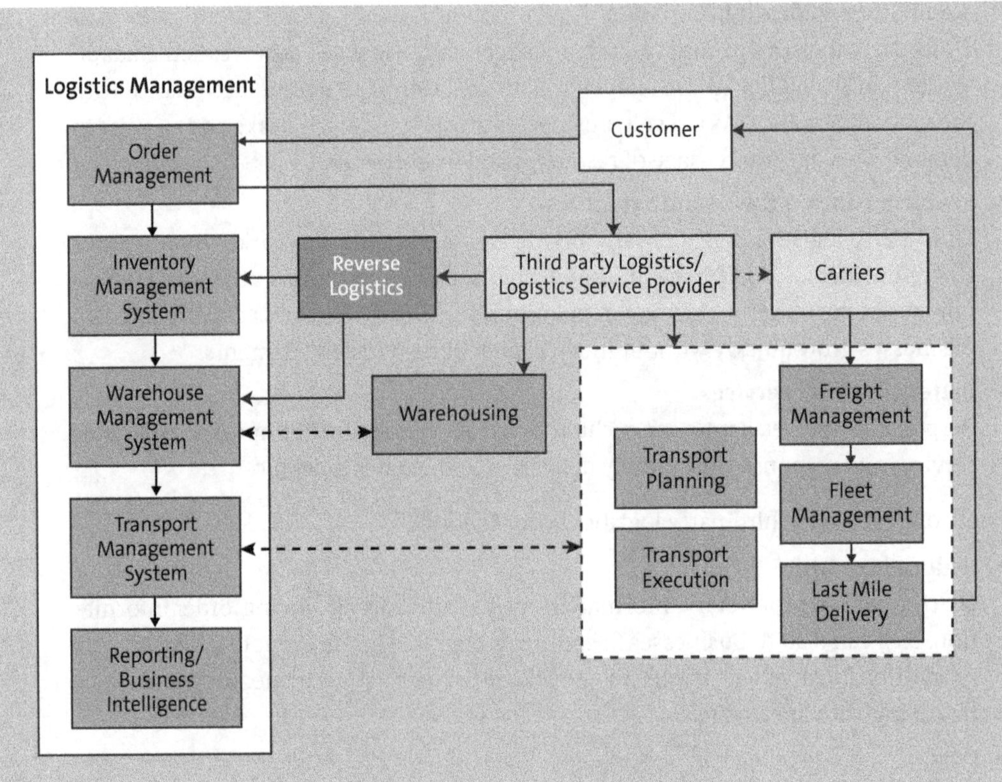

Figure 1.3 Service Covered by Third-Party Logistics Integration

The following sections elaborate on the nature and benefits of third-party logistics integration within three key functional areas involving order processing, outbound and inbound logistics, and inventory.

1.3.1 Order-to-Cash

Order fulfillment encompasses the entire process of receiving an order for shipment to customer, picking and packing, and finally shipping the order to the consignee. During order processing, the third-party logistics team locates the item in the warehouse, packages it, and then carries out the process of shipping it either via an external carrier or using its own freight to its destination.

Some of the benefits of letting third-party logistics service providers manage order fulfillment are as follows:

- Time and cost savings for all order processing, warehousing and inventory management, and shipping
- Fast and efficient order shipments

- Access to a network of shipping carriers and providers, along with potentially better rates
- Last mile connectivity

Order management integration involves message exchanges from your enterprise resource planning (ERP) system to third-party logistics, informing them about impending outbound shipments through outbound delivery notifications and receiving confirmation messages from third-party logistics service providers containing details of picking, packing, and goods issue to shipment, either as a single massage with all details together or through separate messages at each stage of pick, pack, and goods issue. Figure 1.4 depicts a typical order-to-cash process integration with third-party logistics.

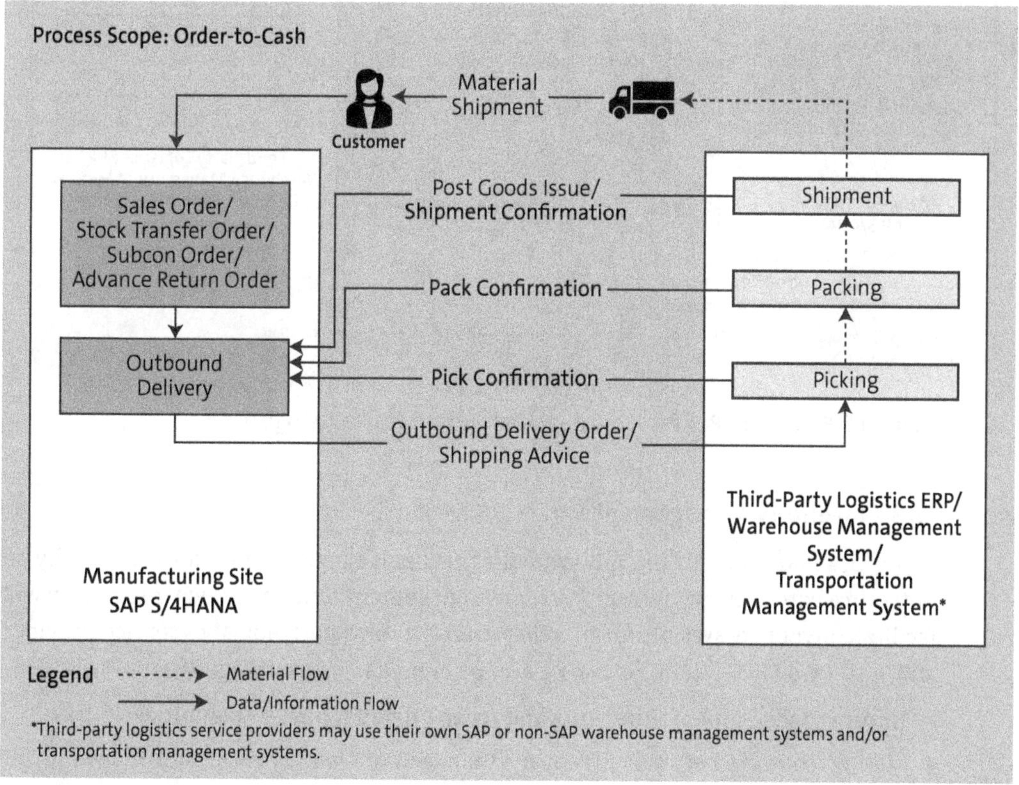

Figure 1.4 Order-to-Cash Process Integration with Third-Party Logistics

1.3.2 Procure-to-Pay

Procure-to-pay integration with third-party logistics synchronizes inbound shipment requests, receipts, freight, and handling of imports and exports with resources and expertise available at third-party logistics service providers. This improved visibility leads to more accurate data for your procurement, inventory, and distribution processes.

Such integration involves message exchanges from your ERP system to the third-party logistics system, informing them of impending inbound shipments through inbound delivery notifications and receiving confirmation messages from the third-party logistics service provider containing details of goods receipt upon receipt of the stock from the delivery communicated earlier. Figure 1.5 depicts a typical procure-to-pay integration with third-party logistics.

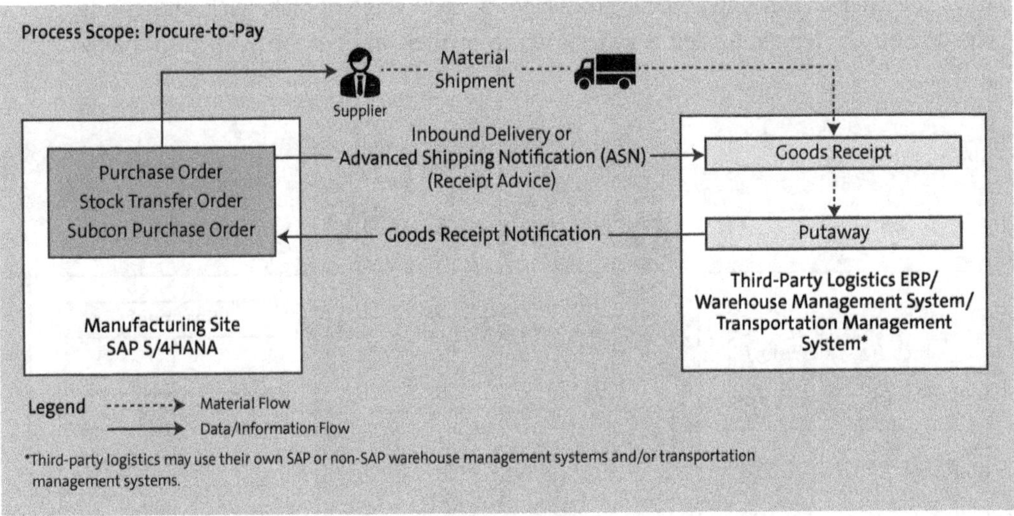

Figure 1.5 Procure-to-Pay Process Integration with Third-Party Logistics

1.3.3 Inventory Management

Inventory management integration with third-party logistics provides the ability to outsource part or all of your supply chain management functions. Such integration will facilitate inventory synchronization between the third-party logistics service provider and your own ERP system. Following are the benefits of such integration:

- Improved visibility of shipments and receipt detail for better planning.
- Tighter controls between inventory management and the execution system for greater transparency for third-party logistics or warehousing activities
- Monitoring of inventory accuracy

Inventory management integration involves message communications between a third-party logistics system and a company's ERP systems. A message communication at a regular interval, as decided by the business (usually daily at off hours), will bring in an inventory snapshot from the third-party logistics system and post it to the company's ERP system. Similarly, any stock status change that can be carried out either at the third-party logistics system or your own ERP system needs to be communicated across to the other system. Figure 1.6 depicts a typical inventory management process with third-party logistics.

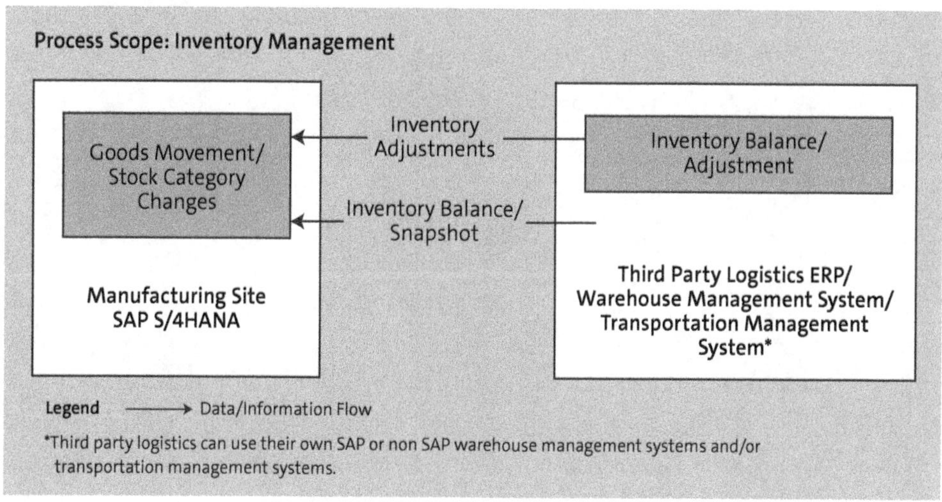

Figure 1.6 Inventory Management Process Area Integration with Third-Party Logistics

1.4 Summary

This chapter covered the details of supply chain integration and process improvement needs that can be provided by external logistics service providers specializing in logistics functions. Different scenarios involving logistics service providers were discussed, and we looked at key value drivers to substantiate business cases involving outsourcing full or part of the logistics supply chain to third-party logistics. Finally, the nature and key business benefits of third-party logistics integration in three major process areas—order fulfillment, procurement, and inventory management—were discussed in some detail.

Chapter 2
Design and Implementation Criteria

The global business environment, with its constantly advancing technology, evolving world markets, and greatly expanded supply chains, makes it increasingly complex and demanding on supply chain professionals to make crucial decisions affecting profitability. With the globalization of the economy, challenges of providing seamless supply chain solutions across geographical and cultural boundaries have increased exponentially.

In the recent past, we've seen how natural calamities and global pandemics can quickly disrupt established supply chains, compelling businesses to rethink their supply chain strategies. In such a demanding situation, selecting the right third-party logistics service provider can be a challenge. There is increased focus on business continuity planning, technology and automation, supplier diversity, and logistics service provider reliability. After all, third-party logistics will be an integral part of the logistic process, and its success will enhance brand reputation as well as customer satisfaction.

With all of this in mind, you need to consider your options carefully before committing to a third-party logistics relationship. Third-party logistics must have proper alignment with your business objectives, growth plans, and customer service culture. This is in view of the tenure of relationship specific to services such as third-party warehousing contracts, which are generally executed on a long-term basis. Consider all selection aspects carefully and thoroughly because it's very difficult to disengage once third-party logistics is onboarded. There should be mutual trust and respect between the business and the logistics service provider, and third-party logistics service provider should be expected to act as a solutions provider and a trusted adviser. They should foster the culture of innovation in the supply chain process and ensure that your products are brought to market quicker and in a cost-effective way. To do so, they should have complete visibility of your present and future fulfillment plans and needs.

Before embarking on plans to onboard a third-party logistics service provider, businesses should prepare a solid third-party logistics strategy to make sure that the chosen third-party logistics service provider is the right fit for their business. Some of the starting considerations are explained in following sections.

2.1 Determine Business Needs

Each business has its own unique process-specific key performance indicators (KPIs) and long-term objectives. Each process area that is relevant for outsourcing to third-party logistics may need to be scrutinized and detailed cost benefit analysis performed before any conclusion is reached in favor of outsourcing. The process areas impacted by such decision are detailed in the following sections.

2.1.1 Order Fulfillment

Figure 2.1 shows the sales and distribution-specific processes covered by logistics service provider integration.

Figure 2.1 Third-Party Logistics Order Fulfillment Process

To measure the performance of third-party logistics, businesses may consider following KPIs for order fulfillment, which can be included as part of a Service-Level Agreement (SLA) to monitor third-party logistics performance:

- **Order accuracy**
  The shipment contains the right items and batches per sales order.
- **On-time shipping**
  Packages are shipped and delivered on time.
- **Shipment accuracy**
  Shipment is made to the correct destination or end customer.
- **Delivery time**
  Times are correct for order pickup from the warehouse or distribution center and for delivery to the end customer.
- **Delivery cost**
  Indicator is used to contain cost, referring to the total cost of shipment divided by the number of packages shipped. Of course, there is a subjectivity as there may be high-value items delivered to customers in distant locations.

2.1.2 Storage and Warehousing

Inventory management and warehousing are important functions to control the cost of acquisition and holding. If stock is kept for long or if you have nonmoving stock, additional cost factors may crop up, such as holding, depreciation, spoilage, and so on. For any third-party logistics partner, this is one area they want to streamline through automation, best practices, and innovations.

Figure 2.2 depicts the common activities for third-party logistics under inventory management and warehouse management (WM).

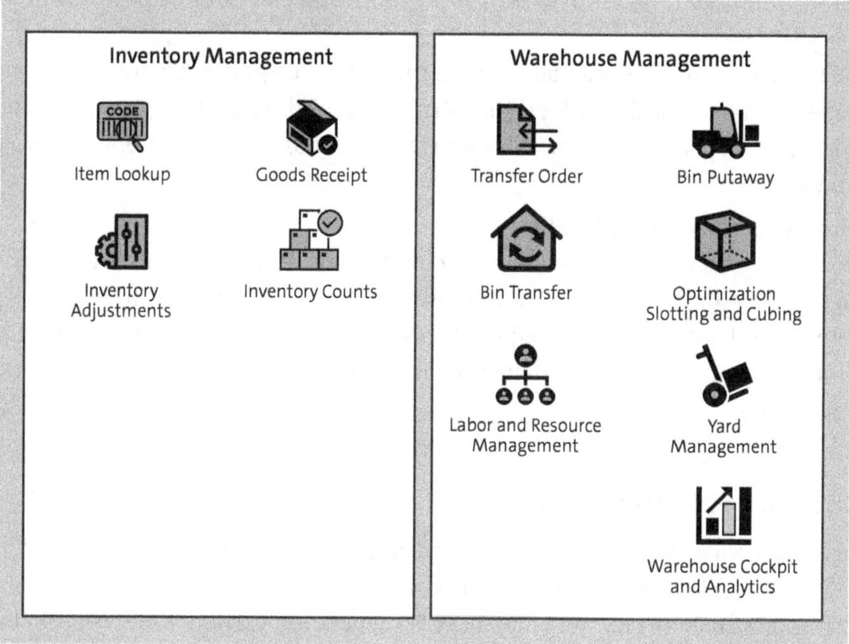

Figure 2.2 Third-Party Logistics Inventory Management and Warehousing Function

Some of the KPIs are listed here:

- **Inventory accuracy**
  This is the percentage of counted inventory against book inventory.
- **Inbound receipt**
  This refers to the time to move the inbound receipt items from the receiving dock to stock. More time means a shortage of stock for the outbound shipment, resulting in lost sales.
- **Inventory turnover**
  If the product sits in inventory for a longer period, it may incur the additional cost of carrying as well as storage conditions and policies around storing the product.
- **Inventory carrying cost**
  These are the costs of storing and holding that inventory, which may include storage

cost, handling, labor, insurance and taxes, transportation, loss or shrinkage, and depreciation.

Other warehousing requirements that may determine the selection of third-party logistics include the following:

- **Public vs. contract warehouse**
  Contracted space may have certain benefits over public space due to the fact that the warehouse space is always available and sometimes comes along with better picking and packing facilities. Third-party logistics agencies may offer both, however, a cost-benefit analysis may need to be performed before making such a decision.
- **Specialty storage**
  Special storage conditions may be required for high-value perishable items such as from the pharmaceutical and food industries. Not all third-party logistics service providers have capacities such as cold storage, hazardous storage, wet bonded storage, and so on.
- **Warehouse monitoring and audit**
  Third-party logistics service providers may have their own IT setup with warehouse management systems (WMSs) for warehouse operations and monitoring. However, a careful consideration may be required to evaluate their capabilities to control, monitor, and audit warehouse activities for accurate receipts, distribution, and stock level management.

2.1.3 Distribution and Transportation

Many logistics service providers are asset-based and have their own fleet of transports. However, for certain legs of shipment, such as ocean or air shipment or even for intermodal transports such as rail, they may have to depend on carriers for execution of shipment. They may reserve space or act as a freight forwarder in such cases and have payment liability to the carrier.

Logistics service providers may use their own SAP or non-SAP transport management systems for planning, execution, and monitoring of transports. Figure 2.3 provides a snapshot of transportation management features and functions.

Capabilities of third-party logistics providers may be assessed for distribution services; some of these are listed here:

- **Distribution methods**
  Drayage, customs brokerage, full container load (FCL), less than container load (LCL), less than truckload (LTL), and truckload (TL).
- **Fleet optimization**
  Efficient delivery of goods based on artificial intelligence (AI) mapping and modeling.

- **Tracking of goods**
  Using GPS and Internet of Things (IoT) devices.
- **Intermodal transport**
  Arranging intermediate movement of goods between different methods of transports.
- **Freight forwarding**
  Organizing and managing transportation by liaising with carrier companies, transport documentation, renting out space in a distribution fleet, and so on.
- **Freight payment and accounting**
  Payment to carrier and relevant accounting
- **IT capabilities of third-party logistics**
  Use of transport management systems and ease of integration with your business.

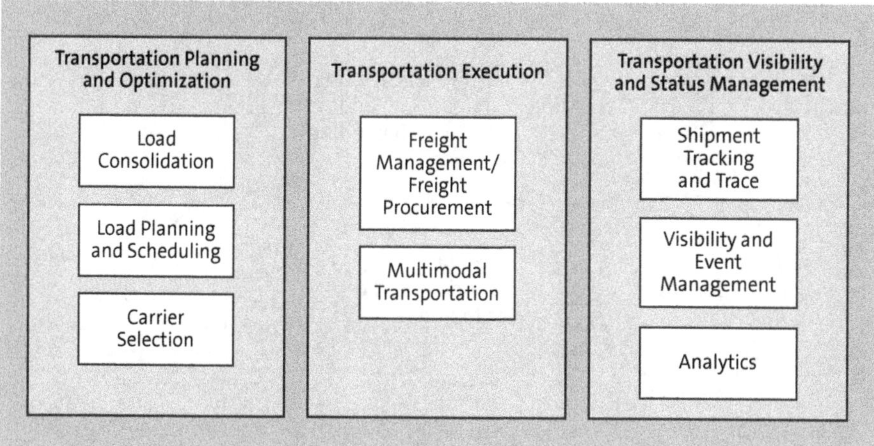

Figure 2.3 Transport Management Functions

2.1.4 Value-Added Services

Value-added services may be a key factor in selecting the right third-party logistics provider for your business. Common value-added services provided by third-party logistics players are as follows:

- **Packaging and assembly**
  Packaging the products in handling units per shipment specifications. Assembling the products when the transport of parts being assembled is convenient and cheaper. Packaging also provides protection against damage and improves product appearance.
- **Kitting**
  Combining related products and putting them in one kit to be sold as a unit or SKU.

- **Customer returns/reverse logistics**
  Inspecting, reworking for minor issues, or sending back customer returns to the manufacturer for replacement.
- **Export/import management**
  Ensuring transport documentation, compliance with international regulations, and cost-effective shipment of goods.
- **Cross-docking**
  Consolidating and deconsolidating goods.

There are other value-added services provided by different third-party logistics providers depending on the nature and extent of their operations. Figure 2.4 provides a glimpse of the value-added services provided by third-party logistics.

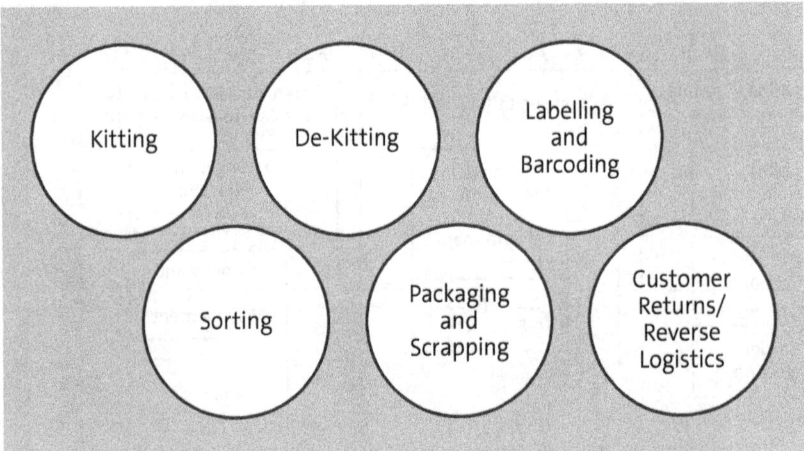

Figure 2.4 Value-Added Service Options from Third-Party Logistics Service Providers

Having understood the options provided by logistics service providers, it's essential to understand the challenges presented by a typical third-party logistics implementation initiative.

2.2 Assess Current State and Future Challenges

A strategy is required for selecting a third-party logistics partner that is based on an assessment of business requirements, the current state of the business, and future business objectives. The following section describes the various action points for an organization or company contemplating outsourcing their logistics services to a third-party logistics partner and important techno-commercial attributes a third-party logistics partner should be evaluated against for qualification.

2.2.1 Create a Cross-Functional Team for Regular Interactions

To start with, businesses can choose a project team consisting of different supply chain functional areas. This team must be led by a third-party logistics project leader, preferably a supply chain or logistics function lead. Selection of team members is critical because key teams need to be involved, such as the following:

- IT (for current IT landscape, cross-system integration, and management reporting requirements)
- Finance and costing (for evaluation of internal costs versus third-party logistics implementation and running costs)
- Shipping and warehousing
- Supply chain
- Procurement
- Sales and marketing
- Customer service
- Quality

2.2.2 Finalize Scope of Services and Future Road Map

Prepare a detailed list of services with key performance measures. Include a future road map like you would for any planned scaling up or expansion in business. This may also include assessing the customer base, market geography, and current strengths and weaknesses of the in-house logistical services (may be measured by a current set of KPIs). Any need for local distribution centers or warehouses close to the customer base, setting up warehouse/distribution center versus hiring warehouse services, and so on can be major cost-saving factors. Another major aspect is distribution and transportation because most third-party logistics carriers are either asset-based or freight forwarders, so they possess knowledge and expertise to deal with international regulations and sophisticated transport facilities.

2.2.3 Assessment of the Current IT Landscape

A detailed study and assessment is required of the current system's effectiveness and accessibility. The following set of data points may reflect the status:

- Age of the current system
- Current number of user licenses and limitations
- Capability of the current system to adapt to business expansion and global reach
- Connectivity to third-party or cloud-based systems
- Reporting and visibility into KPIs, financials, and processes

- Visibility into the supply chain, including operations, production, order status, and shipping

This kind of study helps identify gaps in efficiencies and uncovers bottlenecks in processes and technology.

2.2.4 Evaluation of Third-Party Logistics Capabilities

After due deliberation of all the preceding factors, it may boil down to the crucial and most important step of evaluating possible third-party logistics partners for your business. Following are a few pointers to selecting the right third-party logistics partner:

- **Technological capabilities**
  To boost supply chain visibility and keep track of all operations, a top-notch third-party logistics provider has access to a number of technical solutions, including cloud-based systems, for example, WMS, inventory management system, transport management system, and others. This increases efficiency by assisting logistics providers in obtaining real-time data and reducing supply chain faults. Businesses should ensure that the capabilities of the third-party logistics system are compatible with their automation and growth requirements.
- **Scalability**
  The imminent question is, can third-party logistics demonstrate proven reliability during unplanned spikes in business? The logistics service provider should be able to meet both your present needs and any future needs your company may have as it expands. Ideally, you should pick a third-party logistics services provider that has solid networks and can quickly handle more inventory to meet changing customer demands. On the contrary, they may also need to scale down when demands are low, while remaining equally cost efficient.
- **Financial health**
  Your ideal third-party logistics provider should have the financial stability to deal with significant disruptions and be in a solid position to employ the best practices and take advantage of innovations that will protect and flourish your business.
- **Cost of service**
  Before choosing a third-party logistics supplier, it's vital to examine the prices and services of each potential provider. Select a third-party logistics company that is within your budget without sacrificing service quality, as doing so could harm your company in the long term. A trustworthy third-party logistics partner is transparent in its pricing and level of service.
- **Expertise in business**
  Ask providers if they have expertise in your business vertical. Some third-party logistics providers may have the right business expertise and capabilities to match

your requirements. For example, your business may need intercontinental shipment of hazardous goods in a specific temperature condition. This may require understanding of international regulations pertaining to hazardous goods along with the means to transport such goods.

- **Management and culture**
  Third-party logistics partners should be able to employ best practices, support sustainable processes, and be open to the culture of innovations in technology and business trends.

2.2.5 Integration Approach

A business may need to integrate one of multiple third-party logistics partners. Based on the situation, the following design factors may provide leads to the design of future integration solutions:

- Business case for selection of the probable third-party logistics partner
- The scope and cost of services
- Geographical coverage in terms of location, proximity to customer base, accessibility to nearest production or distribution centers, and so on
- Communication standards and protocols at third-party logistics
- Volume of data and key transactions to be handled
- Technical landscape of the third-party logistics partner systems
- Integration mechanism, for example, electronic data interchange (EDI)/IDoc, XML/IDoc or application programming interface (API), and so on
- Master data upload and synchronization requirements (vendor, customer, material, and any other master data)
- Inventory reconciliation between the business and third-party logistics systems
- System security considerations
- System monitoring requirements, error handling, and escalation matrix
- Third-party logistics requirements documentation
- Readiness of the third-party logistics provider for technical changes and availability

Once the approach is finalized and approved, it may need a multidimensional or techno-functional approach to realize the solution. The following sections highlight some areas that require a dedicated effort to implement the solution.

Mapping Key Business Processes

Functional specification documents are prepared that contain detailed process maps in key areas of order fulfillment, inventory management, WM, transportation, and other related areas. Each of the functional specifications usually contains the following:

- Business process flow/maps
- Assumptions and constraint
- Sample maps from third-party logistics (XML, EDI, or any other format)
- Data maps between IDoc fields and logistics service provider–specific maps
- Data conversion requirements for specific fields of the transaction
- SAP HANA field value-specific limitations, changes, or alternatives to overcome those
- Technical change requirements
- Basis requirements
- Security requirements
- Data dependencies

Use Case: SAP S/4HANA Integration with External Third-Party Logistics Warehouse Management Systems

Some of the use cases for process mapping with third-party logistics systems you may consider include master data replication, inbound delivery processing, outbound delivery processing, and stock category adjustments. Key master data may need to be uploaded to the logistics service provider system for successful processing of transactional IDocs (Figure 2.5).

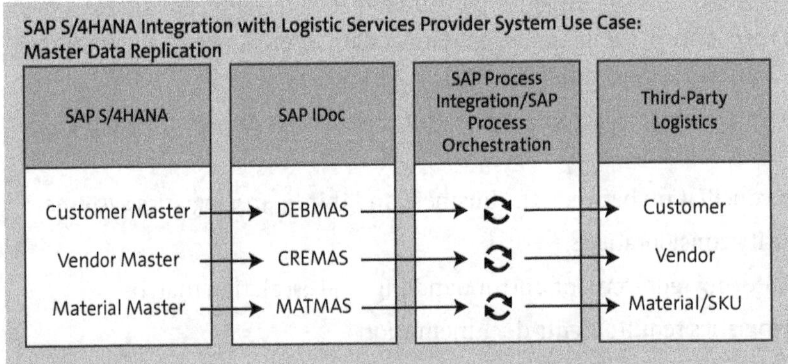

Figure 2.5 Master Data Replication to Third-Party Logistics Systems

Inbound delivery processing may involve delivery notifications to the logistics service provider and, once the material is physically received, confirmation from the logistics service provider regarding goods receipt (Figure 2.6).

Outbound delivery processing may involve delivery notifications to the logistics service provider and, once the material is physically issued and shipped to the customer, confirmation from the logistics service provider regarding delivery confirmation (Figure 2.7).

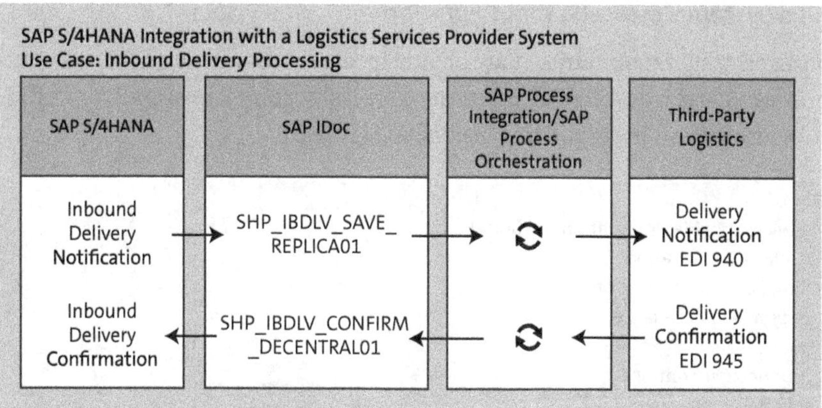

Figure 2.6 Inbound Delivery Processing Using a Third-Party Logistics WMS System

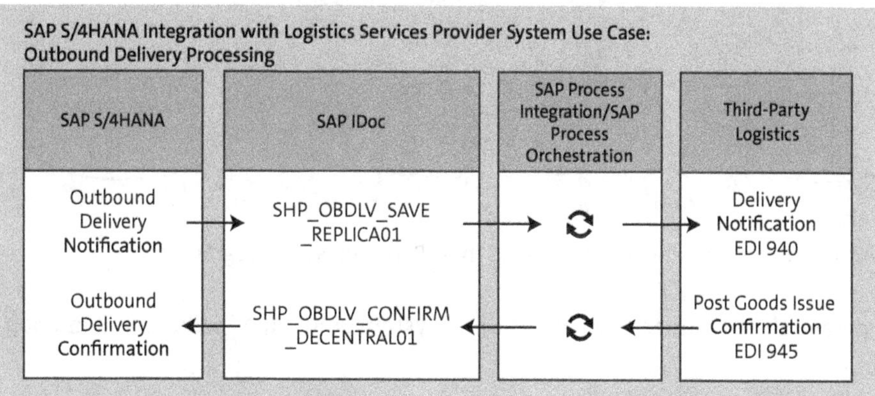

Figure 2.7 Outbound Delivery Processing Using a Third-Party Logistics WMS System

Any adjustments in stocks at the third-party logistics warehouse needs to be recorded to the SAP S/4HANA system immediately to keep the system stocks in sync at both ends (Figure 2.8).

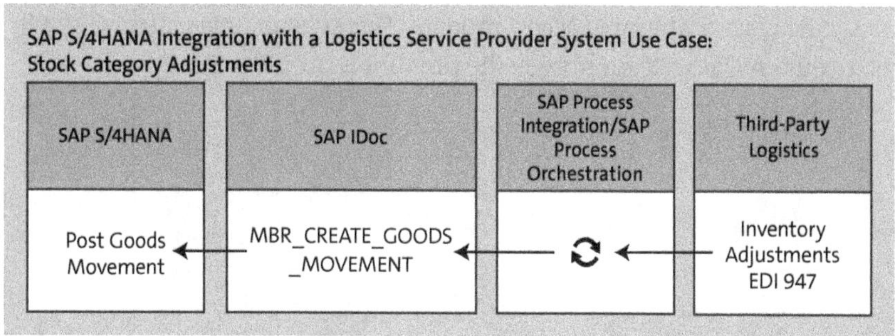

Figure 2.8 Stock Category Adjustments in a Third-Party Logistics WMS System

2.2.6 Design System Landscape for Integration

System landscape design requirements are of vital importance and may need to be vetted by technical experts from all relevant areas. A typical system landscape from third-party logistics integration may look like the one in Figure 2.9.

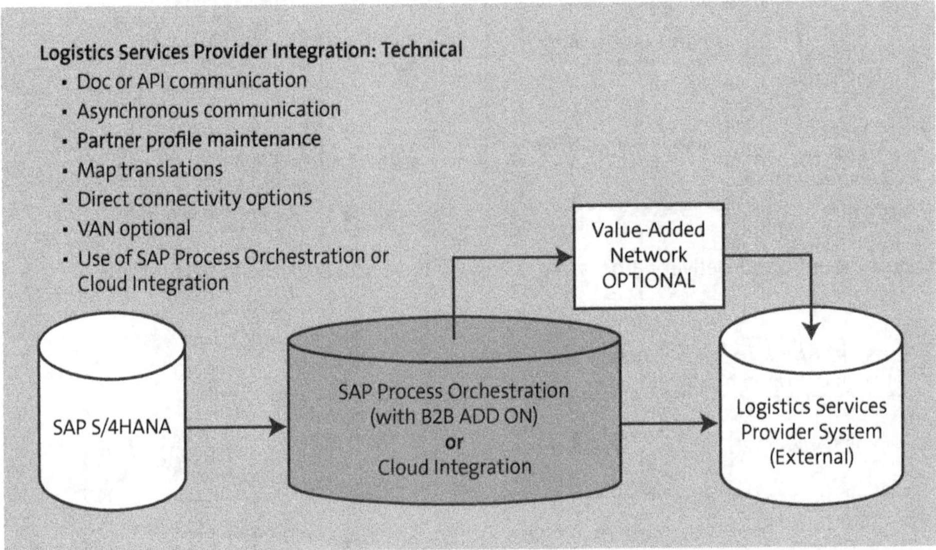

Figure 2.9 Schematic System Landscape for Third-Party Logistics Integration

More details about connectivity options and system landscape can be found in Chapter 10.

2.3 Summary

Implementing third-party logistics integration solutions requires a very meticulous approach to carefully inspect and evaluate every detail of transactional field logics, platform integration considerations, and the extent of technical changes the third-party logistics partners are willing to accommodate. These factors may affect the technical design at your end, as well as increase the timeline of implementation.

Chapter 3
Organizational Entities and Master Data

Setting up necessary organizational elements enables smooth and efficient supply chain processes that impact third-party logistics integration. These organizational elements belong to multiple functional areas and may be configured as part of the relevant functional area configuration. However, specific requirements may surface during interface design, which we'll discuss in this chapter. In addition, master data elements in SAP S/4HANA aren't specific to the third-party logistics process but are usually set up as part of the functional areas as well (mainly procurement, sales, and logistical functions). The SAP S/4HANA business partner approach to centrally manage master data for customers and vendors has made redundant the previous SAP ERP model of vendor and customer masters.

The enterprise structure of an organization in SAP S/4HANA is subdivided based on functional areas, legal requirements (global or country-specific), adherence to audit compliances, and reporting purposes.

In supply chain management, key organizational units are purchasing organizations and groups, sales organizations and sales areas, plants, storage locations, and shipping points. The following section describes the role of organizational entities and how to maintain them in an SAP system.

Once the organizational structure and setup are finalized, the next major action point is determining business partner roles (supplier, sold-to-party, ship-to-party, carrier, invoicing party, etc.) and specific field requirements pertaining to ordering and delivery business processes. This should then be followed by other master data requirements for seamless end-to-end business process execution in SAP. These master data requirements are specific to the material master (key fields relevant to third-party logistics processes), purchasing, and sales.

3.1 Organizational Entities

Enterprise structure is the key to success for business transactions, and a good amount of time and effort is spent deliberating every business requirement impacting the organizational structure. From a design perspective, it may be important to keep the third-party logistics design team involved in enterprise design decisions, especially regarding sales, purchase, and logistics organizational entities.

The following sections deal with the details of organizational units that play a major part in transactions related to third-party logistics.

3.1.1 Purchasing Organization and Groups

Procurement from external vendors as well as intracompany and intercompany stock transfers will require setting up purchasing organizations and purchasing groups. Purchasing organizations have a controlling impact on purchase transactions; however, purchasing groups are independent entities used more for reporting purposes. We'll discuss both purchasing organizations and purchasing groups in the following sections.

Purchasing Organization

A purchasing organization is the highest level of organizational unit in SAP procure-to-pay (SAP Material Management [MM] in SAP ERP) that is responsible for procurement of material and services, responsible for negotiation of conditions with prospective suppliers, and is legally accountable for relevant transactions with external vendors such as tendering and agreement, purchase, and settlement of invoices.

Depending on the organizational policies and procurement framework, the purchasing organization can be maintained in the following ways:

- **Centralized**
  Centralized purchasing has one purchasing organization catering to multiple company codes. Some advantages include better control of vendor negotiations and purchases, ability to leverage the discounts with bulk purchases, and tighter control of purchase spends.
- **Decentralized**
  Decentralized purchasing brings in a few benefits due to reduced bureaucracy, local sourcing and associated price benefits, and accelerated purchase processes. However, this option limits the control while checking the adequacy of suppliers for specific products as well as spend visibility.

Figure 3.1 represent variations in a purchasing organization's assignments when moving from centralized to partially decentralized and finally to decentralized/distributed purchasing.

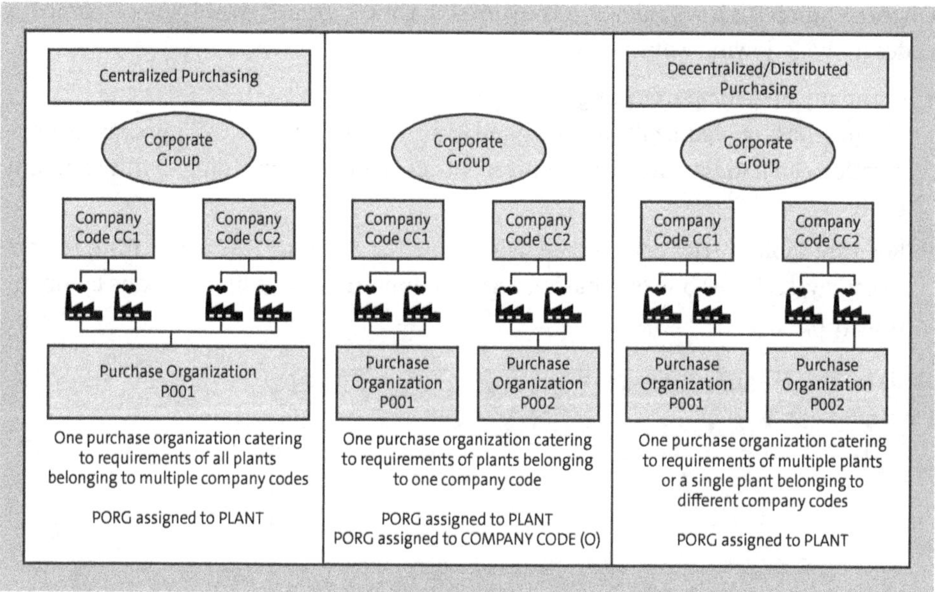

Figure 3.1 Centralized versus Decentralized Purchasing Scenarios

Now that you understand the business context, let's look at how you maintain the purchasing organization in SAP S/4HANA and how it's assigned to company codes and plants.

> **Note**
> More specific details can be found in *Sourcing and Procurement with SAP S/4HANA* (SAP PRESS, 2023) by Justin Ashlock.

Purchasing organizations in SAP S/4HANA can be created using the IMG reference guide via menu path **SPRO • SAP Reference IMG • Enterprise Structure • Definition • Materials Management • Maintain Purchasing Organization** (or use Transaction OX08), as shown in Figure 3.2.

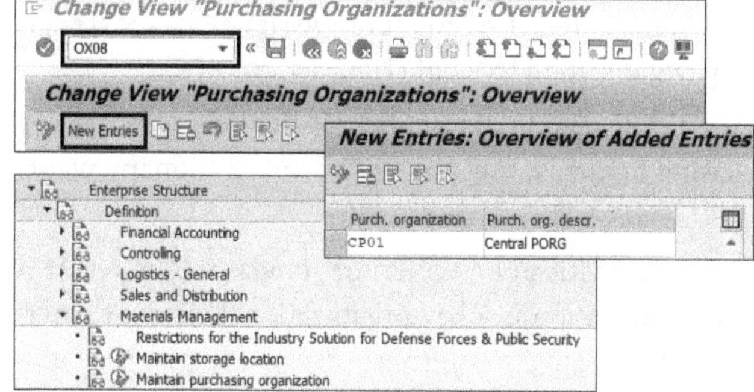

Figure 3.2 Maintain Purchasing Organization in SAP S/4HANA

Once the purchasing organization is maintained, it can be assigned to plant/company codes in the following ways:

- **Company-specific purchasing**
  The purchasing organization is assigned to a company code. Such a purchasing organization can handle purchase requirements of all plants pertaining to this company code.

  Relevant table entries can be seen using table `T024E` (Figure 3.3), where **POrg** is a primary key, and thus one purchasing organization can only be linked to one company code (1:1).

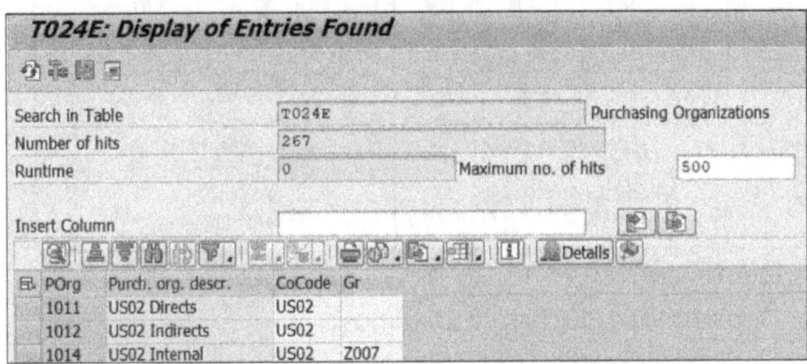

Figure 3.3 Purchasing Organization Assignment to Company Code

- **Plant-specific purchasing**
  The purchasing organization is assigned to plants (irrespective of which company code this plant belongs to). In this case, the company code for purchase is derived from the plant, as each plant must belong to a specific company code. Such a purchasing organization can handle all purchase requirements from all linked plants (Figure 3.4).

- **Cross-plant purchasing**
  This is similar to the plant-specific scenario. In this case, the purchasing organization is linked to all plants across companies. However, it's not linked to any company code (Figure 3.4).

You can assign a purchasing organization to a plant using the IMG reference guide via menu path **SPRO • SAP Reference IMG • Enterprise Structure • Assignment • Materials Management • Assign Purchasing Organization to Plant** (Transaction OX17).

Relevant table entries can be seen using table `T024W`, where both **POrg** and **Plnt** are primary keys, and thus one purchasing organization can only be linked to many plants (1: N) and vice versa.

In logistics service provider–specific scenarios of a typical or standard SAP S/4HANA implementation, the purchasing organization is a key organizational element determining the pricing procedures for purchasing, master records such as purchase info records, purchase contracts, suppliers, and invoicing or settlement processes.

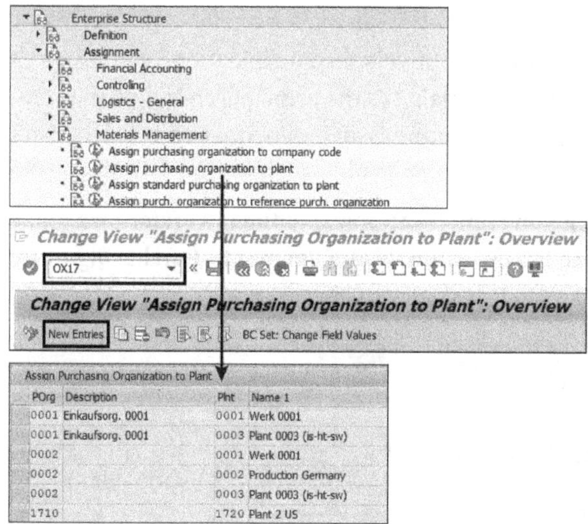

Figure 3.4 Purchasing Organization Assigned to Multiple Plants

Note

While using SAP S/4HANA with embedded SAP TM (i.e., side-by-side deployment), as shown in Figure 3.5, the materials management purchasing organization ❶ is linked to the SAP TM purchasing organization ❷. The purchasing organization in SAP TM is responsible for freight quotations and agreements, freight orders and freight bookings, and forwarding settlement. During transfer of the forwarding settlement document to materials management, the purchasing organization creates a service purchase order on the linked purchasing organization.

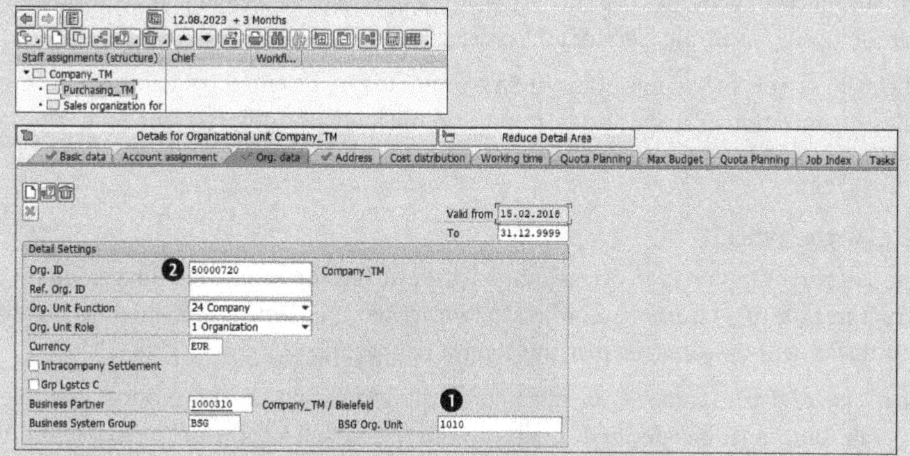

Figure 3.5 Linking the SAP S/4HANA Purchasing Organization to the SAP TM Purchasing Organization

The SAP TM process will be discussed in greater detail in Chapter 8.

Purchasing Group

A group represents a specific buyer or group of buyers responsible for procurement of specific materials or types and classes of materials. It's the principal channel for a company's dealings with its suppliers. A purchase group is an independent entity that isn't assigned to any other organizational units.

Purchasing groups can be maintained in configuration using the IMG reference guide via menu path **SPRO • SAP Reference IMG • Materials Management • Purchasing • Create Purchasing Group** (Transaction OME4), as shown in Figure 3.6.

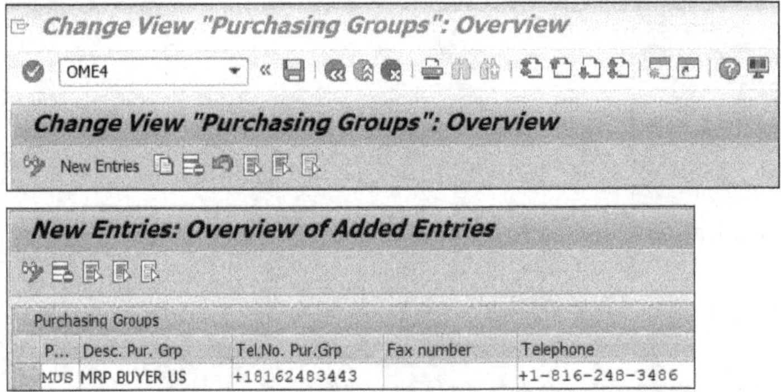

Figure 3.6 Creation of a Purchasing Group

3.1.2 Sales Organization and Sales Area

The sales organization is a foundation for effective sales planning and sales policies. Systematic execution of the plans, policies, and programs of a sales organization control all the sales activities. As such, it ensures maximum efficiency and profitability without losing consumer service and satisfaction.

The sales area is more important in SAP where it's represented by a combination of sales organization, distribution channel, and division. Ensuing sections focus on the details of sales-specific organizational entities.

Sales Organization

The sales organization is an organizational unit within supply chain management that structures the organization according to its sales requirements. It's the highest organizational unit in the sales and distribution functional area.

As a legal entity responsible for sale and distribution of goods and services, the sales organization must be specified in all sales documents. It's therefore available for all basic functions of sales and distribution (e.g., pricing, availability, etc.).

Sales organizations in SAP S/4HANA can be created using the IMG reference guide via menu path **SPRO • SAP Reference IMG • Enterprise Structure • Definition • Sales and Distribution • Define, Copy, Delete, Check Sales Organization** (Transaction OVX5).

Figure 3.7 represents the process of maintaining sales organizations via the preceding menu path or directly through the transaction.

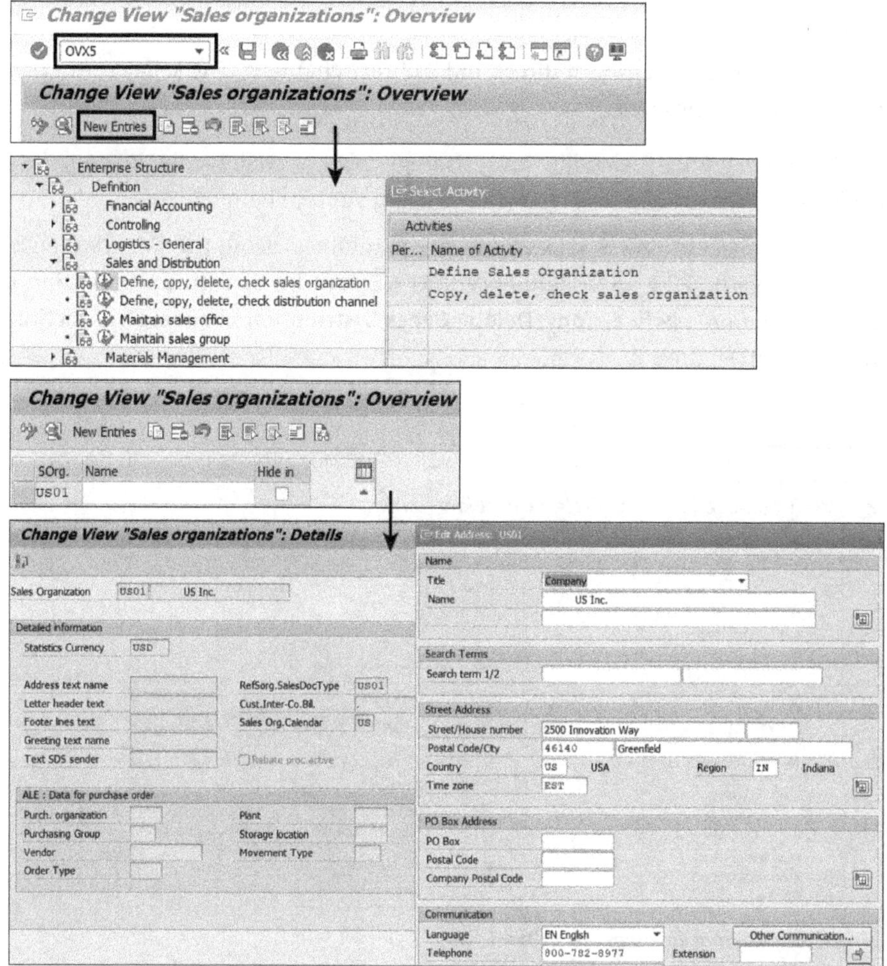

Figure 3.7 Creation of Sales Organization with Address Details

Distribution Channel

The distribution channel is the channel through which saleable materials or services reach the customer. In a business context, distribution channels are the paths that products and services take on their way from the manufacturer or service provider to the end consumer. Distribution channels are also referred to as marketing channels or marketing distribution channels.

There are three types of distribution channels:

- **Direct**
  Involves direct selling to the end customer. The vendor might also maintain its own sales force or sell its products or services through e-commerce.

- **Indirect**
  Uses multiple distribution partners or intermediaries to distribute goods and services from the seller to customers.
- **Hybrid**
  Combines the characteristics of direct and indirect channels. The seller uses both direct and indirect methods.

The distribution channel selection and strategy play a major role in supplier/producer's to-market strategy and results in efficient supply chain management.

Distribution channels in SAP S/4HANA can be maintained using the IMG reference guide via menu path **SPRO • SAP Reference IMG • Enterprise Structure • Definition • Sales and Distribution • Define, Copy, Delete, Check Distribution Channel** (Transaction OVXI), as shown in Figure 3.8.

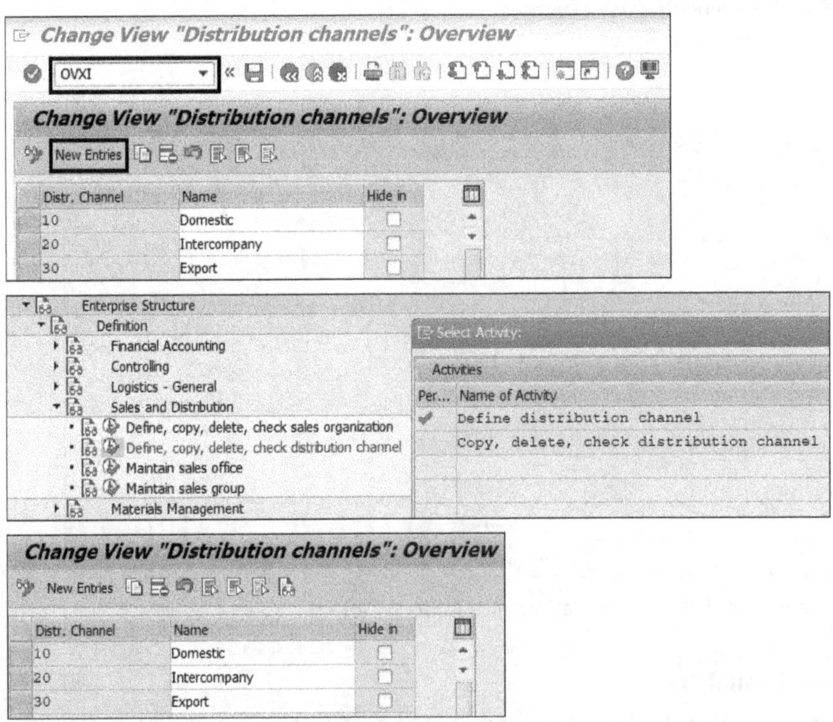

Figure 3.8 Maintain Distribution Channels in Sales

Division

A division in sales and distribution represents grouping or categorization of products and services based on similar attributes. Table 3.1 represents divisions in a typical IT company.

Division	Name
01	IT Products
02	Software
03	Hardware
04	IT Services

Table 3.1 Sales Divisions Based on Product Attributes

Figure 3.9 represents the process of maintaining divisions.

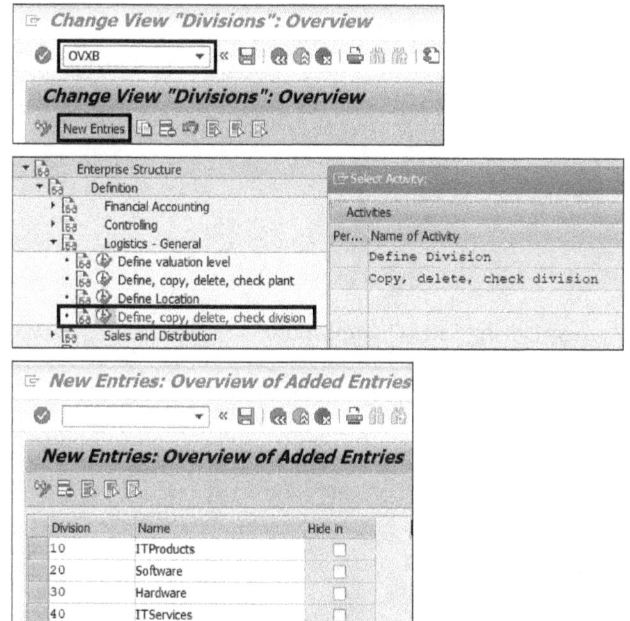

Figure 3.9 Sales Divisions for Different Categories of Goods

Sales Area

A sales area is a combination of three organizational units: sales organization, distribution channel, and division. It determines which distribution channel can be used to sell the products from one division in a sales organization. Every document created in the sales and distribution module must belong to a particular sales area.

To setup a sales area, the following prerequisite steps are mandatory:

- Assign the sales organization to a company code.
- Assign the distribution channel to the sales organization.
- Assign the division to the sales organization.

Again, we won't go into each individual assignment type as this is beyond scope of this topic. However, you'll see how the sales area is defined in configuration. The sales area plays a major part in domestic, export, and intercompany sales orders and billing.

As shown in Figure 3.10, Figure 3.11, and Figure 3.12 maintaining the correct sales area is key to success of sales transactions, and all the SKUs used in sales need to be extended to relevant sales areas.

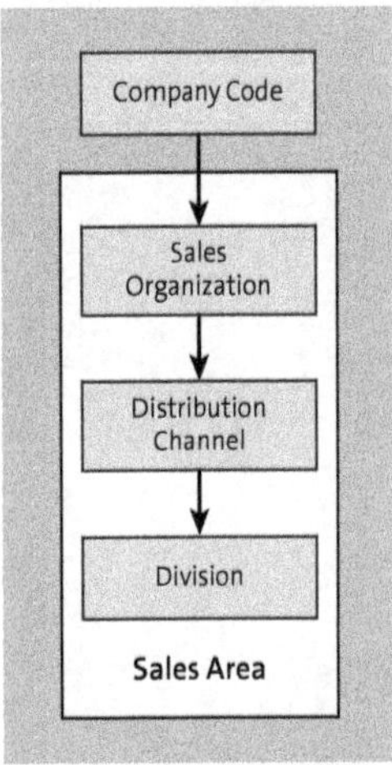

Figure 3.10 Setting Up the Sales Area by Linking Sales Organization, Distribution Channel, and Division

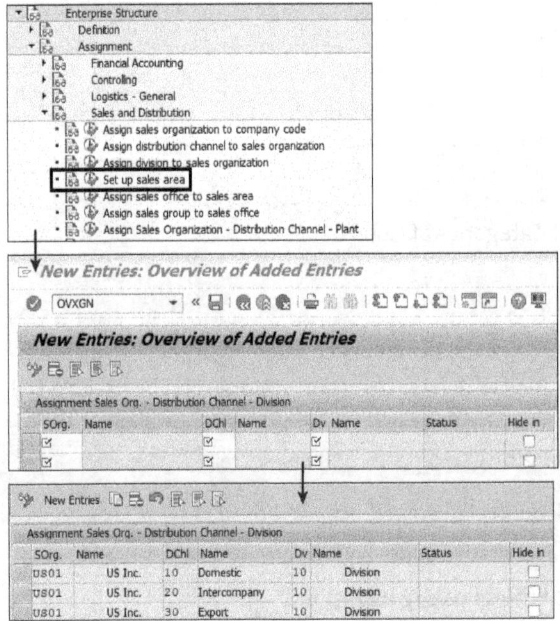

Figure 3.11 Setting Up the Sales Area by Linking Sales Organization, Distribution Channel, and Division Workflow

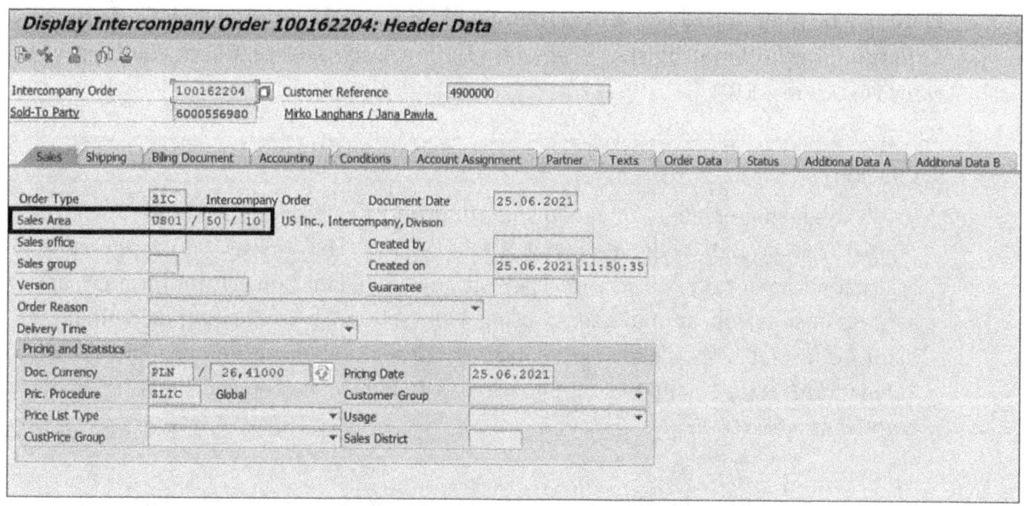

Figure 3.12 Sales Area Used as Mandatory Input in the Header of the Sales Order

Sales areas can be maintained using IMG menu path **SPRO • SAP Reference IMG • Enterprise Structure • Assignment • Sales and Distribution • Set Up Sales Area** (Transaction OVXGN).

3.1.3 Plants, Storage Locations, and Shipping Points

In SAP S/4HANA, the plant is part of the logistics enterprise structure and is the highest entity in the logistics organization. The storage location is maintained under materials management organization, and the shipping point under logistics execution. All of them play major parts in logistics-relevant transactions involving goods movement, and we'll discuss them in more detail in the following sections.

Plant

A plant is an operational facility within a company. In SAP S/4HANA, it represents an organizational unit from the viewpoint of production, procurement, maintenance, materials planning, and distribution of goods and services. From an inventory maintenance perspective, it's the highest organizational entity accounting for inventory checks and balances as well as adjustments and write-offs.

In SAP S/4HANA, a plant is defined as a valuation area if the valuation is at the plant level. Most commonly, this is the approach because plants may be located across cities, countries, or continents, and material valuation may differ based on location. Another approach is to use valuation at the company code level when the size of the organization is small and not too geographically dispersed. In SAP, a plant is defined based on the factory calendar, country keys, and region keys.

Plant can be maintained using IMG menu path **SPRO • SAP Reference IMG • Enterprise Structure • Definition • Logistics - General • Define, Copy, Delete, Check Plant** (Transactions EC02 and OX10).

> **Caution**
>
> It's always recommended to create plants by copying a reference plant rather than creating it from scratch. There are about 142 dependent table entries with assignments that get automatically copied with Transaction EC02 (Plant Copy). After the copy, any address change can be maintained using Transaction OX10. Creating or defining a plant from scratch should always be avoided for missing entries in dependent table. Refer to SAP Note 1754880 - Configurations Missing When Copying/Creating Plants with Transaction OX10.

Figure 3.13 shows how the plant can be created by copying a reference plant:

❶ Copy the reference plant (Transaction EC02).

❷ Maintain the address and other details using Transaction OX10.

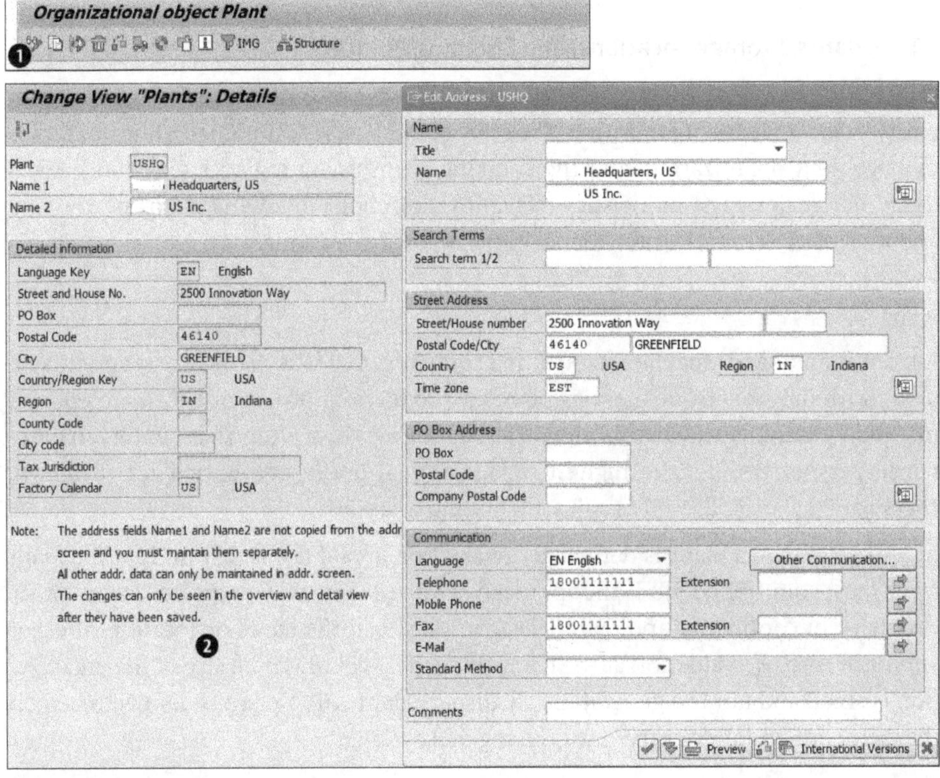

Figure 3.13 Plant Creation

The plant in SAP S/4HANA is always assigned to a company code to form an organizational structure. One plant can be assigned to only one company code. Plant is the highest organizational entity for physical inventory and stock balances.

Storage Location

A storage location is an organizational unit that helps differentiate various stock types stored in a plant. Storage location is a place where stock is kept physically. A plant can consist of multiple storage locations. For a particular storage location, the relevant inventory data is stored at a storage location level in SAP S/4HANA. Addresses of the storage locations need to be maintained if they are geographically at different locations.

Storage locations in SAP S/4HANA can be defined using configuration menu option **SPRO • SAP Reference IMG • Enterprise Structure • Definition • Materials Management • Maintain Storage Location** (Transaction OX09).

As shown in Figure 3.14, all storage locations for a plant can be maintained with or without storage location address details. If a storage location address is maintained, the same gets defaulted in the purchase order.

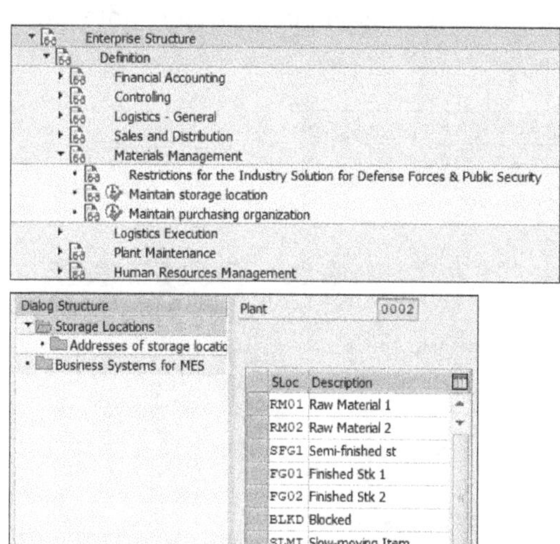

Figure 3.14 Storage Location Maintenance

> **Note**
>
> For third-party logistics integration scenarios, you can define storage locations specific to the locations of the providers, where they receive stock, consolidate/deconsolidate it, use value-added services such as packaging, and ship it to the end customers.

Note that one plant can have many storage locations per the requirements of an organization, but one storage location can be assigned to only one plant.

If you're not using a warehouse management system (WMS), the storage location is the lowest level of inventory management in the system. In the inventory management component, the storage location is the location of physical stock in a plant.

In a WMS or extended warehouse management (EWM) system in a plant, individual warehouses (high-rack storage, block storage, picking area, etc.) are maintained as storage types within a warehouse complex and are grouped together under a warehouse number. As a rule, it's not necessary to define several storage locations for a plant because the WMS warehouse number is assigned to a storage location from inventory management. We'll discuss EWM scenarios in detail in Chapter 7.

Shipping Points

A shipping point is an independent organizational unit where processing and monitoring of deliveries and goods issues are carried out. A delivery is processed by one shipping point only. A shipping point is the highest organizational entity for a shipping organization that can be determined for each order item.

Shipping points can be maintained using IMG menu path **SPRO • SAP Reference IMG • Enterprise Structure • Definition • Logistics Execution • Define, Copy, Delete, Check Shipping Point** (Transaction OVXD). Figure 3.15 represents shipping point maintenance address and other relevant details.

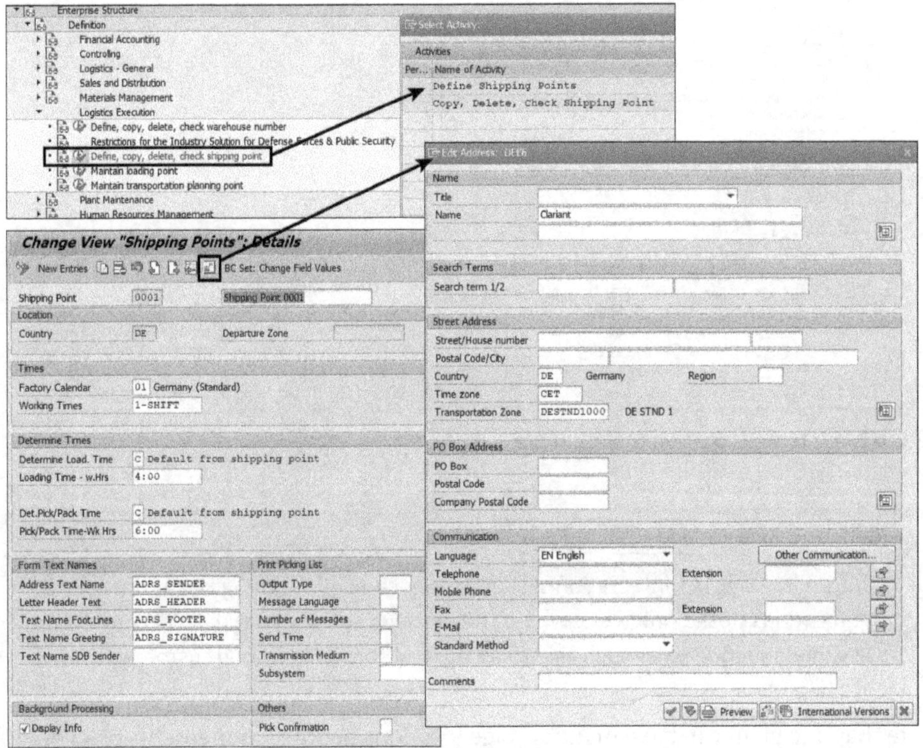

Figure 3.15 Shipping Point Creation with Address Details

The next step is to assign the shipping point to a plant, as shown in Figure 3.16. Multiple shipping points can be assigned to one plant and vice versa.

Shipping points can be assigned to a plant using IMG menu path **SPRO • SAP Reference IMG • Enterprise Structure • Assignment • Logistics Execution • Assign Shipping Point to Plant** (Transaction OVXC).

3

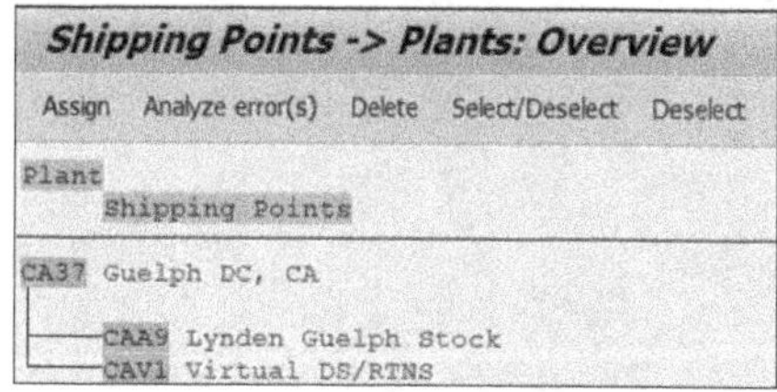

Figure 3.16 Shipping Point Assignment to a Plant

3.2 Business Partners

A business partner is an organization (firm, branch office), person, or group of persons or organizations in which your company has a business interest. Figure 3.17 shows the various possible roles of a business partner. This section covers the different aspects of business partners in more detail.

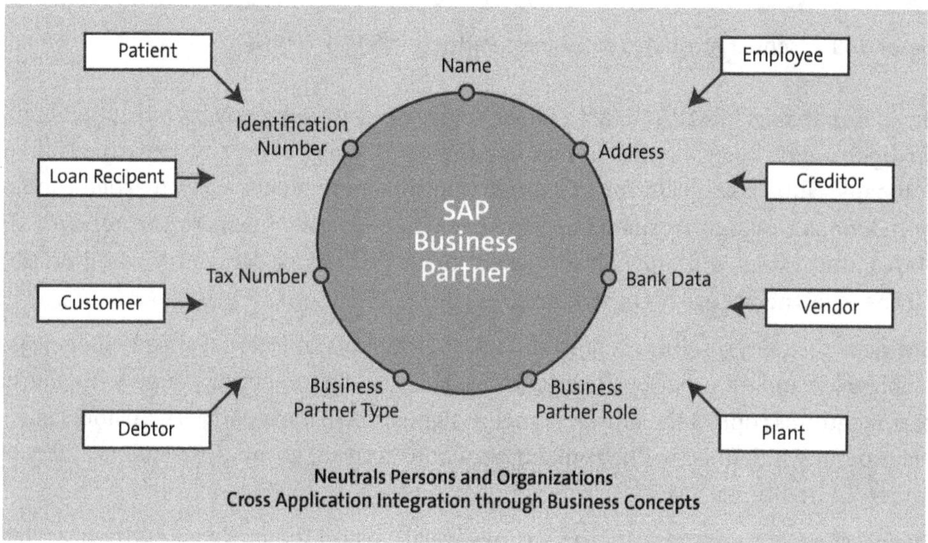

Figure 3.17 Business Partner Roles Based on the Business Interaction with the Company

3.2.1 Characteristics and Classification of Business Partners

In SAP S/4HANA, the approach is to define and manage a central database for both customers and vendors. This is achieved by maintaining business partners with a general data set representing the name, address, and other common features. The same is then extended based on the roles it assumes such as vendor, customer/sold-to party, manufacturer, payer, carrier, and so on. All role-specific parameters are maintained for individual roles upon extension of the business partner.

Business partners in SAP are structured with three distinct entities: business partner categories, groupings, and business partner roles, as shown in Figure 3.18.

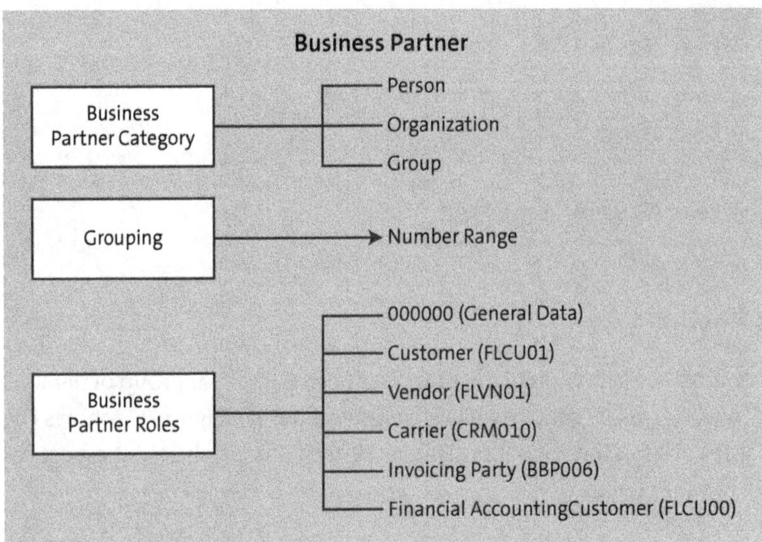

Figure 3.18 Essential Attributes of Business Partners in SAP S/4HANA

Business partners in SAP S/4HANA can be categorized as a *person* (a private individual), an *organization* (a legal entity or part of a legal entity, e.g., company, department in a company, club, or association), or a *group* (multiple individuals, e.g., a married couple or individuals living in a shared apartment). While creating business partners in SAP HANA, the system will only allow relevant fields for master data entry based on the business partner category chosen.

Business partner groupings represent logical groupings of internal stakeholders (e.g., employees) and external business entities, such as customers or vendors. Business partner groups control the number range assigned to business partners. Business partner grouping is defined in Customizing similar to account groups for vendors and customers in traditional SAP ERP.

Business partner roles classify the business relationship the business entity has with your organization and what functions are performed by that role. The role defines what data needs to be maintained for the business partner. SAP S/4HANA comes with predefined roles that are sufficient for most business functions. However, the system allows you to maintain custom roles as well, based on specific business requirements.

Following are examples of business partner roles, as shown in Figure 3.19:

- **000000 Business Partner (Gen.)**
- **FLCU01 Customer**
- **FLCU00 Customer (Fin.Accounting)**
- **FLVN01 Supplier**
- **FLVN00 Supplier (Fin.Accounting)**
- **CRM010 Carrier**
- **BBP006 Invoicing Party**

While creating a business partner master record, an appropriate role needs to be selected based on the role this entity is supposed to play. It could be vendor, customer, carrier, and so on.

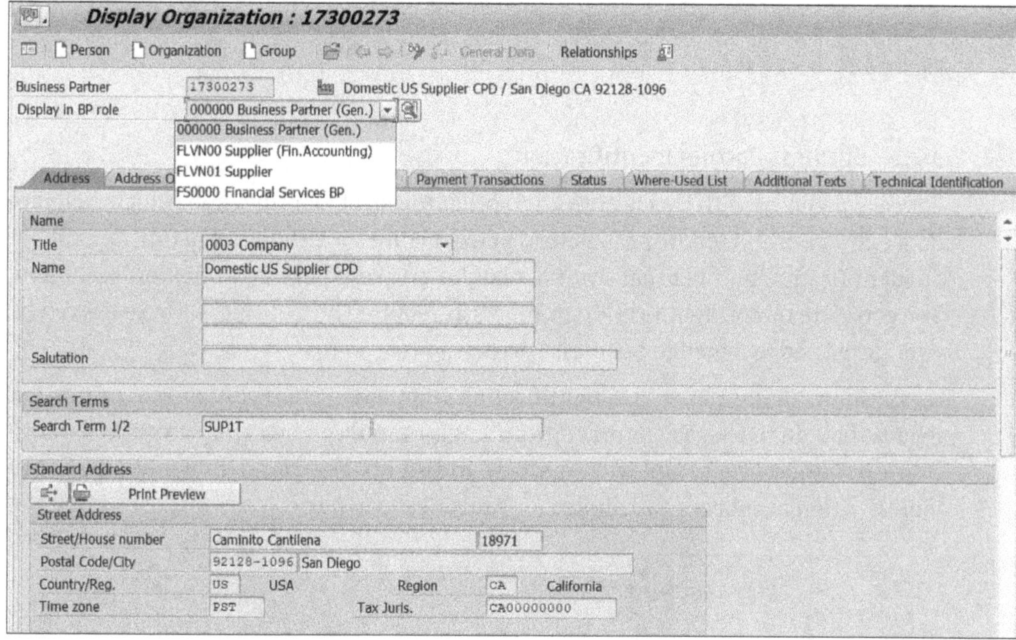

Figure 3.19 Business Partner Roles

The menu path for defining business roles is **SPRO • SAP Reference IMG • Cross-Application Components • SAP Business Partner • Business Partner • Basic Settings • Business Partner Roles • Define BP Roles.** Figure 3.20 shows business partner roles and their attributes. If the **Hide** button is selected, this role doesn't appear in the dropdown menu while creating business partner master data (refer to Figure 3.19).

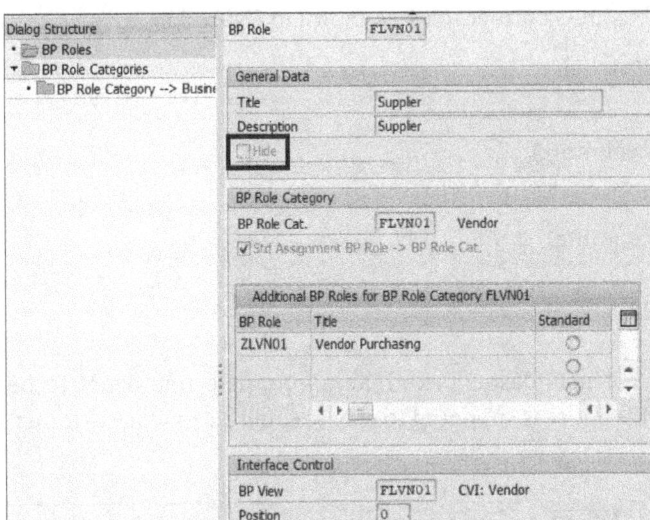

Figure 3.20 Business Partner Role and Attributes

3.2.2 Business Partner Identification

For easy mapping with legacy entity and other forms of identification, which facilitate data extraction and reporting, business partner comes with additional functionality of an identification number that stores alphanumeric keys provided by external sources. You can store many identification numbers as required by a business. These IDs can be configured and assigned to business partners.

As shown in Figure 3.21, in the **Identification Numbers** section contains identification details that can be a legacy number (from single or multiple systems), any other identification number (e.g., employee number) and so on. You can define any amount of identification numbers as attributes for a business partner.

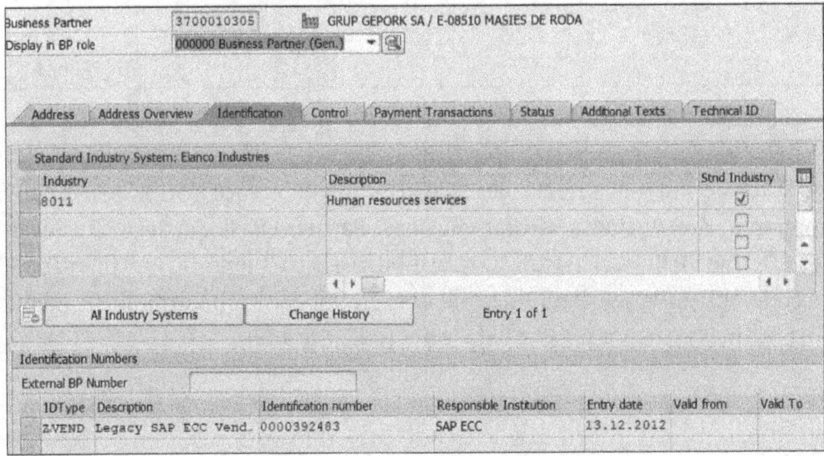

Figure 3.21 Legacy or External Identification with Validity Maintained in the Business Partner

The ID numbers are classified by the identification types defined in Customizing. While maintaining business partner master data, several identification numbers can be maintained for one type.

Now, each identification type is linked to one identification category. Using the identification category, a program can determine the identification type that is assigned to the category in Customizing, and thus access the identification numbers for a business partner. Categories are maintained in Customizing (Figure 3.22 and Figure 3.23).

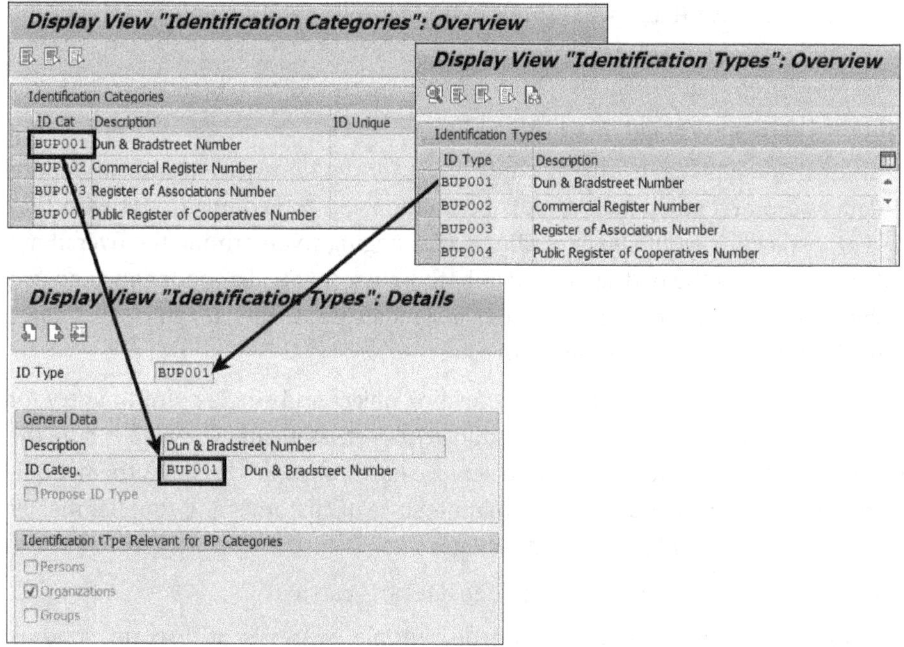

Figure 3.22 Maintenance and Linking of Identification Category and Identification Type

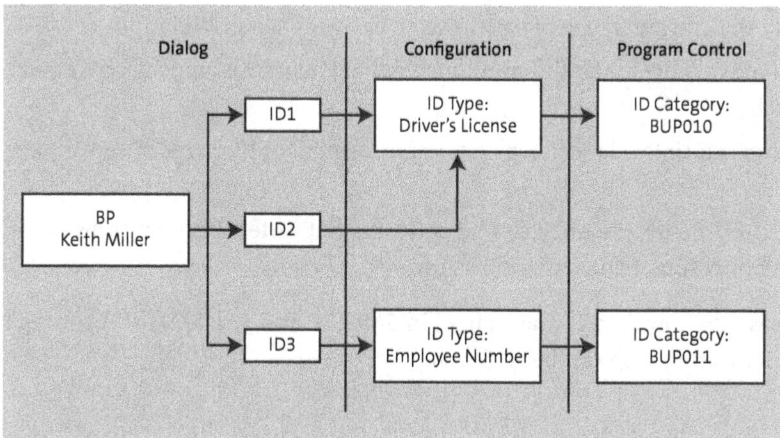

Figure 3.23 Schematic Representation of Identification Category and Type and Corresponding Linkage

Configuration of business partner identification category and type can be maintained through menu path **SPRO • SAP Reference IMG • Cross-Application Components • SAP Business Partner • Business Partner • Basic Settings • Identification Numbers • Define Identification Categories and Define Identification Types.**

3.2.3 Customer-Vendor Integration

In this section, we'll look at the facts and benefits that drive the need for customer-vendor integration (CVI) and what it is. Then, we'll walk through an overview of the steps to configure CVI in SAP S/4HANA.

Benefits of Customer-Vendor Integration

In traditional SAP ERP, having many data objects for the same party results in redundant data and omits key details about the relationship the company has with that party. For instance, it would be exceedingly challenging to determine the overall net spend for a partner if the partner served as both a customer and a vendor and had two data objects (one for each role). It becomes a major data issue when mapping such cases that involve hundreds of thousands of partners.

In SAP S/4HANA, business partner is the leading object and single point of entry for maintaining business partner, customer, and vendor master data. This is to ensure the ease of master data maintenance and to achieve harmonization between them. Compared to classical SAP ERP transactions, maintenance of customer and supplier master data via business partner has many advantages:

- A sole legal entity is represented with one business partner.
- Business partner allows maintenance of multiple addresses with corresponding address usages.
- Business partner can be assigned multiple roles.
- Maximal data sharing and reuse of data leads to easier data consolidation.
- General data is available for all different business partner roles, and specific data is stored for each role.
- Maintenance of multiple relationships can be performed for the same business partner.
- Maintenance of time dependency can be performed at different subentities, roles, addresses, relationships, bank data, and so on.

Figure 3.24 represents how data is maintained in SAP ERP and SAP S/4HANA through which CVI data harmonization can be achieved.

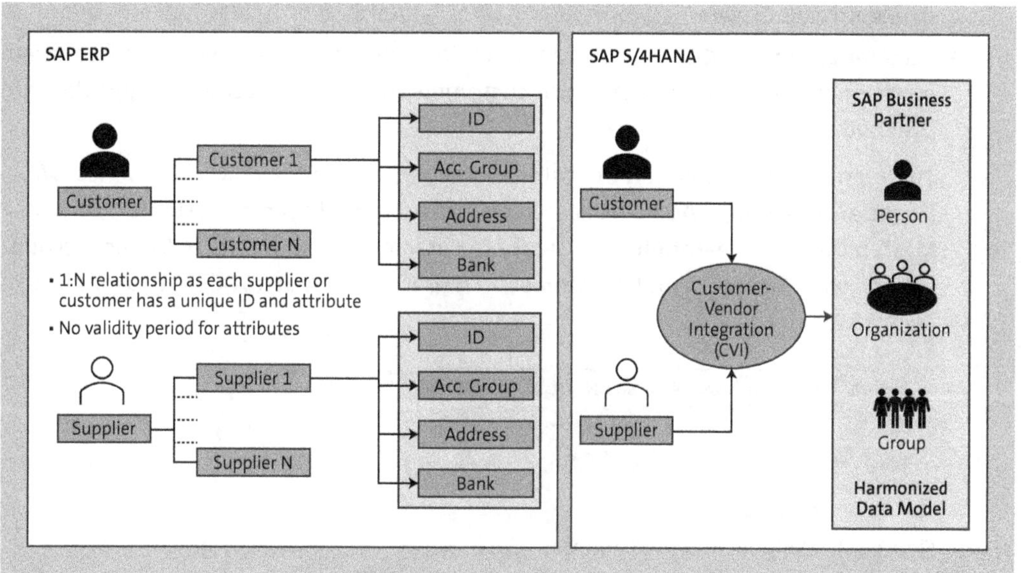

Figure 3.24 Customers and Suppliers in Traditional SAP ERP versus SAP S/4HANA with Data Harmonization via CVI

Thus, SAP CVI is essentially a prerequisite step for enterprises willing to transition to SAP S/4HANA.

Business partner is the leading master data object in SAP S/4HANA. New or greenfield implementations of SAP S/4HANA will require mandatory setup and maintenance of business partners. However, clients migrating from legacy SAP ERP or choosing to maintain a hybrid landscape of both SAP ERP and SAP S/4HANA side by side may choose to go for CVI.

CVI is bidirectional, so business partner master data is converted into customer and vendor data automatically and vice versa, populating data from customer/vendor processing to the business partner. Whenever a new business partner is created or an existing business partner is amended, relevant tables for customer (table KNA1) and vendor (LFA1) are updated accordingly. The process and parameters for auto-conversion are defined in configuration.

Configuring Customer-Vendor Integration in SAP S/4HANA

Before configuring CVI, you must ensure that all the enterprise structures pertaining to finance, sales and distribution, materials management, logistics, and logistics execution have been set up properly. In addition, all general ledger accounts, payment terms, incoterms, or any other mandatory inputs for customer/vendor master fields must be set up in system.

The following sections discuss the setup needed for CVI in sequential order. Some of these may already exist in the system.

Define Account Group

Account groups need to be created for vendors and customers. You can differentiate between different type of vendors or customers through the account groups they are assigned to.

The menu path for creation of customer account groups is **SPRO • Reference IMG • Financial Accounting • Accounts Receivable and Accounts Payable • Customer Accounts • Master Data • Preparations for Creating Customer Master Data • Define Account Groups with Screen Layout (Customer)** (Transaction S_ALR_87003301), as shown in Figure 3.25.

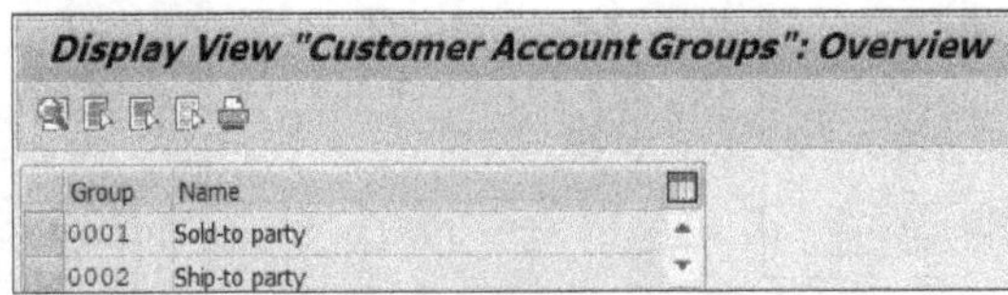

Figure 3.25 Defining Account Groups for Customers

The menu path for creation of vendor account groups is **SPRO • Reference IMG • Financial Accounting • Accounts Receivable and Accounts Payable • Vendor Accounts • Master Data • Preparations for Creating Vendor Master Data • Define Account Groups with Screen Layout (Vendor)** (Transaction S_ALR_87003126), as shown in Figure 3.26.

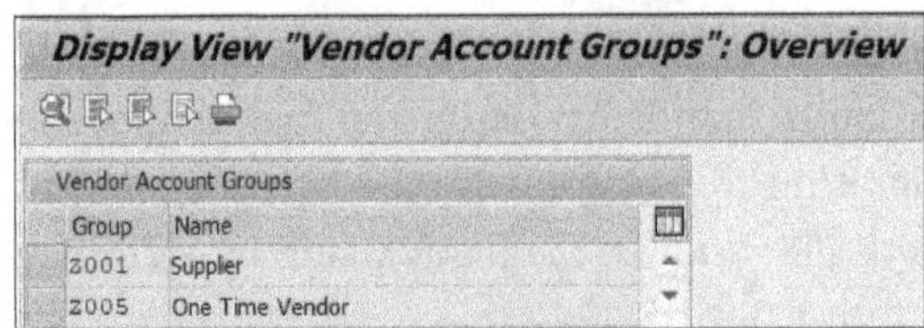

Figure 3.26 Defining Account Groups for Vendors

Create Number Ranges

Number ranges for customer and vendor account groups need to be created for assignment. The menu path for creation of number ranges for customer account groups is **SPRO • SAP Reference IMG • Financial Accounting • Accounts Receivable and Accounts Payable • Customer Accounts • Master Data • Preparations for Creating Customer Master Data • Create Number Ranges for Customer Accounts** (Transaction S_ALR_87003434 or Transaction XDN1, with the number range object DEBITOR for the customer), as shown in Figure 3.27.

Edit Intervals: Customer, Object DEBITOR

Number Range No.	From No.	To Number	NR Status	External
05	0050000000	0059999999	0	☑

Figure 3.27 Creating Number Ranges for Customer Accounts

The menu path for creation of number ranges for vendor account groups is **SPRO • Reference IMG • Financial Accounting • Accounts Receivable and Accounts Payable • Vendor Accounts • Master Data • Preparations for Creating Vendor Master Data • Create Number Ranges for Vendor Accounts** (Transaction S_ALR_87003103 or Transaction XKN1, with the number range object KREDITOR for the vendor), as shown in Figure 3.28.

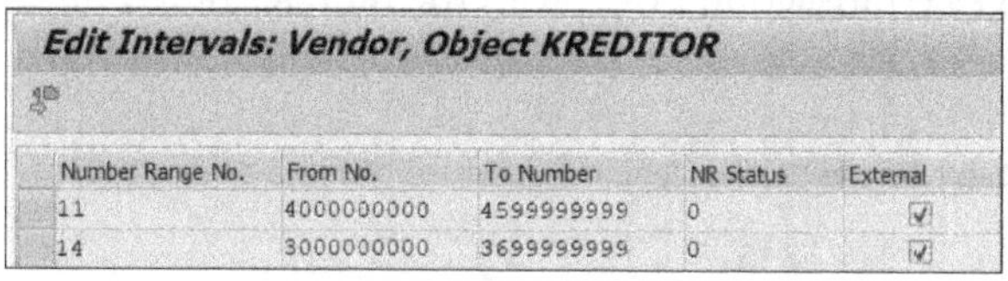

Edit Intervals: Vendor, Object KREDITOR

Number Range No.	From No.	To Number	NR Status	External
11	4000000000	4599999999	0	☑
14	3000000000	3699999999	0	☑

Figure 3.28 Creating Number Ranges for Customer Accounts

Assign Number Ranges

Each of the customer and vendor groups created needs to be assigned to a number range. This number range is consumed when a customer or vendor is created against this account group. The menu path for number range assignment to customer account groups is **SPRO • Reference IMG • Financial Accounting • Accounts Receivable and Accounts Payable • Customer Accounts • Master Data • Preparations for Creating Customer Master Data • Create Number Ranges for Customer Accounts** (Transaction S_ALR_87003428), as shown in Figure 3.29.

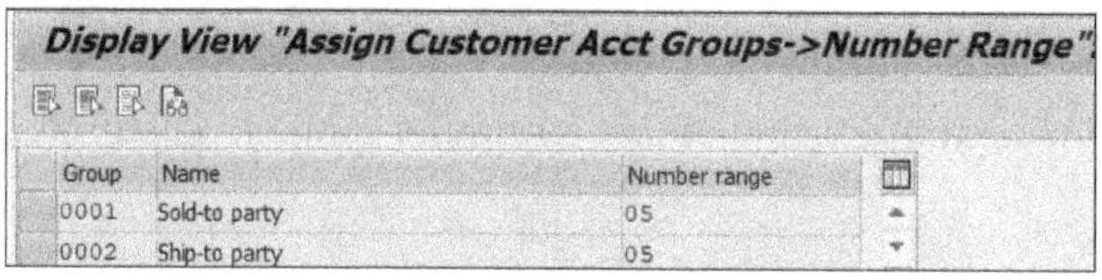

Display View "Assign Customer Acct Groups->Number Range"

Group	Name	Number range
0001	Sold-to party	05
0002	Ship-to party	05

Figure 3.29 Assigning Number Ranges to Customer Account Groups

The menu path for number range assignment to vendor account groups is **SPRO • Reference IMG • Financial Accounting • Accounts Receivable and Accounts Payable • Vendor Accounts • Master Data • Preparations for Creating Vendor Master Data • Assign Number Ranges to Vendor Account Groups**, (Transaction S_ALR_87003201), as shown in Figure 3.30.

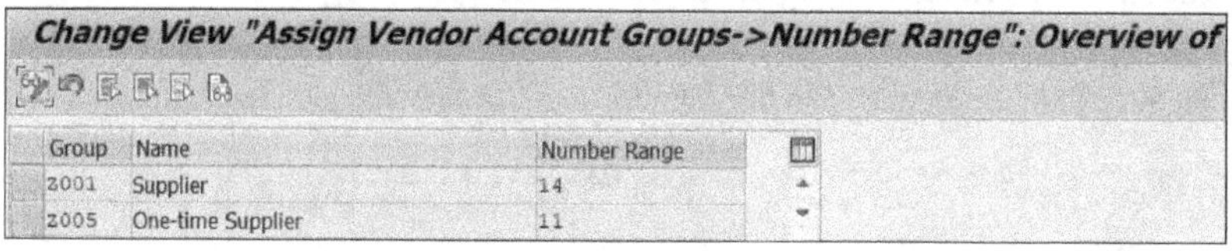

Change View "Assign Vendor Account Groups->Number Range": Overview of

Group	Name	Number Range
Z001	Supplier	14
Z005	One-time Supplier	11

Figure 3.30 Assigning Number Ranges to Vendor Account Groups

Business Partner Number Ranges and Grouping

Each business partner must be assigned to a grouping, which determines how an entry is made in the **Business Partner Number** field. The menu path for creation of number ranges for business partner groups is **SPRO • Reference IMG • Cross-Application Components • SAP Business Partner • Business Partner • Basic Settings • Number Ranges and Groupings • Define Number Ranges** (Transaction BUCF, with the number range object BU_PARTNER), as shown in Figure 3.31.

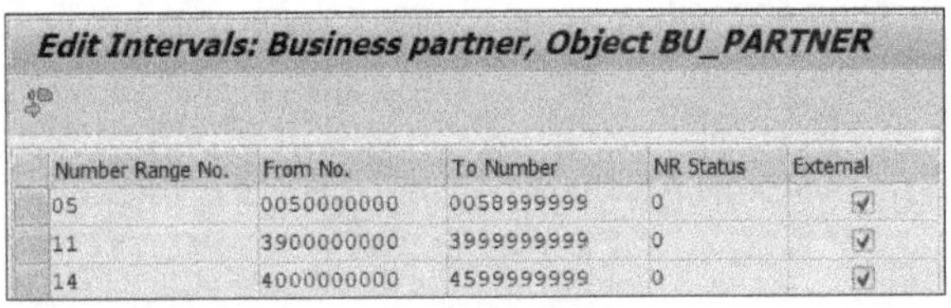

Edit Intervals: Business partner, Object BU_PARTNER

Number Range No.	From No.	To Number	NR Status	External
05	0050000000	0058999999	0	☑
11	3900000000	3999999999	0	☑
14	4000000000	4599999999	0	☑

Figure 3.31 Maintaining Number Ranges for Business Partners Groups

Once number ranges are maintained, they need to be assigned to business partner groups. The **External** standard grouping checkbox must be selected. Here, you must assign the number range used for your actual customer numbers (which is assigned to the customer account group). The same logic applies to vendor account groups as well.

The menu path for creation of number range assignment is **SPRO • Reference IMG • Cross-Application Components • SAP Business Partner • Business Partner • Basic Settings • Number Ranges and Groupings • Define Groupings and Assign Number Ranges.**

Figure 3.32 shows how to define a grouping and assign a number range to customers and vendors:

❶ **External**
This indicates internal or external number ranges.

❷ **Int.Std.Grping**
Selection indicates internal grouping used as standard grouping.

❸ **Hide**
This controls the availability of the group for selection during F4 help.

❹ **Ext.Std Grping**
Selection indicates external grouping used as standard grouping.

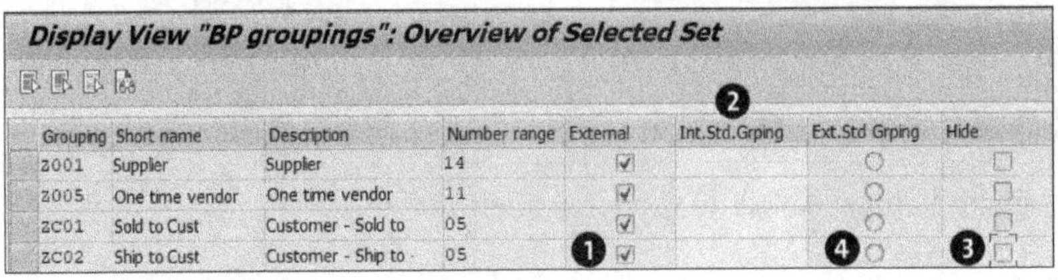

Display View "BP groupings": Overview of Selected Set

Grouping	Short name	Description	Number range	External ❶	Int.Std.Grping ❷	Ext.Std Grping ❹	Hide ❸
Z001	Supplier	Supplier	14	☑		○	☐
Z005	One time vendor	One time vendor	11	☑		○	☐
ZC01	Sold to Cust	Customer - Sold to	05	☑		○	☐
ZC02	Ship to Cust	Customer - Ship to	05	☑		○	☐

Figure 3.32 Defining Grouping and Assigning Number Range (Customer and Vendor)

Settings for Master Data Synchronization

In this configuration setting, you can display synchronization objects provided as standard for master data synchronization and, if required, create custom synchronization objects per the business requirements. This IMG activity is mainly used for system administration to get detailed information on the synchronization objects provided by SAP.

> **Note**
>
> SAP recommends using those synchronization objects that are provided as standard. Creating your own synchronization objects in the customer namespace requires a lot of additional programming.

You'll start configuration with the following steps:

1. **Activate synchronization between the business partner and customer/vendor.**
   The menu path is **SPRO • Reference IMG • Cross-Application Components • Master Data Synchronization • Synchronization Control • Activate Synchronization Options**, as shown in Figure 3.33.

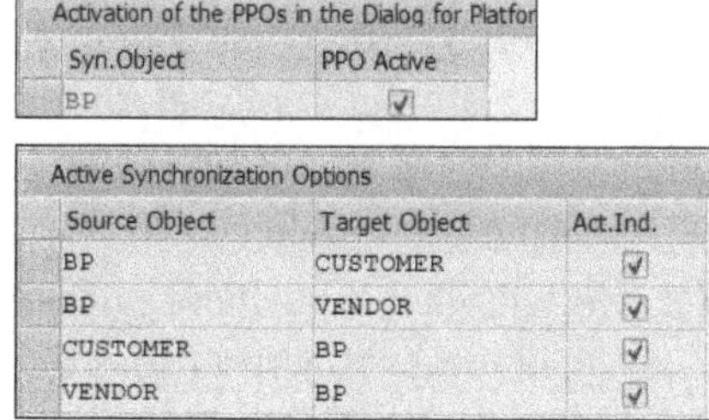

Figure 3.33 Activating Synchronization between Business Partner and Customer/Vendor

2. **Set the business partner role category for direction business partner to customer/vendor.**
   In this configuration setting, you can define which business partner role categories enable customer or vendor integration in the direction from the business partner to the customer/vendor. You can determine how the system creates a corresponding customer or a vendor in financial accounting when a business partner is created.

The menu path to CVI configuration is **SPRO • SAP Reference IMG • Cross-Application Components • Master Data Synchronization • Customer/Vendor Integration • Business Partner Settings • Settings for Customer Integration/Settings for Vendor Integration**, as shown in Figure 3.34.

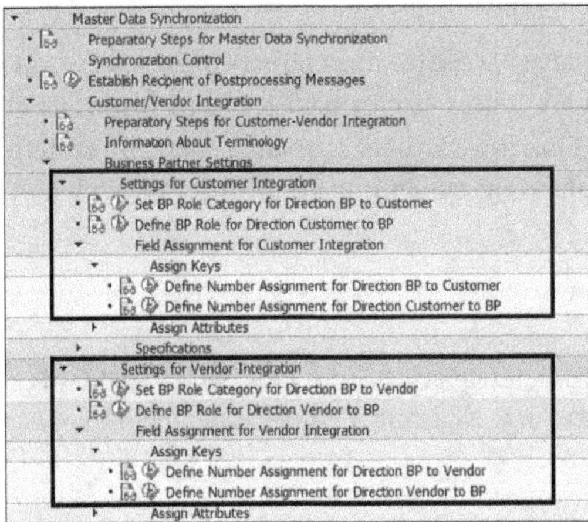

Figure 3.34 Configuration Menu Path for CVI

Let's now look at the settings needed for customer master (in financial accounting) and business partner. We'll skip the vendor integration part as the setups are similar in nature. Follow these steps:

1. **Set business partner role category for direction business partner to customer.**
   This is used to define which business partner role categories enable customer integration in the direction from the business partner to the customer. It will determine how the system creates a corresponding customer in financial accounting when a business partner is created in SAP S/4HANA. The following is shown in Figure 3.35:
   - If the **Mandatory for Customer** radio button is selected, when a business partner is maintained in this role category or in the assigned business partner role, the corresponding customer is automatically processed.
   - If the **Optional for Customer** radio button is selected, when you process the business partner in this role category or in the assigned business partner role, you have the option to decide whether the corresponding customer is created.

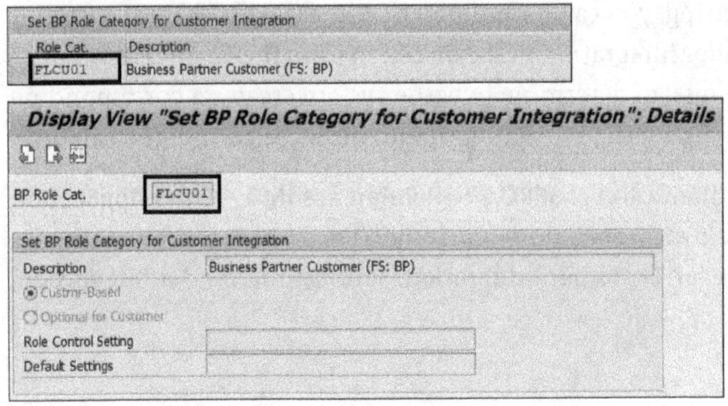

Figure 3.35 Configuration for Auto-Creation of Customer Based on Processing of Business Partner Data

2. **Define the number assignment for direction business partner to customer**
   In this IMG activity, you assign **Account groups for the customer master record** to **Business partner groupings**. This ensures that the customer is updated at the same time as the business partner is processed as part of customer integration. With this assignment, you can choose whether the customer master record is created with an account group with *internal* or *external* number assignment or with *identical* numbers.

Figure 3.26 shows the configuration for this node. If you want to use the same number as the business partner number that is currently being processed when you create the customer master record, set the **Same Numb...** indicator.

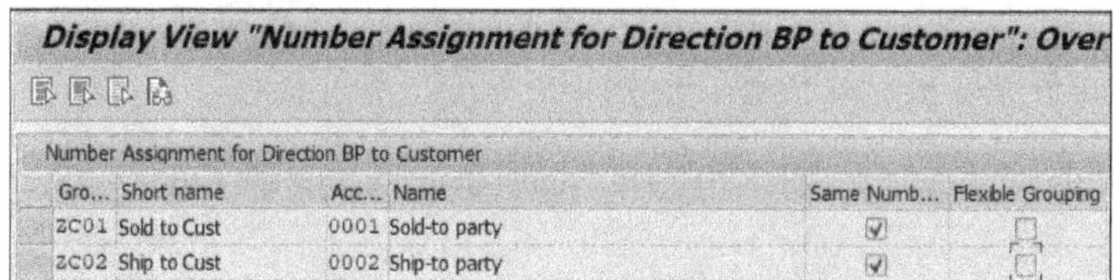

Figure 3.36 Linkage between Business Partner Account Group and Customer Account Group That Determines Customer Creation with the Same Number Range or a Different Number

Now, if a business partner is processed in SAP S/4HANA, it should create a customer number automatically in financial accounting with the same number (see Figure 3.37).

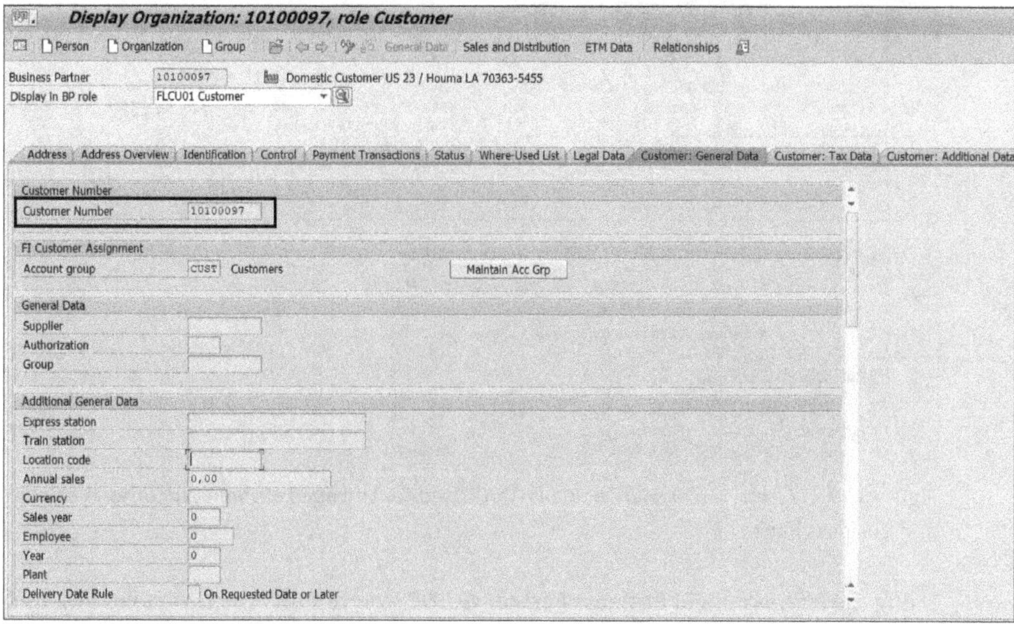

Figure 3.37 Customer Business Partner Reflecting the Customer Number (Highlighted) Created in Financial Accounting

Link between Business Partner and Vendor/Customer

Figure 3.38 is a pictorial representation of how CVI updates the database tables.

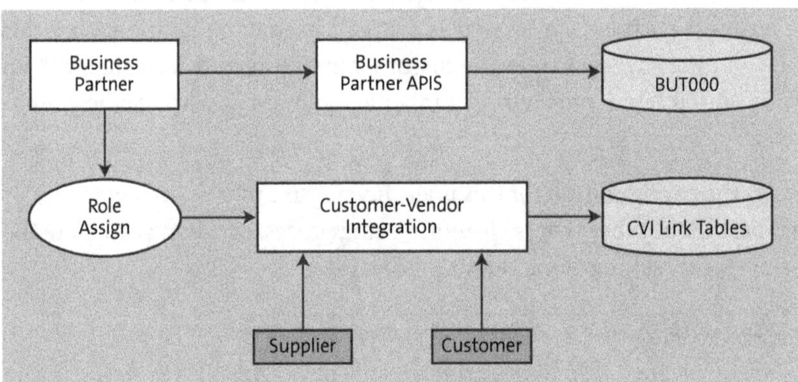

Figure 3.38 Schematic Representation of SAP S/4HANA Table Updates through CVI

Upon creation of business partner in SAP S/4HANA, table BUT000 is updated. The primary key in this table is PARTNER_GUID.

Next, when customer and supplier roles are maintained for this business partner, it passes through relevant APIs and updates link tables CVI_CUST_LINK and CVI_VEND_LINK, as shown in Figure 3.39.

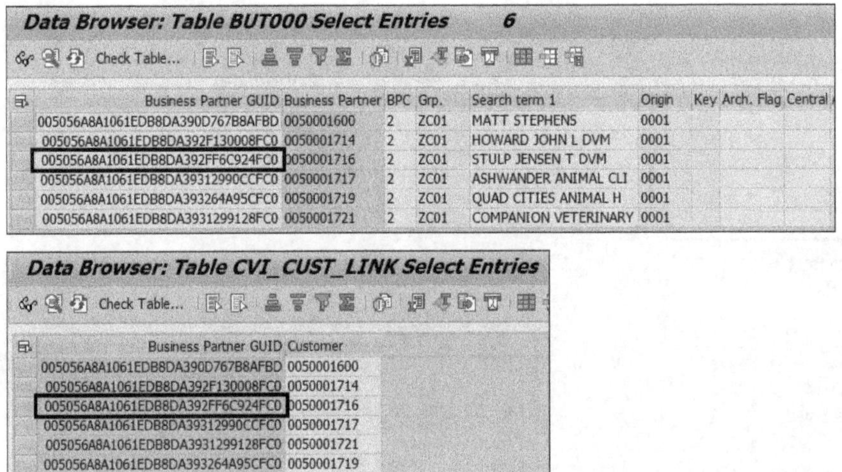

Data Browser: Table BUT000 Select Entries 6

Business Partner GUID	Business Partner	BPC	Grp.	Search term 1	Origin	Key	Arch. Flag	Central
005056A8A1061EDB8DA390D767B8AFBD	0050001600	2	ZC01	MATT STEPHENS	0001			
005056A8A1061EDB8DA392F130008FC0	0050001714	2	ZC01	HOWARD JOHN L DVM	0001			
005056A8A1061EDB8DA392FF6C924FC0	0050001716	2	ZC01	STULP JENSEN T DVM	0001			
005056A8A1061EDB8DA39312990CCFC0	0050001717	2	ZC01	ASHWANDER ANIMAL CLI	0001			
005056A8A1061EDB8DA393264A95CFC0	0050001719	2	ZC01	QUAD CITIES ANIMAL H	0001			
005056A8A1061EDB8DA3931299128FC0	0050001721	2	ZC01	COMPANION VETERINARY	0001			

Data Browser: Table CVI_CUST_LINK Select Entries

Business Partner GUID	Customer
005056A8A1061EDB8DA390D767B8AFBD	0050001600
005056A8A1061EDB8DA392F130008FC0	0050001714
005056A8A1061EDB8DA392FF6C924FC0	0050001716
005056A8A1061EDB8DA39312990CCFC0	0050001717
005056A8A1061EDB8DA3931299128FC0	0050001721
005056A8A1061EDB8DA393264A95CFC0	0050001719

Figure 3.39 CVI Link Tables and Table BUT000 Used for Linkage between Customer/Vendor and Business Partner

Using the common field **Business Partner GUIDE**, you can link the two tables and find the relation between business partner and customer/vendor. Table 3.2 lists the important tables for business partner.

SAP Table	Description
BP000	Business Partner Master (General Data)
BUT000	BP: General Data I
BUT001	BP: General Data II
BP001	FS-Specific Attributes; Organization
BP1000	BP: Business Partner - Role
BUT100	BP: Roles
BUT020	BP: Addresses
BUT021	BP: Address Usages
BUT051	BP Relationship: Contact Person Relationship
BUT0BANK	Business Partner: Bank Data
BUT0BK	BP: Bank Details
DFKKBPTAXNUM	Tax Numbers for Business Partner
BD001	Business Partner: Assignment Customer – Partner
BP030	BP: Business Partner - Address
BC001	Business Partner: Assign Vendor - Partner

Table 3.2 Standard SAP Tables Pertaining to Business Partner–Specific Details

3.3 Master Data

Maintaining master data correctly and extending that data to third-party logistics/logistics service provider's systems are important tasks in all third-party logistics design frameworks. Such master data includes material masters, customer and vendor masters, batch masters, and so on. If the master data can't be extended, relevant conversion parameters or mapping tables may need to be used.

In this section, we'll discuss a few critical master data requirements and functions, as well as features of a few key attributes.

3.3.1 Material Master

From a logistics point of view, the material master is the most important and central master data. Material is used in planning, purchasing, subcontracting, stocking and inventory taking, sales and distribution, and so on.

For third-party logistics purposes, the following three types of materials are most common:

- **Finished products**
  Typically, these are goods that the company produces and sells to clients. FERT is the material type for a finished good.
- **Semifinished products**
  These are used as components for finished or other semifinished products. They are also used as child item components in subcontracting. Semifinished products are represented by material type HALB.
- **Raw materials**
  These are materials used for making semifinished or finished products and are procured from external suppliers. Material type ROH represents such materials.

Let's briefly look at material master views and at which organizational level they are maintained in Table 3.3.

Material Master View	Organizational Level
Basic Data	Client level
Purchasing, MRP, QM, and Costing Data	Plant level
Sales Data	Sales organization level (**Tax** indicator is at country level)
Accounting Data	Valuation area level
Storage Data	Storage location level
Warehouse Data	Warehouse number level

Table 3.3 Material Master Views and Their Organizational Level

We won't look at every detail of the material master because that isn't in the scope of this topic. However, it's important to understand the different process areas or module-specific views of the material master, as well as some of the key fields from the logistics or third-party logistics standpoint.

Figure 3.40 shows the role of a few key fields in the **Basic data 1** view of the material master:

❶ **Base Unit of Measure** (table MARA-MEINS)
Unit of measure (UoM) in which material stock is maintained (table T006A).

❷ **GenItemCatGroup** (table MARA-MTPOS_MARA)
The general item category group used in item category determination during sales and stock transfer (table TPTM).

❸ **Gross weight** (table MARA-BRGEW) used in capacity check and packaging/handling unit (HU) calculations.

❹ **Net weight** (table MARA-NTGEW) also used in capacity check and packaging/HU calculations.

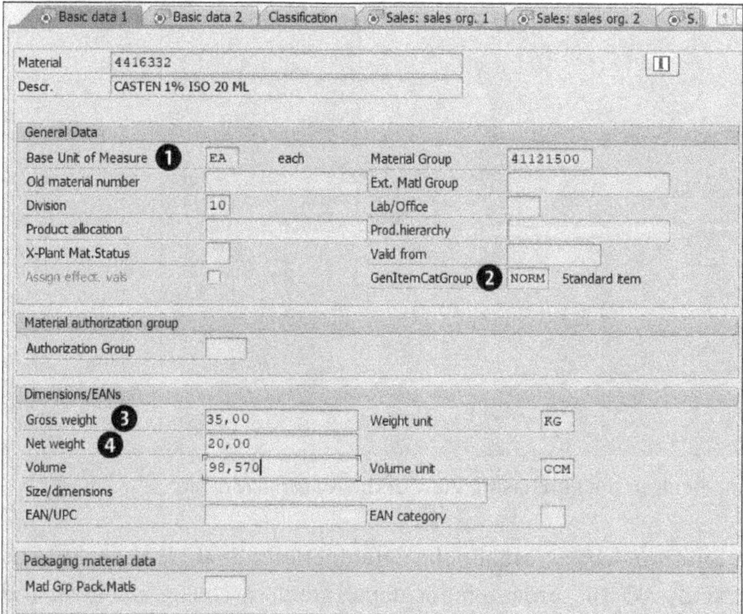

Figure 3.40 Important Fields in the Basic Data View of the Material Master

3.3.2 Procurement-Specific Master Data

Each functional area of businesses or departments are responsible to update the specific view(s) of the material master so that the concerned department will know the business trends, nuances, best practices, and process variables.

Figure 3.41 shows the **Purchasing** view of the material and its key fields:

❶ **Order Unit (table MARA-BSTME)**
Specifies the purchase UoM used in ordering of material during procurement.

❷ **Purchasing Group (table MARC-EKGRP)**
Key for a buyer or a group of buyers who is/are responsible for procurement of a material or a class of materials. This field value is also used to categorize purchase spend.

❸ **GR processing time (days) (table MARC-WEBAZ)**
Number of workdays required after receiving the material for inspection and placement into storage.

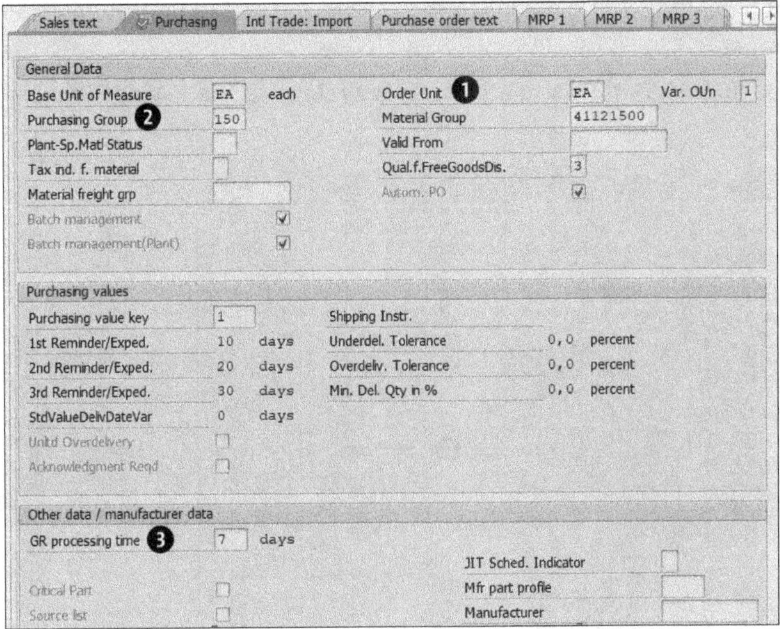

Figure 3.41 Important Fields in the Purchasing View of the Material Master

During data migration from legacy, usually the data migration team looks at holistic data requirements and asks the procurement department to collate, sanitize, and arrange data files for the following business objects:

- Material master **Purchasing** view
- Purchase-specific fields in other views of materials (e.g., **Planning** or **MRP** views, **QM** view, etc.)
- Business partner with vendor role (FLVN00 and FLVN01)
- Purchase info records
- Source list
- Pricing conditions for purchase

We've gone in depth regarding the business partner concept in Section 3.2. Maintaining business partners and purchase master data isn't in the scope of the current topic. However, you need to understand the main fields to look at while investigating or analyzing an issue with communication to third-party logistics or for that matter any external party. Knowledge of SAP tables and data types may also help in understanding the maps with XML, IDocs, EDI, and so on.

3.3.3 Sales-Specific Master Data

Like purchase master data, there are several views in the material master that control sales-specific transactions. In addition, another key master data item is the customer business partner with business partner roles FLCU00 and FLCU01 that need to be maintained for relevant sales areas.

Figure 3.42, Figure 3.43, and Figure 3.44 depict various sales-specific key data fields in the material master. Figure 3.42 shows the **Sales: sales org. 1** view:

❶ **Sales Org. (sales organization; table VKORG)**
Determines the way in which products or services reach the customer.

❷ **Distr. Chl (distribution channel; table VTWEG)**
Determines the way in which products or services reach the customer. Examples of distribution channels are wholesale, retail, or direct sales.

❸ **Division (table MARA-SPART)**
Defines grouping of materials, products, or services. This, along with sales organizations and distribution channels, helps in determining the sales areas for a material, product, or service.

❹ **Delivering Plant (table MVKE-DWERK)**
Plant from which the goods should be delivered to the customer. The value proposed in the sales order item is from this field, if maintained. Otherwise, it can be from the customer master record of the goods recipient or from the material master record.

❺ **Tax Category (table TATYP)**
Identifies the condition that the system uses to automatically determine country-specific taxes during pricing.

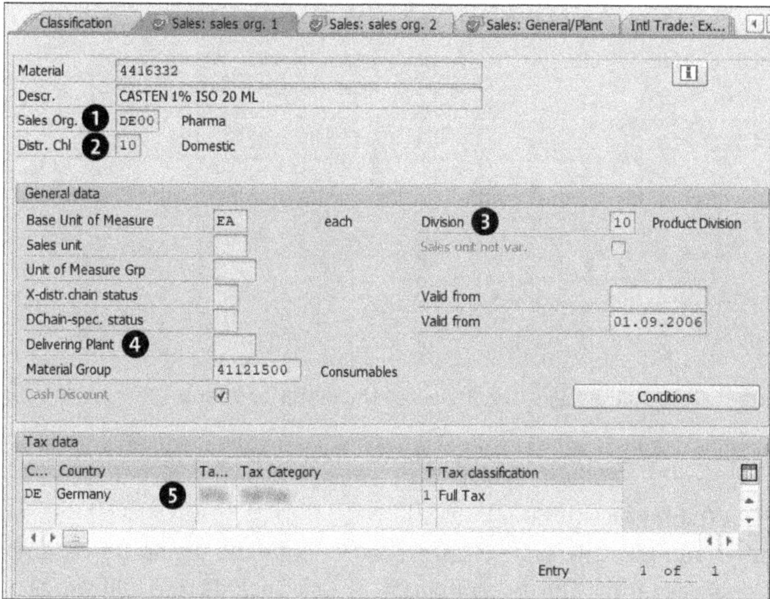

Figure 3.42 Important Fields in the Sales Org. 1 View of the Material Master

Figure 3.43 shows the **Sales: sales org. 2** view with the following fields:

❶ **Acc Assmt Grp Mat. (account assignment group for material; table MVKE-KTGRM)**
Group of materials with the same accounting requirements. Often different types of materials require different accounting requirements (finished goods, trading goods, and services require different accounting during billing).

❷ **Gen. item cat. grp (general item category group; table MARA-MTPOS_MARA)**
Materials grouping that helps the system to determine item categories during sales and stock transfer order (STO) processing. It's based on material type settings in configuration.

❸ **Item Category Group (from material master; table MVKE-MTPOS)**
This is copied from the **Gen. item cat. grp** field, but this is the category group that is assigned to a specific distribution channel. It can also be different from the **Gen. item cat. grp** field, and this field value will have priority if the corresponding activity is distribution channel–specific. However, it requires a business decision to exercise this option.

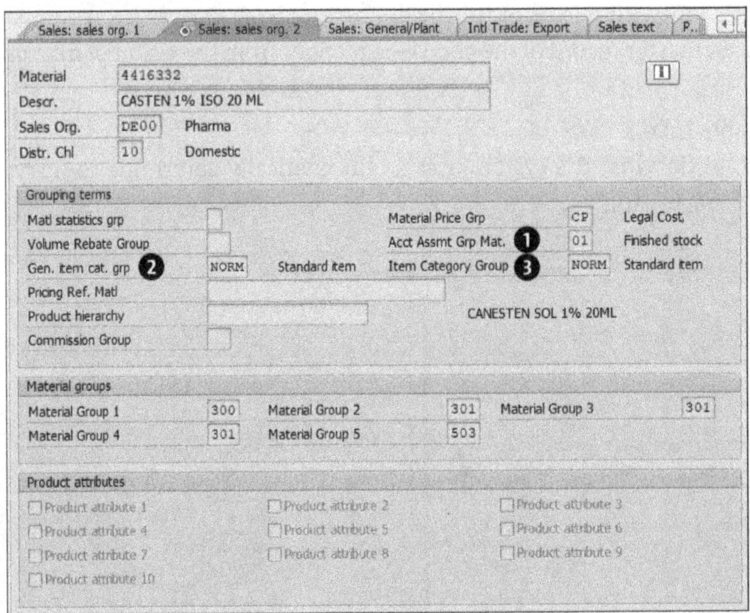

Figure 3.43 Important Fields in the Sales Org. 2 View of the Material Master

Figure 3.44 shows the **Sales: General/Plant** view:

❶ **Availability check (table MARC-MTVPF)**
Specifies whether and how the system checks availability and generates requirements for materials planning. The determination process includes the following:

- Which transaction documents to consider for the availability check
- Period till the end of the replenishment lead time or over the entire period

- Whether the system generates individual requirements or summarized requirements

❷ **Trans. Grp (transportation group; table MARA-TRAGR)**
Grouping of materials that share the same route and transportation requirements.

❸ **LoadingGrp (loading group; table MARC-DWERK)**
Grouping of materials that share the same loading requirements. The system uses the loading group, as well as the shipping conditions (from the customer master) and the delivering plant, to automatically determine the item's shipping point.

❹ **Profit Center (table MARC-PRCTR)**
Assigns an adequate profit center for the material for controlling module purposes. This means that profit made from this material will be assigned to the profit center set in this field.

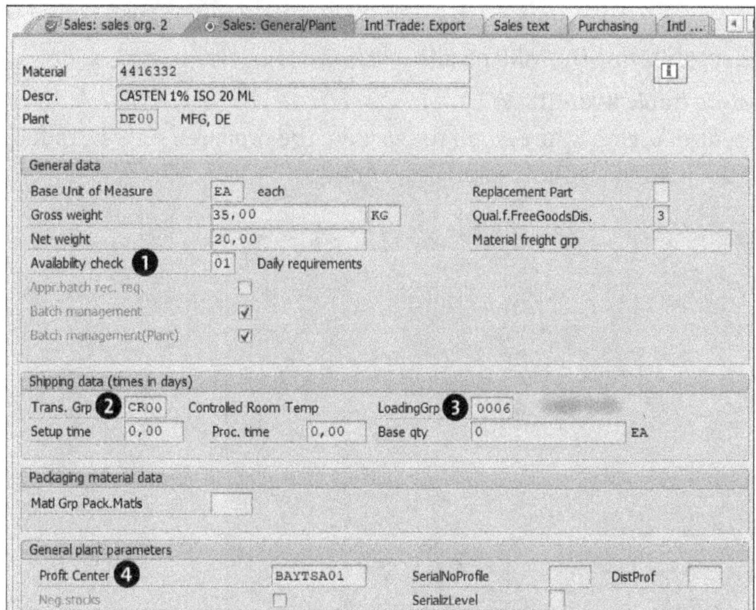

Figure 3.44 Important Fields in the General/Plant View of the Material Master

3.3.4 Other Supply Chain-Relevant Master Data

If we look at the end-to-end supply chain cycle, other business functions such as planning (for sales, purchase, and stock transfer), warehouse management (WM), and quality management play equally important parts in the value chain. However, to avoid too much diversion, we'll quickly look at some of these process-specific key fields in the material master.

Figure 3.45 shows the **MRP 1** view:

❶ **MRP Type (table MARC-DISMM)**

Key that determines whether and how the material is planned. The following standard options are available, but you can also have custom material requirements planning (MRP) types:

- Manual reorder point planning
- Automatic reorder point planning
- Forecast-based planning
- MRP with forecasts for unplanned consumption
- Master production scheduling (MPS)
- Demand-driven replenishment

❷ **MRP Controller (table MARC-DISPO)**

Specifies the number of the MRP controller or group of MRP controllers responsible for material planning for the material. MRP controller manages all the data in MRP views, along with monitoring the MRP results.

❸ **Lot Sizing Procedure (table MARC-DISLS)**

Defines the procedure used by the system to calculate the procurement or production quantity of the material within materials planning.

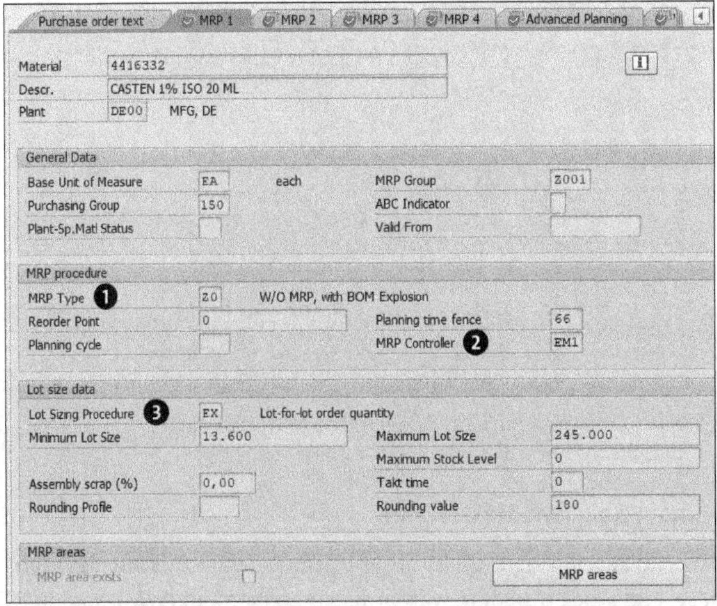

Figure 3.45 Important Fields in the MRP 1 View of the Material Master

Figure 3.46 shows the **MRP 2** view:

❶ **Procurement Type (table MARC-BESKZ)**

Value in this field determines how the material is procured. The following procurements types are possible:

- E: Material is produced in-house.
- F: Material is procured externally.
- X: Material can be both produced in-house and procured externally.
- Blank: No procurement.

Procurement type is defined in the configuration of the material type.

❷ **Special procurement (table MARC-SOBSL)**

The value in this field allows you to override the **Procurement Type** field or further narrow down selections based on the **Procurement Type** field value. For example, procurement type **F** or **X** may have a **Special Procurement Type** of "30", indicating subcontracting.

❸ **Prod. stor. location (issue storage location; table MARC-LGPRO)**

This storage location is copied to the planned order, production order, or run schedule quantity for in-house production of material. For component material, it's the issuing storage location to which a backflush is posted. When the material is produced, it's the receiving storage location to which the receipt of the material is posted.

❹ **Storage loc. for EP (default storage location for external procurement; table MARC-LGFSB)**

Storage location that is proposed in the purchase requisition in materials planning for subsequent storage of the material. For external procurement, this storage location is proposed in the planned order in materials planning.

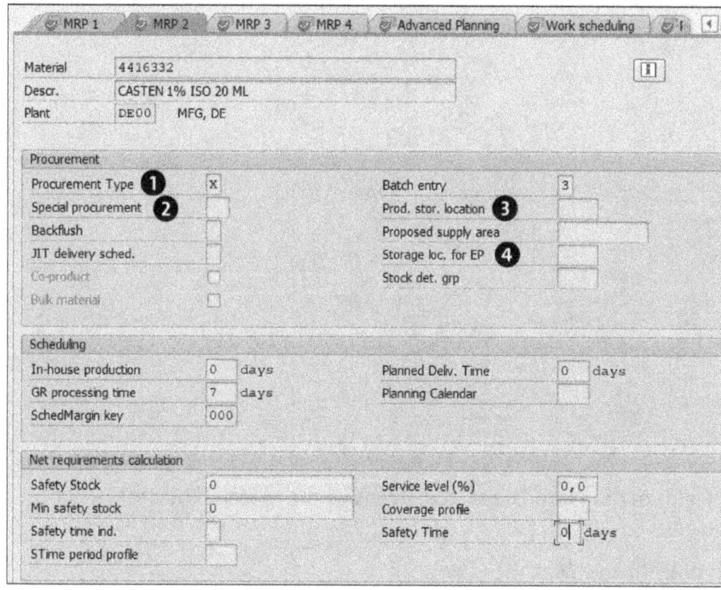

Figure 3.46 Important Fields in the MRP 2 View of Material Master

Figure 3.47 shows the **Plant data/stor. 1** view:

❶ **Temp. conditions (temperature conditions indicator; table MARA-TEMPB)**
Temperature conditions in which the material must be stored.

❷ **Storage conditions (table MARA-RAUBE)**
Indicates storage conditions required by the material. This field value is also used in shipping point determination.

❸ **Mn. Rem. Shelf Life (minimum remaining shelf life; table MARA-MHDRZ)**
Minimum remaining time for the material to be allowed to be stored (if the remaining time is shorter than the minimum, the system will deny goods receipt). It's defined in combination with the period (day, week, month, year) as the **Period Ind. for SLED** field. The shelf life is checked upon goods receipt only if this check has been activated in Customizing for inventory management in the activity **Set Expiration Date Check.**

❹ **Total shelf life (table MARA-MHDHB)**
Period from the date of production to the shelf-life expiration date. If the value is maintained in the material master, the system calculates the shelf-life expiration date from the date of production (goods receipt date) plus the total shelf life. If no value is maintained, the system assumes that the date entered upon goods receipt is the material's shelf-life expiration date.

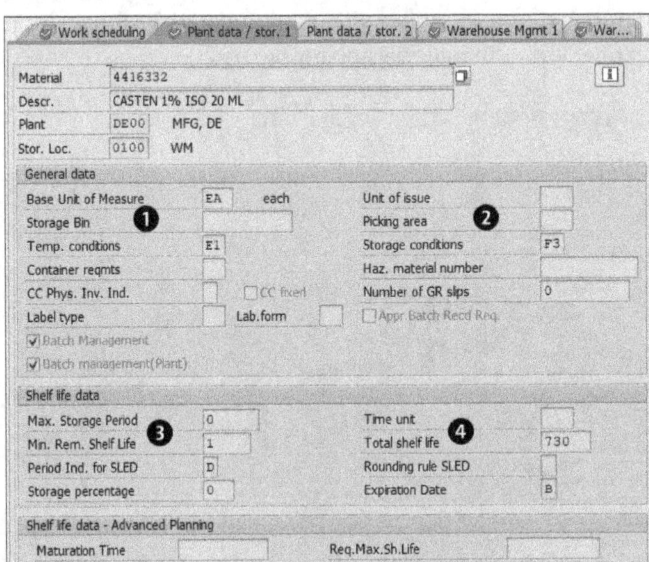

Figure 3.47 Important Fields in the Plant Data/Stor. 1 View of the Material Master

Figure 3.48 shows the **Warehouse Mgmt 1** view:

❶ **Whse. No. (warehouse number; table LGNUM)**
All activities within a warehouse, for example, goods movements and physical inventory, are assigned to a specific warehouse number.

❷ **WM unit (WM UoM; table `MLGN-LVSME`)**
UoM in which WM manages the stock of material. If maintained, the WM document uses this UoM instead of the base UoM. Only maintain this field if you want to use a UoM that differs from the base UoM or the alternative UoMs.

❸ **Stock removal (table `MLGN-LTKZA`)**
This storage type indicator determines the storage types selected by the system when removing stock and the priority with which they are accessed. Up to 10 storage types can be predefined in the hierarchical order. The stock removal, or picking strategy, is defined in Customizing.

❹ **Stock placement (table `MLGN-LTKZE`)**
Storage type indicator that determines the storage types to be selected by the system and the priority in which they are accessed for stock placements. Putaway strategy can be set in Customizing that allows the system to choose the right storage type for stock placement or putaway.

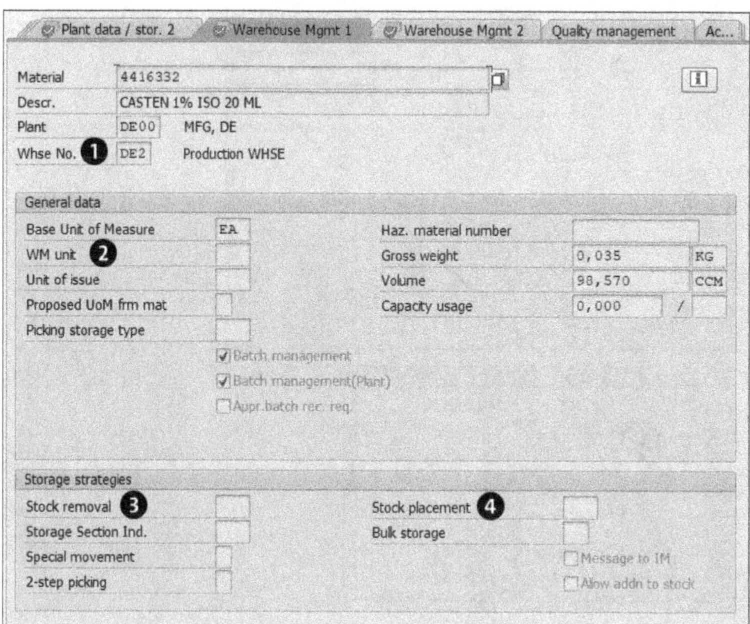

Figure 3.48 Important Fields in the Warehouse Mgmt 1 View of the Material Master

Figure 3.49 shows the **Quality management** view:

❶ **Inspection Setup (table `MARC-QMATV`)**
If checked, indicates that at least one inspection type has been set up for the material and plant.

❷ **QM in Procur. Active (quality management in procurement is active; table `MARA-QMPUR`)**
If maintained, this will activate quality during procurement/ordering of material.

❸ **QM Control Key (control key for quality management in procurement; table MARC-SSQSS)**
This key specifies conditions in quality inspection in procurement. One can set a delivery block, invoice block, and other types of blocks until material inspection is done at the source. Possible entries are predefined in Customizing.

❹ **Inspection Types (table QPART)**
The inspection type defines how an inspection is performed. Several inspection types can be assigned to a material. Inspection type will work only when set to **Active**. Inspection types can be set in Customizing.

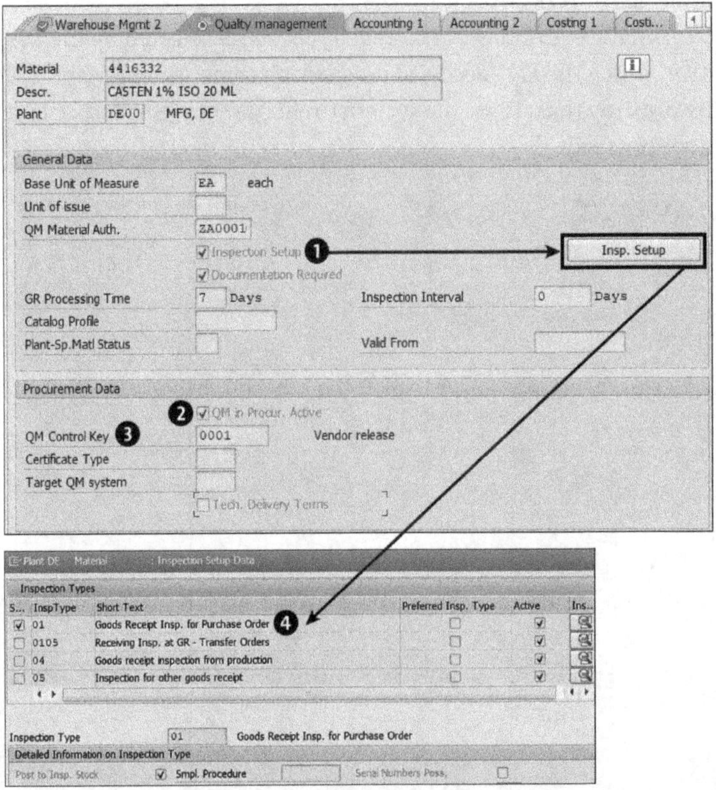

Figure 3.49 Important Fields in the Quality Management View of the Material Master

3.4 Summary

In this chapter, you saw three main areas that contribute to the success of supply chain transactions: organizational entities, business partners, and master data. The intention was to familiarize you with a few key areas such as business partners and the importance of data fields that may help you in analyzing the issues with logistical transactions, interpretation of data maps between SAP and third-party solutions, and designing output or communication channels with external agencies.

Chapter 4
Packaging and Traceability

Customer service and product quality are the two cornerstones on which business reputations are built and destroyed. Even if a consumer receives a flawed product, the company still has a chance to win them over by quickly and seamlessly replacing the flawed item. Batch tracking, an inventory management strategy, is vital in this situation because it has an impact on both product quality and customer satisfaction. Supply chain efficiency and economy of operation necessitates innovative collaboration between packaging, logistics, and supply chain operations. Effective packaging ensures the safety of goods and materials during transit, distribution, and storage.

This chapter emphasizes the basics of batch management and handling unit management (HUM) in SAP and how these are relevant to supply chain, shipping, and delivery transactions. A batch is characterized by a subset of inventory that was produced during a typical production run and has homogenous units with unique specifications. Batch management establishes the traceability across the supply chain from sourcing to manufacturing and then from sales to end customers. In industries such as pharma, mining, fast-moving consumer goods (FMCG), retail, automotive, aerospace, and engineered goods, batch management plays a crucial role in determining product traceability, root cause analysis (RCA), or product fault analysis and may also help in product recall for substitution. Implementing batch management is necessary for a variety of reasons, such as the traceability of quality problems in a batch that needs to be recalled, the need for a certain ISO certification, the approaching expiration date of items, and so on.

In SAP S/4HANA (or in SAP ERP), HUs represent a combination of material or a group of materials packed into a single logistical unit using a packing material. In the logistics process of an organization, materials are packed together with packing materials, such as cartons, pallets, containers, and so on, and moved together within the supply chain or stored in an inventory as a group. All the information contained in the product items, for example, batches, weight, and so on, is retained in the HUs and is always available.

As shown in Figure 4.1, batch management plays an important role in the product value chain and cuts across multiple modules of supply chain and logistics.

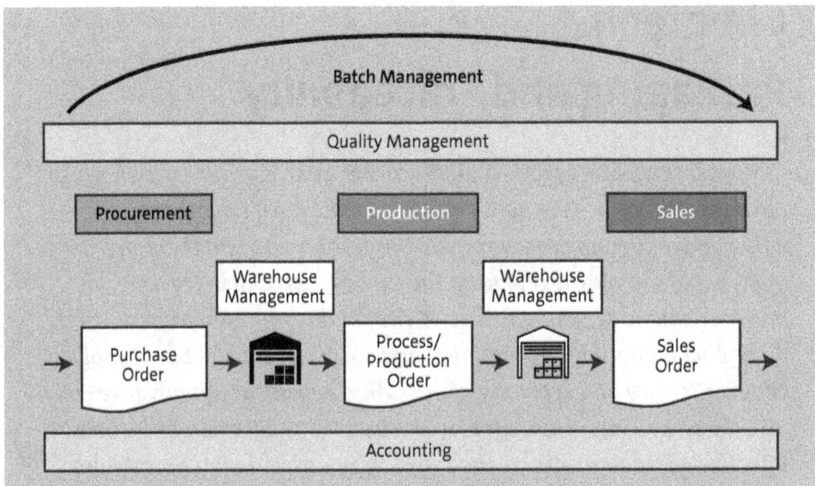

Figure 4.1 Integration of Batch Management with All Applications of the SAP System

4.1 Batch Management

Batch management aims to create consistency from product to product and run to run, as well as making the best use of your supplies, minimizing waste, and streamlining manufacturing processes. It's a concept closely tied to both quality assurance and operational efficiency. In this section, we'll look at how batch management contributes to the value chain in traceability and various other regulatory practices. In addition, it may be essential to understand the integration points with various supply chain modules or processes.

4.1.1 Benefits and Integration Points

There are various reasons to use batch management:

- Defect tracing, callback activities, and regression requirement
- Legal requirements (e.g., the guidelines set out by Good Manufacturing Practice [GMP] or regulations on hazardous materials)
- Need for differentiated quantity- and value-based inventory management (e.g., due to heterogeneous yield/result qualities or varying constituents in production)
- Differences in usage and the monitoring thereof in materials planning in sales and distribution and production
- Requirements for the production or the process (e.g., settlement of material quantities based on different batch specifications)

From a third-party logistics perspective, following are the integration points with respect to how the batch number forms a part of relevant transactions:

- **Sales order and outbound delivery**
  Batch determination perform a batch analysis together with a batch usability and availability check. The system decides if there is enough stock to fulfil the sales order. Outbound delivery created against the sales order is then communicated to logistics service provider for further transactions.
- **Purchase order and material receipts**
  If there isn't shortfall in raw material with desired specifications, the eligible vendor is asked to supply necessary quantity with the correct specifications. Batch number is created internally at the receiving system and assigned either to inbound delivery or to Goods-receipt postings. Similar to outbound delivery, inbound delivery information (external purchase or subcontracting) is communicated to logistics service provider for arrange for material pick-up and delivery to destination.
- **Quality management**
  For quality-managed materials, the goods received from external suppliers or the lots produced internally will have batch numbers assigned and are subject to quality inspection. After the inspection, results recorded, and usage decision posted, the materials become available for batch determination. Inventory snapshots on a daily basis, at the minimum, need to be communicated to the logistics service provider for reconciliation purpose. If needed, this could be arranged for shorter intervals as well.
- **Warehouse management (WM)**
  Using the transport order, the batch of raw material is placed into stock at the storage bin level. For materials managed via WM/extended warehouse management (EWM), warehouse transactions are always performed during inbound and outbound movements of stock. However, logistics service provider integration happens usually at the inventory management level as many logistics service providers run their own warehouse management systems (WMSs; SAP/non-SAP) and integration at the delivery level provides some flexibility to both clients and service providers.

4.1.2 Define Batch Level and Batch Status Management

This is one of the crucial design decisions while adopting batch management and deciding what level of granularity is needed for traceability through batches. Batch numbers can be distinctive or unique at the following levels:

- **Plant level**
  During selection at the plant level, the batch number is exclusive to that particular material and plant combination. The batch number is assigned and recognized in a plant where the material was created or procured. During stock transfer to another plant, the recipient plant may not recognize the batch number, so the system has to re-create the same batch number and then assign it to the material received. There

is always the possibility of having the same batch number with different specifications in different plants.

In addition, as an alternative, a different internal batch number is generated at goods receipt in that plant, and the user has an option to store different batch characteristics values (batch specifications) for the batch in the receiving plant. This setting is useful in the absence of inter-plant transfer of materials.

- **Material level**
  The same batch number has the same meaning for the material in all plants. One batch number can't have different specifications in different plants.

  During stock transfer to another plant, the specification of the batch is known in the destination plant even if the batch is placed into storage in that plant for the first time.

- **Client level**
  The batch number is unique in the whole client. A batch number can only be assigned once in the client. The same batch number can't be assigned to different materials.

As shown in Figure 4.2, the menu path for configuration is **SPRO • SAP Reference IMG • Logistics - General • Batch Management • Specify Batch Level and Activate Status Management • Batch Level** (Transaction OMCE).

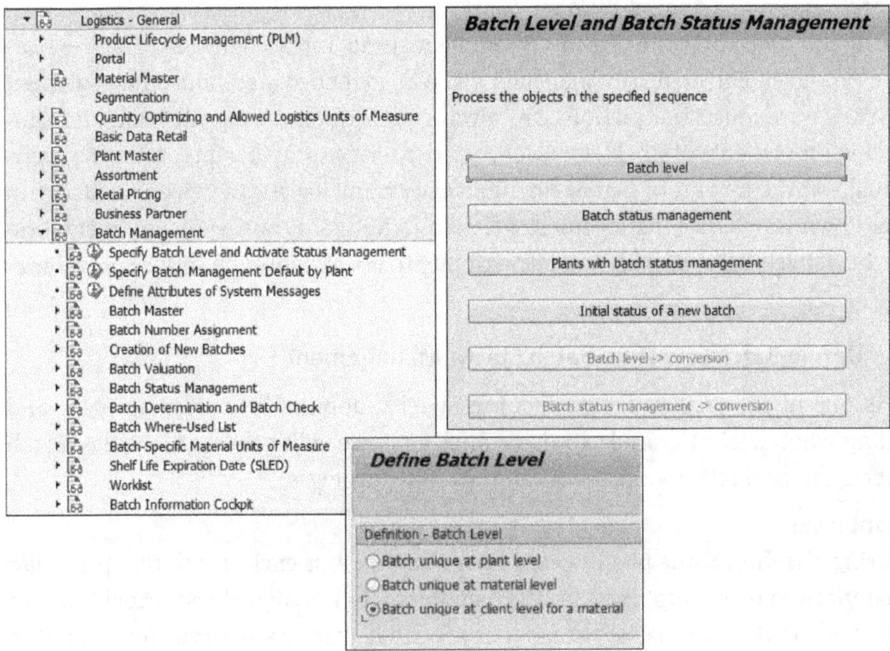

Figure 4.2 Specifying Batch Level in Configuration

Note

Table MCHA stores information of batches at the plant level whereas table MCH1 stores batch information at the material or client levels. In the standard delivery, batches are uniquely defined at the material level.

For batch-level conversion, SAP provides a standard report called RM07CHDX. When changing the batch level from plant to material, there are a few prerequisites listed in SAP Note 891902 that must also be taken into consideration.

Another useful feature of batch status management is provided to indicate whether a batch is usable or not. This allows you to prevent goods movements for batches that aren't usable, as well as their delivery to customers and their use in production. This way, standard batch handling can be supplemented by batch status management.

When a new batch is received in inventory, the status is first set to **Unrestricted Use**. However, the following situations allow for restricting the use of a batch:

- Quality management specifies the status **Restricted Use** in its usage decision.
- The user sets the status to **Restricted Use** using Transaction MSC2N (Change Batch). This setting tags the material batch stock as **Unrestricted** or **Restricted**, which is an added advantage to stop the batch from being used even when the batch stock lies in unrestricted stock type (**Restricted Batch** status).

Batch status can be looked at as an added stock type over and above the standard stock types in SAP such as the unrestricted stock, blocked stock, and quality stock. This means that if a batch stock is in quality stock type or blocked stock type and tagged with **Unrestricted Batch** status, then it's still counted as quality stock.

If a batch stock is in quality stock type or blocked stock type and tagged as **Restricted Batch** status, then it's still counted as quality stock, though when the stock moves from quality to unrestricted stock type, the batch stock would still not be available for use because the **Restricted Batch** status is still stopping it from consumption. When a batch stock is in unrestricted stock type and tagged with **Restricted Batch** status, the batch stock is unavailable for consumption.

The selected batch level determines the plants in which status management is used for a material to be handled in batches, as shown in Table 4.1.

Batch Level	Status Management
Material or client level	Automatically in all plants
Plant level	Can be chosen for individual plants

Table 4.1 Management of Batch Level Status

4.1.3 Specify Batch Management Default by Plant

This setting indicates whether batch management is turned on automatically at the plant level. It provides the option to activate or deactivate batch management by default for specific plants. Such use cases may include distribution centers where no batch management may be required for any materials; however, for manufacturing plants, detailed batch management handling may be required.

As depicted in Figure 4.3, the **Not Batch Management** option default setting for plants is enabled where batch management isn't required. Otherwise, it's turned off for plants that require it.

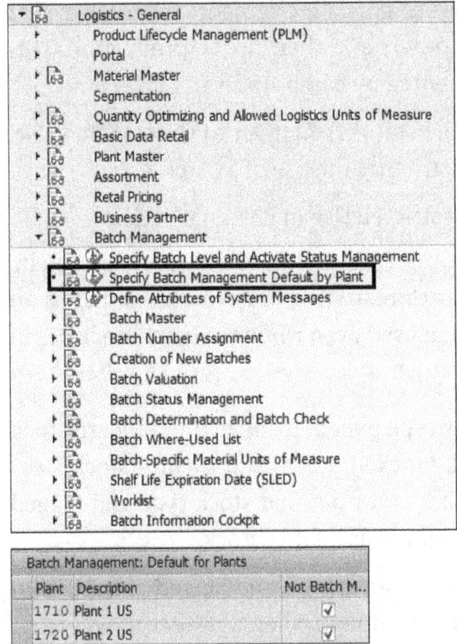

Figure 4.3 Deactivate Batch Management for Specific Plants

Enabling the **Not Batch Management** setting can help avoid repetitive tasks for specific plants, and batch management–relevant fields will be suppressed in the plant view of the material master for a batch-managed material.

The default settings for plant-specific batch management can be maintained using menu path **SPRO • SAP Reference IMG • Logistics - General • Batch Management • Specify Batch Management Default by Plant** (Transaction OM54).

4.1.4 Batch Number and Assignments

Numbering of batches is an important factor as it can be purely numeric and system generated. On the other hand, businesses may also decide to have external numbering with numeric or alphanumeric characters or digits.

Configuration for number ranges can be achieved using menu path **SPRO • SAP Reference IMG • Logistics - General • Batch Management • Batch Number Assignment • Activate Internal Batch Number Assignment** (Transaction OM54), as shown in Figure 4.4.

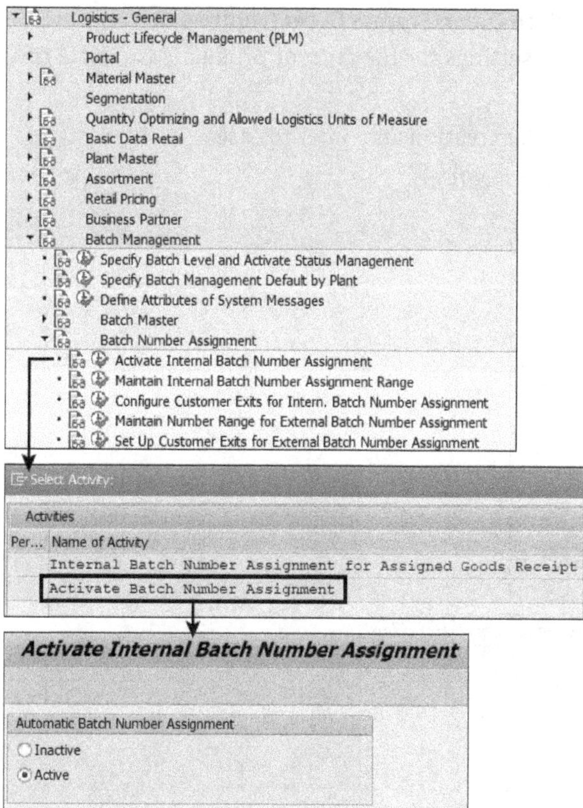

Figure 4.4 Activating Internal Batch Number Assignment

Next, internal and external number ranges are defined by following menu path **SPRO • SAP Reference IMG • Logistics - General • Batch Management • Batch Number Assignment • Maintain Internal Batch Number Assignment Range / Maintain Number Range for External Batch Number Assignment** (Transaction OMAD), as shown in Figure 4.5.

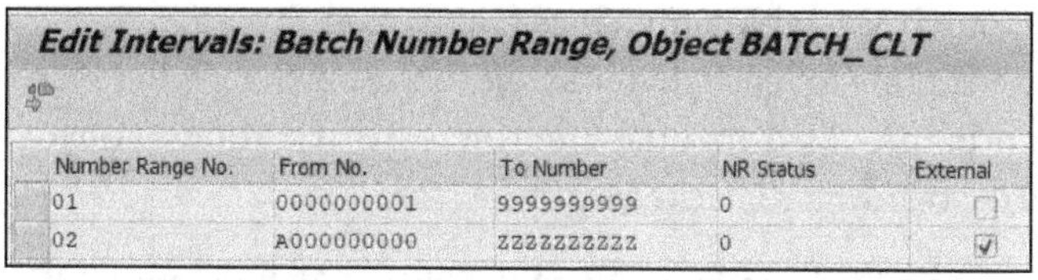

Edit Intervals: Batch Number Range, Object BATCH_CLT

Number Range No.	From No.	To Number	NR Status	External
01	0000000001	9999999999	0	☐
02	A000000000	ZZZZZZZZZZ	0	☑

Figure 4.5 Creation of Internal and External Number Ranges

4.1.5 Creation of New Batches

In this configuration step, you can decide whether the batch creation is manual or automatic, per user exit, and so on. The system provides the following options:

- **Define Batch Creation for Production Order/Process Order (Figure 4.6)**
  The following batch management settings for the current production control profiles can be made in this step:
  - Automatic batch creation on order creation and order release
  - Automatic classification on batch creation

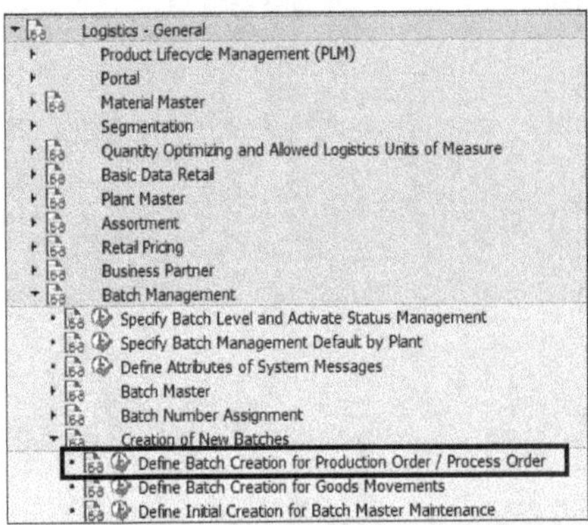

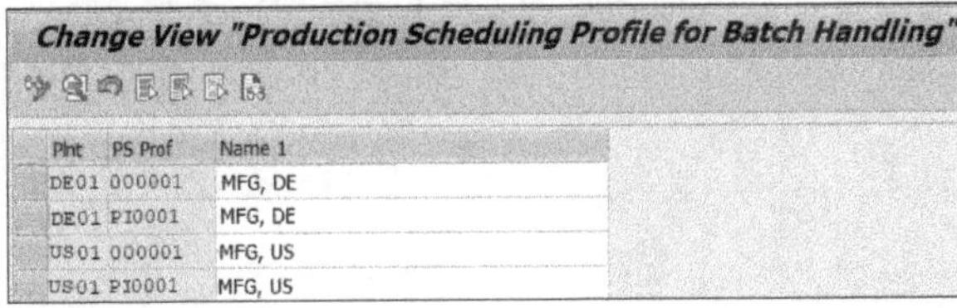
Change View "Production Scheduling Profile for Batch Handling"

Plnt	PS Prof	Name 1
DE01	000001	MFG, DE
DE01	PI0001	MFG, DE
US01	000001	MFG, US
US01	PI0001	MFG, US

Figure 4.6 Plant-Specific Settings for Auto Batch Creation Based on the Production Schedule Profile

- **Define Batch Creation for Goods Movements (Figure 4.7)**
  Allows you to define for each movement type whether and how a new batch is created for goods movements. To issue batch numbers automatically, you must set the internal batch number assignment. To do this, refer to Section 4.1.4.
- **Define Initial Creation for Batch Master Maintenance (Figure 4.8)**
  This function can be used to define for each material type if the batch number is created automatically or can be provided manually during batch creation via batch master maintenance and, if so, how it's checked by the system.

The preceding three options are available under menu path **SPRO • SAP Reference IMG • Logistics - General • Batch Management • Creation of New Batches**.

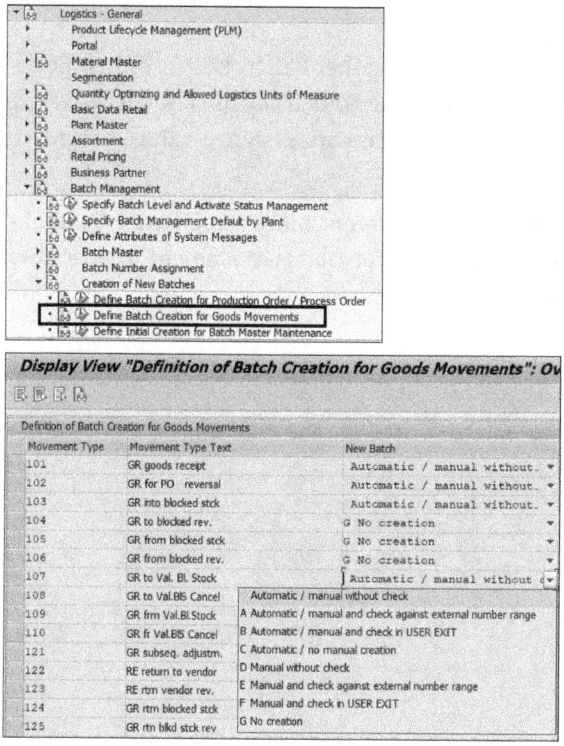

- Logistics - General
  - Product Lifecycle Management (PLM)
  - Portal
  - Material Master
  - Segmentation
  - Quantity Optimizing and Allowed Logistics Units of Measure
  - Basic Data Retail
  - Plant Master
  - Assortment
  - Retail Pricing
  - Business Partner
  - Batch Management
    - Specify Batch Level and Activate Status Management
    - Specify Batch Management Default by Plant
    - Define Attributes of System Messages
    - Batch Master
    - Batch Number Assignment
    - Creation of New Batches
      - Define Batch Creation for Production Order / Process Order
      - Define Batch Creation for Goods Movements
      - Define Initial Creation for Batch Master Maintenance

Display View "Definition of Batch Creation for Goods Movements": Ov

Definition of Batch Creation for Goods Movements

Movement Type	Movement Type Text	New Batch
101	GR goods receipt	Automatic / manual without...
102	GR for PO reversal	Automatic / manual without...
103	GR into blocked stck	Automatic / manual without...
104	GR to blocked rev.	G No creation
105	GR from blocked stck	G No creation
106	GR from blocked rev.	G No creation
107	GR to Val. Bl. Stock	Automatic / manual without
108	GR to Val.BIS Cancel	
109	GR frm Val.Bl.Stock	
110	GR fr Val.BS Cancel	
121	GR subseq. adjustm.	
122	RE return to vendor	
123	RE rtrn vendor rev.	
124	GR rtrn blocked stck	
125	GR rtn blkd stck rev	

Automatic / manual without check
A Automatic / manual and check against external number range
B Automatic / manual and check in USER EXIT
C Automatic / no manual creation
D Manual without check
E Manual and check against external number range
F Manual and check in USER EXIT
G No creation

Figure 4.7 Specifying the Method of Batch Creation for Each Movement Type

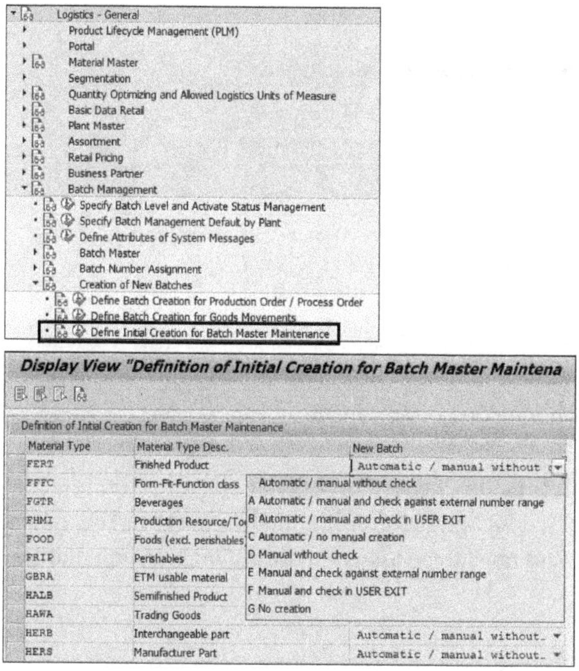

- Logistics - General
  - Product Lifecycle Management (PLM)
  - Portal
  - Material Master
  - Segmentation
  - Quantity Optimizing and Allowed Logistics Units of Measure
  - Basic Data Retail
  - Plant Master
  - Assortment
  - Retail Pricing
  - Business Partner
  - Batch Management
    - Specify Batch Level and Activate Status Management
    - Specify Batch Management Default by Plant
    - Define Attributes of System Messages
    - Batch Master
    - Batch Number Assignment
    - Creation of New Batches
      - Define Batch Creation for Production Order / Process Order
      - Define Batch Creation for Goods Movements
      - Define Initial Creation for Batch Master Maintenance

Display View "Definition of Initial Creation for Batch Master Maintena

Definition of Initial Creation for Batch Master Maintenance

Material Type	Material Type Desc.	New Batch
FERT	Finished Product	Automatic / manual without
FFFC	Form-Fit-Function class	
FGTR	Beverages	
FHMI	Production Resource/To	
FOOD	Foods (excl. perishables	
FRIP	Perishables	
GBRA	ETM usable material	
HALB	Semifinished Product	
HAWA	Trading Goods	
HERB	Interchangeable part	Automatic / manual without...
HERS	Manufacturer Part	Automatic / manual without...

Automatic / manual without check
A Automatic / manual and check against external number range
B Automatic / manual and check in USER EXIT
C Automatic / no manual creation
D Manual without check
E Manual and check against external number range
F Manual and check in USER EXIT
G No creation

Figure 4.8 Batch Creation Options Based on Material Types

4.1.6 Batch Valuation

In this step, you specify for each movement type whether batch classification is possible during goods movements and how it's to be carried out. If classification is activated, the object characteristics contained in the batch class are assigned values automatically.

The menu path for configuration is **SPRO • SAP Reference IMG • Logistics - General • Batch Management • Batch Valuation • Valuation for Goods Movement in Inventory Management • Activate Batch Classification for Goods Movements in IM** (Transaction OMCV), as shown in Figure 4.9.

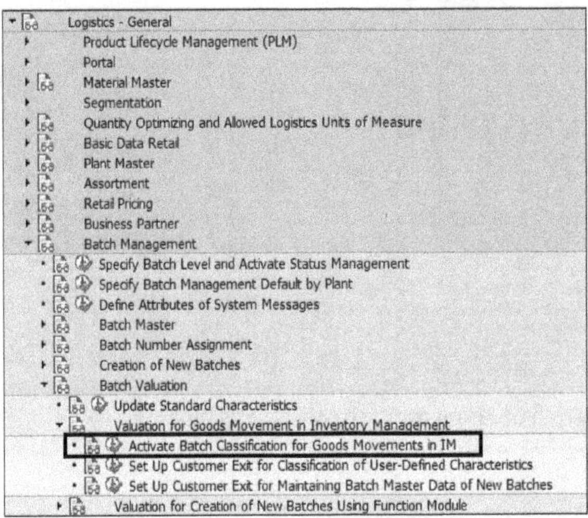

Display View "Classify Batches Within Goods Movement": Overview of Sel

M...	Classify batches	Ext.class.	Copy Chars	Copy Chars
101	Classification Carried Out in Background (All Batches)	☑	Do not copy characteri...	Do not copy charact
201	Classification Carried Out in Background (All Batches)	☑	Do not copy characteri...	Do not copy charact
303	Classification Carried Out in Background (All Batches)	☑	Do not copy characteri...	Do not copy charact
543	Classification Carried Out in Background (All Batches)	☑	Do not copy characteri...	Do not copy charact
643	Classification Carried Out in Background (All Batches)	☑	Do not copy characteri...	Do not copy charact

Do not copy characteristics
1 Copy only classification characteristics
2 Copy only configuration characteristics
3 Characteristics frm configuration (prio) and classification
4 Characteristics frm classification (prio) and configuration

Figure 4.9 Categorization of Batches during Goods Movement

For use of the batch determination function, ensure that batch classification is active for all inward movement types so that all batches are classified. Otherwise, the values of the characteristics in the batch class need to be assigned during master data maintenance.

As shown in Figure 4.9, the **Ext.class.** (extended classification) column is selected that uses a customer function call (CFC) to assign values to characteristics of a batch to be specified as a background job. For example, you can use a customer exit in the background to assign values to characteristics that aren't from quality management (e.g., time of goods receipt). All required goods movement data is available in the customer exit interface. The data is controlled according to the movement type. This can be defined using the configuration shown in Figure 4.10.

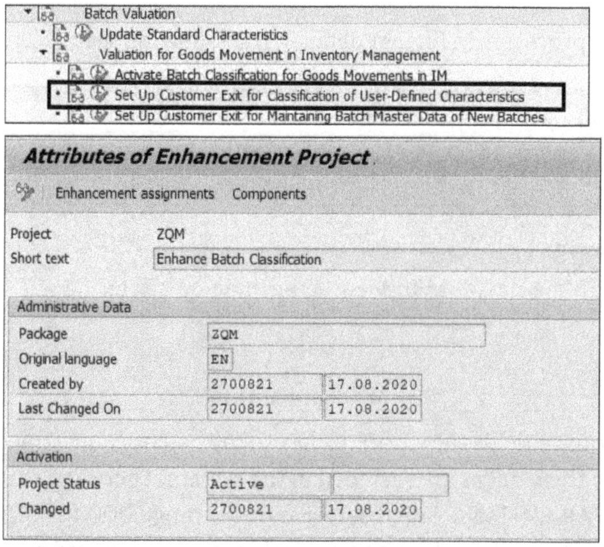

Figure 4.10 Use of Customer Function Call to Assign Values of Characteristics

4.1.7 Batch Determination and Batch Check

Batch determination is a key feature in batch management that allows automatic determination or identification of a batch of materials during a business transaction. For instance, batch determination can be used to automatically recommend the batch of finished items that must be delivered to a client or the raw materials that must be used during production. Batch management is a common practice among manufacturing and distribution companies.

Automatic batch determination is used in inventory management, production, sales and distribution, and WM to select batches and allocate them in a manufacturing order, an outbound delivery, or a goods issue/good receipt/goods transfer in inventory management.

Configuration for batch determination is very similar to the pricing procedure in purchasing. To configure the batch determination process, the following sequence of configuration steps need to be completed:

1. Create condition tables.
2. Maintain the batch determination access sequence.
3. Define the appropriate batch strategy type (such as production, sales and distribution, warehouse management, etc.).
4. Maintain the batch search procedure.
5. Allocate the batch search procedure to organizational entities (such as the sales area or warehouse number) and activate the batch search procedure.

We'll walk through each of these processes in further detail in the following sections.

Prerequisites for SAP Batch Determination

The prerequisites for batch determination are as follows:

- Batch level needs to be setup in configuration.
- Materials must be batch-managed.
- Material must be created with the classification view in the material master.

Condition Table

Condition tables are combinations of fields that form the key to the batch search strategy. These fields determine how the batches are searched in the system. There are preconfigured condition tables in SAP S/4HANA. SAP reserves number range 000 to 500, and the following number areas are reserved for condition tables in the standard system:

- 00–09 Sales and Distribution
- 10–19 Warehouse Management
- 20–29 Inventory Management
- 30–39 Production/Manufacturing

For custom creation of condition tables, number range 501 onward can be used.

The condition table can be created using menu path **SPRO • SAP Reference IMG • Logistics - General • Batch Management • Batch Determination and Batch Check • Condition Table**. Alternatively, you can use transaction codes to create condition tables in individual process areas such as Transaction OMA1 (for inventory management), Transaction OPLB (for process and production orders), Transaction V/C7 (for sales and distribution), and Transaction OMK4 (for WM).

Figure 4.11 shows a custom condition table configured in the sales and distribution process area.

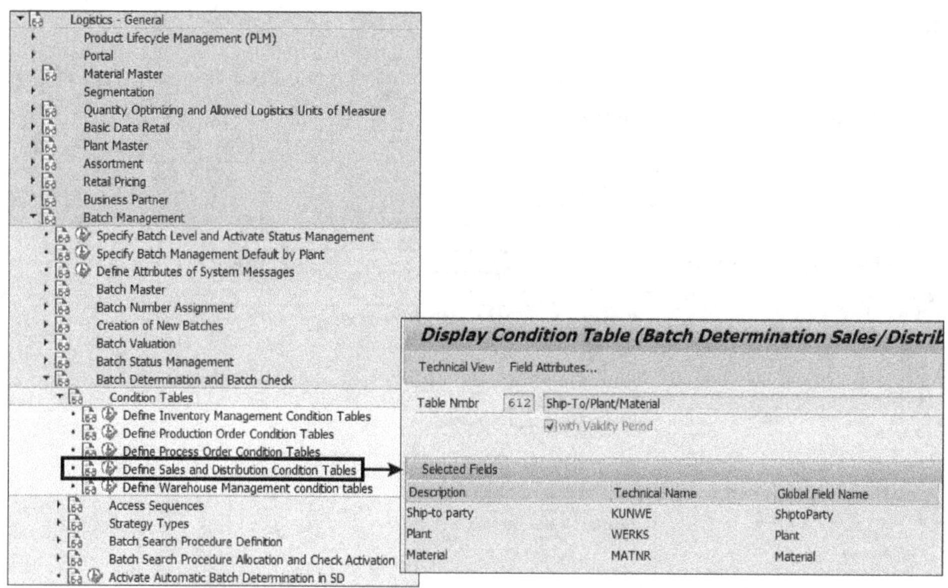

Figure 4.11 Condition Table for Batch Determination in Sales and Distribution (Transaction V/C7)

Batch Determination Access Sequence

The batch determination access sequence is associated with the batch search strategy. The primary purpose of the access sequence is to enable sequential access to the condition tables by the strategy type. The strategy type with access sequence is automatic in nature. Users can't manually enter or edit the rate or value. Users can manually edit the rate or value for those without an access sequence.

SAP S/4HANA provides a number of access sequences by default. If necessary, you can use these or create a new custom access sequence. The condition table chosen in the previous stage has to be entered in the access sequence in the proper sequence. The access sequence may contain one or more condition tables.

Configuration for access sequence follows menu path **SPRO • SAP Reference IMG • Logistics - General • Batch Management • Batch Determination and Batch Check • Access Sequences.** Alternatively, you can use transaction codes to maintain the access sequence in individual process areas such as Transaction OMCX (for inventory management), Transaction OPLF (for process and production orders), Transaction V/C2 (for sales and distribution), and Transaction OMKU (for WM).

Figure 4.12 represents a typical configuration in sales and distribution starting from the access sequence that encapsulates sequential access of condition tables:

❶ **Access Sequence**

❷ Two "accesses" beneath **Access Sequence**

❸ Condition **Table Number**

❹ **Int** indicator to consider NULL value in document of condition records

These condition tables are collections of fields that fetch the field values or rates from the corresponding SAP transaction.

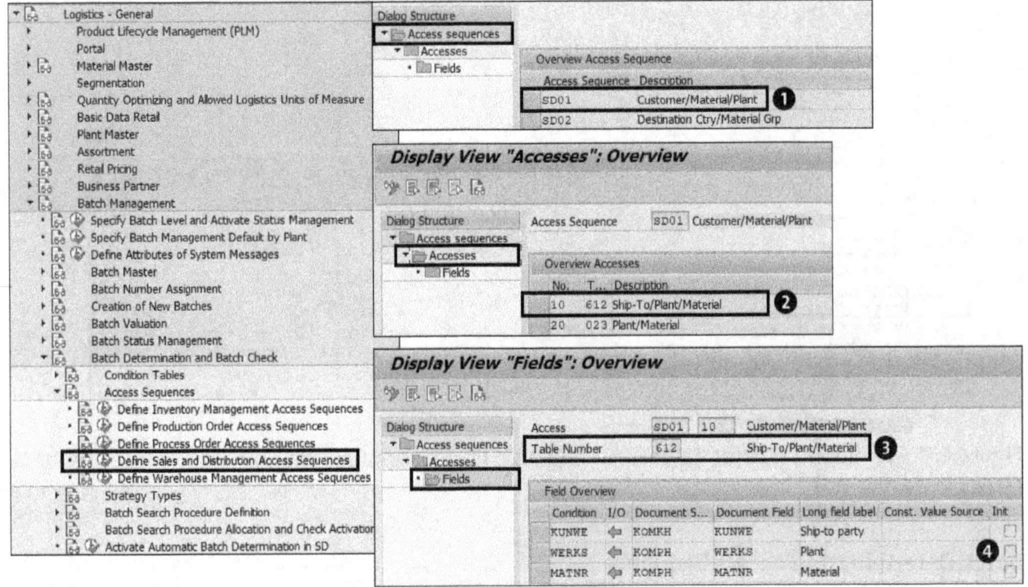

Figure 4.12 Access Sequence for Sales and Distribution

Strategy Types

Strategy type determines the type of conditions and criteria applied during the batch determination process. The selection criteria and the sort rules used in batch determination procedures are stored in the configuration of the strategy type. Selection criteria or the search class are used to search for batches according to their characteristic values.

Strategy types are used ultimately to create the batch search master data or batch search condition record so that the system knows the exact level at which the search for batches is to be carried out along with the selection/sort rules of these batches. The search class isn't assigned to the material but is a separate batch class that needs to be created and assigned to the material. This batch class will have the characteristics, and when the batch is created, those characteristic values will be populated.

The selection class (Transaction BMC1/BMC2/BMC3) is used to help the system choose a certain batch based on the characteristics and their fixed values. Sort rules (Transactions CU70/CU71/CU72) are used to help the system sort the batches for the given class characteristics.

Figure 4.13 shows the assignment of the selection class ❶ and sort rule for the batch search strategy ❷. In the sort rule ❸, selecting the **Ascending** checkbox for characteristics represents first expiry first out (FEFO), whereas selecting **Descending** represents the last in first out (LIFO) option.

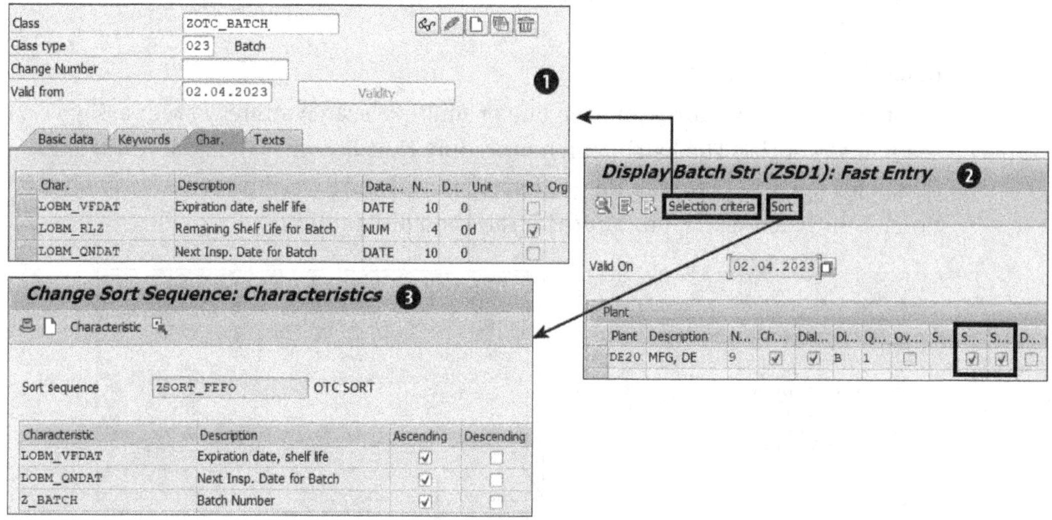

Figure 4.13 Assignment of Selection Class and Sort Rule to the Batch Search Strategy

The configuration menu path for maintaining batch search strategy types is **SPRO • SAP Reference IMG • Logistics - General • Batch Management • Batch Determination and Batch Check • Strategy Types**. Alternatively, you can use transaction codes to maintain strategy types: Transaction OMCW (for inventory management), Transaction OPLE (for process and production orders), Transaction V/C1 (for sales and distribution), and Transaction OMKT (for WM).

Figure 4.14 represents the configuration of the batch determination strategy in sales and distribution.

Application V Sales/Distribution
Condit. type ZSD1 OTC Batch Strategy Access seq. SD01 Customer/Material/Pla

Selection
Class Type 023 Batch
Class ZOTC_BATCH OTC Batch Search Maint.
Selection type

Sorting
Sort sequence ZSORT_FEFO OTC SORT Maint.

Batch split
No. of splits 9 ☑ Changes allowed ☐ OverDel allowed

Quantity proposal
Display UoM B ☑ Dialog batch determin. Qty proposal 1 Maint.

Presentation
Screen number 3000 Status CUA_V

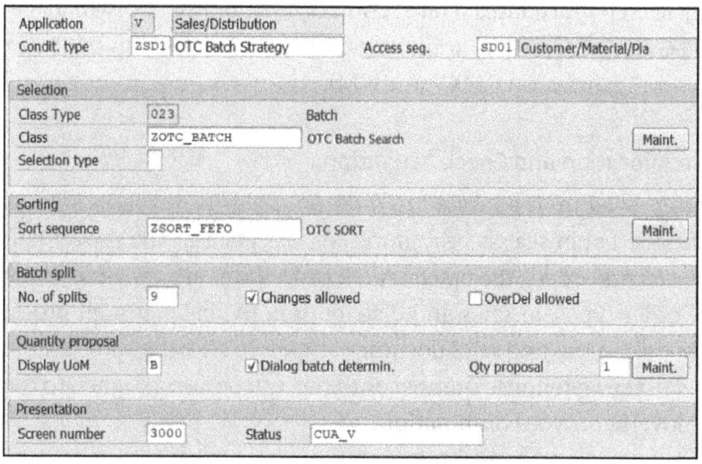

Figure 4.14 Configuration for the Batch Strategy Type

You then assign the selection class to the strategy and use the same in the condition record master data. In the condition records, you need to enter the required fixed values against any class characteristics that you want to use to search the batches.

Batch Search Procedure

A batch search procedure can have one or multiple search strategy types assigned in sequential priority. The batch search procedure thus becomes the final configuration that holds the total search configuration together. Figure 4.15 represents the configuration for the batch search procedure for sales and distribution.

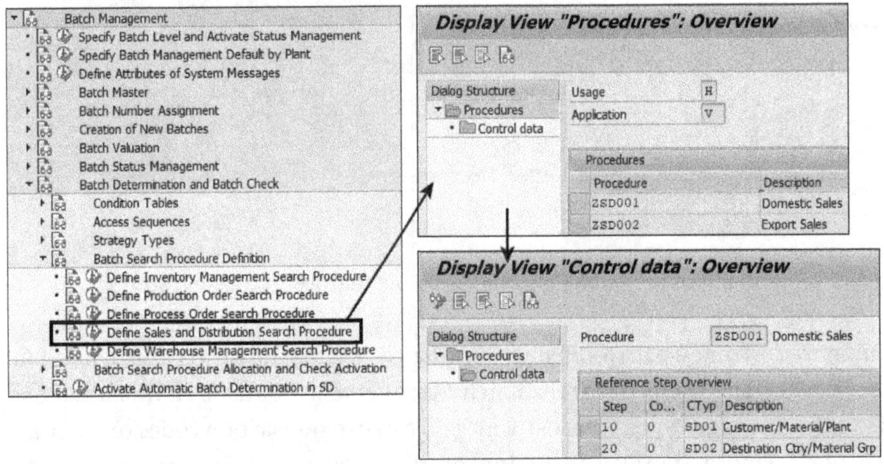

Figure 4.15 Batch Search Procedure Configuration

The configuration menu path for maintaining batch search procedures is **SPRO • SAP Reference IMG • Logistics - General • Batch Management • Batch Determination and Batch Check • Batch Search Procedure Definition.** Alternatively, you can use transaction codes to maintain batch search procedures: Transaction OMCY (for inventory management), Transaction OPLG (for process and production orders), Transaction V/C3 (for sales and distribution), and Transaction OMKV (for WM).

Batch Search Procedure Allocation and Check Activation

Finally, the batch search procedure is allocated to different fields per process area. In inventory management, the batch search procedures are assigned to the movement types. In production or process orders, the batch search procedures are assigned to the production plants and order types. In sales and distribution, the batch search procedures are assigned to the sales areas and sales document types. In WM, the batch search procedures are assigned to the warehouse number and they can be also assigned to the warehouse number + movement type combination.

The configuration menu path for maintaining batch search procedure allocations is **SPRO • SAP Reference IMG • Logistics - General • Batch Management • Batch Determination and Batch Check • Batch Search Procedure Allocation and Check Activation.** Alternatively, you can use transaction codes to maintain batch search procedure allocations: Transaction OMCG (for inventory management), Transaction OPL8 (for process and production orders), Transaction V/C5 (for sales and distribution), and Transaction OMK1 (for WM).

Figure 4.16 represents the assignment in the sales and distribution process area.

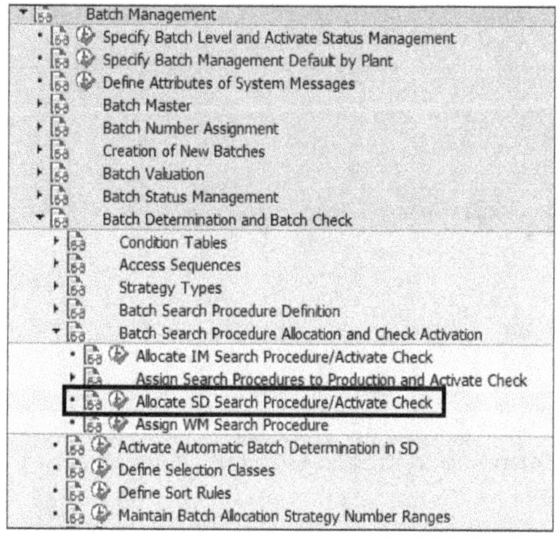

Sales org.	Distr. Chl	Division	SalesDocTy	Search Proced.	Check Batch
DE20	10	10	OR	ZSD001	☑
DE20	20	10	CBIC	ZSD002	☑

Figure 4.16 Assignment of the Batch Search Procedure and Batch Check to the Sales Area and Sales Document Type

Activate Automatic Batch Determination in SD

In this step, automatic batch determination for order or delivery item categories can be activated. When entering an order item or a delivery item, batch determination is triggered automatically.

Configuration menu path for maintaining Batch Search Strategy is **SPRO • SAP Reference IMG • Logistics - General • Batch Management • Batch Determination and Batch Check • Activate Automatic Batch Determination in SD** (Transaction V/CA), as shown in Figure 4.17.

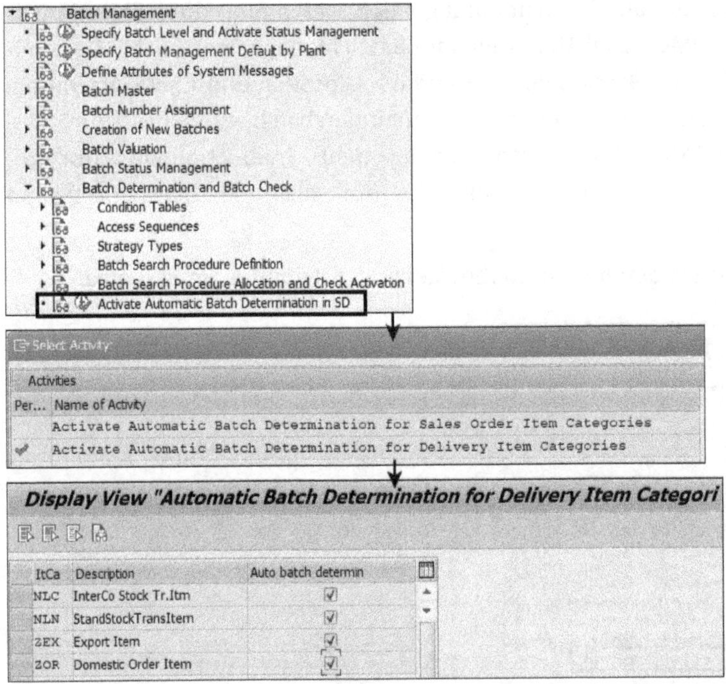

Figure 4.17 Setup for Auto-Triggering Batch Determination in Sales Order or Delivery

4.1.8 Batch Where-Used List

The information from inventory postings and reversal postings in inventory management is used to create batch where-used entries. The function covers transfers between different batches, but not transfers from one plant to another or from one storage facility to another. The various batches in production planning are linked via production orders, process orders, and subcontract orders.

A batch where-used list represents the list of objects where the batch was produced from and the objects where the batch was issued to. If the batch is under inspection with an inspection lot, the list also shares the inspection lot number. The function only displays batches for which a receipt posting or issue posting has been carried out. Batches that were chosen when an order was released aren't displayed.

The scope of the batch where-used list is as follows:

- Inventory management: Transfer postings
- Production planning: Production orders, process orders

The configuration menu path for maintaining batch where-used lists is **SPRO • SAP Reference IMG • Logistics - General • Batch Management • Batch Where-Used List • Make Settings for Batch Where-Used List** (Transaction OMBB), as shown in Figure 4.18.

- Batch Management
  - Specify Batch Level and Activate Status Management
  - Specify Batch Management Default by Plant
  - Define Attributes of System Messages
  - Batch Master
  - Batch Number Assignment
  - Creation of New Batches
  - Batch Valuation
  - Batch Status Management
  - Batch Determination and Batch Check
  - Batch Where-Used List
    - Make Settings for Batch Where-Used List
    - Set Up Customer Ext for Batch Where-Used List
    - Business Add-In: Enhancement to Automatic Plant Determination
    - Business Add-In: Access to Customer-Specific Batch Usage Data

Plnt	Name 1	BaWU synchron.postng	BaWU deactivated
AUP1	MFG, AU	☑	☐
DEP1	MFG, DE	☑	☐

Figure 4.18 Configuring Batch Where-Used Lists

Prerequisite and Constraints

Prerequisites for and constraints of the batch where-use list include the following:

- All materials that are part of the batch where-used list must be subject to management in batches.
- Creation of a batch where-used list is possible if the receipt records and issue records are stored in the batch where-used file (Transaction MB57).
- The batch where-used list function can't be used for purchase orders or production orders with account assignment because goods receipts are entered without specifying a batch number.

Figure 4.19 displays the batch where-used list for a specific material and batch number, including the following:

❶ Delivery number

❷ Top-down or bottom-up hierarchy

❸ Inspection lot number

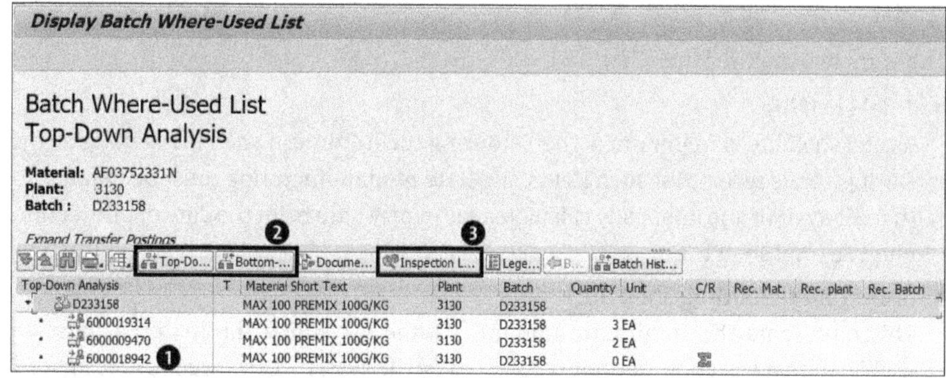

Figure 4.19 Batch Where-Used List Display

SAP standard table CHVW captures the details of batch where-used lists, which are most used for batch traceability, that is, understanding and analyzing the history of batches when customer complaints arise.

Unlike SAP ERP, in SAP S/4HANA, the material master view has the **Batch management(Plant)** indicator for batch management in the plant (in addition to the **Batch management** indicator). The flexibility it provides is that you can even extend the material to another plant without the indicator. Figure 4.20 shows the material master view with batch management indicators.

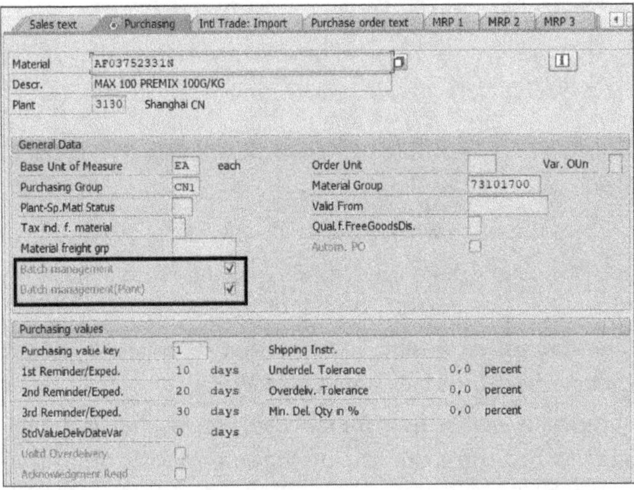

Figure 4.20 Batch Management Indicators in the Material Master in SAP S/4HANA

4.1.9 Shelf-Life Expiration Date

Shelf life denotes the length of time after which materials may no longer be used, consumed, or sold. It typically applies to perishable items in food, pharma, or chemical industries. If the shelf-life check expiration date check is active, you must enter the shelf-life expiration date or the production date of the material at the time of a goods receipt.

There are two kinds of shelf life:

- **Total shelf life**
  If total shelf life is maintained, then entering the minimum shelf life is mandatory. During goods receipt for such items, the **Date of manufacturing** must be entered so that the system automatically calculates the expiry date from the date of production plus the total shelf life.
- **Minimum remaining shelf life**
  This represents the minimum amount of time for which the material must be usable, so that the goods receipt is accepted by the system. During goods receipt of a material, the system checks the length of time between the goods receipt date and

the date of manufacture (as entered by the user during goods receipt). If the calculated period is equal to or more than the "minimum remaining shelf life," the system allows for goods receipt posting; otherwise, it throws an error.

In this configuration step, the material shelf-life expiration date check is set for goods receipts for each plant and movement type.

The menu path for shelf-life configuration is **SPRO • SAP Reference IMG • Logistics - General • Batch Management • Shelf Life Expiration Date (SLED) • Set Expiration Date Check** (Transaction OMJ5). Figure 4.21 represents the shelf-life activation for **Plant** and **Movement Type**.

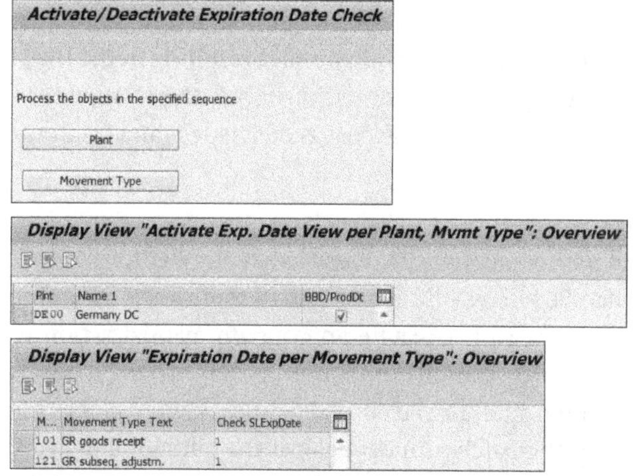

Figure 4.21 Activating the Shelf-Life Expiration Date for Plant and Movement Types

Figure 4.22 represents a material master maintained with shelf-life data.

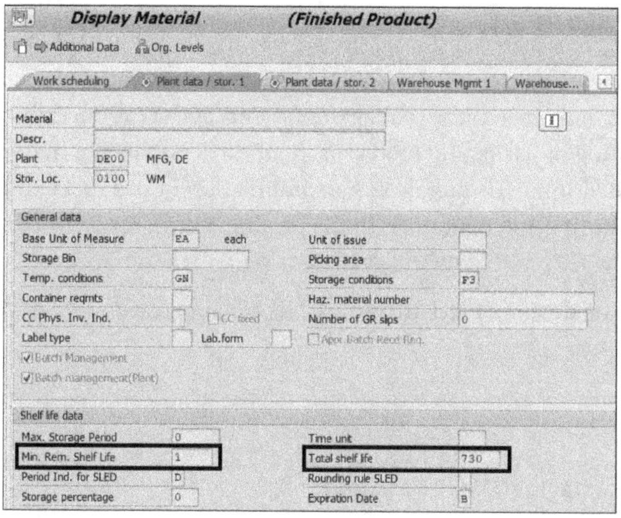

Display Material (Finished Product)

Additional Data | Org. Levels

Work scheduling | Plant data / stor. 1 | Plant data / stor. 2 | Warehouse Mgmt 1 | Warehouse...

Material
Descr.
Plant DE00 MFG, DE
Stor. Loc. 0100 WM

General data

Base Unit of Measure EA each | Unit of issue
Storage Bin | Picking area
Temp. conditions GN | Storage conditions F3
Container reqmts | Haz. material number
CC Phys. Inv. Ind. | CC fixed | Number of GR slips 0
Label type | Lab.form | Appr.Batch Rec. Req.
Batch Management
Batch management(Plant)

Shelf life data

Max. Storage Period 0 | Time unit
Min. Rem. Shelf Life 1 | Total shelf life 730
Period Ind. for SLED D | Rounding rule SLED
Storage percentage 0 | Expiration Date B

Figure 4.22 Material Master Plant/Storage View Data

4.1.10 Batch Management in a Third-Party Logistics Scenario

So far, we've seen how batch management is activated and how setup is carried out in sequence matching with business requirements. We also looked at positive differences SAP S/4HANA brings in while maintaining the material master. Now, we'll look into the complexities in the business scenarios that sometimes make it difficult to map the process with standard settings causing the need for additional technical development. Some business examples are described next.

Usually, logistics service providers are informed of customer deliveries when an outbound delivery is created. Upon saving the document, an outbound message gets triggered that informs the logistics service providers about delivery details, including batch numbers to be delivered to customer. It may so happen that a different batch number gets delivered to the customer due to availability issues resulting in the logistics service provider sending the post goods issue information back with a mismatch in batch numbers. This causes an error while posting the post goods issue against the outbound delivery.

In another variation of the preceding scenario, a single batch quantity may be split into two containing the original batch number along with a new batch number for the split quantity. Such situations can be handled in two different ways: (1) routing the incoming error message through a user workflow to necessitate appropriate manual action to change the outbound delivery and retriggering the outbound message to the logistics service provider again, or (2) performing additional technical development to accept the changed batch number that updates the batch number against the outbound delivery programmatically in the background. Complex technical development is required for the second option. Split batch can be handled using a suitable qualifier in IDoc messages.

4.2 Handling Unit Management

For warehouse or stores managers using SAP S/4HANA, the HU is a potent tool that helps reduce inventory errors, enhance stock management, and boost productivity. HUs are physical units consisting of packaging materials (load carriers/packing material) and the goods contained within. HUs are always a combination of products and packaging materials. All the information contained in the product items, for example, about batches, is retained in the HUs and is always available.

The following section covers the details of HU attributes, HU advantages, and how to configure HUs in SAP S/4HANA.

4.2.1 Overview

Following is a summary of advantages HUM has to offer:

- Inventory accuracy and tracking
- Shipping accuracy
- Unique label and compliance
- Nesting of HUs
- Efficient stock management
- Traceability throughout the supply chain
- Compatible to future technology (radio frequency identification [RFID] and mobility solutions)

HUs can be nested (see Figure 4.23), and a new HU can be created from several other HUs and can be used in all processes within the supply chain. Each HU has a unique, scannable identification number that can be constructed according to industry standards such as European Article Number 128 (EAN 128) or the Serial Shipping Container Code (SSCC) number.

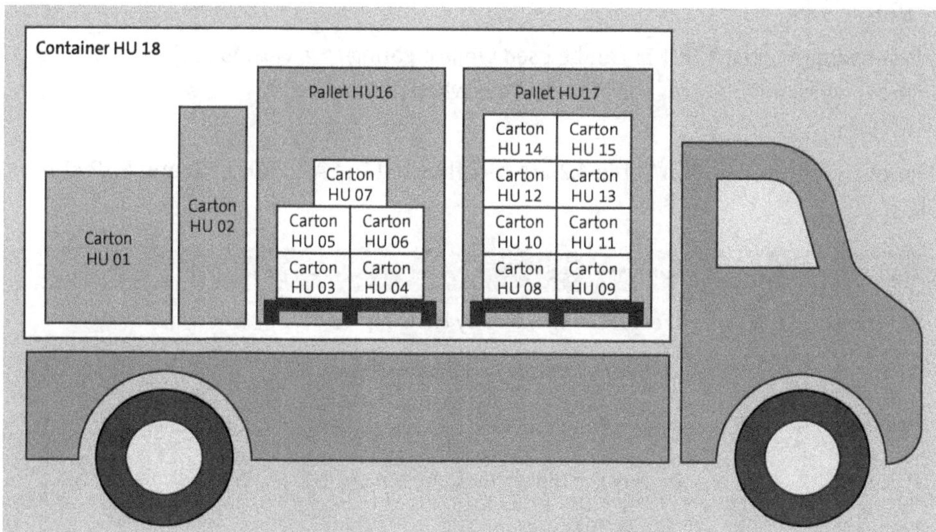

Figure 4.23 Example of Nested Handling Units

The HU object contains the following attributes:

- Identification number
- Dimensions
- Weight
- Volume
- Status
- Materials and quantities
- Packaging materials
- Packing instructions

HUs consist of a HU header and HU items. For identification purposes, it's given a unique number. The header data of a HU is divided in the following way and appears as different tabs:

- Weight/volume/dimensions
- Additional data, such as sales and distribution data and dangerous goods data
- General data
- Transaction history
- Output details

HU items can consist of material items, packaging materials, or other HUs that appear in an overview. The status HU reflects various physical states of HUs such as weighed, loaded, blocked or posted goods issue, and so on. User-specific procedures can be used to implement a status HU per requirements, as explained in the following sections.

Note

User parameter HU_ST_INIT can be used while creating HUs without object references to set which initial status the HU is to have when it's created.

The transactions listed in Table 4.2 are mostly used in SAP S/4HANA and SAP ERP for HUs.

Transaction	Purpose
HU01/HU02/HU03	Create/Change/Display Handling Unit
HUMO	Monitor and Process Handling Units
HUIBD	Display Handling Unit Details Specific to Inbound Delivery
HUOBD	Display Handling Unit Details Specific to Outbound Delivery
HUTRA	Display HU for Shipment
COP1/COP2	Pack/Process Work Orders
VLMOVE	Goods Movement for Handling Units and Materials
HUPAST	Packing Station in the Warehouse (Create HUs with Object Reference Type Packing-Station)
HUCOWA	Components for Work Orders
HUCOWE	Finished Product for Work Order
LT09	Putaway, Picking, and Stock Transfers at WM HU

Table 4.2 SAP Transactions Pertaining to Handling Units

Figure 4.24 shows HU header, item and status level details:

❶ Item-level details

❷ **Weight**, **Volume**, and **Dimensions** areas

❸ **HU Identification** field

❹ **HU Identification 2** field

❺ **System Status** reflecting goods issue posted

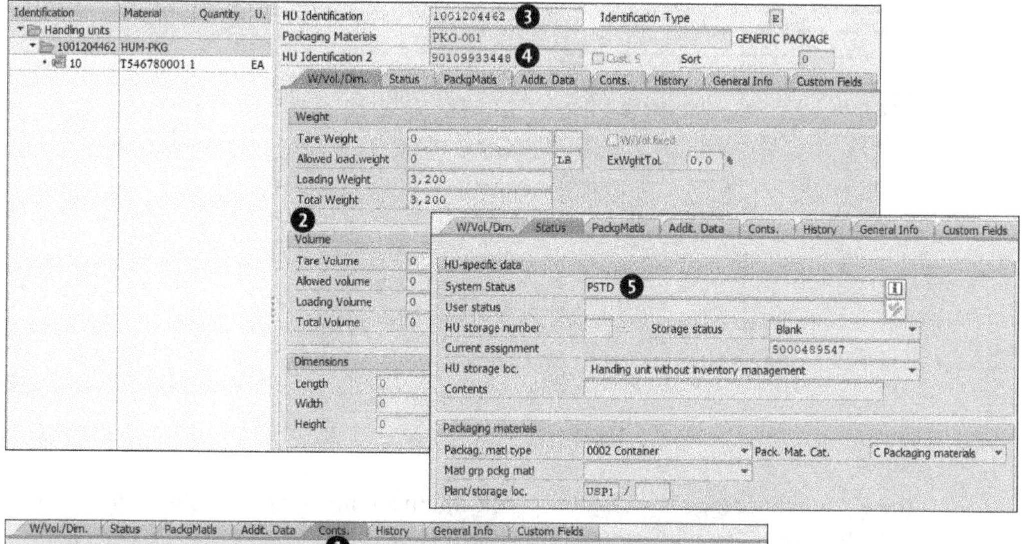

Figure 4.24 Handling Unit Details

Transaction documents impacted by HUs include inbound and outbound deliveries, material documents, work orders, and transfer orders.

The following sections elaborate on the necessary configuration settings for HUM.

4.2.2 Define Number Ranges for Handling Units

When creating HUs, the system assigns a number from the internal number range interval, or it can be maintained manually. The **Number Range Object** is **RV_VEKP**. Number ranges for HUs can be maintained using configuration menu path **SPRO • SAP Reference IMG • Logistics - General • Handling Unit Management • Basics • Technical Basics • Define Number Ranges for Handling Units** (Transaction VNKP), as shown in Figure 4.25.

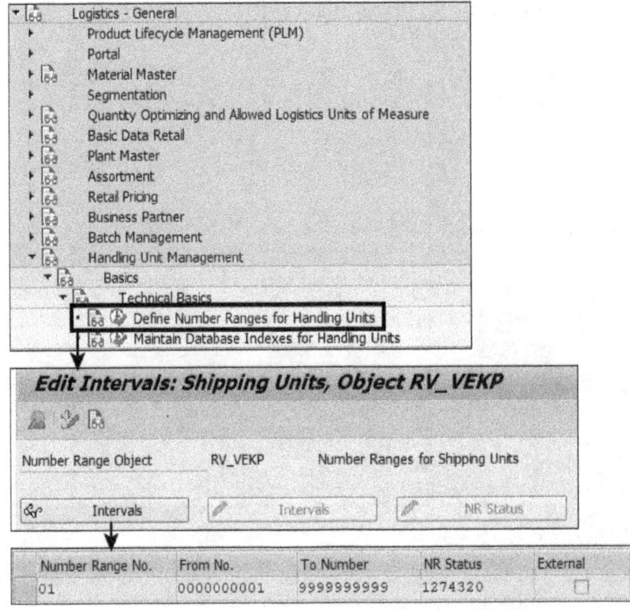

Figure 4.25 Maintain Number Range for Handling Units

> **Note**
>
> The HU number range is a system-specific internal number and is visible only at the SAP tables for HU header, namely table VEKP (HU Header) and table VEPO (HU Item).

In the following sections, we'll walk through how to define number ranges for HUs.

Unique External Identification of Handling Units

HUs can be identified when assigned a unique number. If multiple HUs have been assigned the same HU identification, the system can't determine exactly which HU is being requested, thereby leading to failure of scanner-supported processing, for example.

This configuration step establishes that the assignment of HU identification is unique at the client level. The menu path is **SPRO • SAP Reference IMG • Logistics - General • Handling Unit Management • External Identification • Set Unique Number Assignment for HU Identification**, as shown in Figure 4.26.

> **Note**
>
> It's only possible to indicate a storage location as HU-managed if the **HU ID Unique** indicator is set. More precisely, a link from the HU to the storage unit isn't possible if there is no assurance of unique number assignment. The standard system contains unique number assignment.

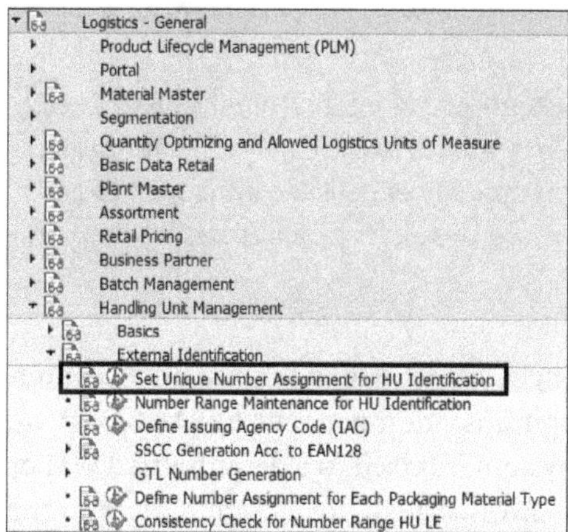

Figure 4.26 Activation of Unique Number Assignment for HU

Number Range Maintenance for HU Identification

In this step, a number range is maintained for external identification for the HU. The number range object is **HU_VEKP**, as shown in Figure 4.27, along with the configuration menu path and settings.

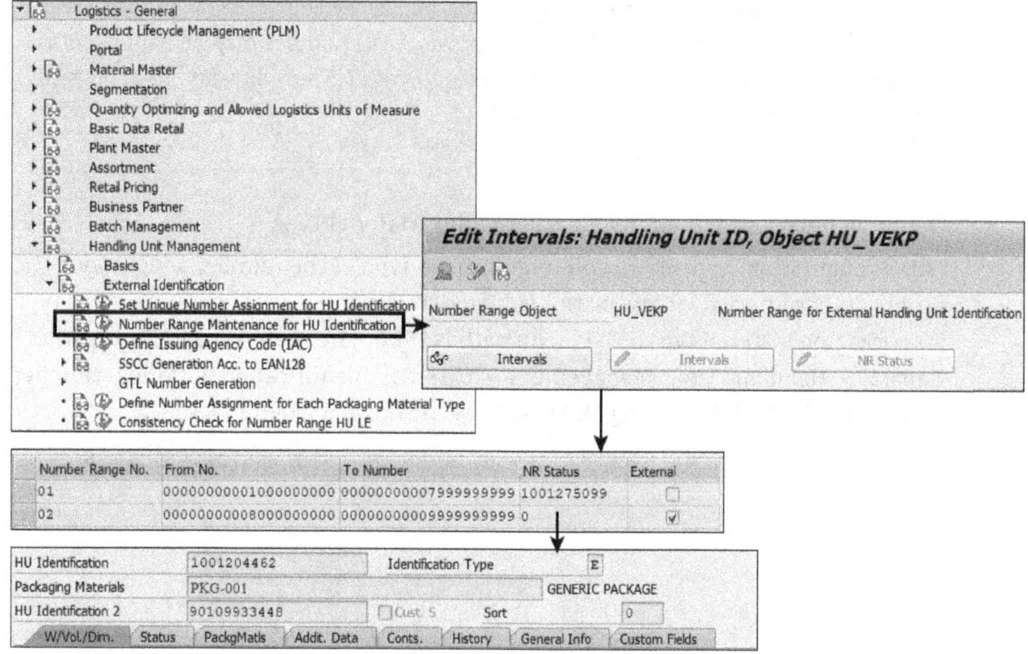

Figure 4.27 External Identifier for Handling Units

Note

You shouldn't mix this number up with the system-specific internal number associated with number range object **RU_VEKP**. The one maintained as the external identification number is visible to users as the HU identification number depicted in Figure 4.25 and Figure 4.27. To edit existing HUs, you should use this number for accessing the HU details.

As you can see in Figure 4.27, two number ranges are defined in the system. The internal number range is used when the packing transaction happens in SAP S/4HANA. There may be case when the SAP S/4HANA system is linked to a decentralized WM or EWM system, and packages generated in the EWM system are communicated to the SAP S/4HANA system. In such a scenario, the external number range is used in SAP S/4HANA.

Note

From the third-party logistics perspective, the HU identification field value can be sent to logistics service providers as an identifier for the HU. This is applicable in cases where the shipper has packed the material and communicated to the logistics service provider about the impending shipment with HU details.

In some cases, the logistics service provider may do the packaging and may generate the HU identifier at their end. The HU number generated by the logistics service provider may not match the number range configured at the shipper's end. In such a scenario, the communicated logistics service provider HU number may be populated as the **HU Identification 2** value, as shown in Figure 4.27. The value can be populated through inbound IDocs or manually as well.

Define Number Assignment for Each Packaging Material Type

This configuration step allows you to define the type of the number assignment, the number range interval for internal and external number assignments, and the HU type for packaging material types. The menu path is **SPRO • SAP Reference IMG • Logistics - General • Handling Unit Management • External Identification • Define Number Assignment for Each Packaging Material Type**, as shown in Figure 4.28.

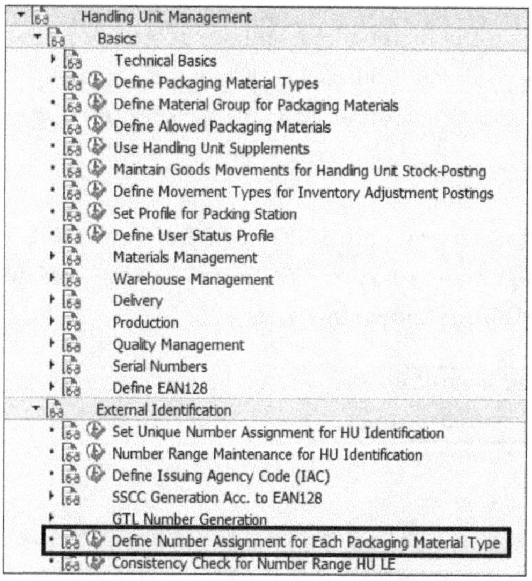

- Handling Unit Management
  - Basics
    - Technical Basics
    - Define Packaging Material Types
    - Define Material Group for Packaging Materials
    - Define Allowed Packaging Materials
    - Use Handling Unit Supplements
    - Maintain Goods Movements for Handling Unit Stock-Posting
    - Define Movement Types for Inventory Adjustment Postings
    - Set Profile for Packing Station
    - Define User Status Profile
    - Materials Management
    - Warehouse Management
    - Delivery
    - Production
    - Quality Management
    - Serial Numbers
    - Define EAN128
  - External Identification
    - Set Unique Number Assignment for HU Identification
    - Number Range Maintenance for HU Identification
    - Define Issuing Agency Code (IAC)
    - SSCC Generation Acc. to EAN128
    - GTL Number Generation
    - Define Number Assignment for Each Packaging Material Type
    - Consistency Check for Number Range HU LE

PkMtT	Description	Num...	Ty.	IntInterv	ExtInterv
0001	Pallets	B	1	01	02
0002	Container	B	2	01	02
0003	Ship	B		01	
RAIL	RAILCAR	B		01	

Figure 4.28 Number Range Assignment for Packaging Material Types

4.2.3 Settings for Packaging Materials

The purpose of packaging is to contain or secure the materials that will be packed. To put it another way, the material that needs to be packaged might be placed within or on top of the packaging. The packaging material can be a load carrier. The most important packaging materials include crates, boxes, containers, wire baskets, and pallets.

There is another type of packing material called auxiliary packaging material. Auxiliary packaging material and packaging material together are used to pack materials safely. Auxiliary packaging material includes equipment, safety, and protective packaging, such as lids, foil, lining, and anything used to wrap or protect materials. Auxiliary packaging material can be packaging material but can't be used as a load carrier.

Note

Material master data for each packaging material needs to be maintained with material type VERP (packaging) or any other customer-specific material type defined in Customizing. Material type VERP contains additional packing-relevant fields (in the **Sales: General/Plant Data** view).

> The gross weight and the volume entered in the material master have special significance for packing. While creating HUs, the system automatically copies the packaging material's master data and uses the gross weight and volume as the packaging material's tare weight and tare volume.

Configuring packaging material types follows menu path **SPRO • SAP Reference IMG • Logistics - General • Handling Unit Management • Basics • Technical Basics • Define Packaging Material Types** (Transaction VHAR), as shown in Figure 4.29.

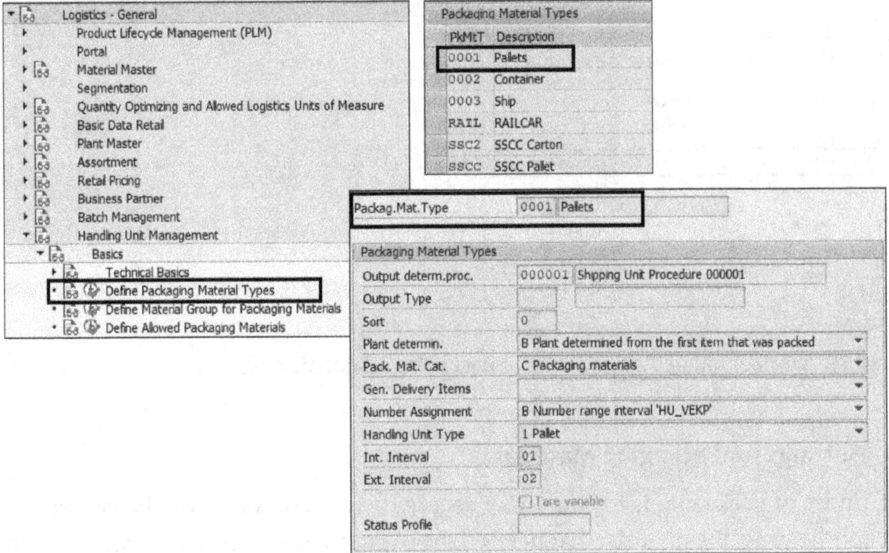

Figure 4.29 Setup for Packaging Material Types

In the following sections, we'll walk through the settings for configuring packaging material.

Define Material Group for Packaging Materials

In this configuration step, material groups for packaging materials are defined. Similar packaging materials are grouped together. The material grouping for packaging materials is entered in the material master record of the shipping material (**Basic data 1** view). For example, you can group together one-way or two-way pallets under the material group called Pallets.

As shown in Figure 4.30, the full configuration menu path is **SPRO • SAP Reference IMG • Logistics - General • Handling Unit Management • Basics • Technical Basics • Define Material Group for Packaging Materials** (Transaction VEGR).

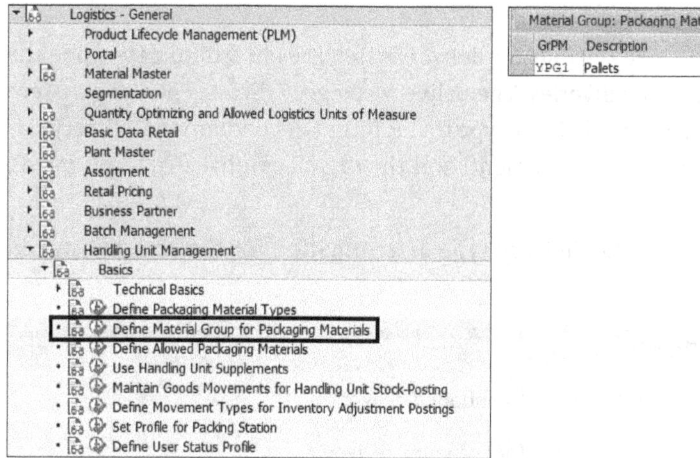

Figure 4.30 Define Packaging Material Groups

Define Allowed Packaging Materials

When packing materials or HUs, the system checks whether the packaging materials for the HU to which the items are to be packed is allowed for packing. When the packaging material type is assigned to a specific material group for packaging materials, only those packaging materials that belong to this group are allowed for packing. By assigning shipping material types to material groups for shipping materials, you define which shipping materials are allowed for packing.

The allowed packaging material type can be maintained using configuration menu path **SPRO • SAP Reference IMG • Logistics - General • Handling Unit Management • Basics • Technical Basics • Define Allowed Packaging Materials** (Transaction VHZU). Figure 4.31 shows the menu path and configuration to define allowed packaging materials.

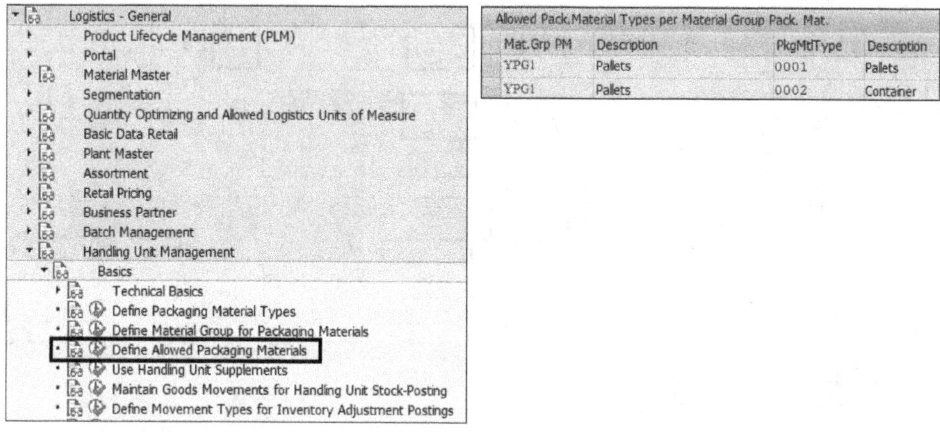

Figure 4.31 Assignment of Allowed Packaging Material Types to the Packaging Material Group

Delivery Type Determination

This step allows you to make settings for delivery categories in SAP to determine the delivery type in HU-related deliveries. The delivery category dictates how the system determines a delivery type. The delivery type that is found depends on the delivery category (classification and goods movement) and the organizational structures (plant, storage location).

In this activity, you can set the delivery type determination for the objects listed in Table 4.3.

Delivery Category	Description
ID	Delivery category for inbound deliveries
OD	Delivery category for outbound deliveries
TP	Delivery category for stock transfer deliveries/posting changes
RL	Delivery category for return delivery to supplier
LB	Delivery category for subcontracting
RK	Delivery category for customer returns

Table 4.3 Delivery Categories in SAP S/4HANA

The configuration menu path is **SPRO • SAP Reference IMG • Logistics - General • Handling Unit Management • Basics • Delivery • Delivery Type Determination.**

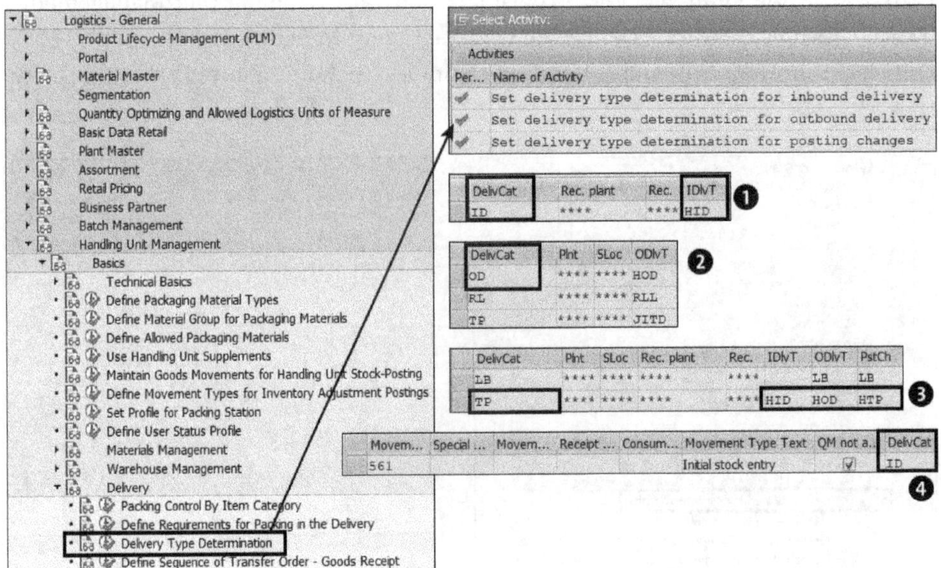

Figure 4.32 Delivery Type Determination of Delivery Categories for HU Movements

Figure 4.32 shows the following:

❶ Delivery type determination for inbound delivery (**HID**)

❷ Delivery type determination for outbound delivery (**HOD**)

❸ Delivery type determination for posting changes (**HTP**)

❹ The prerequisite of goods movement type configuration (**ID**)

Prerequisite

The delivery type for each movement type must be set for the delivery type determination to take place.

Material Master Settings

The following setups or assignments are required in the material master while maintaining packaging materials, as shown in Figure 4.33 and Figure 4.34:

❶ Create the material master with **Material type** "VERP Packaging" for packaging material.

❷ Maintain the **Packaging material data** for the packaging material.

❸ Assign the **Matl Grp Pack.Matls** (material group for packaging material) for shipping materials or materials to be packed.

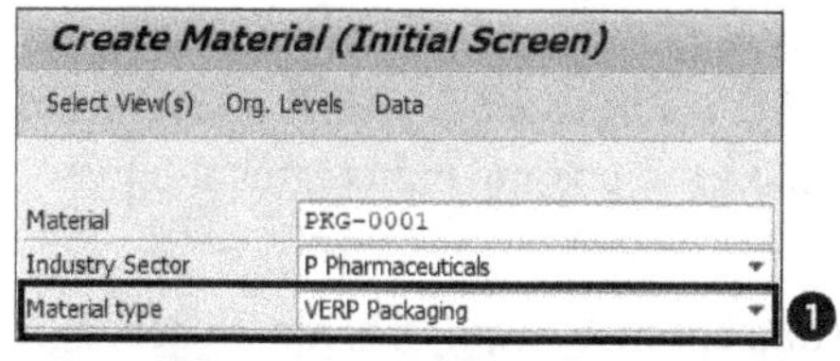

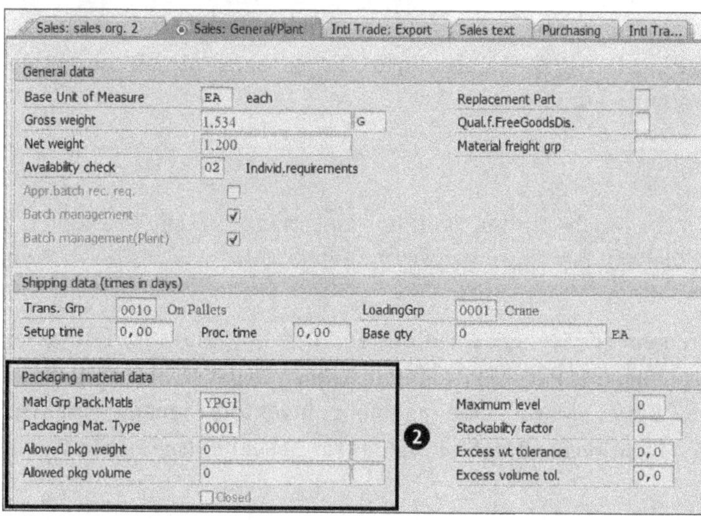

Figure 4.33 Material Master Settings for Packaging and Shipping Materials: Part 1

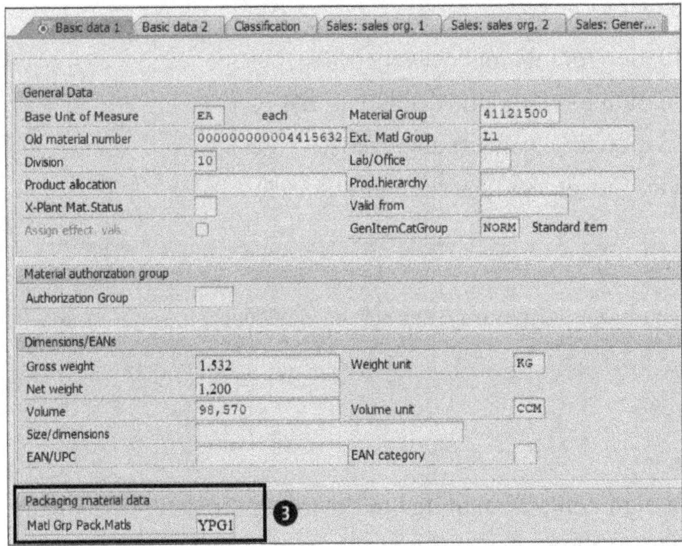

Figure 4.34 Material Master Settings for Packaging and Shipping Materials: Part 2

Handling Unit Requirement for Storage Locations and Default Values for Deliveries

In this step, for each storage location, check for HU requirements and that the default values for creating a delivery are maintained. Default settings at the plant level can be overridden with more specific default settings at the storage location level.

When you set the HU requirement indicator (**HU req.**) for a storage location, you can only make stock postings if you specify the HU. If you don't specify the HU, the system doesn't create a material document when you make the posting but creates a delivery document.

Because the stock in a storage location that is subject to HU requirements is packed in principle, you must maintain a partner storage location for transactions such as unpacking. Use a storage location in the same plant as the partner storage location that has the same or no warehouse number.

> **Note**
>
> You're only allowed to change the **HU req.** indicator for storage locations that don't contain any stock and that aren't assigned to a decentralized WMS.

Configuring **HU Requirement for Storage Locations and Default Values for Deliveries** can be achieved using menu path **SPRO • SAP Reference IMG • Logistics - General • Handling Unit Management • Basics • Material Management • Inventory Management • HU Requirement for Storage Locations and Default Values for Deliveries** (see Figure 4.35).

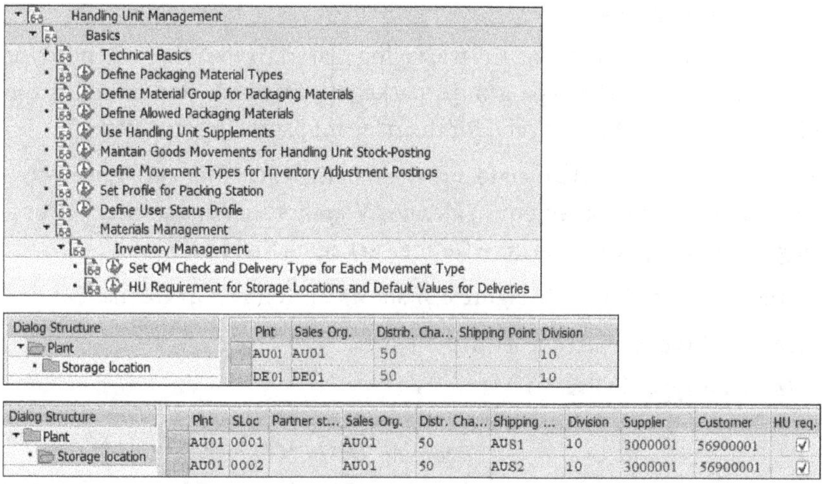

- Handling Unit Management
  - Basics
    - Technical Basics
    - Define Packaging Material Types
    - Define Material Group for Packaging Materials
    - Define Allowed Packaging Materials
    - Use Handling Unit Supplements
    - Maintain Goods Movements for Handling Unit Stock-Posting
    - Define Movement Types for Inventory Adjustment Postings
    - Set Profile for Packing Station
    - Define User Status Profile
    - Materials Management
      - Inventory Management
        - Set QM Check and Delivery Type for Each Movement Type
        - HU Requirement for Storage Locations and Default Values for Deliveries

Dialog Structure
- Plant
  - Storage location

Plnt	Sales Org.	Distrib. Cha...	Shipping Point	Division
AU01	AU01	50		10
DE01	DE01	50		10

Dialog Structure
- Plant
  - Storage location

Plnt	SLoc	Partner st...	Sales Org.	Distr. Cha...	Shipping ...	Division	Supplier	Customer	HU req.
AU01	0001		AU01	50	AUS1	10	3000001	56900001	☑
AU01	0002		AU01	50	AUS2	10	3000001	56900001	☑

Figure 4.35 Defining HU-Specific Storage Locations

4.2.4 Handling Units with Extended Warehouse Management

With EWM, HUs are controlled at the storage type level, and within the storage type, it can be set whether the use of HUs is mandatory, optional, or not required. This section covers HU-specific configuration details in an SAP S/4HANA EWM system.

Define Handling Unit Types in SAP S/4HANA Extended Warehouse Management

HU types in EWM can be defined using menu path **SPRO • SAP Reference IMG • Integration with Other SAP Components • Extended Warehouse Management • Additional Material Attributes • Attribute Values for Additional Material Master Fields • Define Handling Unit Type**, as shown in Figure 4.36.

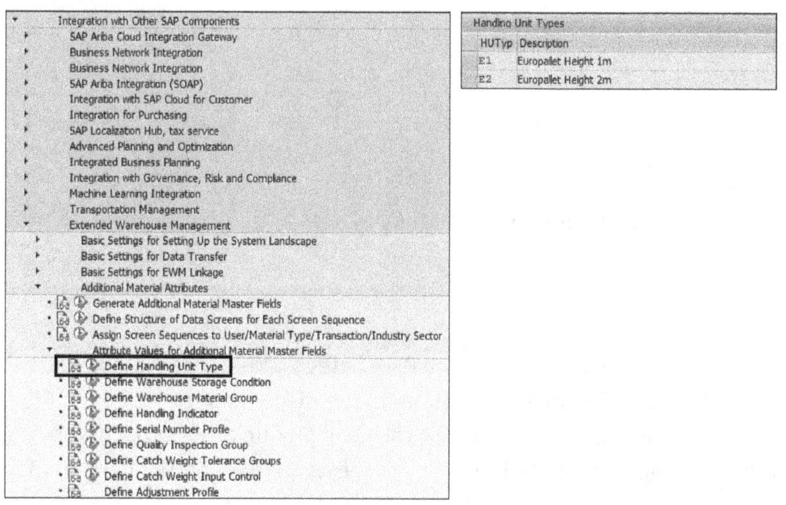

Figure 4.36 Define Handling Unit Types in Extended Warehouse Management

Define Packaging Material Types

In this Customizing activity, you define the packaging material types that you need for packing. The packaging material type groups packaging materials and contains the main control characteristics that are valid for the relevant packaging material.

Packaging material types can be defined using menu path **SPRO • SAP Reference IMG • SCM Extended Warehouse Management • Extended Warehouse Management • Cross-Process Settings • Handling Units • Basics • Define Packaging Material Types.**

In addition to the configuration path, Figure 4.37 shows the following screens:

❶ **Packaging Material Types** in inventory management

❷ **Packaging Material Types in EWM**

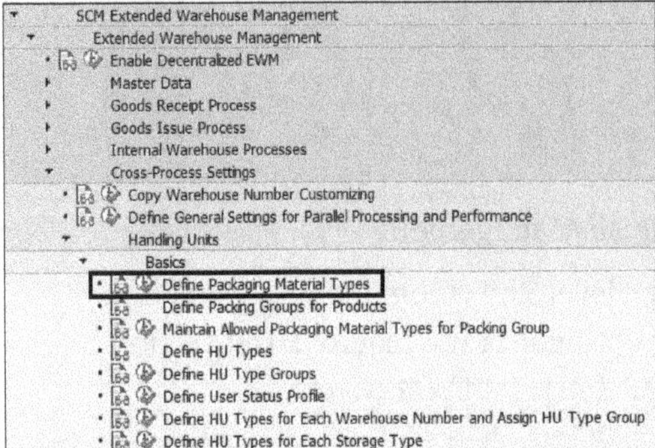

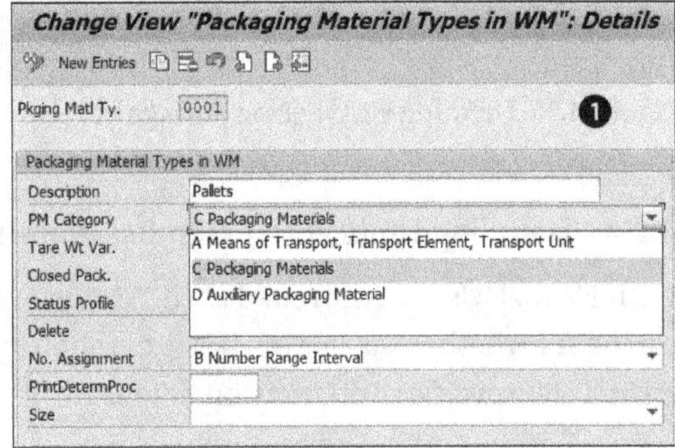

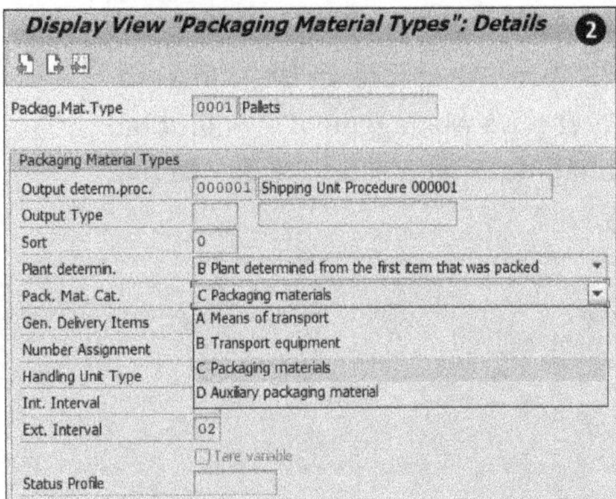

Figure 4.37 Defining Packaging Material Types in EWM

> **Note**
>
> Means of transport and transport equipment (**PM Category**) has been grouped in EWM. Number ranges for HUs in EWM act as external number ranges for SAP S/4HANA; that is, if HUs are generated in EWM, they will pick up an external number range in SAP S/4HANA when distributed. Conversely, the ID number for HUs generated in inventory management will be treated as external numbers in EWM.

Define Handling Unit Types for Each Storage Type

In this IMG activity, for each storage type, you define the permitted HU types that are used by the HU type check. Alternatively, you can also specify valid HU type groups for each storage type. EWM can manage storage bins of different sizes within a storage type. The storage bins are determined by means of storage bin type determination, for example, which consists of the HU type check and the storage bin type check.

> **Note**
>
> The HU type check and the storage bin type check are only performed if you've activated the attribute **HU Type Check** for the storage type (configuration activity: **Define Storage Types**).

Assignment of the HU type to a specific storage type can be maintained using menu path **SPRO • SAP Reference IMG • SCM Extended Warehouse Management • Extended Warehouse Management • Cross-Process Settings • Handling Units • Basics • Define HU Types for Each Storage Type** (see Figure 4.38).

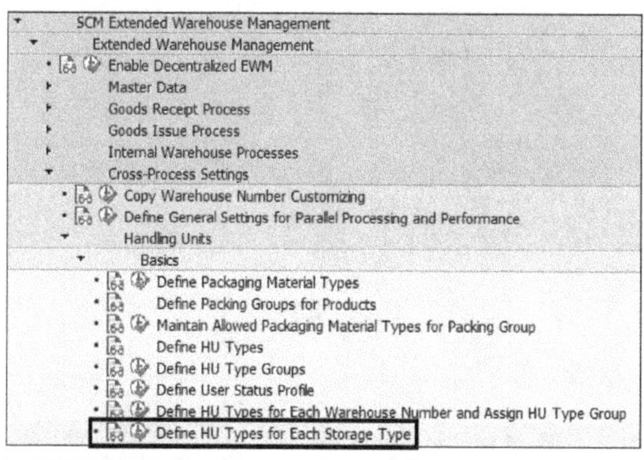

HU Types per Storage Type

WhN	Ty.	HU Type	HU Type	HU Type	HU Type	HU Type	HU Type	HU Type	HU Type	HU Type	HU Type
0001	0010	E1	E2								
0001	0040	E2									
0001	0050	E1	E2								
0001	0070	E1	E2								
0001	0080	E1	E2								

Figure 4.38 Permitted HU Types for Each Storage Type

4.3 Summary

The chapter focused on the details of batch management configuration in SAP S/4HANA with relevant pointers to the benefits of batch management and its integration across supply chain modules. Batch management aims to create consistency from product to product and in production cycles, as well as making the best use of your supplies, minimizing waste, and streamlining manufacturing processes.

The second half of this chapter dealt with another submodule of logistics: handling unit management (HUM). HUM simplifies operations related to units moving between supply chains and warehouses. In any logistics process, material is packed and moved together with cartons, pallets, and similar packaging materials, or it's stored together. HUM reflects packing-based logistics structures in the SAP system. We looked at the details of HUs, their configuration and tracking, and how they're maintained within EWM. Using HUs, it's much easier to track the movements of the entire HU and the materials contained within rather than tracking each material individually.

Chapter 5

Shipping and Delivery

Shipping is an essential part of a supply chain and a key process in the success of a business. A seamless and streamlined shipping process can elevate a company's supply chain management and help set them apart from the rest if they are educated on the ins and outs of essential logistical/shipping processes.

Shipping refers to the transportation of goods from a company or a producer to the consumer. It's a value chain that necessitates identifying essential steps in logistics, their interlinkages, and management of stakeholders for each step. Shipping represents an overall process of many individual transactions in which both internal and external interfaces are involved.

Many business organizations avail the services of third-party logistics for shipping and logistics management to benefit from facilities such as modern warehousing, transportation, latest trends and innovation in supply chain practices, competitive rates, and so on. An efficient shipping strategy focuses on building experiences that delight the customer while keeping costs low.

In the beginning of this chapter, we'll take a quick look at a few key aspects of business processes involving shipping and logistics. What the shipping organization looks like and configuration-specific details will be dealt with next. This will be followed by delivery-specific details and setups, such as availability check and output controls, that play a significant role in third-party logistics scenarios.

5.1 Shipping and Logistics Processes

Shipping management in logistics comprises a wide range of functions from resource planning, organization, execution, and transportation of shipment. When these activities are outsourced to third-party logistics service providers, either in full or in part, responsibilities may transform from fully owned to shared, as explained here:

- A *shipper* takes the order and prepares it for shipping as cargo. The business entity itself or the freight forwarders may act as the shipper.
- Once an order has shipped, it's the *carrier's* job to ensure that it's delivered safely to the customer. Once the order is with the carrier, order tracking can be used in real

time to notify the customer of their order process, as well as give them an estimated delivery time.

- *Third-party logistics service providers* offer both inbound and outbound logistics services and are responsible for taking care of the shipping process. They may also be involved in supply chain management activities.

Due their close connections with shipping carriers, third-party logistics service providers can negotiate better rates for cost-effective and efficient transportation. Upon shipment, third-party logistics service providers can control the entire process of managing and tracking all orders in real time and keep the customer up-to-date about delivery status. Another major contribution from third-party logistics service providers is transport documentation as they have the requisite expertise of international and local laws and regulations.

Next, let's discuss the three major categories or types of logistics processes:

- **Inbound logistics**
  This involves processes that move products from a manufacturer or external vendor to a warehouse. It covers transport, storage, delivery, and inspection of goods, including sourcing products and production and delivery lead times.
- **Outbound logistics**
  This refers to operations necessary to move the products from the warehouse to end customers and involves activities such as sales order processing, picking and packing, post goods issue, and shipment to customer.
- **Reverse logistics**
  This includes processing of customers returns and exchanges, involving both inbound and outbound processes such as inspection of returned products, processing refunds, shipping out exchanged products, and so on. An efficient reverse logistics process helps in reducing costs and resources.

In SAP S/4HANA, shipping consists of several individual process steps in a specific sequence. These steps, as listed here, are interconnected and depend on other steps:

- Inbound/outbound delivery creation
- Picking or transfer order confirmation in warehouse management (WM), putaway for inbound delivery
- Packing and creation of packing list
- Post goods issue/receipts
- Transportation planning and execution

Table 5.1 provides an overview of general shipping terms.

Term	Description
Letter of indemnity (LOI)	A document protecting the contract partners (e.g., shippers and carriers) and owner from losses. In other words, the party or parties are indemnified against a possible loss by some third party, such as an insurance company.
Consignee	Person or entity that is the shipment receiver and generally the owner of the shipped goods.
Incoterms	Set of 11 internationally recognized rules that define the responsibilities of sellers and buyers. Incoterms specify who is responsible for paying for and managing the shipment, insurance, documentation, customs clearance, and other logistical activities. The following incoterms are applicable for any mode of transport: ■ EXW: Ex Works (place of delivery) ■ FCA: Free Carrier (named place of delivery) ■ CPT: Carriage Paid to (place of destination) ■ CIP: Carriage and Insurance Paid To (place of destination) ■ DAP: Delivered at Place (named place of destination) ■ DPU: Delivered at Place Unloaded (place of destination) ■ DDP: Delivered Duty Paid (place of destination). ■ The following four incoterms are applicable for sea and inland waterway transport: ■ FAS - Free Alongside Ship (name of port of loading) ■ FOB - Free on Board (named port of loading) ■ CFR - Cost and Freight (named port of destination) ■ CIF - Cost Insurance and Freight (named port of destination)
Bill of lading	A bill of lading is a legal document that has a few important functions in shipping and logistics. It's a contract between the shipper, carrier, and consignee stating what goods are being shipped, where the shipment is coming from, and where it's headed to.
Importer of record	The party responsible for ensuring that imported goods comply with all customs and legal requirements of the country of import.
Port of loading	Port where the cargo is loaded onto the shipping vessel or container. It's the starting point of the shipping journey and is an important factor in determining shipping costs and transit time.

Table 5.1 Shipping Terms Used in Logistics Business Processes

Term	Description
Port of discharge	Place where a vessel discharges or unloads some or all of its shipments. The shipments are, thereafter, dispatched to their various consignees. A port of discharge is also known as the port of unloading.
Freight costs	Also termed as freight charges or freight rates, this refers to the amount paid to a carrier company for the transportation of goods from the point of origin to an agreed location.
Full container load (FCL)	Full container load;, that is, the client's goods are transported in a filled and sealed container.
Less than container load (LCL)	Grouped shipment, that is, the client' goods aren't numerous enough to fill a container, so the goods of several clients are put into one container.
Full truckload (FTL)	This allows for more precise planning of shipment and delivery because shipping plans need not be grouped with others. Shipment is usually faster, but cost may be a disadvantage.
Less than truckload (LTL)	This allows shippers to pay only for what they occupy, which makes it a less expensive and more environmentally conscious method to ship. However, it usually takes longer to reach the destination.

Table 5.1 Shipping Terms Used in Logistics Business Processes (Cont.)

5.2 Logistics Execution

Logistics execution or basic shipping is the link between procurement and distribution, regardless of whether the processes involved are internal or related to third parties (vendors, customers, or service providers):

- Inbound delivery connects the procurement process with the WM process to facilitate putaway and goods receipt.
- Production processes are integrated with WM to facilitate both staging of the raw materials required for production and receipt from production.
- Outbound delivery links the sales and distribution process with WM to execute picking, packing, and shipping of products to customers and distribution centers.

Figure 5.1 shows the integration points of logistics execution with purchasing and with sales and distribution operations.

In the following sections, we'll discuss the differences in the SAP S/4HANA process flow with or without deliveries.

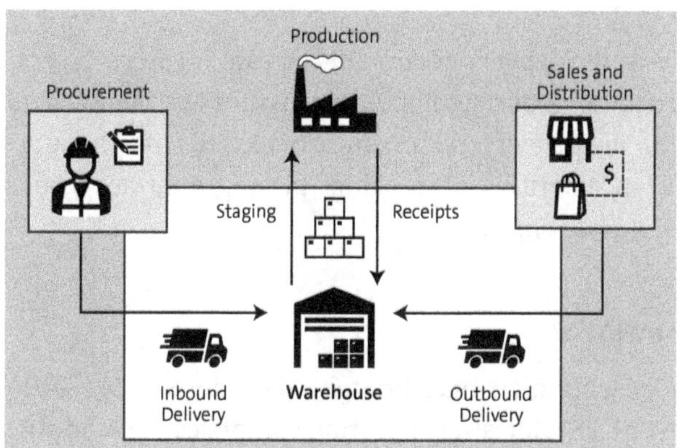

Figure 5.1 Basic Functions of Logistics Execution in the End-to-End Supply Chain Process

5.2.1 Mapping Logistics Execution Processes

There are two ways to map the processes in the logistics execution area (see Figure 5.2 for a comparison of logistics execution processes, including WM, with and without delivery options):

- **Without delivery**
  The without delivery option is simpler and doesn't involve inbound or outbound delivery. However, in this case, there is no option in the system to manage and record real-time activities between ordering and stock postings. In addition, this option isn't suitable from a third-party logistics perspective, where delivery is an essential integration point.

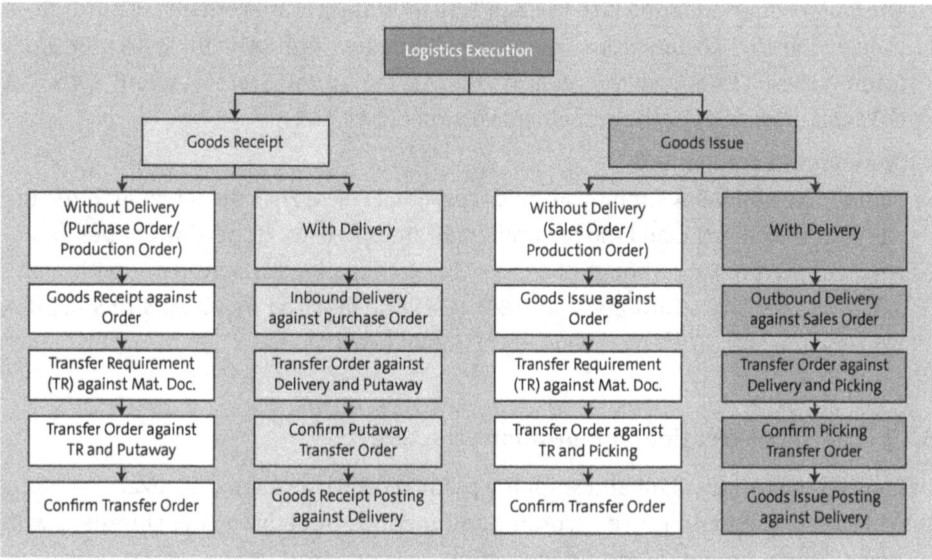

Figure 5.2 Logistics Execution Processes Mapping with and without Delivery

- **With delivery**
  In process mapping with delivery, the delivery document is created against the purchase order/sales order. Subsequent goods receipts/goods issues are then posted against the delivery document. The delivery document is equipped with the functionalities to capture details and statuses of packing, picking/putaway, goods receipts/issues, transportation, and so on.

5.2.2 SAP ERP versus SAP S/4HANA

We'll take a look at the logistics execution changes brought about by SAP S/4HANA when compared with SAP ERP. In SAP ERP, logistics execution comprises the following components:

- Shipping
  - Inbound deliveries
  - Outbound deliveries
- Transportation (LE-TRA)
- Warehouse management (LE-WM)

SAP S/4HANA retains the concept of logistics execution like that of SAP ERP, however, with a different future road map:

- **Shipping**
  SAP S/4HANA also uses the outbound and inbound delivery, and the respective configuration nodes are the same as in SAP ERP.
- **Transportation management**
  The SAP recommendation is to use SAP Transportation Management (SAP TM). Even though the LE-TRA functionality is still retained within SAP S/4HANA, per SAP's future vision, LE-TRA will be replaced by SAP TM in SAP S/4HANA. SAP S/4HANA 1709 and onward offer the embedded version of SAP TM.
- **Warehouse management**
  Per the SAP S/4HANA simplification document, LE-WM isn't the target architecture within SAP S/4HANA, and the recommendation is to use extended warehouse management (EWM). Even though the LE-WM functionality is still available within SAP S/4HANA, per SAP's future vision, LE-WM will be replaced by EWM. SAP S/4HANA 1610 and onward offer an embedded version of EWM.

5.2.3 Logistics Execution Organization Structure

In logistics execution, plant and shipping point are the two main organizational units for categorizing an enterprise according to functions of delivery and shipment with respect to procurement, sales, and inventory management.

Shipping Points

A shipping point is the top level in the organization for shipping and indicates the place or location that goods are shipped from or received. It can be allocated to a single plant or to multiple plants. A delivery belongs to exactly one shipping point; that is, all items of a delivery belong to one shipping point. Groups of deliveries may also belong to exactly one shipping point. A shipping point has an address and is defined with a four-character alphanumeric key. A shipping point is used as a selection criterion for processing deliveries such as picking or goods issue. In addition, it can be used as a selection criterion for lists of deliveries and the delivery worklist.

Shipping points can be created using configuration menu path **SPRO • SAP Reference IMG • SAP Customizing Implementation Guide • Enterprise Structure • Definition • Logistics Execution • Define, Copy, Delete, Check Shipping Point** (Transaction OVXD).

Figure 5.3 provides screenshots of the configuration menu path, shipping point fields, and address-specific details.

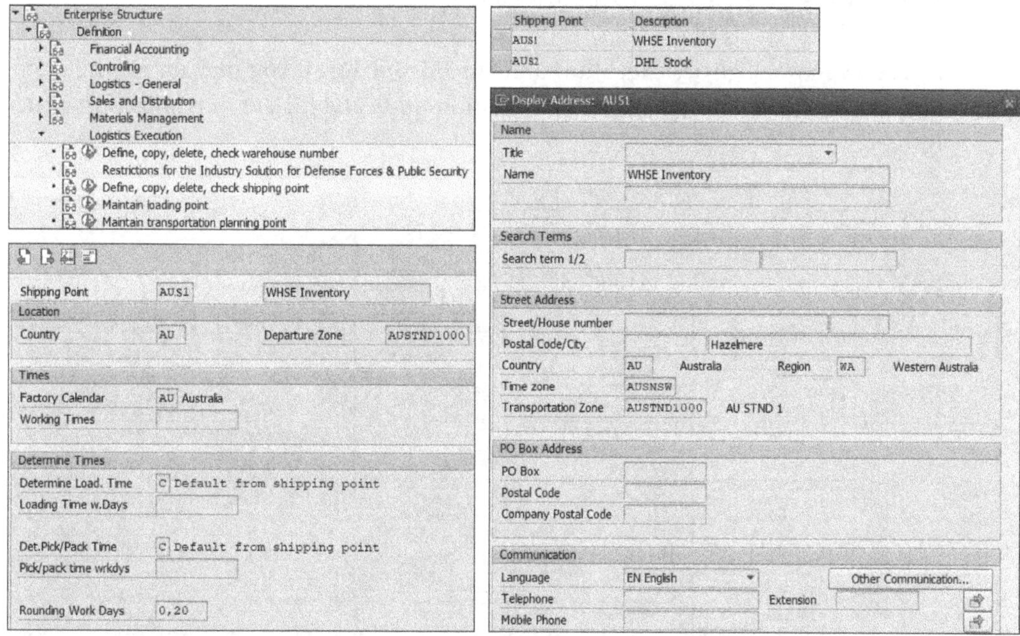

Figure 5.3 Define Shipping Points with Address Details

Shipping Point Assignment to Plant

Single or multiple shipping points can be allocated to a plant. In addition, one shipping point can belong to multiple plants.

The configuration menu path for assigning Shipping Point to Plant is **SPRO • SAP Reference IMG • SAP Customizing Implementation Guide • Enterprise Structure • Assignment • Logistics Execution • Assign Shipping Point to Plant** (Transaction OVXC).

Figure 5.4 shows the menu path for configuration and assignment of shipping point to plant.

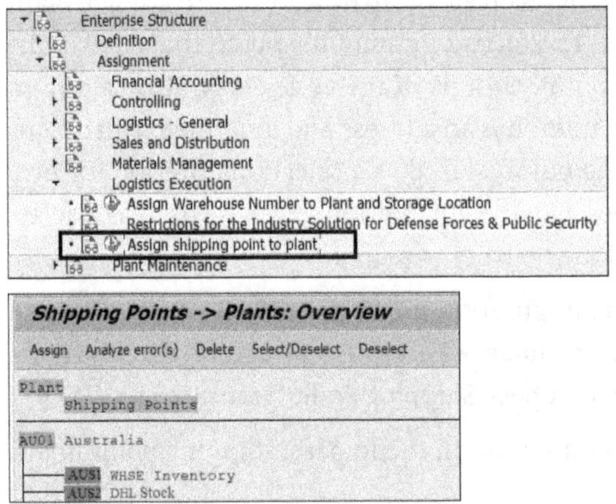

Figure 5.4 Assigning Shipping Points to a Plant

Figure 5.5 shows logistics execution units in SAP S/4HANA. You may notice the inter-linkages between different logistics execution units and logistics organization structure.

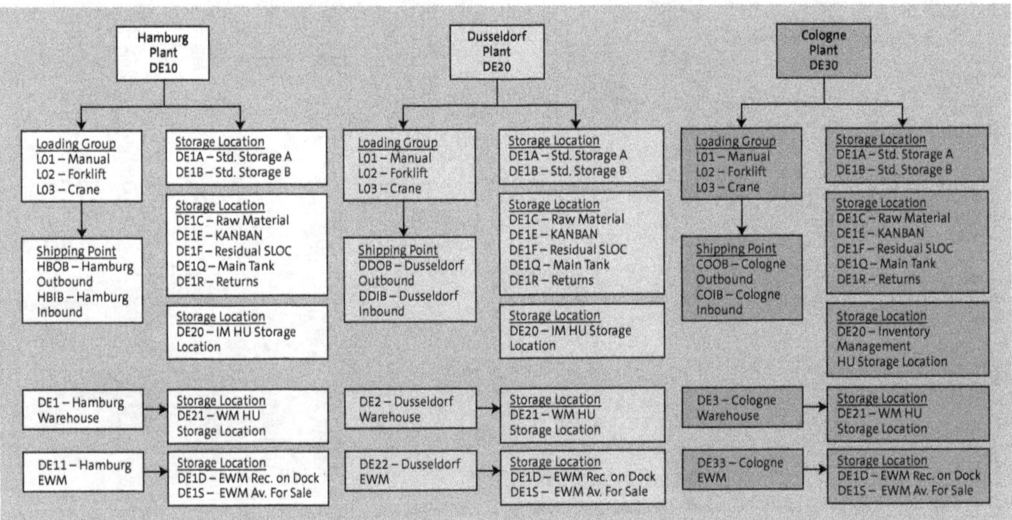

Figure 5.5 Sample Logistics Execution Organization Structure

5.2.4 Shipping-Specific Configurations

Next, you'll see the major setups for enabling the shipping function in SAP S/4HANA.

Define Shipping Condition

The shipping condition represents conditions required for goods to be shipped. It's maintained as a two-character alphanumeric key along with a description for the shipping conditions. Shipping condition is maintained in customer master record in the sales data for each sales area. The system copies the shipping condition into the sales document header. Shipping point determination can only be carried out by the system if a shipping condition is contained in the sales document.

Shipping condition can be maintained using menu path **SPRO • SAP Reference IMG • SAP Customizing Implementation Guide • Logistics Execution • Shipping • Basic Shipping Functions • Shipping Point and Goods Receiving Point Determination • Define Shipping Conditions** (Transaction OVSF).

Figure 5.6 shows the configuration menu path and example shipping conditions

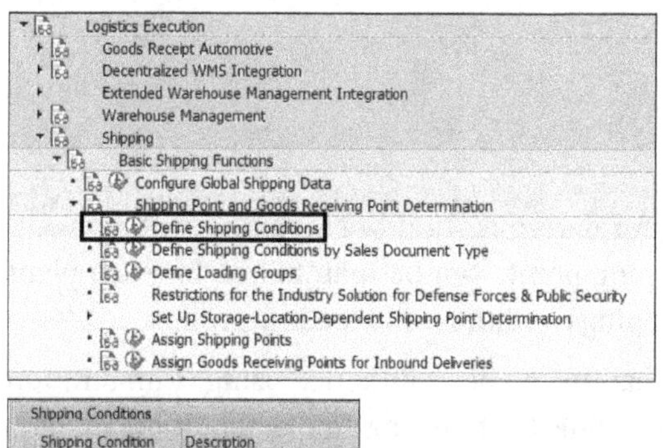

Shipping Conditions

Shipping Condition	Description
01	Normal
02	Express Delivery
03	Direct Ship
04	Returns

Figure 5.6 Defining Shipping Conditions

> **Note**
>
> Shipping condition can also be defined for each sales document type. This shipping condition then supersedes the one specified in the customer master record. This setting can be maintained in the **Determine shipping condition by sales document type** node.

Define Loading Group

Loading groups is an essential element in shipping point determination. The value for this field is maintained in the material master in the **Sales: General/Plant** view for the corresponding plant. In other words, it categorizes the material master data by the criterion to load these goods. An alphanumeric key with up to 4 digits and a description are maintained for the loading group.

Loading groups can be defined using configuration menu path **SPRO • SAP Reference IMG • SAP Customizing Implementation Guide • Logistics Execution • Shipping • Basic Shipping Functions • Shipping Point and Goods Receiving Point Determination • Define Loading Groups** (Transaction OVSX). Figure 5.7 shows loading group examples along with the configuration menu path.

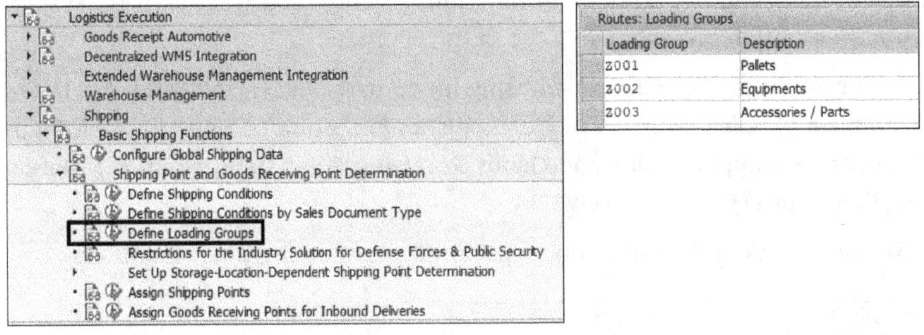

Routes: Loading Groups

Loading Group	Description
Z001	Pallets
Z002	Equipments
Z003	Accessories / Parts

Figure 5.7 Defining Loading Groups

Shipping Point Determination

In this configuration step, the criteria for determination of shipping points for outbound deliveries are maintained. Shipping points can be maintained for each plant based on different combinations of shipping conditions and loading groups.

The criteria for shipping point determination can be maintained using configuration menu path **SPRO • SAP Reference IMG • SAP Customizing Implementation Guide • Logistics Execution • Shipping • Basic Shipping Functions • Shipping Point and Goods Receiving Point Determination • Assign Shipping Points** (Transaction OVL2; table TVSTZ). Figure 5.8 shows field combinations for the determination of shipping points based on **Shipping Condition, Loading Group**, and **Plant**.

Logistics Execution
Goods Receipt Automotive
Decentralized WMS Integration
Extended Warehouse Management Integration
Warehouse Management
Shipping
Basic Shipping Functions
Configure Global Shipping Data
Shipping Point and Goods Receiving Point Determination
Define Shipping Conditions
Define Shipping Conditions by Sales Document Type
Define Loading Groups
Restrictions for the Industry Solution for Defense Forces & Public Security
Set Up Storage-Location-Dependent Shipping Point Determination
Define Rule for Determination of Shipping Point
Define Rule for Determination of Shipping Point for Subcontracting
Assign Shipping Points According to Storage Location
Assign Shipping Points
Assign Goods Receiving Points for Inbound Deliveries

Shipping Point Determination

Shipping Condition	Loading Group	Plant	Proposed Shipping Point	Manual Shipping Point	Manual Shipping Point
01	Z001	AUP1	AUS1	AUV1	AUS5
01	Z002	AUP1	AUS2	AUV1	AUS6
01	Z003	AUP1	AUS2	AUV1	AUS6

Figure 5.8 Maintaining Parameters for Shipping Point Determination

> **Note**
>
> The value maintained in the **Proposed Shipping Point** column is the one determined by the system. However, this can be changed by the user while creating purchase or sales documents. The F4 or dropdown option in the field will show the values maintained in the manual shipping point columns.

As shown in Figure 5.9, the shipping point in the stock transfer order (STO) is determined based on **Supplying Plant**, **Shipping Point**, **Shp.Cond.** (shipping condition), and **Loading Group** (all marked).

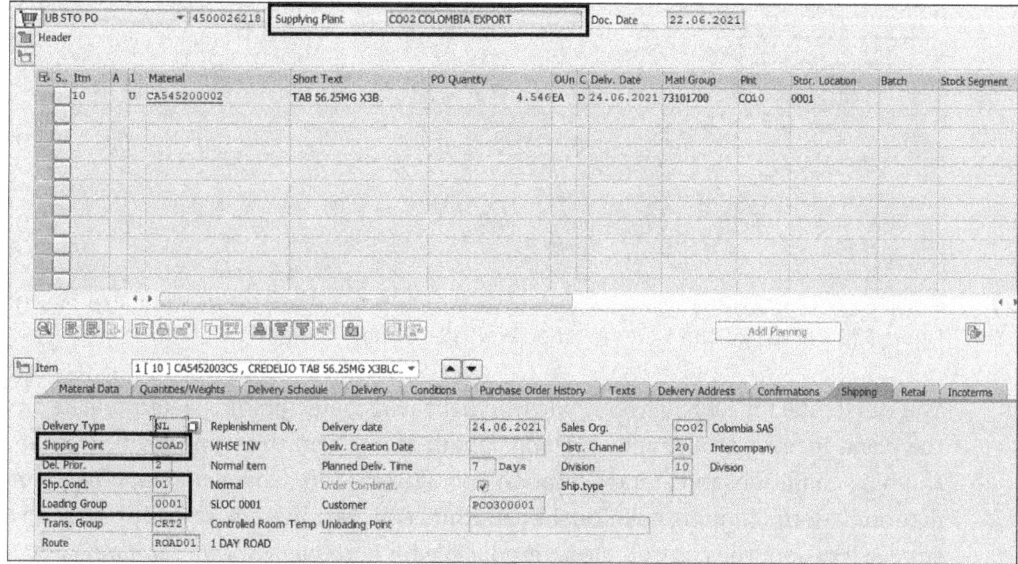

Figure 5.9 Shipping Point Determination in Stock Transfer Order (Intracompany) Purchase Order

The automatic determination of the shipping point depends on the following criteria:

- Shipping condition from the ship-to party (can be overwritten by shipping condition from sales document type, if configured per sales document type)
- Loading group from the material master
- Plant from the purchase order/sales order item/delivery plant
- Storage location from the purchase order/sales order item (if the shipping point determination is based on the storage location)

In a sales order, the shipping point is determined based on the shipping conditions from the customer master (sold-to party) record, loading group from the material master record (**Sales/Plant** data screen), and the delivering plant. (Figure 5.10 shows the resulting **Shipping Point** in a sales order **Shipping** tab at the item level.)

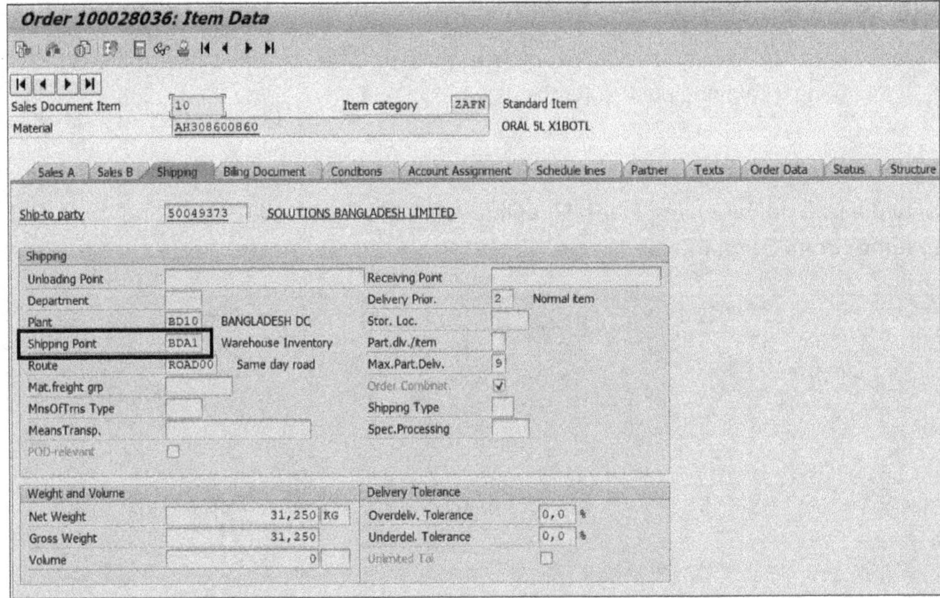

Figure 5.10 Shipping Point Determination in a Sales Order

You can define for each delivery type whether the shipping point is determined from the plant, loading group, and shipping condition, or from the plant, loading group, shipping condition, and storage location (storage-location-dependent shipping point determination). Shipping point can be determined in two different ways: by plant or by storage location. The control rule is responsible for both plant-specific shipping point determination and storage-location-specific shipping point determination.

> **Note**
>
> Storage-location-dependent shipping point determination can be used in the stock transfer between storage locations scenario, for example.

The control for plant- or storage-location-dependent shipping point determination can be set up using menu path **SPRO • SAP Reference IMG • SAP Customizing Implementation Guide • Logistics Execution • Shipping • Basic Shipping Functions • Shipping Point and Goods Receiving Point Determination • Set Up Storage-Location-Dependent Shipping Point Determination • Define Rule for Determination of Shipping Point.**

Figure 5.11 shows the configuration menu path along with the screen for controlling shipping point determination based on either the plant or storage location.

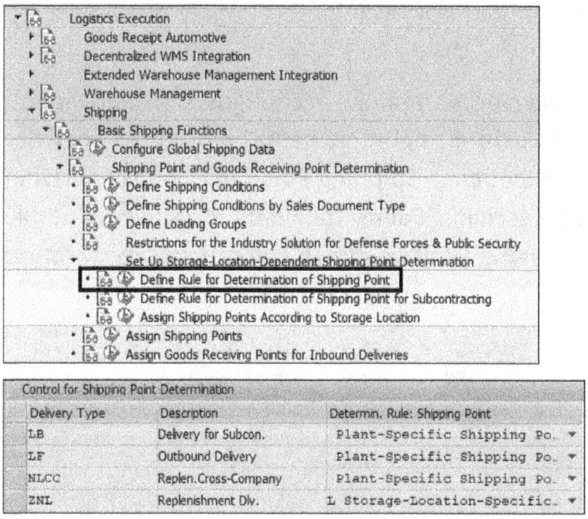

Delivery Type	Description	Determin. Rule: Shipping Point
LB	Delivery for Subcon.	Plant-Specific Shipping Po…
LF	Outbound Delivery	Plant-Specific Shipping Po…
NLCC	Replen.Cross-Company	Plant-Specific Shipping Po…
ZNL	Replenishment Dlv.	L Storage-Location-Specific…

Figure 5.11 Control for Determination of Shipping Point (Plant or Storage Location)

Setting Up Storage-Location-Dependent Shipping Point Determination

Shipping point determination at the storage location level can be set up using menu path **SPRO • SAP Reference IMG • SAP Customizing Implementation Guide • Logistics Execution • Shipping • Basic Shipping Functions • Shipping Point and Goods Receiving Point Determination • Set Up Storage-Location-Dependent Shipping Point Determination • Assign Shipping Points According to Storage Location**. Figure 5.12 shows the configuration for the storage-location-specific shipping point determination rule.

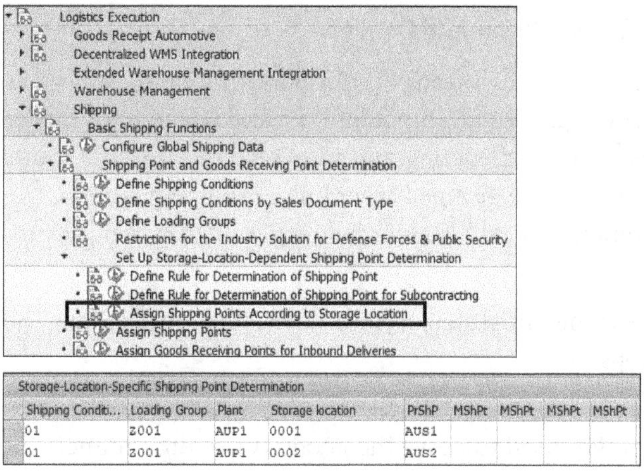

Shipping Condit...	Loading Group	Plant	Storage location	PrShP	MShPt	MShPt	MShPt	MShPt
01	Z001	AUP1	0001	AUS1				
01	Z001	AUP1	0002	AUS2				

Figure 5.12 Setup for Storage-Location-Dependent Shipping Point Determination

Assign Shipping Point as Goods Receiving Points for Plant/Storage Location

In this configuration step, the criteria for shipping point determination for inbound deliveries are maintained. Assign one or several goods receiving points to the combination plant/storage location. For multiple assignments of receiving locations, you can set a priority.

Shipping points for inbound deliveries can be maintained in configuration using menu path **SPRO • SAP Reference IMG • SAP Customizing Implementation Guide • Logistics Execution • Shipping • Basic Shipping Functions • Shipping Point and Goods Receiving Point Determination • Assign Goods Receiving Points for Inbound Deliveries** (table TWLVZ). Figure 5.13 shows different receiving locations for plant and storage location combinations.

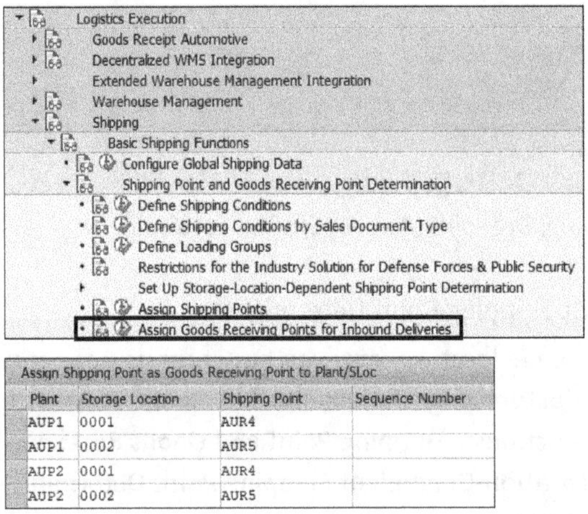

Plant	Storage Location	Shipping Point	Sequence Number
AUP1	0001	AUR4	
AUP1	0002	AUR5	
AUP2	0001	AUR4	
AUP2	0002	AUR5	

Figure 5.13 Setup for Shipping Point as Goods-Receiving Locations (for Inbound Deliveries)

5.2.5 Route Determination in Sales Order and Delivery

The route determines the itinerary and the means of transport in shipping, as well as influences transportation scheduling. Specific shipment type and selection of carrier will have a bearing on transportation scheduling and may require a specific route. Based on routes, sales order items can be grouped according to shipping criteria. The system determines routes automatically for a sales order item and can repeat the procedure for a delivery.

Establishing a shipment's route is the initial step of route determination. Defining the route requires the following steps:

1. **Define modes of transport.**
   These are types of transportation used to move the goods. A two-digit numeric or alphanumeric is used to define the modes of transport.

   Modes of transport can be maintained in configuration using menu path **SPRO • SAP Reference IMG • SAP Customizing Implementation Guide • Logistics Execution • Shipping • Basic Shipping Functions • Routes • Define Routes • Define Modes of Transport** (Transaction OVTB).

As shown in Figure 5.14, four different modes of transport are defined in configuration.

Mode of Transport			
ShTy	Description	SType	Description
01	Road	1	Road
02	Rail	2	Rail
03	Sea	4	Sea
04	Inland Waterway	3	Inland waterway

Figure 5.14 Modes of Transport

Note

Mode of transport is assigned to a shipping type. The shipping type serves as a classification of modes of transport in dangerous goods management.

2. **Define shipping types.**
   Shipping types are the actual vehicle or carrier used to transport the material. A two-digit alphanumeric key represents various types of shipment methods, as shown in Figure 5.15. Here, one or several modes of transports can be assigned to each shipping type, along with a shipping type procedure group.

Shipping Types				
PT	Description	MdTr	Description	STPG
01	Truck	01	Road	0001
02	Mail	06	Postal Service	0002
03	Sea	03	Sea	0003
04	Rail	02	Rail	0004
05	Airplane	05	Air	

Figure 5.15 Shipping Types and Linkage with Mode of Transport and Shipping Type Procedure Group

The configuration for assignment of shipping type to mode of transport and shipping type procedure group is through menu path **SPRO • SAP Reference IMG • SAP Customizing Implementation Guide • Logistics Execution • Shipping • Basic Shipping Functions • Routes • Define Routes • Define Shipping Types** (Transaction OVTA).

Note

Each shipping type is assigned to a shipping type group, which serves to determine the pricing procedure for calculating the shipping costs.

3. **Define routes and route stages.**
   Routes and stages are more relevant for shipment and transportation, so they may be important to understand from a transportation perspective. In this step, routes with route stages are maintained. Additionally, it also provides the option of maintaining the transportation connection point. All relevant details, such as shipping

type, preliminary and subsequent legs of transport, duration, calendar, and distance, are maintained here.

Configuration for routes and route stages can be done using menu path **SPRO • SAP Reference IMG • SAP Customizing Implementation Guide • Logistics Execution • Shipping • Basic Shipping Functions • Routes • Define Routes • Define Routes and Stages** (Transaction OVTC).

Figure 5.16 shows the configuration details of routes and transportation connection points.

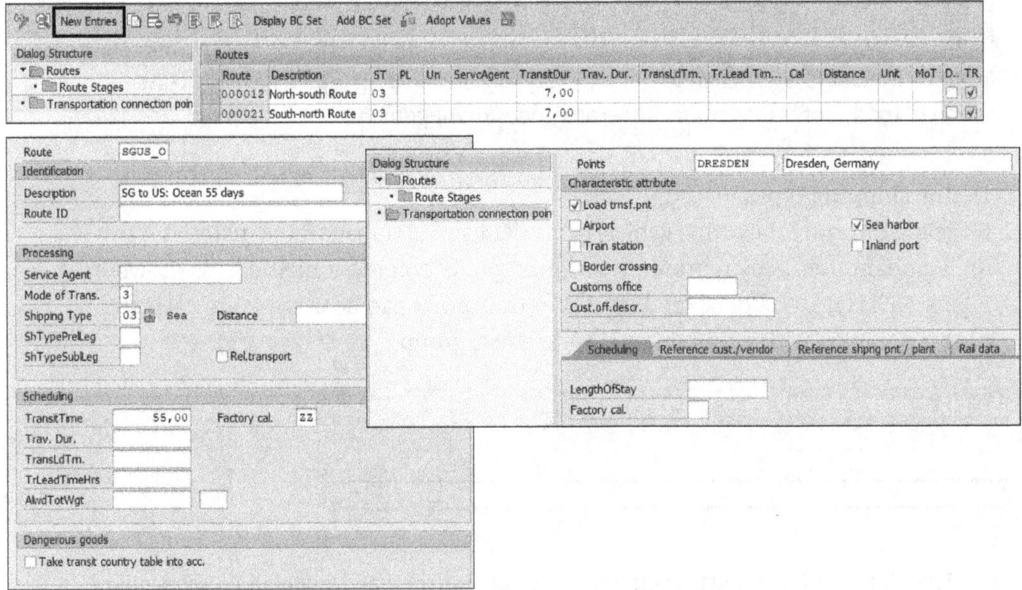

Figure 5.16 Setup for Routes and Transportation Connection Points

Maintaining route stages may be important while using the transportation component of basic shipment (LE_TRA). Figure 5.17 provides a snapshot of a sample route stage.

Dialog Structure
Routes
Route Stages
Transportation connection poin

Route 000012 North-south Route

Route Stages

Itn.	Cat.	Cat	Dep.point	Description	ST	ServcAgent	Distance	Unit	Tot.durat.	Trav. Dur.	Work Time	Cal	LegId	SpPI	ShCRel	Proced..
1	1		0000000001		03		0								☐	
2	1		0000000002		01		0								☐	
3			0000000003				0								☐	

Figure 5.17 Maintenance of Route Stages for a Specific Route

4. **Route determination.**

   The SAP system determines routes automatically for each sales document item. It determines the itinerary and mode of transport in shipping. Routes are determined depending on the following criteria:

- Country and transportation zone (departure zone) of the shipping point, based on the configuration maintained
- Shipping conditions, with the default value based on the customer master of the sold-to party (business partner sales area view: **Shipping**) or from the sales document type (if configured)
- Transportation group, with the default value from material master in the **Sales: General/Plant** view
- Country and transportation zone (receiving zone) of the ship-to party, with the default value from the customer master of the ship-to party
- Weight group, if route determination is carried out again at the delivery note level, based on the configuration maintained

The SAP system copies the route from the sales document item into the delivery at header level.

Note

You can manually overwrite the route that is determined. The route can be redetermined in outbound delivery based on the weight (weight group). The configuration of the delivery type dictates whether the route is redetermined.

Figure 5.18 is a schematic representation of logical or criterial route determination.

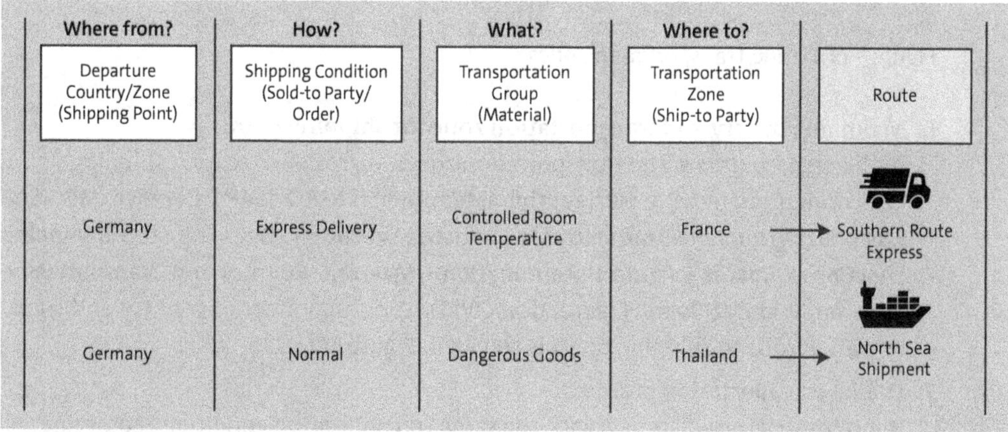

Figure 5.18 Criteria for Route Determination

5. **Define the transportation zone.**
   Transportation zones can be defined per country and depending on your business requirements. Transportation zones represent delivery recipient (ship-to party) regions or delivery shipper (supplier) regions. You define the transportation zones for each country and enter the corresponding description. Multiple transportation zones can also be assigned to a country.

Transportation zones can be defined using configuration menu path **SPRO • SAP Reference IMG • SAP Customizing Implementation Guide • Logistics Execution • Shipping • Basic Shipping Functions • Routes • Route Determination • Define Transportation Zones** (Transaction OVR1). Figure 5.19 shows country-specific transportation zones along with the configuration menu path.

Customers: Transportation Zones

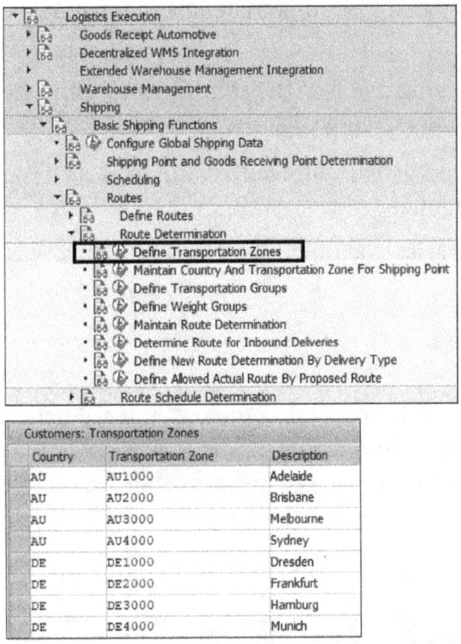

Country	Transportation Zone	Description
AU	AU1000	Adelaide
AU	AU2000	Brisbane
AU	AU3000	Melbourne
AU	AU4000	Sydney
DE	DE1000	Dresden
DE	DE2000	Frankfurt
DE	DE3000	Hamburg
DE	DE4000	Munich

Figure 5.19 Define Transportation Zones

6. **Maintain country and transportation zone for shipping point.**
   In this step, countries and transportation zones are assigned to each shipping point. The configuration menu path for this assignment is **SPRO • SAP Reference IMG • SAP Customizing Implementation Guide • Logistics Execution • Shipping • Basic Shipping Functions • Routes • Route Determination • Maintain Country and Transportation Zone for Shipping Point** (Transaction OVL7). The assignment of country and transportation zone to shipping point is shown in Figure 5.20.
7. **Define transportation groups.**
   It groups together materials with the same transportation condition requirements. You define the transportation groups with a four-digit alphanumeric key along with a description. Transportation group is maintained in the **Sales: General/Plant** view of the material master data to enable route determination in shipping.

Note

The system can only carry out route determination if the transportation group is included in the material master record. As the transportation group is a prerequisite for route determination, select the transportation group in the material master record as a required-entry field to ensure the entry.

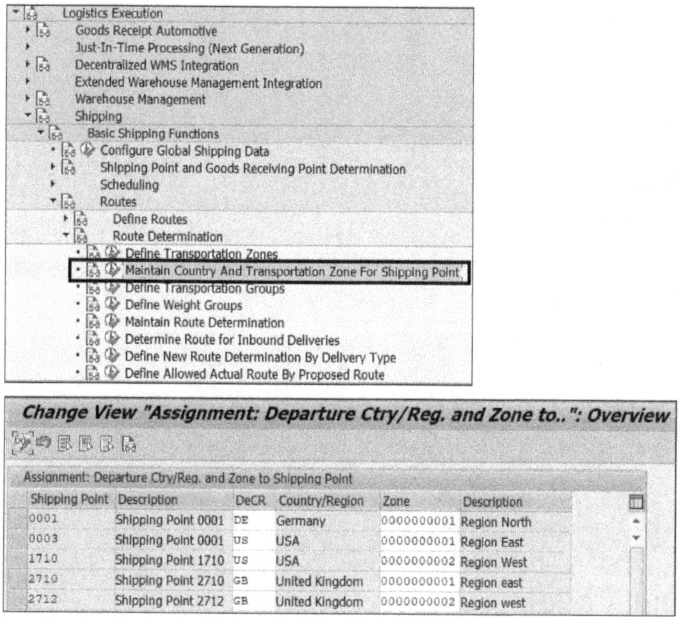

Change View "Assignment: Departure Ctry/Reg. and Zone to..": Overview

Assignment: Departure Ctry/Reg. and Zone to Shipping Point

Shipping Point	Description	DeCR	Country/Region	Zone	Description
0001	Shipping Point 0001	DE	Germany	0000000001	Region North
0003	Shipping Point 0001	US	USA	0000000001	Region East
1710	Shipping Point 1710	US	USA	0000000002	Region West
2710	Shipping Point 2710	GB	United Kingdom	0000000001	Region east
2712	Shipping Point 2712	GB	United Kingdom	0000000002	Region west

Figure 5.20 Assigning Transportation Zones and Country to Shipping Point

Transportation groups can be defined using configuration menu path **SPRO • SAP Reference IMG • SAP Customizing Implementation Guide • Logistics Execution • Shipping • Basic Shipping Functions • Routes • Route Determination • Define Transportation Groups** (Transaction OVSY). Figure 5.21 provides a snapshot of the configuration menu path and transportation group details.

8. **Maintain route determination.**
   This configuration step deals with setting the route determination for a sales order or a delivery. Route determination in a sales order is dependent on the following factors:
   - Country of departure/departure zone
   - Country of destination/receiving zone
   - Shipping conditions
   - Transportation group

In a delivery, route determination is dependent on the preceding list and one additional factor, the weight group. The configuration menu path is **SPRO • SAP Reference IMG • SAP Customizing Implementation Guide • Logistics Execution • Shipping • Basic Shipping Functions • Routes • Route Determination • Maintain Route Determination** (Transaction OVRF). Figure 5.22 shows route determination for a sales order.

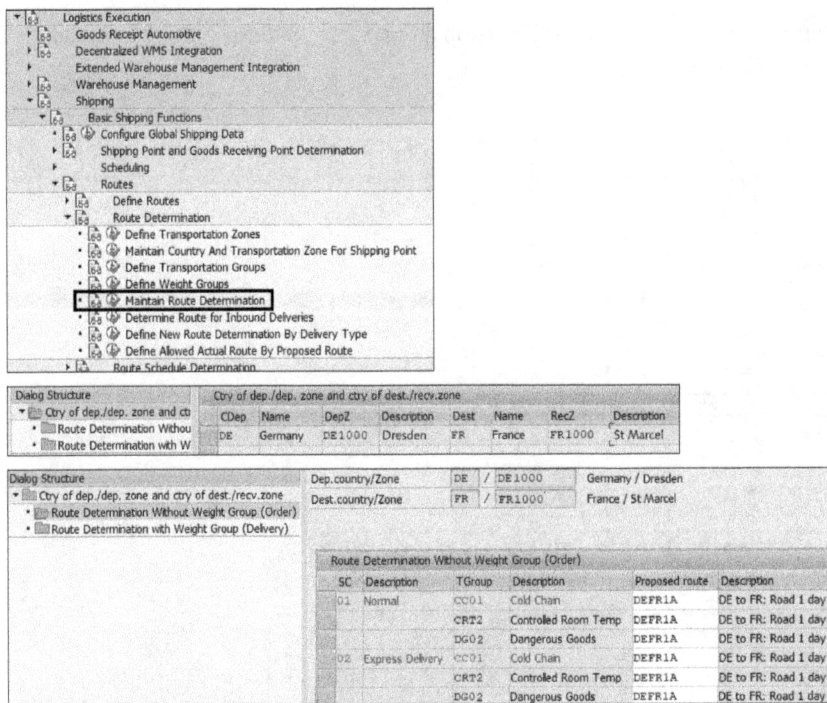

Figure 5.21 Define Transportation Groups

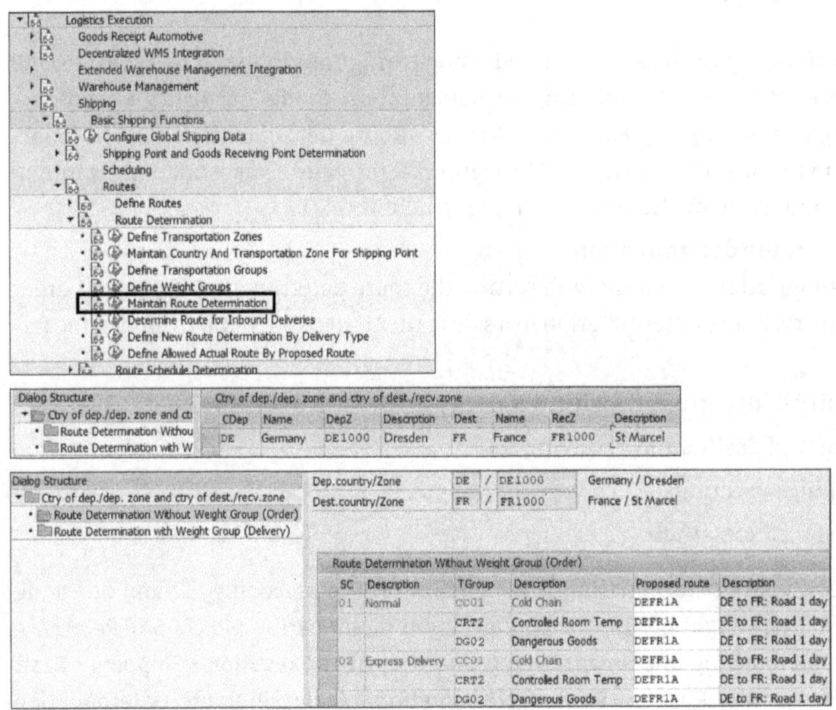

Figure 5.22 Maintain Route Determination for a Sales Order

5.2.6 Inbound and Outbound Deliveries

SAP S/4HANA provides standard delivery types and item categories that can meet the majority of business requirements. However, before implementing custom delivery types, keep in mind that deliveries alone can't be handled in isolation because they are closely linked to sales document types and billing types. Therefore, all the delivery-specific configurations that are listed for sales documents and billing documents should also be considered.

In this section, we'll discuss the details of delivery types and dependent configurations such as item categories, batch creation control, and shipping blocks.

Define Delivery Types

In this step, you can perform the following setup configurations:

1. Define the delivery type and its description.
2. Define attributes of delivery types.

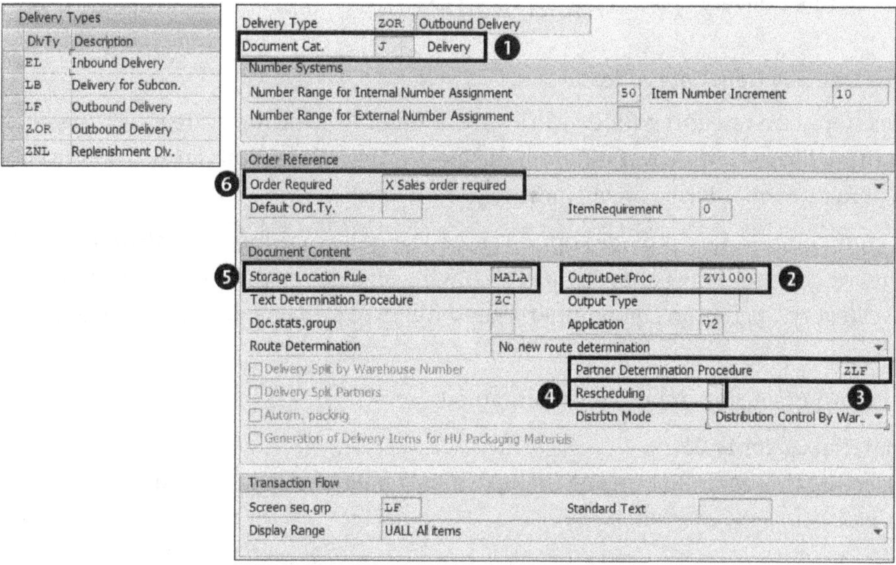

Figure 5.23 Delivery Document Attributes

As shown in Figure 5.23, this includes the following:

❶ **Delivery Type**: Classification for the different types of documents that you can process, for example, quotations, sales orders, deliveries, and invoices

❷ **Output.Det.Proc**: Output determination procedure for header/item, which defines the output categories and their sequence

❸ **Partner Determination Procedure**: Grouping of partner functions that specify which partner functions are allowed for a particular business transaction (e.g., for processing a delivery) and which of those partner functions are mandatory

❹ **Rescheduling**: Enable or disable rescheduling whenever a new delivery is created, and control of whether a redetermination of shipping deadlines should be run for backlogged deliveries using this indicator

❺ **Storage Location Rule**: Rule for determining the storage location for picking

❻ **Order Required**: Requiring a sales order as the basis for delivery

3. Assign the number range of delivery document and item number increment to control the numbering of new items.
4. Choose delivery split options based on partner function or warehouse number.
5. Specify whether the route is based on a sales order document or a new route is determined.
6. Select the **Text Determination Procedure**, that is, the sequence in which the text types appear in the delivery document header.

Delivery types can be defined using configuration menu path **SPRO • SAP Reference IMG • SAP Customizing Implementation Guide • Logistics Execution • Shipping • Deliveries • Define Delivery Types** (Transaction OVLK).

Define Delivery Item Categories

In this step, item categories for deliveries are maintained. The item categories, together with the delivery types, cover different business transactions. You can also define custom item categories based on business requirements.

This configuration step provides options for editing the existing or standard item categories or creating a new one—either by copying an existing one or using a fresh new item category. Specify an alphanumeric key with up to four characters for the item category.

The following settings are usually maintained:

- **Material Quantity 0**
  Selecting this checkbox means the text item material quantity need not be maintained.
- **Quantity**
  The settings in this area allow different system responses to be defined; for example, you can set the system to display an error message if the quantity is zero.
- **Overdelivery**
  This allows you to define the system response as no message/warning/error if the original order quantity is exceeded.
- **Warehouse Control and Packing**
  This area includes relevant controls for picking and packing.

Item categories for delivery type can be maintained using configuration menu path **SPRO • SAP Reference IMG • SAP Customizing Implementation Guide • Logistics**

Execution • Shipping • Deliveries • Define Delivery Item Categories (Transaction OVLP). Figure 5.24 shows the **Delivery Item Categories** screen and various field-level attributes of delivery item categories.

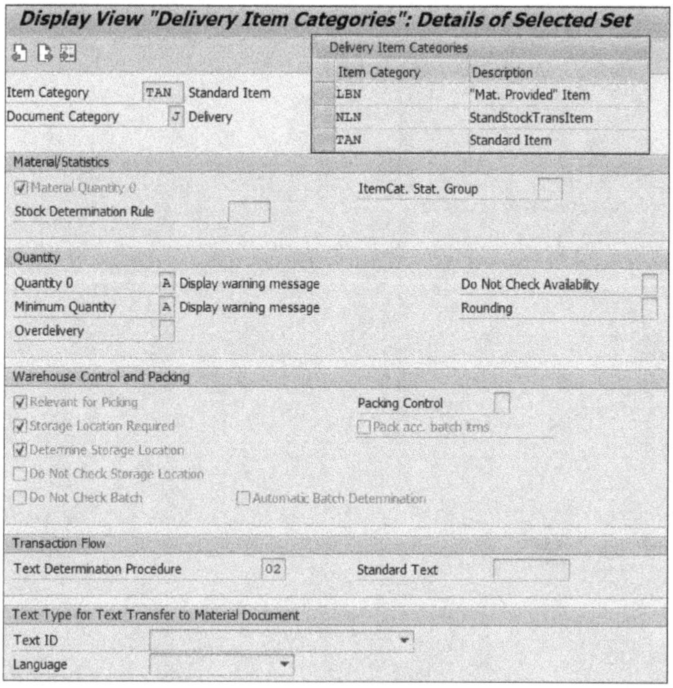

Figure 5.24 Attributes of Item Categories

Define Item Category Determination in Deliveries

In this menu option, you define the item category determination for deliveries. The following applies:

- If an order item or a schedule line is copied into a delivery, the item category is also copied.
- For items independent of orders in the delivery (e.g., packaging material that is entered in the delivery) or deliveries without reference to an order, the item category is determined either by the item category group of the material or based on the sales document type.

For each item category, combinations of allowed delivery types and item category groups are maintained. As an alternative, combinations of item categories and item category groups can be allocated to each delivery type. There is a provision for finer control through another two parameters: **Item usage** and **Item category** of the higher-level item.

The configuration menu path for delivery-specific item category determination is **SPRO • SAP Reference IMG • SAP Customizing Implementation Guide • Logistics**

Execution • Shipping • Deliveries • Define Item Category Determination in Deliveries (Transaction 0184). Figure 5.25 shows the configuration menu path and field combination logic for delivery item category determination.

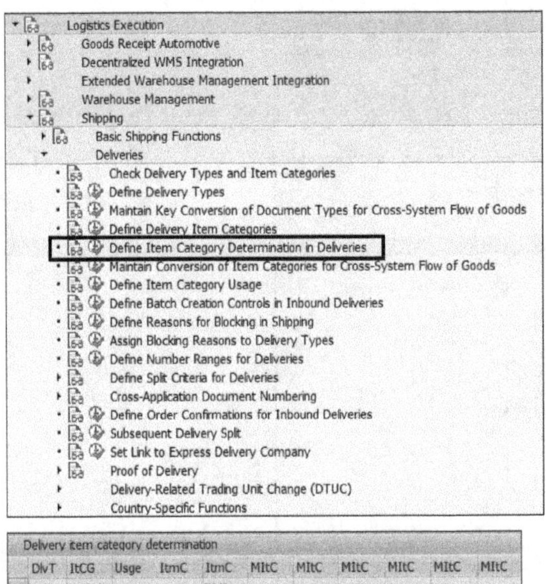

Delivery item category determination

DlvT	ItCG	Usge	ItmC	ItmC	MItC	MItC	MItC	MItC	MItC	MItC
ZOR	NORM			TAN						
ZOR	NORM	C		TAN						
ZOR	NORM	CHSP		TAN						
ZOR	NORM	CHSP	KLN	KLN						
ZOR	NORM	CHSP	TANN	TANN						
ZOR	NORM	PACK		DLN	DLX	DLP	KEN			
ZOR	NORM	PSEL	TAX	TAPS						
ZOR	VERP			DLN						

Figure 5.25 Delivery-Specific Item Category Determination

> **Note**
>
> The SAP system automatically determines the corresponding schedule line category using the item category of a delivery item independent of an order. The information in the schedule line category is necessary for the availability check as well as for quantity postings and value postings in materials management due to the goods movement.

Define Batch Creation Controls in Inbound Deliveries

This step deals with setting up controls for batch creation in the inbound delivery. For each desired delivery item category, you can specify how batch creation, batch classification, or expiration date checks are performed in inbound deliveries. Figure 5.26 shows the standard control options available in the system. The configuration menu path is **SPRO • SAP Reference IMG • SAP Customizing Implementation Guide • Logistics Execution • Shipping • Deliveries • Define Batch Creation Controls in Inbound Deliveries.**

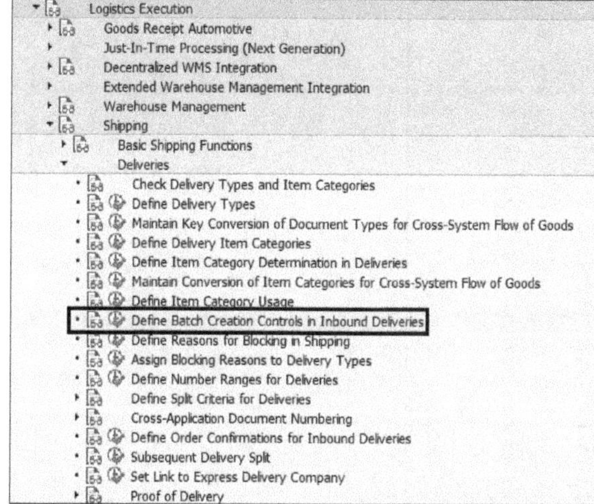

Batch Installation in the Inbound Delivery

Item Category	Description	Batch creation
ELN	Inbound Delivery	3 Batch creation, classification, SLED check

No batch creation, classification, SLED check
1 Batch creation, no classification, SLED check
2 Batch creation, classification, no SLED check
3 Batch creation, classification, SLED check

Figure 5.26 Control Options for Batch Creation in Inbound Delivery

Define Order Confirmations for Inbound Deliveries

Inbound deliveries are either created automatically when shipping notifications are received in the system through IDocs or manually if confirmation control in purchasing is set accordingly. If the shipping notification isn't being used, inbound deliveries can be created for purchase orders manually. The configuration menu path is **SPRO • SAP Reference IMG • SAP Customizing Implementation Guide • Logistics Execution • Shipping • Deliveries • Define Order Confirmations for Inbound Deliveries.**

This setup is necessary for WM because all inbound deliveries for which a WM transfer order is to be created are distributed to the WMS.

> **Note**
>
> Creation of inbound deliveries can be initiated by choosing **Logistics • Logistics Execution • Inbound Process • Goods Receipt for Inbound Delivery • Inbound delivery • Create • Collective processing of Purchase Orders** (Transaction VL34).

Figure 5.27 shows the standard confirmation control options available for inbound delivery.

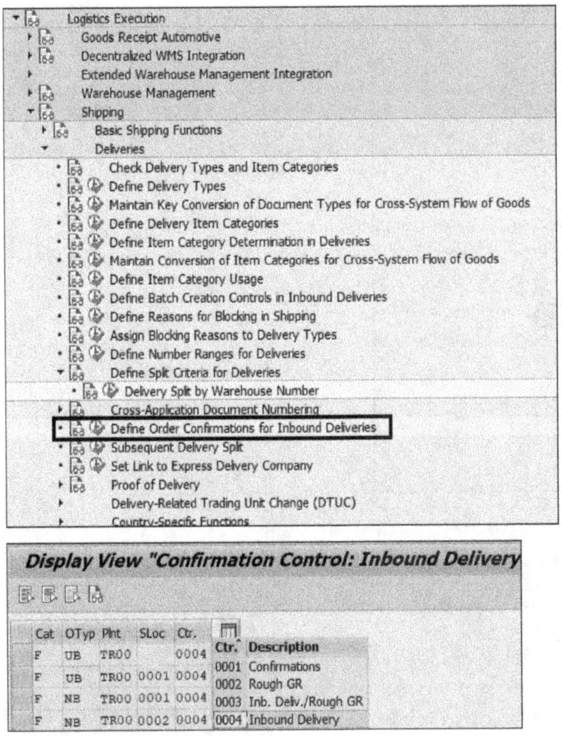

Figure 5.27 Order Confirmation Options for Inbound Delivery

Shipping Blocks

Delivery blocking reasons are used to separate certain transactions from the process flow for special reasons and to avoid further automatic processing. As a result, the transaction is blocked until the concerned authority reviews the situation, makes a decision, and possibly removes the delivery block manually. Following are the steps for maintaining a blocking reason and then assigning it to relevant delivery types:

1. **Define reasons for blocking in shipping.**
   Blocking reasons for delivery processing can be defined based on which activities should be blocked, such as picking or goods issue. This is a two-digit alphanumeric key, and you can define the custom blocking reason as well.

   Reasons for shipping blocks at transaction levels is maintained using configuration menu path **SPRO • SAP Reference IMG • SAP Customizing Implementation Guide • Logistics Execution • Shipping • Deliveries • Define Reasons for Blocking in Shipping** (table TVLS). As shown in Figure 5.28, various blocking reasons are maintained for order, confirmation, delivery due list, picking, and post goods issue.

2. **Assign blocking reasons to delivery types.**
   Based on the blocking reason applicable for each process, the appropriate blocking reason can be assigned to each of the delivery types.

- Logistics Execution
  - Goods Receipt Automotive
  - Decentralized WMS Integration
  - Extended Warehouse Management Integration
  - Warehouse Management
  - Shipping
    - Basic Shipping Functions
    - Deliveries
      - Check Delivery Types and Item Categories
      - Define Delivery Types
      - Maintain Key Conversion of Document Types for Cross-System Flow of Goods
      - Define Delivery Item Categories
      - Define Item Category Determination in Deliveries
      - Maintain Conversion of Item Categories for Cross-System Flow of Goods
      - Define Item Category Usage
      - Define Batch Creation Controls in Inbound Deliveries
      - Define Reasons for Blocking in Shipping
      - Assign Blocking Reasons to Delivery Types

Deliveries: Blocking Reasons/Criteria

DB	Delivery Block Desc.	Order	Conf.	Print	DlvDueListBlock	Pick.Block	GI Block
01	Missing Credit limit	☐	☑	☐	☐	☑	☑
02	Allocation Block	☑	☑	☑	☐	☐	☐
03	Duplicate Items	☑	☐	☐	☑	☐	☐
04	CustomerConfirmation	☑	☐	☐	☐	☐	☐
05	Duplicate PO	☐	☑	☑	☑	☐	☐
06	Block EDI Cust Order	☑	☐	☐	☐	☐	☐
07	Future Delivery Date	☑	☑	☑	☑	☑	☑

Figure 5.28 Define Delivery Blocking Reasons

The configuration menu path is **SPRO • SAP Reference IMG • SAP Customizing Implementation Guide • Logistics Execution • Shipping • Deliveries • Assign Blocking Reasons to Delivery Types** (table TVLSP). Figure 5.29 shows the assignment of shipping block reasons to delivery types and the related configuration menu path.

- Logistics Execution
  - Goods Receipt Automotive
  - Decentralized WMS Integration
  - Extended Warehouse Management Integration
  - Warehouse Management
  - Shipping
    - Basic Shipping Functions
    - Deliveries
      - Check Delivery Types and Item Categories
      - Define Delivery Types
      - Maintain Key Conversion of Document Types for Cross-System Flow of Goods
      - Define Delivery Item Categories
      - Define Item Category Determination in Deliveries
      - Maintain Conversion of Item Categories for Cross-System Flow of Goods
      - Define Item Category Usage
      - Define Batch Creation Controls in Inbound Deliveries
      - Define Reasons for Blocking in Shipping
      - Assign Blocking Reasons to Delivery Types
      - Define Number Ranges for Deliveries

Delivery Blocks

DB	Description	DlvTy	Description
01	Missing Credit limit	NLCC	Replen.Cross-Company
01	Missing Credit limit	ZOR	Outbound Delivery

Figure 5.29 Assignment of Blocking Reason to Delivery Type

So far, we've seen in detail how delivery types are maintained, and the relationship between delivery types and item categories is crucial for the success of shipping-specific transactions. While most of these settings are related to outbound deliveries, we also explored the control settings specific to inbound deliveries for confirmation control and batch creation during delivery processing. Lastly, we saw how shipping blocks are maintained and their importance regarding control or checks over the delivery transactions.

In the next section, you'll see the configuration options available for picking specific transactions.

5.2.7 Picking

The picking process involves taking goods from a storage location and staging the right quantity in a picking area where the goods will be prepared for shipping. Let's look at the specific settings for the picking process in SAP S/4HANA.

Define Relevant Item Categories

This configuration step specifies the picking relevance of item categories in outbound deliveries. The same indicator controls the putaway relevance for inbound deliveries. Delivery item categories relevant for picking can be maintained using the configuration menu path **SPRO • SAP Reference IMG • SAP Customizing Implementation Guide • Logistics Execution • Shipping • Picking • Define Relevant Item Categories** (Transaction OVLP), as shown in Figure 5.30.

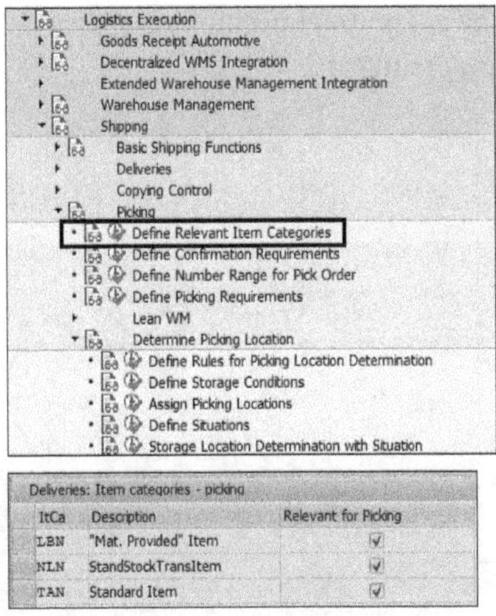

Deliveries: Item categories - picking

ItCa	Description	Relevant for Picking
LBN	"Mat. Provided" Item	☑
NLN	StandStockTransItem	☑
TAN	Standard Item	☑

Figure 5.30 Picking Relevance of Item Categories

Define Number Range for Pick Order

In this step, number ranges for picking requests are defined. The number range object is **SD_PICKING**, as shown in Figure 5.31. Number ranges for picking requests can be maintained using menu path **SPRO • SAP Reference IMG • SAP Customizing Implementation Guide • Logistics Execution • Shipping • Picking • Define Number Range for Pick Order** (Transaction S_ALR_87006706).

Edit Intervals: Picking request, Object SD_PICKING

Number Range No.	From No.	To Number	NR Status	External
01	0000000001	1999999999	21340	☐

Figure 5.31 Number Range for Picking Request

Determine Picking Location

The picking location can be determined automatically by the system when entering a delivery item even if no storage location was specified in the corresponding item of the reference sales document. You can still change the automatically determined picking location manually in the delivery item. In SAP standard, the storage location determination for picking takes place at the time of delivery creation. Automatic determination of the picking location depends on both the picking relevancy of the delivery items and the rules for automatic picking location determination.

Figure 5.32 shows the settings needed for the picking location determination during delivery creation and how the picking rule is assigned to the delivery type.

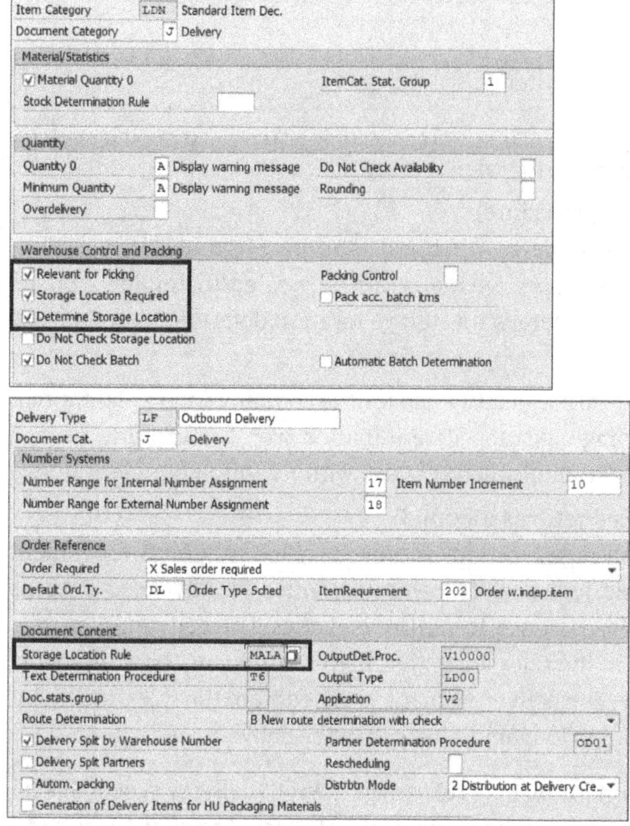

Figure 5.32 Picking Relevance of Delivery Item Category and Picking Rule Assignment to Delivery Type

The following sections elaborate on various picking rules available in SAP S/4HANA that help in determining a suitable picking location.

Define Rules for Picking Location Determination

In this activity, a suitable picking rule is assigned to each delivery type to enable the system to determine the picking location automatically. Each delivery type can have only one picking rule. The picking rule controls how the system determines the picking location.

In SAP S/4HANA, the following standard picking rules are available for determining the picking location (these are noneditable):

- MALA
  This rule determines the picking location according to the following:
  - Storage condition
  - Delivering plant of the delivery item
  - Shipping point of the delivery
- RETA
  This rule determines the picking location according to the following:
  - Storage location
  - Delivering plant of the delivery item
  - Situation

  This rule is particularly important for retail scenarios.
- **MARE, MALA, then RETA (backup procedure)**
  In this case, the system chooses rule MALA first, and if the predetermined storage location can't be applied (e.g., the respective material isn't created/maintained for the concerned storage location, or there is no storage location determined in Transaction OVL3), it switches to rule RETA.

  Next, if there is no storage location maintained for the combination of plant/situation/storage condition, storage location determination won't return any storage location, and the delivery document will be created without this parameter value.
- **MSRE (applicable for advanced returns inbound)**
  In this rule logic, default storage locations can be specified for goods movement postings related to specific logistical follow-up activities in Advanced Returns Management. This is usually assigned to return deliveries. The delivering plant is automatically determined in the sales document item from the material; however, it can be changed manually. The delivering plant is copied from the sales document item into the delivery. It can no longer be changed in the delivery.

The configuration menu option for picking rule assignment to delivery type is **SPRO • SAP Reference IMG • SAP Customizing Implementation Guide • Logistics Execution • Shipping • Picking • Determine Picking Location • Define Rules for Picking Location**

Determination (Transaction OVLQ). Figure 5.33 shows a snapshot of picking rule assignment to delivery type and the relevant configuration menu option.

- Logistics Execution
  - Goods Receipt Automotive
  - Decentralized WMS Integration
  - Extended Warehouse Management Integration
  - Warehouse Management
  - Shipping
    - Basic Shipping Functions
    - Deliveries
    - Copying Control
    - Picking
      - Define Relevant Item Categories
      - Define Confirmation Requirements
      - Define Number Range for Pick Order
      - Define Picking Requirements
      - Lean WM
      - Determine Picking Location
        - Define Rules for Picking Location Determination
        - Define Storage Conditions
        - Assign Picking Locations
        - Define Situations
        - Storage Location Determination with Situation

Deliveries: Dlv.types - Purchase Order Storage Loc

DlvTy	Description	Rule Determ. StLoc.
LB	Delivery for Subcon.	MALA
LF	Outbound Delivery	MALA
NL	Replenishment Dlv.	MALA
NLCC	Replen.Cross-Company	MALA
ZLOR	Outbound Delivery	MALA

Figure 5.33 Assignment of Picking Rule to Delivery Type

Define Storage Conditions

Specific conditions for storage of goods are defined in this step. The conditions defined here are maintained in the material master in the **Plant data/stor. 1** view for each storage location. The storage condition is copied to the delivery item. Storage conditions help in determining the picking location. The storage condition is a two-character alphanumeric key.

Storage conditions can be defined using configuration menu path **SPRO • SAP Reference IMG • SAP Customizing Implementation Guide • Logistics Execution • Shipping • Picking • Determine Picking Location • Define Storage Conditions** (Transaction OMS6). Examples of different storage conditions with the relevant configuration menu path are shown in Figure 5.34.

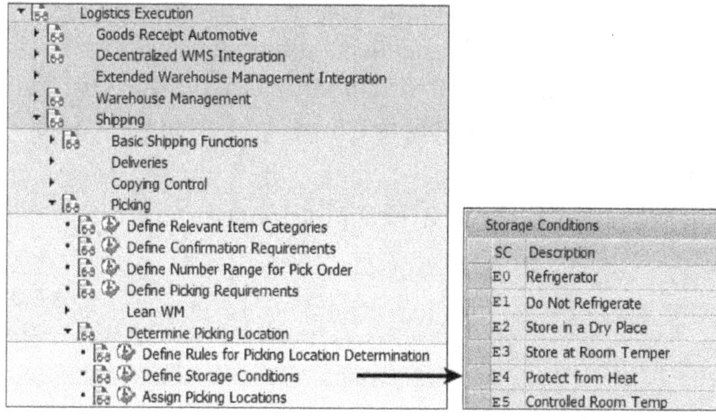

Storage Conditions

SC	Description
E0	Refrigerator
E1	Do Not Refrigerate
E2	Store in a Dry Place
E3	Store at Room Temper
E4	Protect from Heat
E5	Controlled Room Temp

Figure 5.34 Define Storage Conditions

Assign Picking Locations

For each plant, picking locations can be maintained for all combinations of shipping points and storage conditions. As shown in Figure 5.35, pick locations based on specific combinations of plant, shipping point, and storage condition can be maintained using configuration menu path **SPRO • SAP Reference IMG • SAP Customizing Implementation Guide • Logistics Execution • Shipping • Picking • Determine Picking Location • Assign Picking Locations** (Transaction OVL3).

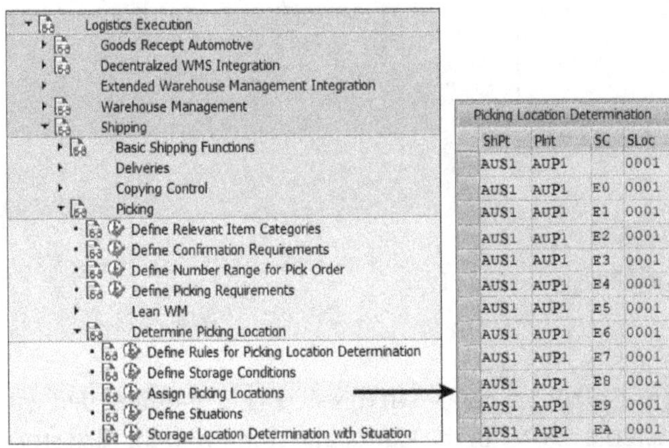

ShPt	Plnt	SC	SLoc
AUS1	AUP1		0001
AUS1	AUP1	E0	0001
AUS1	AUP1	E1	0001
AUS1	AUP1	E2	0001
AUS1	AUP1	E3	0001
AUS1	AUP1	E4	0001
AUS1	AUP1	E5	0001
AUS1	AUP1	E6	0001
AUS1	AUP1	E7	0001
AUS1	AUP1	E8	0001
AUS1	AUP1	E9	0001
AUS1	AUP1	EA	0001

Figure 5.35 Picking Location (Storage Location) Determination Based on Shipping Point and Storage Condition

You may notice that the first line doesn't have a storage condition maintained. This is the default option when material used in delivery doesn't have a storage condition maintained in the material master.

5.2.8 Incompletion Control for Deliveries

While creating and processing deliveries, it's often necessary to check the completeness of the delivery document. A delivery is incomplete if a certain field or set of fields relevant for the delivery haven't been filled in or if certain settings defined in Customizing haven't been considered. The delivery incompletion log ensures that this completion check is executed automatically.

In the delivery, the following levels can be checked for completeness:

- Header data of the delivery
- Item data of the delivery
- Partner data
- Texts

You can display the list of all incomplete deliveries or navigate directly from delivery processing to the incompletion log.

Incomplete outbound deliveries can be displayed using the **SAP Easy Access** menu via **Logistic Execution • Information System • Shipping List of Outbound Deliveries • Incomplete Outbound Deliveries** (Transaction V_UC).

Similarly, incomplete inbound deliveries can be displayed using the **SAP Easy Access** menu via **Logistic Execution • Information System • Goods Receipt • Inbound Delivery Lists • Incomplete Inbound Deliveries** (Transaction V_UC_7).

The delivery incompletion log controls whether the next subsequent functions are allowed in the case of an incomplete delivery. Table 5.2 shows the subsequent activities for a delivery.

Delivery Type	Subsequent activities
Inbound	Putaway Packing Goods received
Outbound	Picking Packing Goods issue Billing

Table 5.2 Possible Follow-Up Actions for Delivery Types

Note

It's possible to set the system in a way that incomplete deliveries can't be saved to the database. In such a case, the delivery is created in a collective processing run, the reasons for the incompleteness are printed in the collective processing log, and the delivery isn't created by the system.

Now, let's look at how to configure the incompleteness procedure and check certain fields for incompleteness.

Define Status Groups

Status groups are defined for representing the state of incompleteness in sales or delivery documents. These status groups are then assigned to the fields in an incompleteness procedure.

In a status group, you can define which functions can be performed for an incomplete sales and delivery document or for items where a field entry is missing. This function is used to block a document for subsequent transactions. In addition, it can also be set to check at the header level whether the general data is complete. Existing status groups can be used, but you can create custom status groups as well.

Status groups for incompletion control can be defined using configuration menu path **SPRO • SAP Reference IMG • SAP Customizing Implementation Guide • Logistics Execution • Shipping • Basic Shipping Function • Incompletion Control for Deliveries • Define Status Group** (Transaction OVAO). Figure 5.36 provides a snapshot of different status groups (**00**, **01**, **02** . . . **A0**, **A1**, **A2**) and relevant incompletion checks at the transaction level.

Incompletion Control: Status Groups

S..	General	Delivery	Billng Doc.	Price	Goods movement	Picking/puta...	Pack
00	☐	☐	☐	☐	☐	☐	☐
01	☑	☐	☐	☐	☐	☐	☐
02	☑	☑	☐	☐	☐	☐	☐
03	☑	☐	☑	☐	☑	☐	☐
04	☑	☑	☑	☐	☑	☐	☐
05	☑	☐	☑	☑	☑	☐	☐
06	☑	☑	☑	☑	☑	☐	☐
16	☑	☐	☐	☐	☐	☑	☐
30	☑	☐	☐	☐	☐	☐	☐
32	☑	☐	☐	☐	☐	☐	☑
58	☑	☐	☑	☐	☑	☑	☑
A0	☑	☐	☐	☐	☐	☐	☐
A1	☑	☐	☐	☐	☑	☑	☑
A2	☑	☐	☐	☐	☑	☑	☐

Figure 5.36 Define Status Group

Define Incompleteness Procedures

Incompleteness procedures specify which fields require mandatory entry while processing delivery documents. Multiple incompleteness procedures can be maintained for groups. These groups stand for objects, for example, headers and items, in delivery documents.

The incompleteness procedure can be maintained using menu path **SPRO • SAP Reference IMG • SAP Customizing Implementation Guide • Logistics Execution • Shipping • Basic Shipping Function • Incompletion Control for Deliveries • Define Incompleteness Procedures** (Transaction OVA2).

As shown in Figure 5.37, Figure 5.38, and Figure 5.39 in this and subsequent sections, the incompleteness procedure has a hierarchy starting from incompleteness group that encompasses several incompleteness procedures. Individual incompleteness procedures are maintained at SAP S/4HANA field levels wherein each relevant field is checked for incompleteness at the transaction level.

Each group is broken down into multiple incompleteness procedures for specific document types, for example, **Delivery header**.

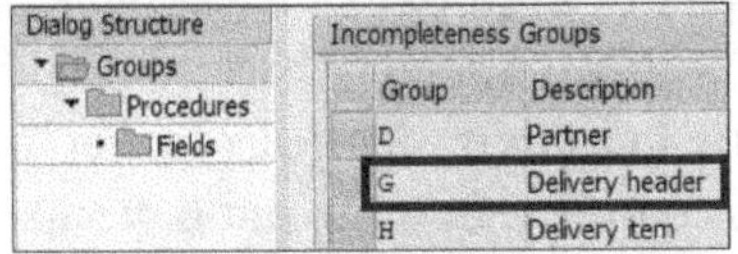

Figure 5.37 Define Incompleteness Procedure Groups

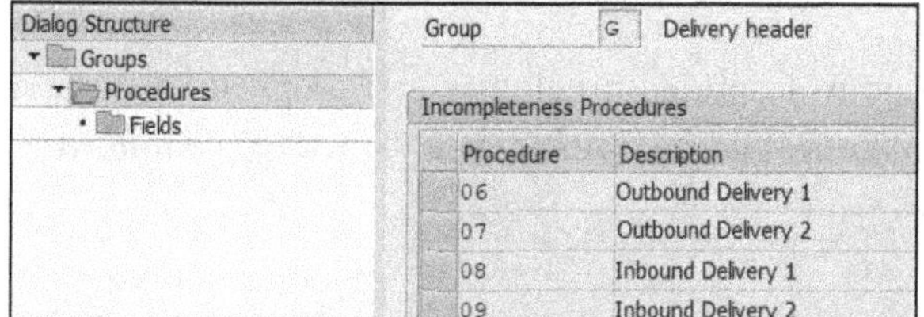

Figure 5.38 Multiple Incompleteness Procedures Assigned to the Delivery Header Group

In an incompleteness procedure, the desired fields that are to be checked for completeness are grouped together. If a user doesn't fill one of these fields in the document, the document is considered incomplete.

These fields are assigned with the appropriate status group, which helps to control, for example, whether the system records a status message for this field or whether certain subsequent activities for the document are blocked.

Note

For every field in the procedure, you can specify whether a warning message will be issued during processing if no data has been entered into the relevant field. However, this function doesn't exist in delivery processing. When you select the control field, it has no further consequences.

As shown in Figure 5.39, the incompleteness procedure is maintained with relevant SAP fields. These fields are assigned to an incompleteness status group, which determines the transactions where these fields need to be checked for completeness of data or value.

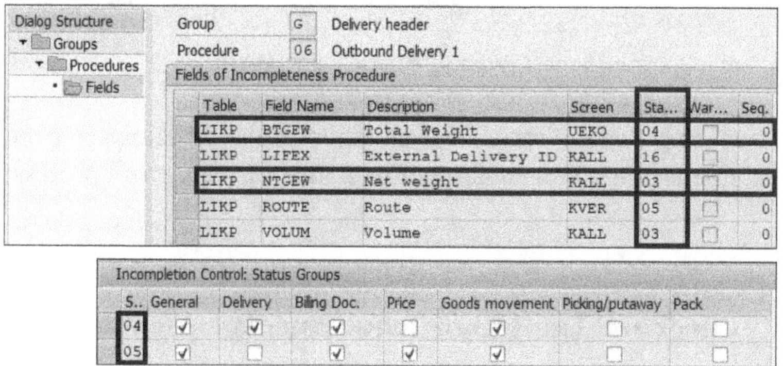

Figure 5.39 Group of Fields to Be Checked while Processing Outbound Deliveries

Once incompleteness procedures are maintained, it needs to be assigned to respective document type.

Assign Incompleteness Procedures

In this step, incompleteness procedures defined previously are assigned to specific object or document types. Incompleteness procedure can be assigned to the following objects;

- Partner functions
- Delivery types
- Delivery item categories

The menu path for assignment of the incompleteness procedure is **SPRO • SAP Reference IMG • SAP Customizing Implementation Guide • Logistics Execution • Shipping • Basic Shipping Function • Incompletion Control for Deliveries • Assign Incompleteness Procedures** (Transaction VUA4). Figure 5.40 provides a snapshot of the assignment of an incompleteness procedure to a delivery type along with the configuration menu path.

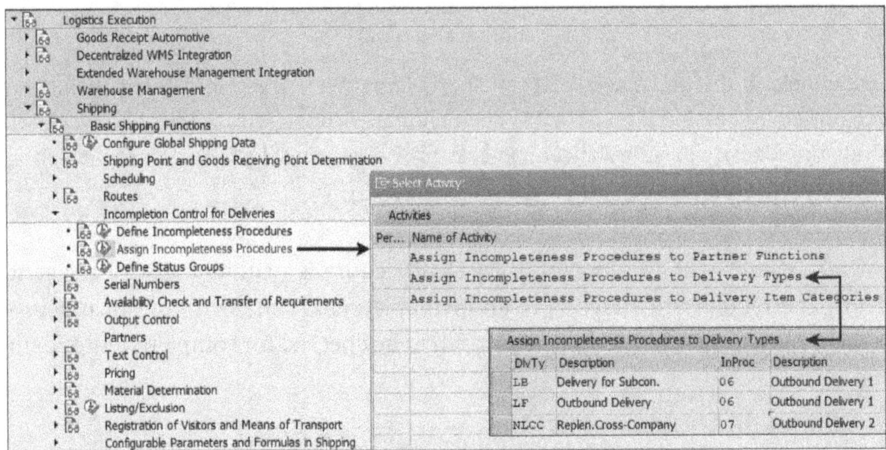

Figure 5.40 Assignment of Incompleteness Procedure to Delivery Types

The system uses the incompleteness procedure to determine which fields appear in the incompletion log when the user doesn't enter information during document processing.

5.3 Availability Check and Transfer of Requirements

Availability check is an essential step in the business process that assesses whether the desired delivery quantity can be met on the specified delivery date for a sales order or STO. The system accounts for the time it takes to generate or collect the item as well as predelivery procedures such as scheduling pickup or packing times. Additionally, it conducts several background tasks, such as backorder processing, rescheduling, and available-to-promise (ATP) checks. Let's review a few terms used in this context:

- **Backorder processing**
  Processing of a sales order that hasn't been fully confirmed or not confirmed at a certain delivery date. Backorders occur when there is insufficient inventory to cover a sales order item's demand. Transaction V_RA or CO06 displays backorders. You can also access them using **SAP Easy Access** menu **Logistics • Sales & Distribution • Sales • Backorders • Backorder Processing**.
- **Rescheduling**
  This is a proposal wherein the confirmed quantities that are already assigned to sales orders can be reassigned to other orders with a high priority. Either Transaction V_V2 (Execute) or Transaction V_R2 (Evaluate) is used for rescheduling. Alternatively, you can also access rescheduling using **SAP Easy Access** menu **Logistics • Sales & Distribution • Sales • Rescheduling • Execute** (Transaction V_V2) or **Evaluate** (Transaction V_R2).
- **Available-to-promise (ATP)**
  Process of checking the available quantities of a material. The equation for ATP is as follows:

  ATP quantity = Warehouse stock + Planned receipts (Future receipts) – Planned issues (Future issues)

  To examine stock on hand, you can use Transaction CO09 or the **SAP Easy Access** menu path **Logistics • Sales & Distribution • Sales • Environment • Availability Overview**.

  Warehouse stock includes unrestricted stock, stock in transfer, stock in quality inspection, blocked stock, restricted use stock, and so on. Planned receipts includes purchase orders, purchase requisitions, planned orders, production orders, and so on. Planned issues includes sales orders, deliveries, reservations and sales requirements, and so on.
- **Replenishment lead time (RLT)**
  The time between when the need of a product is determined and the time it becomes available for use or sale. For materials procured externally, RLT equals the processing time required by the purchasing department plus the planned delivery time and goods receipt processing time. The equation for RLT is as follows:

  RLT = Purchase processing time + Planned delivery time + Goods receipt processing time

In the following sections, we'll look into the criteria that governs availability check, as well as the relevant configuration details.

5.3.1 Basic Criteria for Availability Check

Backward scheduling calculates the material availability date. This is the date on which sufficient material should be available (for activities such as picking, packing, loading, and finally goods issue) to successfully deliver the goods to the customer on his requested delivery date (assuming the customer requires complete delivery). The delivery, therefore, should be created at the latest by this date or by the transportation planning date, whichever is earlier.

Availability check is a function of the sales and distribution component that, along with transfer of requirements, helps in checking if a sufficient quantity of goods is available on the calculated availability date. If a sufficient quantity isn't available on this date, then goods can't be delivered on the customer's requested delivery date. Forward scheduling then calculates the new availability date, and thus a new delivery date is confirmed in the schedule line. The availability check is done at the plant level. Results of the availability check may vary according to RLT, partial delivery constraints, or other such considerations.

Types of availability check include the following:

- Against ATP quantities (with or without RLT)
- Against product allocation
- Rule based

Figure 5.41 explains how the availability check works.

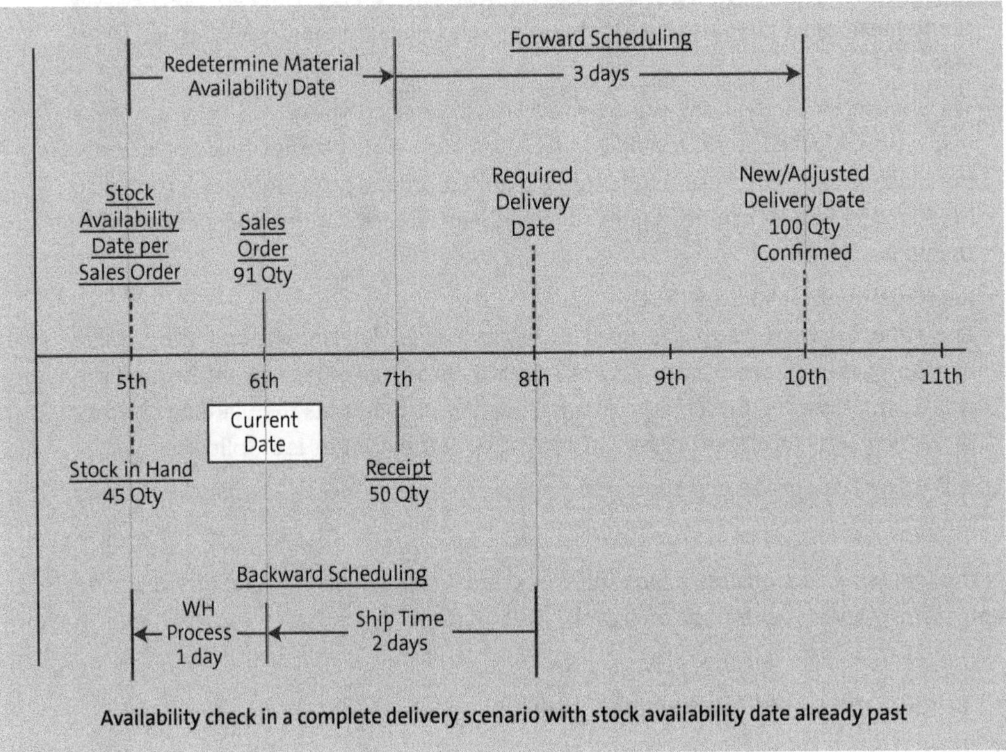

Figure 5.41 Availability Check with Backward and Forward Scheduling Examples

5.3.2 Configuration Steps for Availability Check

In this section, first we'll discuss the general settings for the availability check and then delve deeper into the relevant settings in procure-to-pay and order-to-cash areas. For the sake of brevity, we'll skip the production planning part here as this may not be relevant to third-party logistics integration.

Define Checking Groups

In this IMG step, checking groups are maintained that specify whether and how the system checks availability and generates requirements in material planning when processing sales orders or deliveries. Sales order requirements and delivery requirements can be controlled separately. In SAP S/4HANA, you can create individual requirement records only.

The availability check group can be defined using menu path **SPRO • SAP Reference IMG • SAP Customizing Implementation Guide • Logistics Execution • Shipping • Basic Shipping Functions • Availability Check and Transfer of Requirements • Availability Check • Define Availability Check Group** (Transaction OVZ2).

You can define custom checking groups per business requirements, as shown in Figure 5.42:

❶ Block for several users to be able to process the material simultaneously in different transactions without blocking each other

❷ Individual/summarized requirements

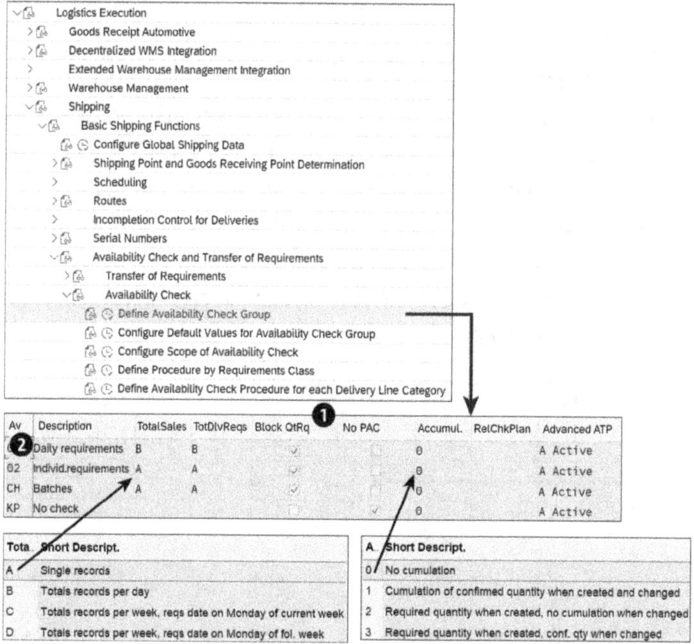

Figure 5.42 Define Checking Groups (Transaction OVZ2)

The **TotalSales** field is for sales documents, and the **TotDLvReqs** field is for deliveries.

Single requirements are stored in table VBBE. Single requirements means that for every single schedule line a record will be created in table VBBE. If you're using summarized requirements, then they are stored in table VBBS. Here you'll get one record for several sales order items on the same date. Table VBBS has no information about the document number, and this table is used in Transaction MD04 and Transaction CO09 to display the requirements of sales and shipping documents. However, in SAP S/4HANA, you can create individual requirement records only. If column **No PAC** (no product availability check) is selected, it switches off the availability check according to ATP logic for the relevant checking group.

The **Block QtRq** field ensures that the same available quantity can't be allocated several times when doing the ATP in parallel sessions. There are two blocking concepts for the availability check:

- **Soft block (Transaction OVZ2)**
  As shown in Figure 5.42, **Block QtRq** is set. This is a block on the material which is only in place during the availability check. For example, if a material A is used in an order, it is blocked during the availability check based on the material master and availability check settings. The quantity reserved for this transaction is recorded in the blocking table. Any other user processing the material at that time receives this information. This complements the information gained from the availability check and gives you a more accurate picture of the current availability situation. After the availability check has been completed, the block is removed. SAP recommends using the soft block.
- **Hard block (Transaction OVZ1)**
  You can set the indicator as **Block** or **X** (TMVFP-VFPSP). This results in blocking the material for ATP check by other users, and the block remains until the transaction has been saved or exited. If you've performed an ATP check in one session, but you haven't saved the order yet, then in parallel sessions, it's not possible to perform an ATP check. That is, parallel processing of the same material will throw an error message (msg. no. V1082). Figure 5.43 shows the hard block setting for individual checking groups. When the hard block in Transaction OVZ1 is set, it overrides the settings against the soft block (Transaction OVZ2).

AvailCheck	Description	Initiator	Block
CH	Batches	A	☑
CH	Batches	B	☑

Figure 5.43 Hard Block on Availability Check

The **No PAC** field switches off the availability check according to ATP logic for the checking group concerned. In addition, the **Accumul.** field is an important setting as it can avoid an over-confirmation situation. Dependent on this setting, the system can behave differently when creating or changing a document. When setting **Accumul.** to a value of **3**, the system will check whether the required quantities of the documents can be covered by the stock and receipt elements. If not, you can't confirm a new sales order when creating.

Define Checking Rules

Checking rules are transaction based. In other words, in a standard SAP system, if you're creating the sales order, then the rule is A; if delivery, then it's B; and if you're creating/releasing a production order, then it's PP. Transaction OPJL is used to define and maintain checking rules, as shown in Figure 5.44. SAP provides standard checking rules for relevant transactions, but you can create your own custom checking rule per business requirements.

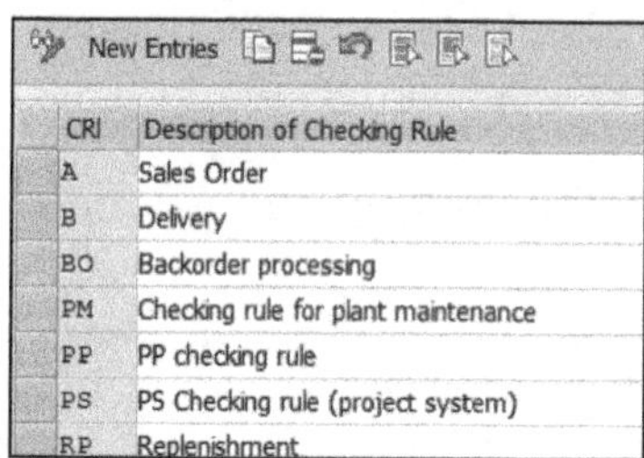
New Entries

CRl	Description of Checking Rule
A	Sales Order
B	Delivery
BO	Backorder processing
PM	Checking rule for plant maintenance
PP	PP checking rule
PS	PS Checking rule (project system)
RP	Replenishment

Figure 5.44 Maintain Checking Rules (Transaction OPJL)

Configure Default Values for Availability Check Groups

In this IMG activity, default values of checking groups can be maintained for material type and plant combinations. This value is suggested while creating the material master record, but you can overwrite the default value.

The configuration menu path for maintaining default values for availability check groups is **SPRO • SAP Reference IMG • SAP Customizing Implementation Guide • Logistics Execution • Shipping • Basic Shipping Functions • Availability Check and Transfer of Requirements • Availability Check • Configure Default Values for Availability Check Group** (Transaction OVZ3).

Figure 5.45 shows the configuration menu path and default value of the availability check as **CH** for the material type and plant combination.

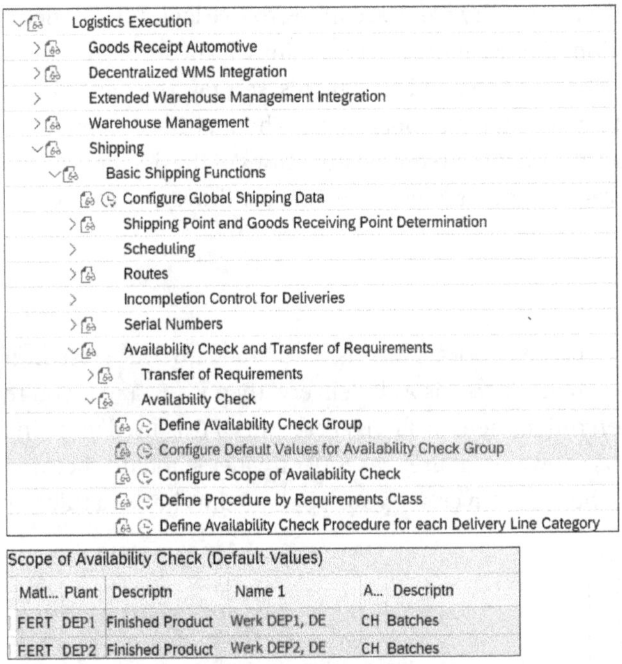

- Logistics Execution
  - Goods Receipt Automotive
  - Decentralized WMS Integration
  - Extended Warehouse Management Integration
  - Warehouse Management
  - Shipping
    - Basic Shipping Functions
      - Configure Global Shipping Data
      - Shipping Point and Goods Receiving Point Determination
      - Scheduling
      - Routes
      - Incompletion Control for Deliveries
      - Serial Numbers
      - Availability Check and Transfer of Requirements
        - Transfer of Requirements
        - Availability Check
          - Define Availability Check Group
          - Configure Default Values for Availability Check Group
          - Configure Scope of Availability Check
          - Define Procedure by Requirements Class
          - Define Availability Check Procedure for each Delivery Line Category

Scope of Availability Check (Default Values)

Matl...	Plant	Descriptn	Name 1	A...	Descriptn
FERT	DEP1	Finished Product	Werk DEP1, DE	CH	Batches
FERT	DEP2	Finished Product	Werk DEP2, DE	CH	Batches

Figure 5.45 Default Value of Availability Check Based on the Material Type and Plant Combination (Transaction OVZ3)

Configure Scope of Availability Check

This setting specifies which stocks, receipts, and issue elements are considered when the availability check is executed. The scope of check is defined using a combination of checking group and checking rule. The checking rule specifies the scope of the availability check for the respective transactions in sales and distribution by specifying precisely which stocks, receipt, and issue elements should be taken into account during the availability check. In addition, the checking rule specifies whether the availability check is to take RLT into consideration.

The menu path for configuring the scope of the availability check is **SPRO • SAP Reference IMG • SAP Customizing Implementation Guide • Logistics Execution • Shipping • Basic Shipping Functions • Availability Check and Transfer of Requirements • Availability Check • Configure Scope of Availability Check** (Transaction OVZ9). As shown in Figure 5.46, the scope of availability check is based on the availability check and checking rule combination.

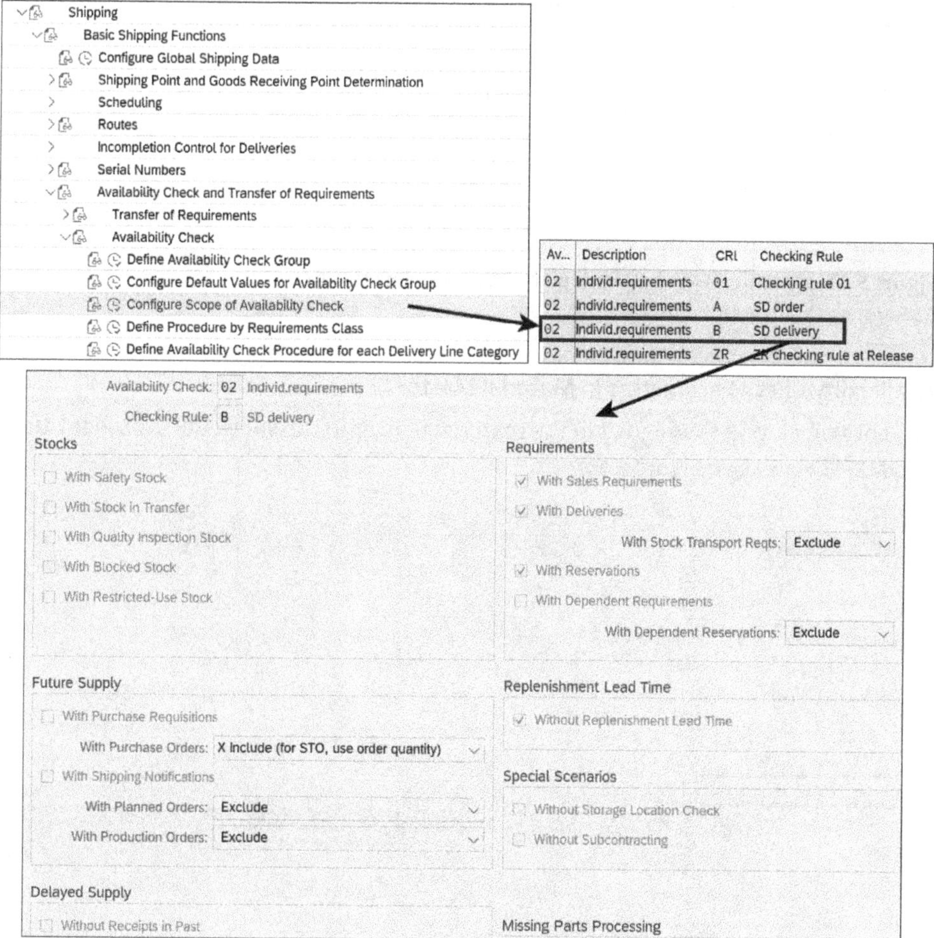

Figure 5.46 Configure Scope of Availability Check (Transaction OVZ9)

Define the Availability Check Procedure for Each Delivery Line Category

In this step, the availability check can be switched off for particular item categories in deliveries. For example, return delivery items don't require an availability check.

The configuration menu path is **SPRO • SAP Reference IMG • SAP Customizing Implementation Guide • Logistics Execution • Shipping • Basic Shipping Functions • Availability Check and Transfer of Requirements • Availability Check • Define Availability Check Procedure for Each Delivery Line Category** (Transaction OVZK). Figure 5.47 shows the provision for switching off the availability check through configuration.

Availability Check for Each Delivery Item Categor

ItCa	Description	Avail.check off
ELCR	Advanced Returns CC	X
ELLR	Advanced Returns IC	X
ELN	Inbound Delivery	X
ELR8	IC SIT Adv. Returns	X
ELSU	Inb.Dely SLS Subitem	X
HIDN	SH - Item	X
HODN	Goods Issue - Item	X

Figure 5.47 Switching Off the Availability Check for Delivery Item Categories (Transaction OVZK)

Availability Check Parameters in Material Master

The checking group is assigned in the material master in the **Sales: General/Plant Data** and **MRP 3** views (see Figure 5.48).

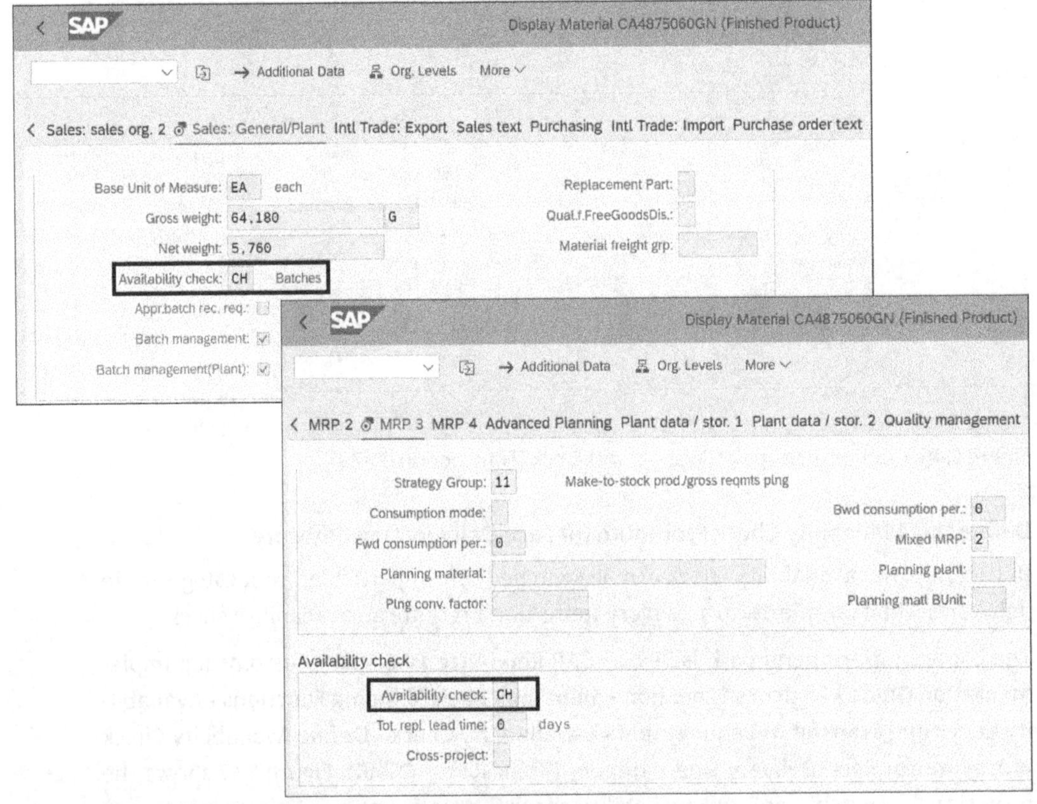

Figure 5.48 Availability Check (Checking Group) in Material Master Views

The **Availability Overview** screen and **Scope of check** tab of material for a specific plant can be viewed using Transaction CO09, as shown in Figure 5.49.

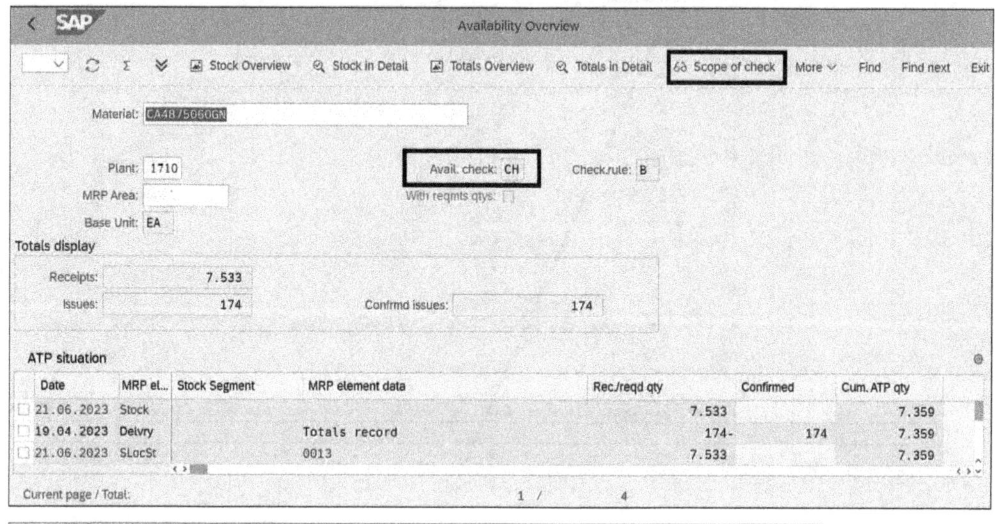

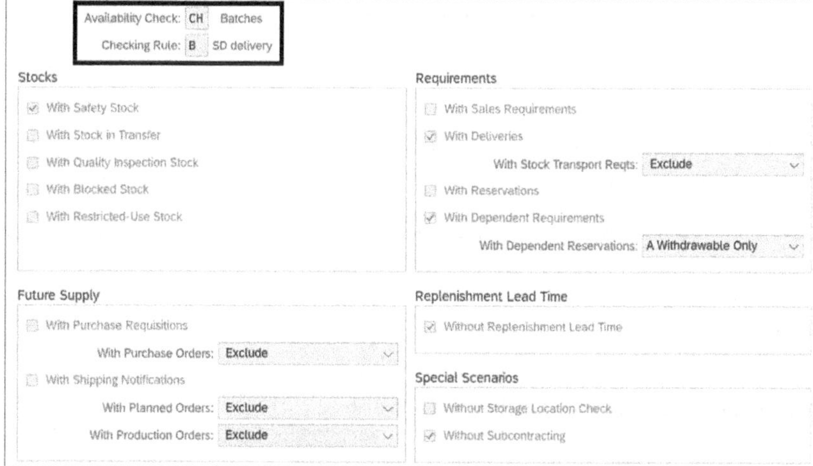

Figure 5.49 Availability Overview of Material Using Transaction CO09

As shown in Figure 5.50, the availability overview of material in an outbound delivery can be viewed using the menu path **Environment • Availability**.

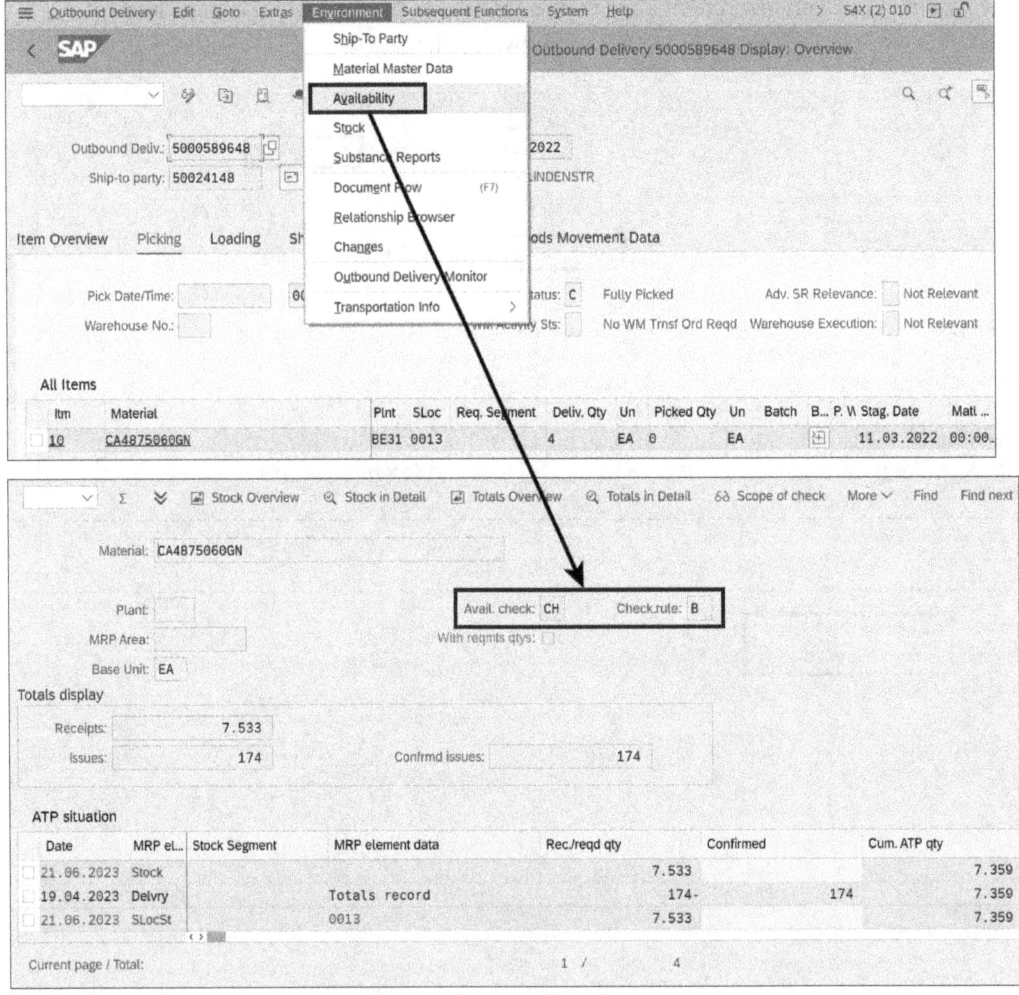

Figure 5.50 Availability Overview in an Outbound Delivery

5.4 Output Control in Shipping

SAP S/4HANA offers new output management using Business Rules Framework plus (BRFplus). The main characteristics of this new output management are as follows:

- Runs in a cloud or an on-premise environment
- Offers a unified solution for all output-related tasks across the product
- Supports ready-to-use output scenarios (predelivered configuration)
- Standard extensibility for configuration and documents (form and email templates)
- Provides central monitoring tools for key users
- Can be natively integrated into SAP Fiori apps (and SAPGUI apps)

Note

The following SAP Notes are relevant while considering BRFplus-based output management for new implementations as well as migration to SAP S/4HANA:

- 3097507 - Output Management in SAP S/4HANA On-Premise
- 2791338 - FAQ: SAP S/4HANA Output Management
- 2267376 - S4TWL - Billing Document Output Management
- 2248229 - SAP S/4HANA Output Management - BRFplus Files for Output Parameter

Output management in the logistics area is supported by a new output control. However, when integration to third-party logistics is concerned, it may involve a much more complex set of outputs to logistics service providers as each of them may be using different means of communication, and the output may involve integration with workflows, events, and special functions.

Per SAP Note 2791338 (FAQ: SAP S/4HANA Output Management), SAP S/4HANA output management doesn't intend to cover the same scope as table NAST. Thus, there are functionalities covered by table NAST not supported by SAP S/4HANA output management.

Functionalities not covered in the new BRFplus-based framework must be handled by alternatives such as table NAST. We list a few standard transmission mediums and their code representations (in parentheses) below:

- Special functions (8)
- Events/SAP Business Workflow (9)
- Distribution/application link enabling (ALE) (A)
- Workflow tasks (T)
- Support for IDocs, which isn't robust enough in BRFplus

For third-party logistics integration, usage of table NAST is the preferred approach, so we'll discuss the various features of communication control and features available with table NAST.

5.4.1 Disabling Business Rule Framework Plus-Based Output Management

The process for issuing output is available through output management. Each application that supports issuing document output decides whether to support SAP S/4HANA output control and/or output determination (table NAST) as the active process for issuing output. If an application provides the option to support both types of output processes to issue output for a given document type, you'll have to decide which output process best suits your requirements. Otherwise, your document output will be issued by the default output process supported by the application.

SAP S/4HANA comes with default BRFplus-based output. You may need to ensure that BRFplus-based output isn't active by default, and, if so, disable it.

As shown in Figure 5.51, activation or deactivation of BRFplus-based output can be done using configuration menu path **SPRO • SAP Reference IMG • Cross-Application Components • Output Control • Manage Application Object Type Activation**.

Activation of Application Object Type

Application Object Type	Text	Status	Default
MANUFACTURING_OP_ACTIVITY	Manufacturing Operation Activity	2 Application Inactive	
SETTLEMENT_DOCUMENT	Settlement Document	2 Application Inactive	2 Application Inactive

Figure 5.51 Maintaining BRFplus-Based Output Disabled for Shipping Output Types

5.4.2 Customizing Output Determination in Shipping

There are two options for output control in sales and delivery: proposal of output from the customer master record and proposal of output using the condition technique. We'll discuss both techniques in the following subsections.

Output Determination Procedure via Customer Master

This uses the customer account group to determine the output determination procedure. Use Transaction V/44 to maintain condition types for output (see Figure 5.52).

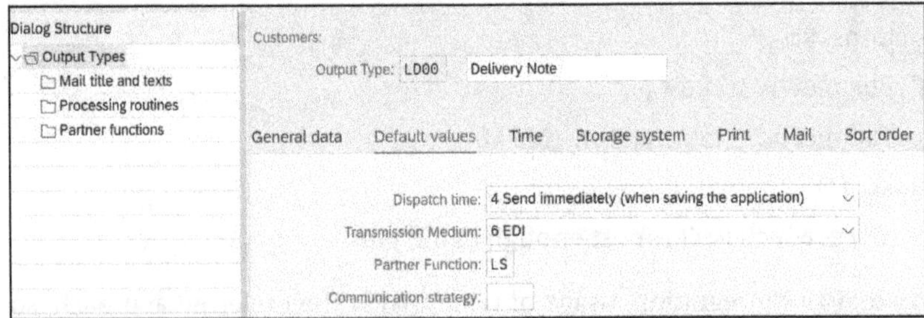

Figure 5.52 Maintain Condition Type for Output Control Using Customer Master

Use Transaction V/46 to maintain the output procedure for the customer master (see Figure 5.53).

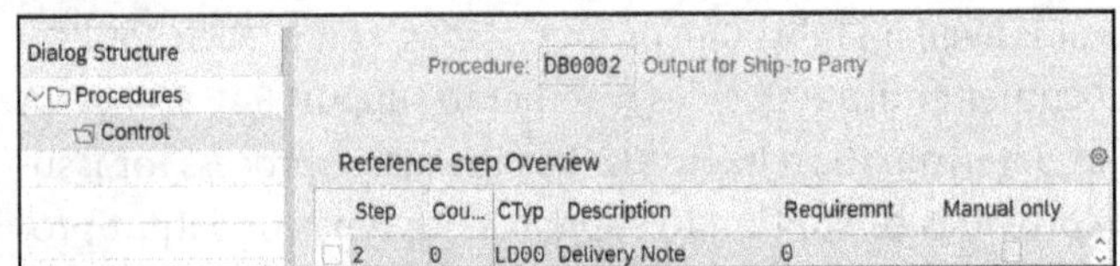

Figure 5.53 Output Procedure for Output Driven by the Customer Master

Use Transaction V/37 to assign the output procedure to the customer account group (see Figure 5.54).

Group	Name	Out.pr	Description
0001	Sold-to party	DB0001	Output for Sold-to Party
0002	Ship-to party	DB0002	Output for Ship-to Party

Figure 5.54 Assignment of the Output Procedure to the Customer Account Group

Alternatively, you can also assign the procedure in the configuration of the customer account group using Transaction OVTO. As shown in Figure 5.55, the customer account group configuration contains the output procedure.

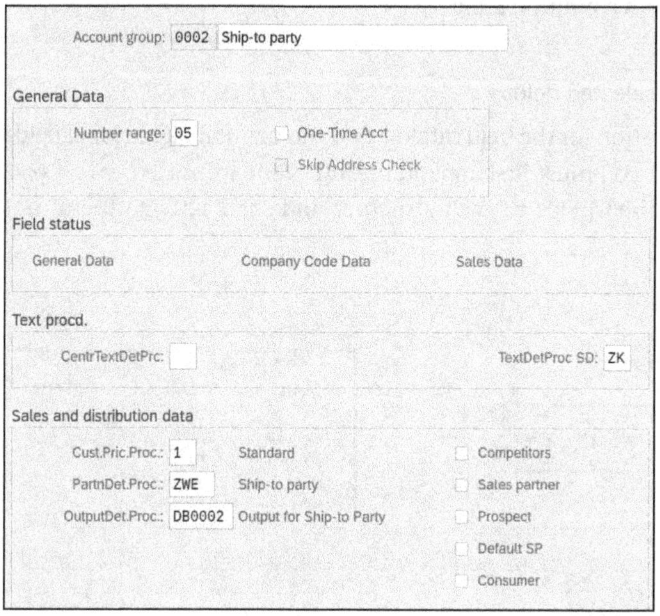

Account group: 0002 Ship-to party

General Data

Number range: 05
One-Time Acct
Skip Address Check

Field status

General Data
Company Code Data
Sales Data

Text procd.

CentrTextDetPrc:
TextDetProc SD: ZK

Sales and distribution data

Cust.Pric.Proc.: 1 Standard
PartnDet.Proc.: ZWE Ship-to party
OutputDet.Proc.: DB0002 Output for Ship-to Party
Competitors
Sales partner
Prospect
Default SP
Consumer

Figure 5.55 Assignment of Output Procedure in Customer Account Group Configuration

Output Determination Procedure Using the Condition Technique

SAP recommends the output proposal using the condition technique as it's more flexible. For example, the condition technique makes it possible to determine the most suitable output from several different types of output. Output control occurs separately in the header and item levels.

The following Customizing steps are carried out in sequence:

1. Maintain the condition table.
2. Maintain the access sequences.
3. Maintain the output types.
4. Maintain the output determination procedure.

We'll discuss each of these steps in more detail in the following sections.

Maintain Condition Table

The condition table consists of a combination of keys fields to identify an individual condition record. A condition record is defined as how the system stores the specific conditions. You can use menu path **SPRO • SAP Reference IMG • Logistics Execution • Shipping • Basic Shipping Functions • Output Control • Output Determination • Maintain Output Determination for Outbound Deliveries • Maintain Condition Table • Maintain Output Condition Table for Deliveries** (Transaction V/60).

Figure 5.56 shows the following important details:

❶ Selected fields in a custom condition table

❷ Field catalog

❸ Technical details of the selected fields

❹ SAP's standard configuration for the field catalog. You can maintain additional fields in this catalog, however, you must first append the SAP standard structures KOMKBV2 and KOMB with the required fields in each structure and then activate them (not shown).

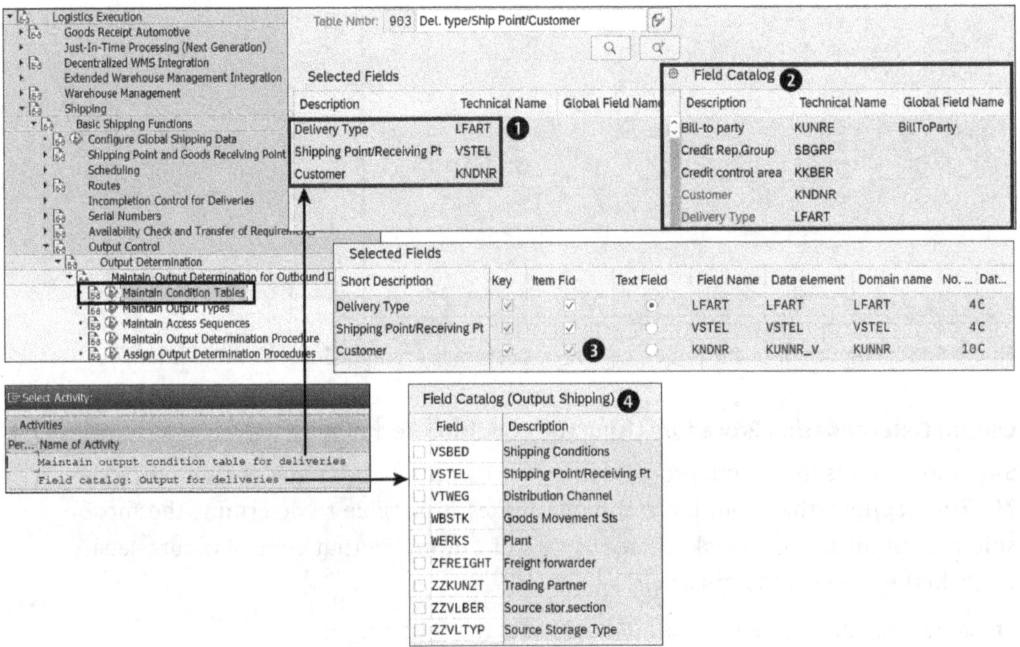

Figure 5.56 Maintain Condition Table

Sometimes, you may need additional fields to be added to the field catalog per business requirements. This can be done using menu path **SPRO • SAP Reference IMG • Logistics Execution • Shipping • Basic Shipping Functions • Output Control • Output Determination • Maintain Output Determination for Outbound Deliveries • Maintain Condition Table • Field Catalog (Output Shipping)** (refer to the lower-right corner of Figure 5.56).

> **Note**
>
> As a prerequisite for catalog field enhancement, communication structure KOMKBV2 and KOMB need to be updated with additional fields.

Maintain Access Sequence

The access sequence can be maintained using menu path **SPRO • SAP Reference IMG • Logistics Execution • Shipping • Basic Shipping Functions • Output Control • Output Determination • Maintain Output Determination for Outbound Deliveries • Maintain Access Sequences**. The **Exclusive** checkbox controls whether the system stops searching for a condition record after the first successful access for a condition type within an access sequence. Figure 5.57 shows the **Access Sequences** configuration with relevant access to table fields.

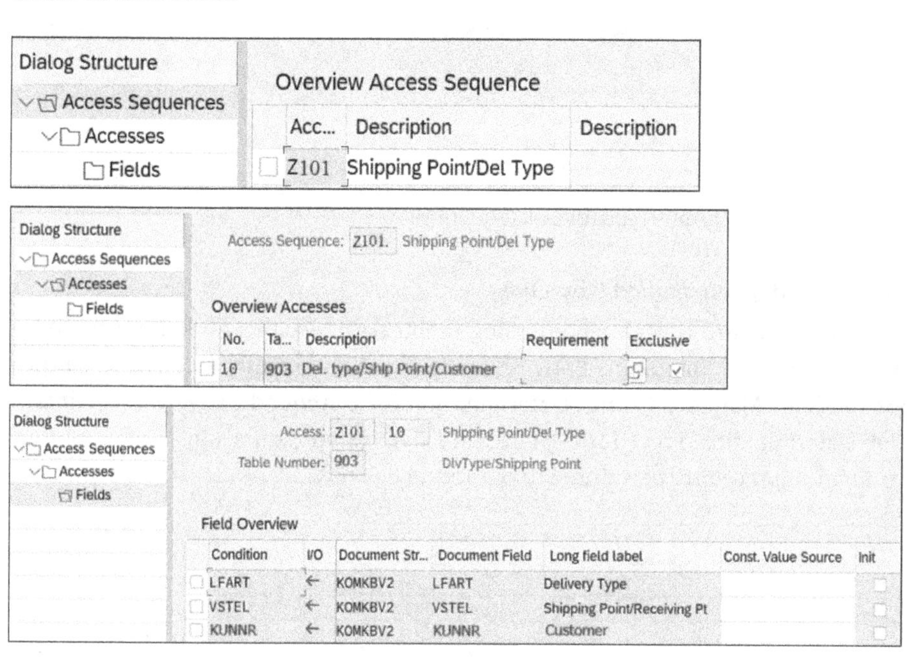

Figure 5.57 Maintain Access Sequence with Relevant Accesses and Fields

Maintain Output Types

Output types are maintained using menu path **SPRO • SAP Reference IMG • Logistics Execution • Shipping • Basic Shipping Functions • Output Control • Output Determination • Maintain Output Determination for Outbound Deliveries • Maintain Output Types** (Transaction V/34). As shown in Figure 5.58, the output type configuration for delivery contains transmission medium, partner function assignment, and processing routine, if applicable.

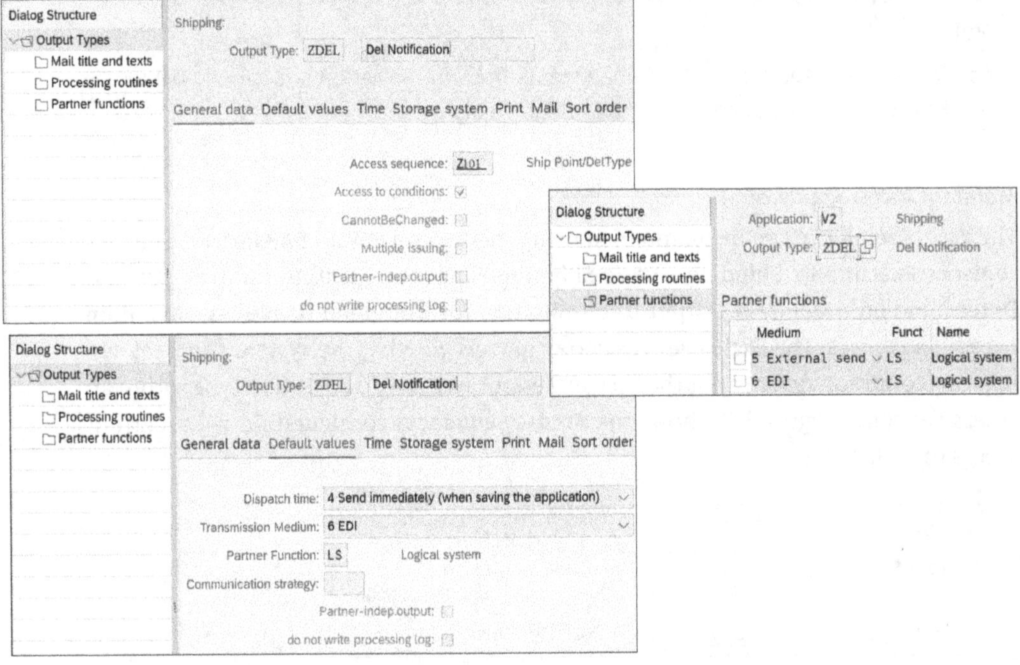

Figure 5.58 Maintain Output Type for Delivery Output

Maintain Output Determination Procedure

The output procedure is maintained using menu path **SPRO • SAP Reference IMG • Logistics Execution • Shipping • Basic Shipping Functions • Output Control • Output Determination • Maintain Output Determination for Outbound Deliveries • Maintain Output Determination Procedure.** Figure 5.59 depicts configuration for the output determination procedure pertaining to outbound delivery.

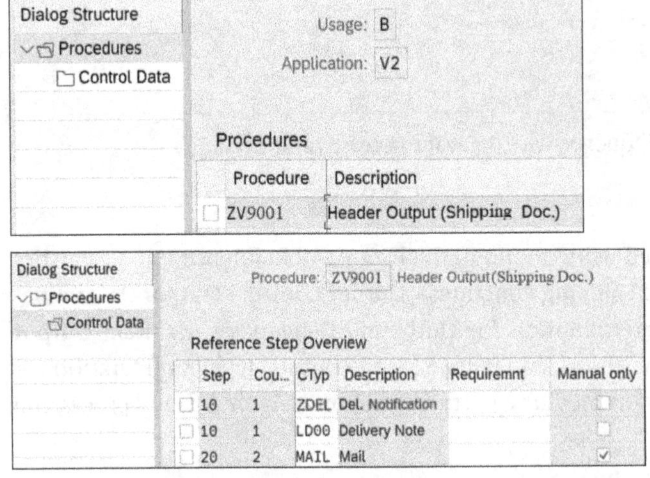

Figure 5.59 Output Determination Procedure for Outbound Delivery

The access sequence for output, output types, and output determination procedures (refer to Figure 5.55, Figure 5.56, and Figure 5.57) can be maintained using Transaction NACE. Figure 5.60 shows the following:

❶ Selecting **Application V2** for **Shipping**

❷ Selecting **Access sequences**, **Output types**, and **Procedures**

❸ Performing condition record maintenance for output type **LD00** (**Delivery Note**)

❹ Maintaining output type configuration

❺ Performing output procedure configuration

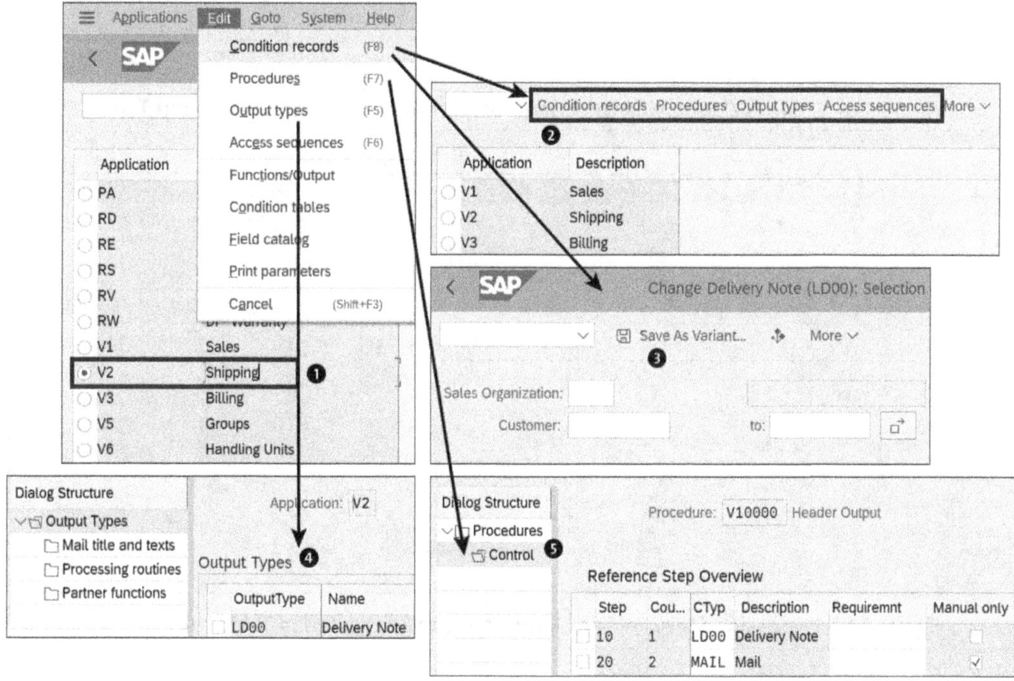

Figure 5.60 Output-Relevant Configuration Settings Using Transaction NACE

Condition records need to be maintained for the condition tables used in the access sequence. As shown in Figure 5.61, use Transaction NACE, and select **Application V2** for **Shipping** ❶. Click **Condition records**, and select the **Output Type** ❷. Then, select the **Key Combination** ❸, and maintain the input value for mandatory field(s) ❹. Maintain the values in the resulting screen ❺.

Once the output message is processed, the processing status of the message is stored in table NACH (Figure 5.62).

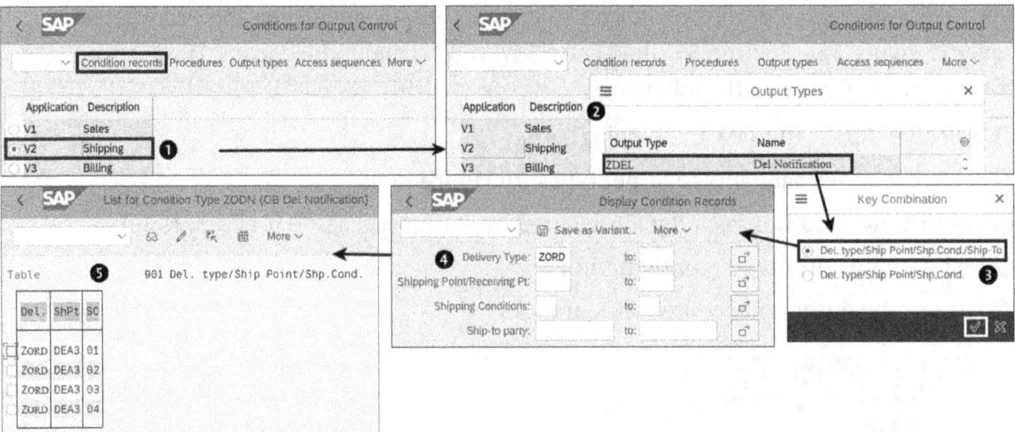

Figure 5.61 Creating/Displaying the Condition Record for the Shipping Output Type

Data Browser: Table NACH Select Entries 200

Table: NACH
Displayed Fields: 21 of 46 Fixed Columns: 2 List Width 0250

Cond.record no.	Created on	Usage	Table Number	Application	Condition Type	Valid From	Valid To	Partner Functn	Partner	Transm. Medium	Number	Date/Time	Comm.strategy
0000001372	21.12.2020	B	901	V2	ZDEL	00.00.0000	00.00.0000	LS		6	0	4	
0000001374	21.12.2020	B	901	V2	ZDEL	00.00.0000	00.00.0000	LS		6	0	4	
0000001376	21.12.2020	B	901	V2	ZDEL	00.00.0000	00.00.0000	LS		6	0	4	
0000001377	21.12.2020	B	901	V2	ZDEL	00.00.0000	00.00.0000	LS		6	0	4	
0000001378	21.12.2020	B	901	V2	ZDEL	00.00.0000	00.00.0000	LS		6	0	4	
0000001379	21.12.2020	B	901	V2	ZDEL	00.00.0000	00.00.0000	LS		6	0	4	
0000001411	21.12.2020	B	901	V2	ZDEL	00.00.0000	00.00.0000	LS		6	0	4	
0000001412	21.12.2020	B	901	V2	ZDEL	00.00.0000	00.00.0000	LS		6	0	4	
0000001413	21.12.2020	B	901	V2	ZDEL	00.00.0000	00.00.0000	LS		6	0	4	
0000001414	21.12.2020	B	901	V2	ZDEL	00.00.0000	00.00.0000	LS		6	0	4	
0000001415	21.12.2020	B	901	V2	ZDEL	00.00.0000	00.00.0000	LS		6	0	4	
0000001416	21.12.2020	B	901	V2	ZDEL	00.00.0000	00.00.0000	LS		6	0	4	
0000001417	21.12.2020	B	901	V2	ZDEL	00.00.0000	00.00.0000	LS		6	0	4	
0000001418	21.12.2020	B	901	V2	ZDEL	00.00.0000	00.00.0000	LS		6	0	4	
0000001419	21.12.2020	B	901	V2	ZDEL	00.00.0000	00.00.0000	LS		6	0	4	

Figure 5.62 Processing Status of the Output Message Reflected in Table NACH

5.4.3 Message Control

The output or follow-up processing of partner-dependent messages is automated via message control. The application then calls the message control using the specified interfaces.

These interfaces are maintained with specific data combinations along with the required processing logic. If one of these data combinations appears during processing, the appropriate processing is started (synchronously or asynchronously). Processing of messages can be defined with several different methods, for example, as a workflow, print output, or as an IDoc dispatch for electronic data interchange (EDI).

Figure 5.63 provides a pictorial overview of message control framework in SAP.

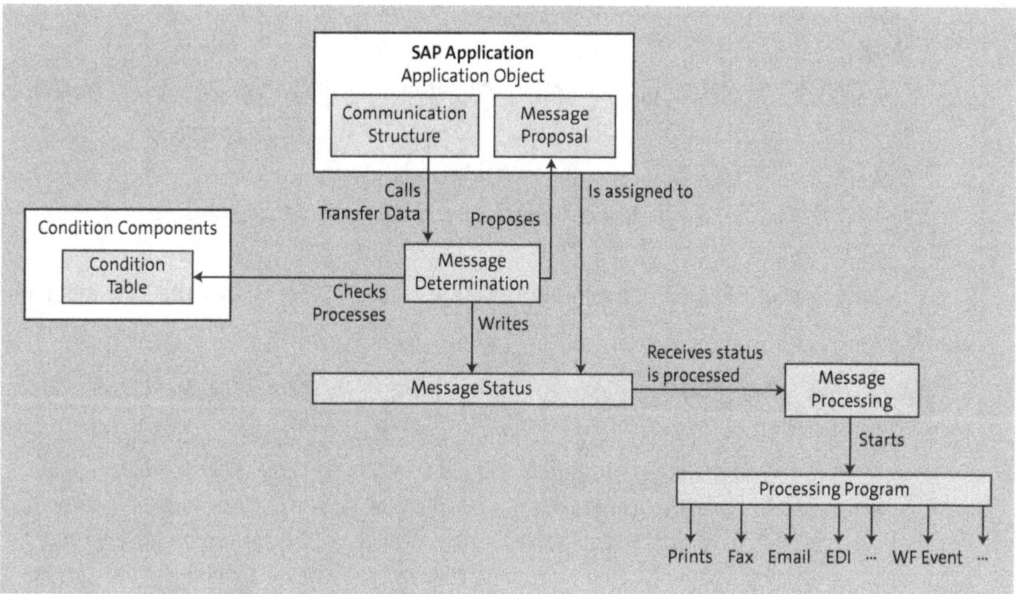

Figure 5.63 Message Control in SAP S/4HANA

5.4.4 Automatic Inbound Delivery Creation during Two-Step Stock Transfers

SAP has provided a standard output type called SPED for auto-creation of inbound delivery upon posting of goods issue during a two-step stock transfer. The process flow is shown in Figure 5.64.

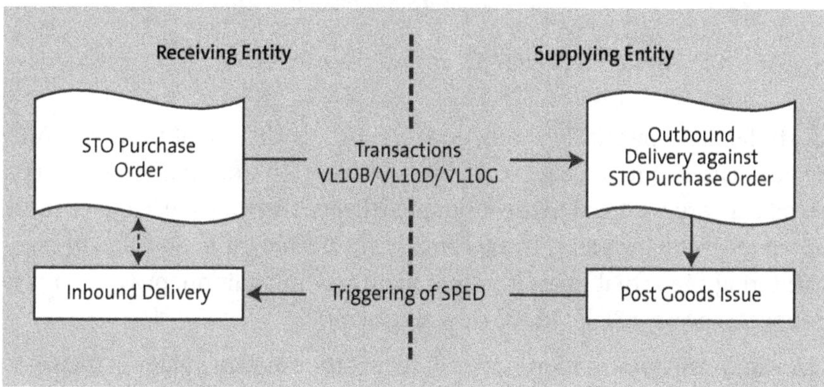

Figure 5.64 Process Flow for Automatic Inbound Delivery Creation during Two-Step Stock Transfer

Note

The following SAP Notes provide details of the functionality and other relevant factors:

- 965176 – Creation of Inbound Delivery during Two-Step Stock Transfers
- 2438735 – SPED Output – Create New Inbound Delivery
- 1119073 – FAQ: Automatic Inbound Delivery Creation for Stock Transfers

However, use of SPED is completely optional. Table 5.3 compares the two available options.

Method/Features	System Dependency	Document Flow	Implementation Effort
SPED	SPED is used when both supplying and receiving entities are using the same system.	Upon successful processing, the SPED output type automatically updates the document flow of outbound delivery with inbound delivery information.	Standard output provided by SAP requires little functional effort to implement. Tracking failure and retriggering messages requires only functional effort.
ALE/IDoc	This approach can be used across systems.	An IDoc automatically updates the document flow of the outbound delivery with inbound delivery information.	This requires substantial functional and technical effort to implement. Tracking of IDoc failures may be troublesome.

Table 5.3 Comparison of Output Methods SPED and ALE/IDoc for STOs

We'll start with the first option, SPED functionality. During the use of the SPED functionality for automatic inbound delivery creation for two-step stock transfers, the data from outbound delivery is copied to the inbound delivery and can be used to post the goods receipt in the receiving plant. The automatically created inbound delivery saves effort from entering the actual quantities manually and, in addition, provides a preview of the expected goods receipt in the receiving plant.

Automatic inbound delivery creation is mandatory if the receiving plant is managed using a decentralized warehouse management system (WMS) or an EWM system; in this case, the inbound delivery is required as a communication document for the decentralized system.

Automatic inbound delivery creation is also required if handling unit management (HUM) is activated for the receiving storage location. In this case, the inbound delivery serves as a reference document for the handling units (HUs) that are transferred by the preceding replenishment delivery.

The configuration for output type SPED can be maintained using menu path **SPRO • Reference IMG • Logistic Execution • Shipping • Basic Shipping Functions • Output Control • Output Determination • Maintain Output Determination for Outbound Deliveries • Maintain Output Types** (Transaction V/34). Figure 5.65 shows the configuration of the SPED output type.

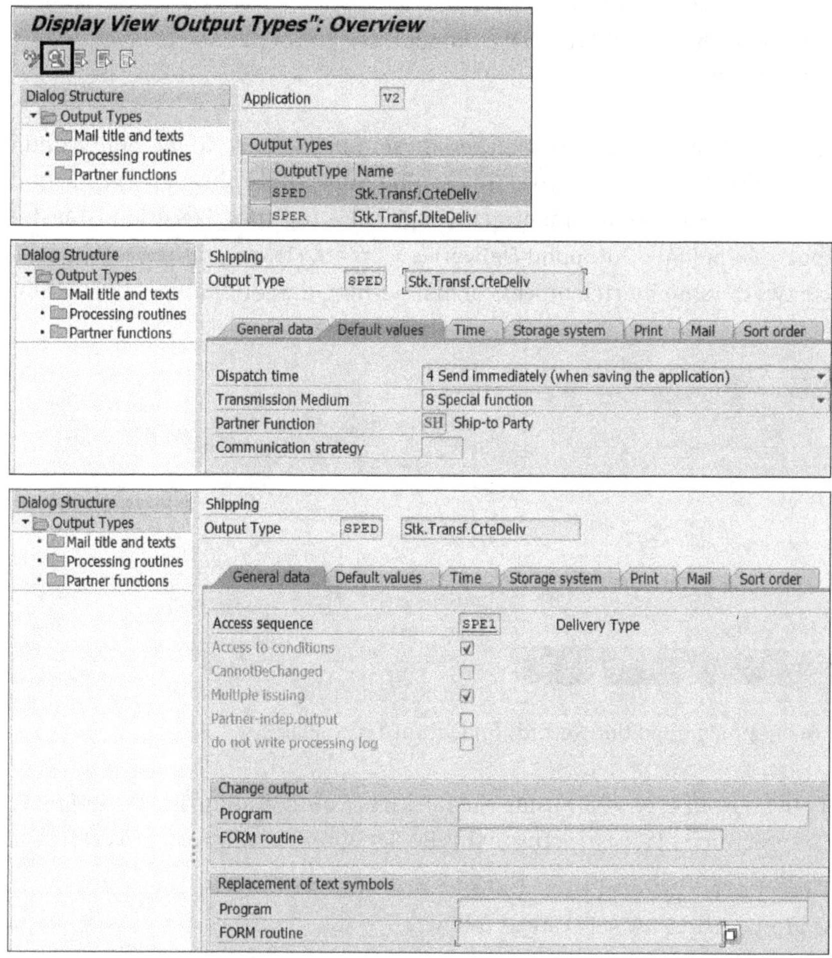

Figure 5.65 Settings for Output Type SPED

Next, assign the output type SPED to the output determination procedure using menu path **SPRO • Reference IMG • Logistic Execution • Shipping • Basic Shipping Functions • Output Control • Output Determination • Maintain Output Determination for Outbound Deliveries • Maintain Output Determination Procedure** (Transaction V/34). As in Figure 5.66, output type SPED is assigned to the document output procedure at the header level of the delivery document.

Dialog Structure
- Procedures
  - Control

Procedure: V10000 Header Output

Reference Step Overview

Step	Cou...	CTyp	Description	Requiremnt	Manual only
10	1	LD00	Delivery Note		☐
20	2	MAIL	Mail		☑
25	0	SPED	Stk.Transf.CrteDeliv	408	☐

Figure 5.66 Assigning the Output Type to the Output Procedure at the Header Level of the Shipping Document

Once the output type and output procedures are set up, you need to maintain condition records for the SPED output type. This can be done either through Transaction NACE or via the **SAP Easy Access** menu path **Logistics • Logistics Execution • Master Data • Output • Shipping • Outbound Deliveries • Create** (Transactions VV21/22/23). Figure 5.67 shows the step-by-step process of maintaining the condition record for output type SPED.

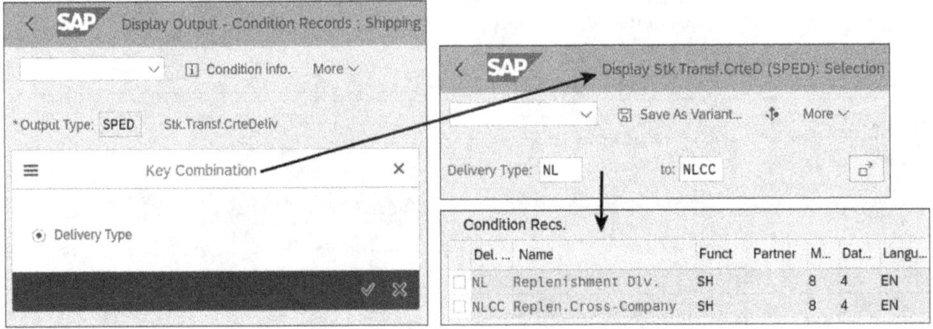

Figure 5.67 Maintaining Condition Records for Output Type SPED

SAP delivers the output type SPED using the Business Configuration Set (BC Set) /SPE/TNAPR. Use Transaction SCPR20 to activate this BC Set. Alternatively, you can create the output type SPED manually.

The process prerequisites for SPED are as follows:

1. The **Conf. Control** (confirmation control key) must be set to **0004**, or **Inbound Delivery** must be maintained in the STO purchase order. Goods receipt assignment and goods receipt relevance must be activated for this confirmation control key. To enable the confirmation control key to be determined automatically in the case of automatically created STOs, one of the following setups may be needed.
   - **Info record**: Maintain the confirmation control key in the info record (see Figure 5.68).
   - Configuration: Default confirmation control (**Ctr.**) can be maintained using configuration menu path **SPRO • Reference IMG • Logistic Execution • Shipping • Deliveries • Define Order Confirmations for Inbound Deliveries** (see Figure 5.69).

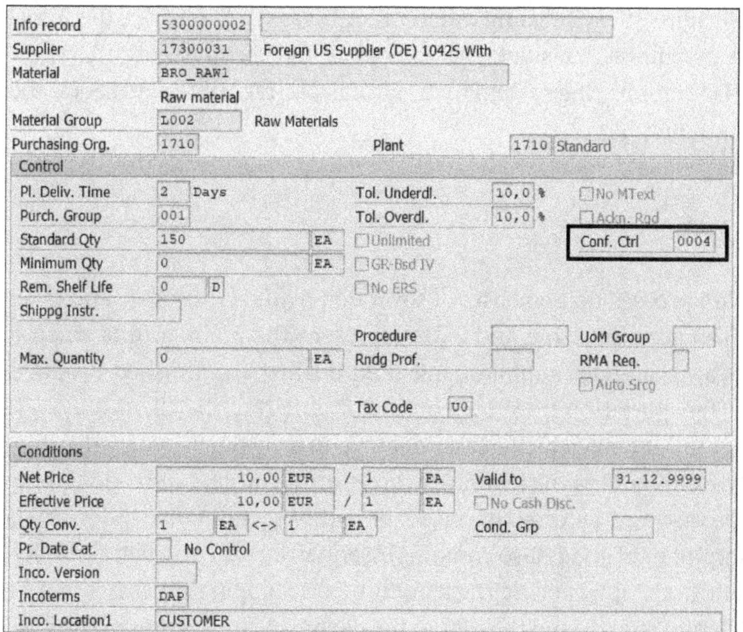

Figure 5.68 Maintenance of Confirmation Control Key in the Info Record

Cat	OTyp	Plnt	SLoc	Ctr.
F	UB			0004
F	UB	ARP1	0001	0004
F	UB	ARP1	0002	0004
F	NB			0004
F	NB	DEP1	0001	0004
F	NB	DEP1	0002	0004

Figure 5.69 Setting the Default Confirmation Control Key (CCK) for the Document Type, Plant, and Storage Location Combination

2. Condition record for output type SPED needs to be maintained for intercompany and intracompany delivery types (refer back to Figure 5.65).

To close this section, we'll discuss the ALE/IDoc approach. In Section 5.4.2, you saw how to configure output determination using the condition technique for EDI/IDoc type messages. In addition to message determination configuration, several other setups such as sender and receiving ports, partner profiles, and connectivity between systems need to be established.

IDoc mapping is another area that needs to be developed carefully because data from the outbound delivery such as batch number, HU details, and customer-to-vendor conversion are required to be mapped to the inbound delivery document.

This approach is well suited if we're using different SAP systems. However, the initial set up of the process requires considerable functional and technical efforts. There should be a proper IDoc monitoring system to monitor IDocs, identify failed IDocs, and correct and retrigger them.

5.5 Summary

Shipping is a complex process that consists of several individual processes and takes numerous factors into account. Time and costs represent the main criteria when it comes to shipping raw materials, supplies, and semifinished and finished products based on an order. Transportation from the contractor (seller) to the client (buyer) is the final step of order fulfillment, which in a company is usually organized by the shipping department. The shipment of the inventory can often be tracked, and the shipping route or shipping process can be directly traced, either by scanning barcodes or via RFID. For this, the shipment of goods also contains essential documents such as a delivery bill, consignment note, or a specific tracking number. Shipping processes can be designed more efficiently overall, and the higher the degree of automation in the individual steps, the greater is the efficiency and accuracy of the shipping process.

Chapter 6
Inventory Management and Reconciliations

Inventory management is the process of storing, ordering, and selling goods and services. This also involves the management of various supplies and processes. One of the most critical aspects of inventory management is managing the flow of raw materials from procurement to finished products. The aim is to minimize overstocks and improve efficiency that results in projects executed on time and within budget, or timely and accurate shipment to customers.

Inventory is the most critical and important aspect of the supply chain. It encompasses all parts of the supply chain from sourcing of raw materials to the sale of finished goods to the end customer. The ideal method of inventory management is to store inventory until needed. Goods or materials are transferred across the distribution or fulfillment network till it's ready for shipment. Inventory shortages may lead to unhappy customers or internal stakeholders (e.g., manufacturing operations). On the other hand, large inventory or overstock has its own liabilities—the cost to store and insure it along with the risk of spoilage, theft, and damage. Organizations with complex supply chains and manufacturing processes must find a way to optimize inventory.

Inventory reconciliation or stock reconciliation refers to the process of comparing and verifying the accuracy of a company's physical inventory with the inventory records in its accounting system. The goal of this process is to ensure that the company's financial records accurately reflect the actual physical inventory it has on hand and to identify and correct any discrepancies that may exist between the two. Discrepancies may include errors in counting or recording the inventory, and mismatches in inventory records due to theft or loss, damage, or obsolescence of inventory items.

Benefits of inventory management include the following:

- Saving money by freeing up cash tied up in overstocking
- Improved cash flow as money is spent on inventory that sells, so cash is always moving through the business
- Satisfied customers due to timely and accurate shipment results

Figure 6.1 provides an overview of inventory management depicting storage and inward/outward movements of stock to various business partners.

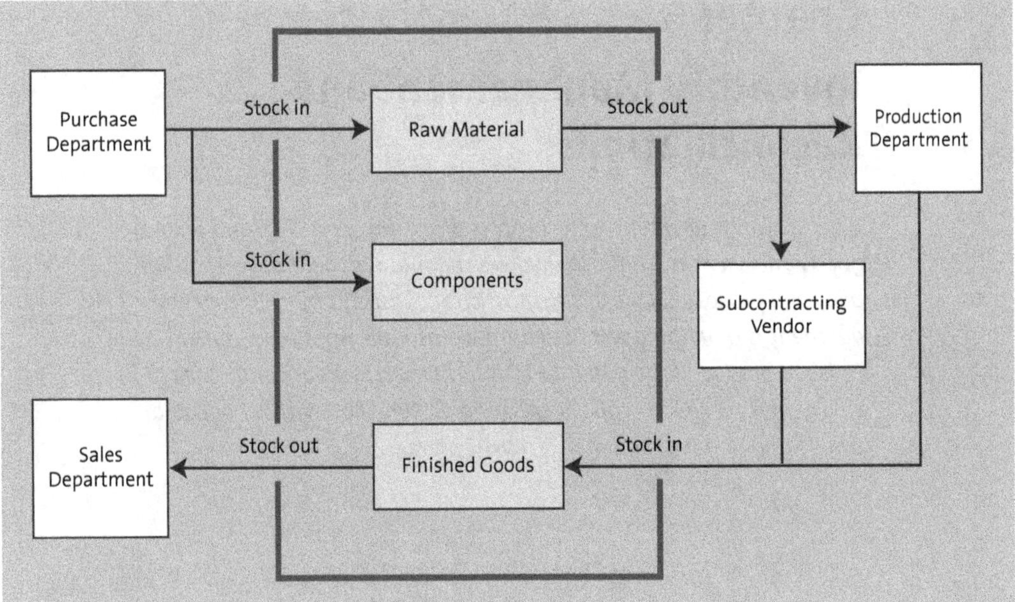

Figure 6.1 Overview of the Inventory Management Process

In the following sections, we'll look at the basic functions of inventory management and its importance from a supply chain perspective. This will be followed by specific SAP S/4HANA inventory management settings required for processes involving third-party logistics integration.

6.1 Goods Movement

Inventory management deals with placing and handling the stock such as receipts from external vendors or manufacturing operations, goods issues to external customers for shipment, or issue of raw material or semifinished products to production or assembly lines for manufacture or assembly of finished goods.

Key activities of inventory management include the following:

- Dealing with the management of stock, either on a value basis, a quantity basis, or both
- Planning, entry, and keeping records of all the movements of goods

In SAP, any goods movement creates a document, known as the material document, that updates all the stock quantities and values in the inventory. A material document is referred to by a document number and document year.

Let's look at few important configuration settings related to goods movement.

6.1.1 Field Selection for Goods Movement

The function called *field selection* provides the option to change the attributes of screen fields dynamically at runtime. With field selection, you can activate or deactivate the following attributes at runtime:

- Input
- Required entry
- Display only
- Invisible
- Highlighted

Using the general field selection, settings for field selection for goods movements initial and header screens can be maintained. A few examples are listed here:

- Define the delivery note and, if required, bill of lading fields as required fields when working with shippers, freight forwarders, or logistics service providers.
- Switch off the print indicators completely or just for certain transactions if not needed.

Note

Field selection for the item screens depends on the movement type and is configured in separate steps pertaining to types of goods movements such as receipts, issues, and transfer postings.

In the standard configuration, all fields that can be configured are defined as optional entries, except for the **Goods Receipt/Issue Slip** field. The **Goods Receipt/Issue Slip** is only active for Italy and is therefore suppressed in the standard configuration.

All screens for field selection are arranged in the **Goods Movements: Initial/Header Screens** screen group. Field selection for goods movement can be maintained using configuration menu path **SPRO • SAP Reference IMG • Materials Management • Inventory Management and Physical Inventory • Field Selection for Goods Movements Initial/Header Screens** (Transaction OMJN).

There are three possible ways the field selection can be maintained in SAP S/4HANA:

- **Maintain the attributes for modifiable fields independently of an influencing value.**
  Modifiable fields are configurable and define how the individual fields are displayed on the screen by using indicators such as ready for input (**Input**), required entry (**Req.**), display only (**Disp.**), hidden/suppressed (**Hide**), and highlighted (**HiLi**). Figure 6.2 provides a snapshot of the configuration of a field selection group for **Screen grp** of **Goods Movements Initial/Header Screens** with the previously mentioned field options.

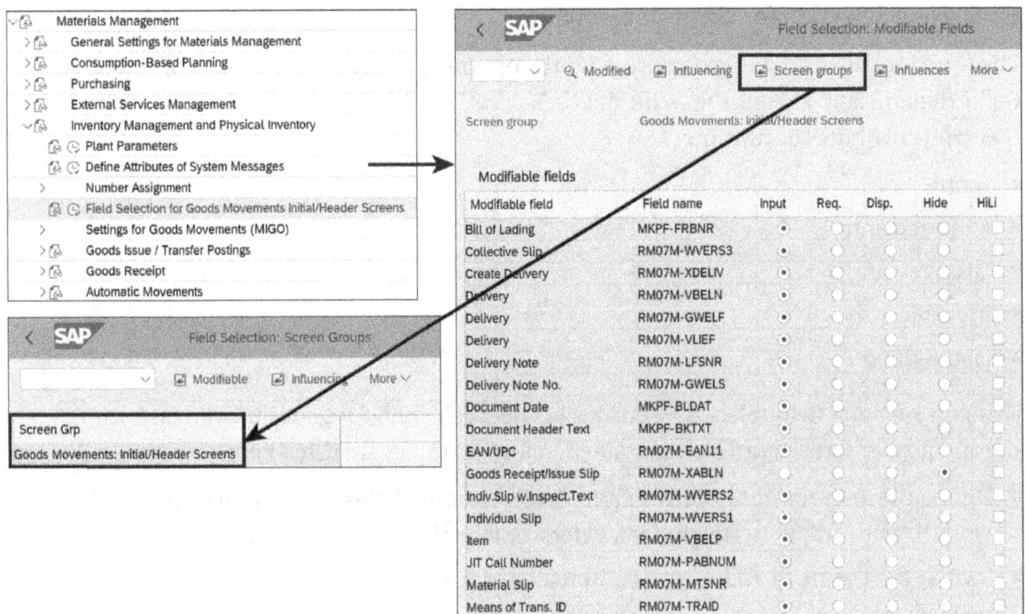

Figure 6.2 Field Selection Group with Modifiable Fields

- **Maintain the attributes of modifiable fields based on influencing field and influencing value.**
  For the influencing fields, you can use indicators to define how the modifiable field is displayed on the screen, depending on a value. For goods movements, you can only maintain the field selection depending on the transaction or transaction fields such as document type and transaction/event type. For example, you can't define the field selection depending on the plant because the plant hasn't yet been entered when you access the transaction.

  Select the Influencing button to display the list of **Influencing** fields. Double-click on the **Influencing** field. The fields that you can modify based on the **Influencing** field appear. Maintain appropriate values in the **Content** field and the corresponding indicators for the modifiable fields. Save your entries.

  Figure 6.3 shows the **Modifiable fields** attributes based on the **Influencing field** of **Document Type** and **Contents** of "WL".

- **Maintain an indicator for a modifiable field, and also determine the values of the influencing fields.**
  This is the third option where you can define the modifiable fields first and then determine the corresponding values of influencing fields for any specific modifiable field:

- Select the **Modifiable** button.
- Set the attributes of **Modifiable field**.
- Double-click on the **Modifiable field**.
- The fields that you can modify based on the influencing value appear.
- Maintain a suitable indicator for the **Influencing field** (for specific modifiable fields).
- Save your entries.

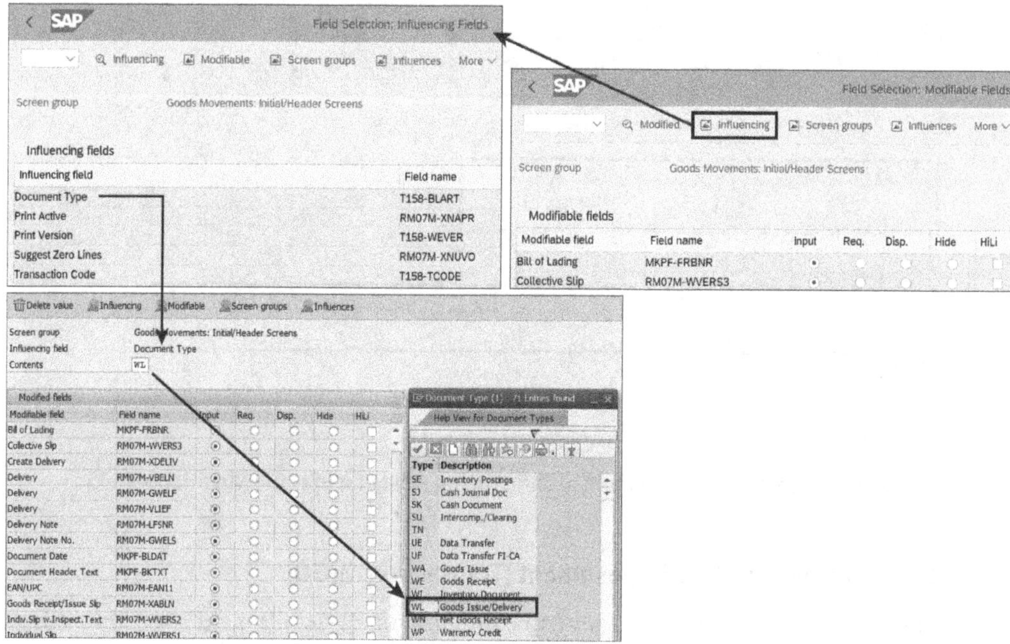

Figure 6.3 Define Field Selection for Influencing Field "Document Type"

Figure 6.4 provides snapshots of the preceding logical steps for the **Modifiable field** value **Bill of Lading**.

> **Note**
>
> If you want to set different field selection definitions for several users (e.g., in different countries), you can copy an SAP transaction and configure the field selection for each transaction.
>
> SAP Note 1645529—OPUK - Customizing of Influencing Field "Level" Isn't Considered—may give you some idea about the field selection–related restrictions in field selection for work breakdown structure (WBS) elements.

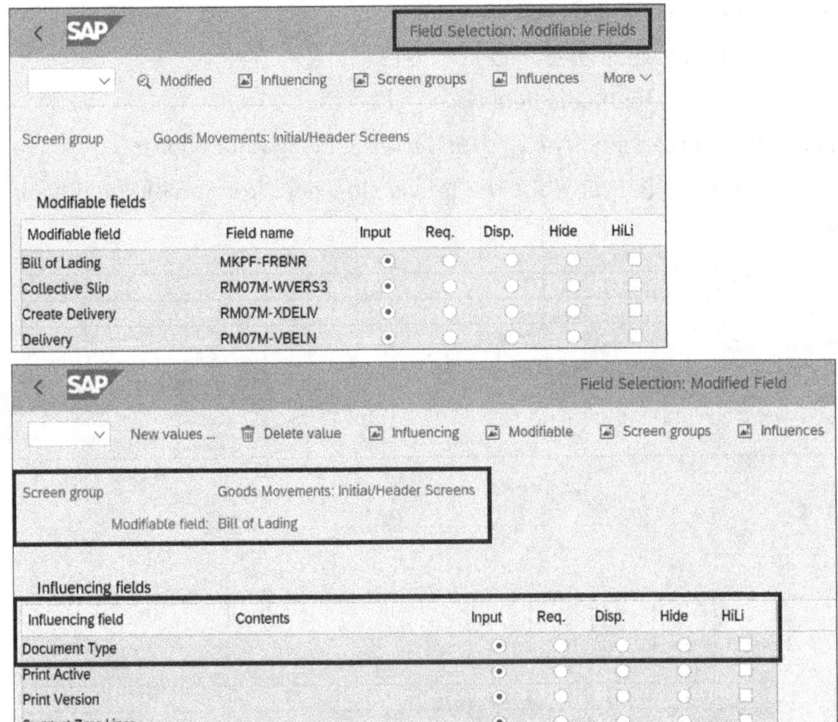

Figure 6.4 Maintain Indicators for the Influencing Field of a Specific Modifiable Field

6.1.2 Settings for Goods Movement (Transaction MIGO)

This section of materials management configuration deals with field selection settings for Transaction MIGO, field selection for movement type, and, finally, for each transaction, defining which transactions and reference documents can be referenced while posting goods movement.

Field Selection for Transaction MIGO

This configuration step deals with field selection settings for Transaction MIGO. Some examples of these settings include the following:

- Hide the header data completely or hide fields that aren't required.
- Hide pushbuttons that aren't required.
- Define the **Delivery note** and **Bill of lading** fields as mandatory entries if you need to track shipments.
- Show the **GR/GI slip** (GOHEAD-XABLN) field as goods receipt/goods issue (GR/GI) slip numbers may be required in some countries.

- Display the **Ordering Plant** (GODYNPRO-PO_WERKS) field for purchase order–related goods movements as a selection criterion in addition to the purchase order number and item number.
- Hide the **Print** indicator if printouts of GR/GI slips aren't needed.

The configuration menu path for field selection attributes for Transaction MIGO is **SPRO • SAP Reference IMG • Materials Management • Inventory Management and Physical Inventory • Settings for Goods Movement (Migo) • Field Selection for MIGO** (Transaction OMJX). Figure 6.5 shows field selection attributes for Transaction MIGO.

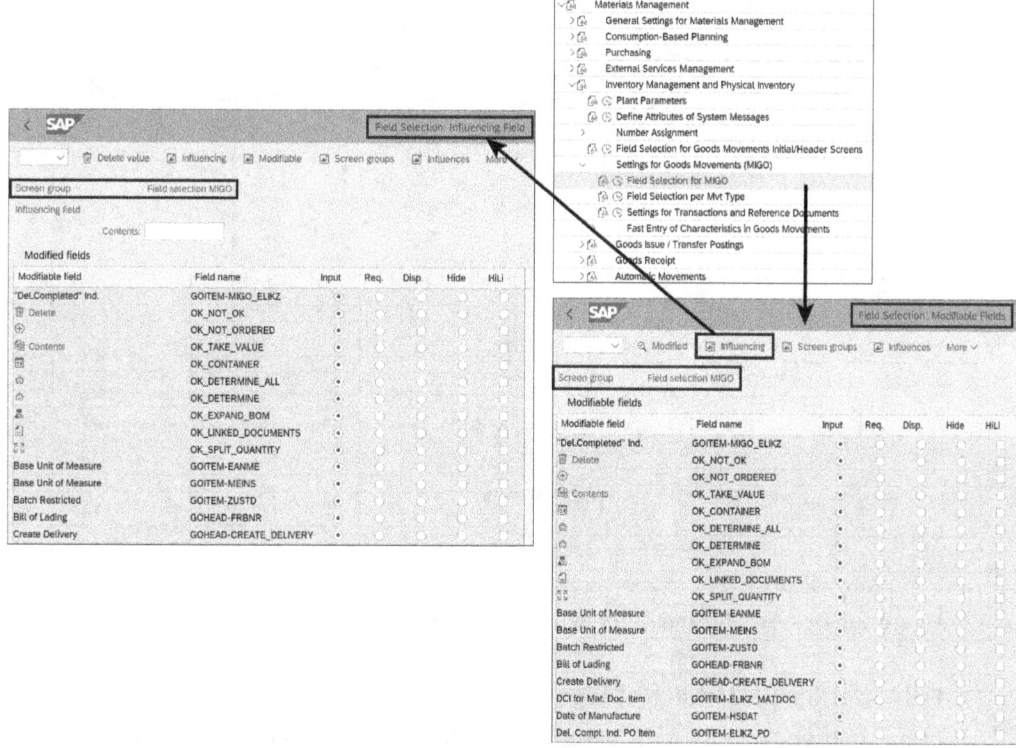

Figure 6.5 Field Selection Settings for Transaction MIGO

In the standard system, all configurable fields are defined as optional entries, except for the **GR/GI slip** (GOHEAD-XABLN) field and the **Plant** (GODYNPRO-PO_WERKS) fields. These fields are hidden in the standard system. You can modify the field settings per business requirements; however, SAP recommends the standard settings.

The characteristic **Mandatory** field is only analyzed for fields with the prefix GOHEAD (header fields) and GOITEM (item fields).

The following exercise will help you understand the correlation between **Modifiable** and **Influencing Fields**:

1. Click on the Influencing button.
2. From the list of **Influencing Fields** that appears, choose **Bill of Lading**, and set the indicators **Req.** and **HiLi**.
3. Click on the Influences button.
4. The resulting screen will show how the **Influencing** and **Modifiable** fields are interrelated (see Figure 6.6).

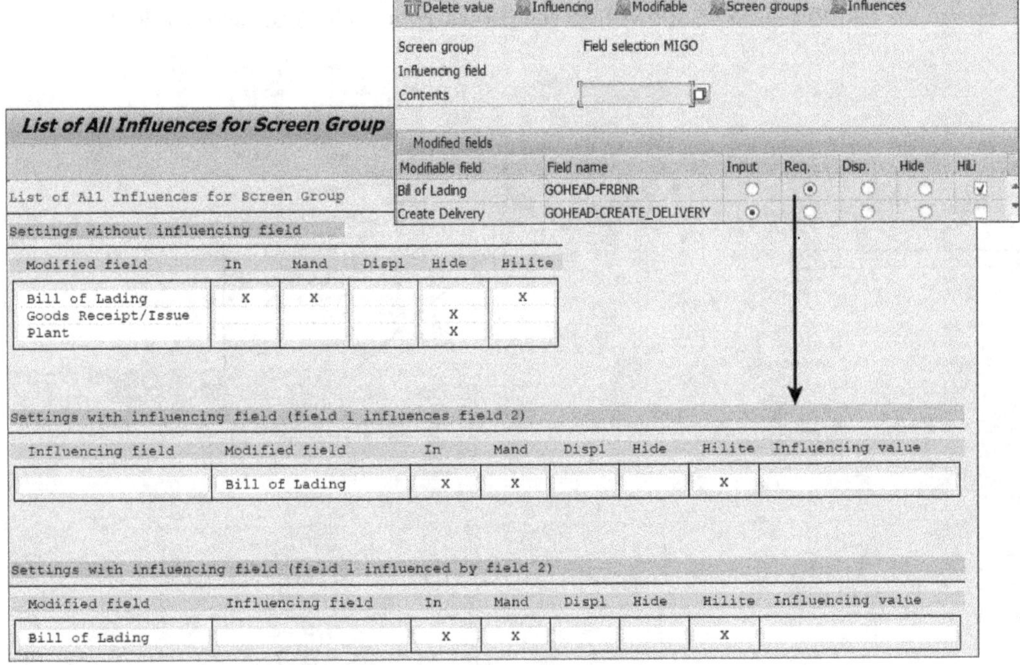

Figure 6.6 Field Setting and Its Influences for a Mandatory Field

Field Selection per Movement Type

In this step, you maintain all the settings for field selection by movement type. You can specify for each field whether it's an optional entry or a mandatory entry. In the table, you see all fields that are visible. You can access the **Field Invisible** setting by deleting the appropriate entry from the table. If you want to add a field for a movement type, create a new entry, enter the movement type, and choose the field you want to add from the list of possible entries.

The configuration menu path for **Field Selection** settings specific to movement type is **SPRO • SAP Reference IMG • Materials Management • Inventory Management and Physical Inventory • Settings for Goods Movement (MIGO) • Field Selection per Mvt Type.** Figure 6.7 shows the configuration menu path and settings of field attributes per movement type.

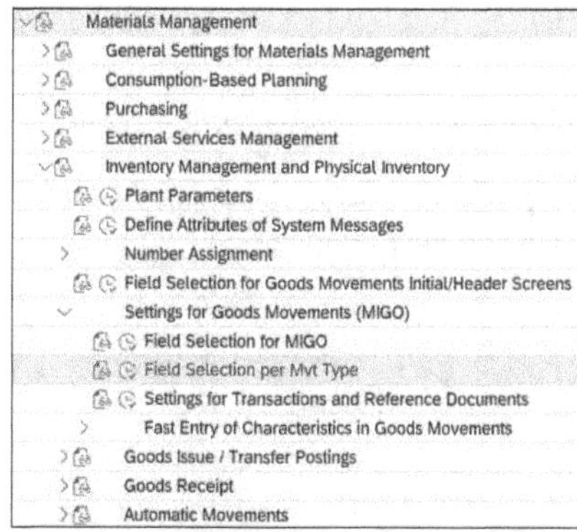

New Entries

MIGO: Field Selection per Movement Type

M...	Field Name	Required Entry	Optional Entry
101	CUSTNAME	○	◉
101	KUNNR	○	◉
101	SGTXT	○	◉
101	WEMPF	○	◉
102	KUNNR	○	◉
102	SGTXT	○	◉
102	WEMPF	○	◉

Figure 6.7 Movement Type Field Selection Settings

Settings for Transactions and Reference Documents

This configuration setting provides the option to limit the number of selectable transactions (goods receipt, goods issue, etc.) and the selectable reference documents (material document, delivery note, etc.). In addition, you can specify the movement type that the system is to propose for every reference document. SAP provides the following preset transactions:

- **Transaction MIGO**
  Used to enter all goods receipts.
- **Transaction MIGO_GR**
  Used to enter goods receipts from external procurement.
- **Transaction MIGO_GO**
  Used to enter goods receipts for production orders.
- **Transaction MIGO_GI**
  Used to enter goods issues.

By default, the **Display material documents** option is always active.

Configuration settings for transactions and reference documents can be maintained using menu path **SPRO • SAP Reference IMG • Materials Management • Inventory Management and Physical Inventory • Settings for Goods Movement (MIGO) • Settings for Transactions and Reference Documents.** Figure 6.8 shows settings for transactions and reference documents.

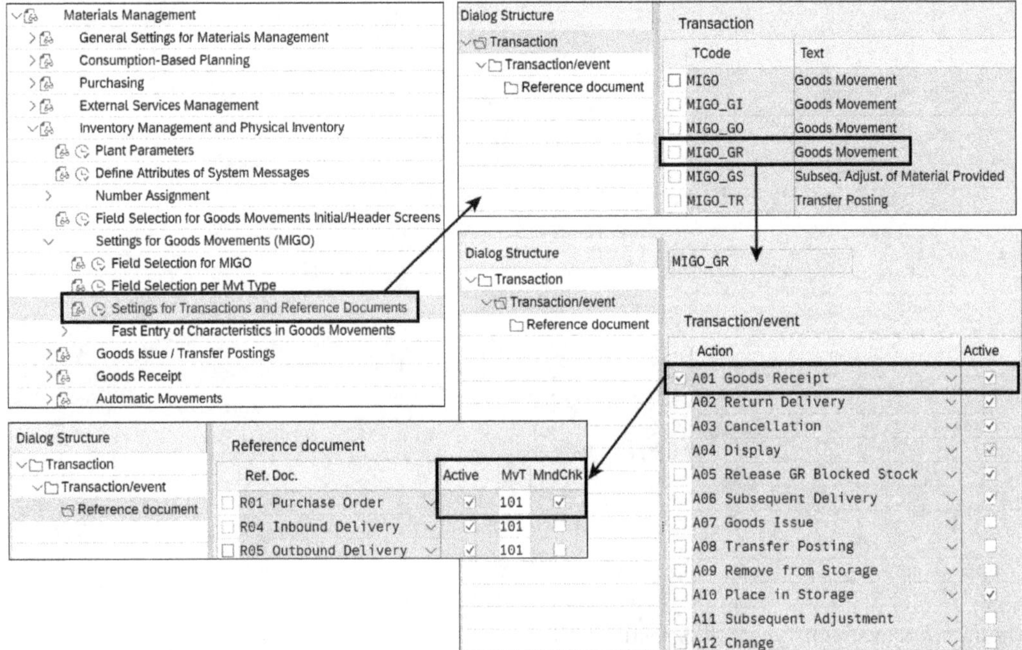

Figure 6.8 Settings for Transactions and Reference Documents

It's possible to restrict the posting of documents unless the check function is executed. For example, if working with subsequent deliveries or goods receipt blocked stock isn't required when goods receipt is posted, these transactions can be removed from the list of possible entries. Such an omission is possible by deselecting the **Active** checkbox. If you work with goods receipts only or with goods issues only, you can choose a transaction that displays only the relevant transactions and reference documents. If you want to ensure that a user also sees the warning and information messages when entering goods receipts for a purchase order, you can select the binding **Mandatory** check indicator for the combination of **Goods Receipt** and **Purchase Order**.

6.1.3 Goods Issue and Transfer Postings

This section of configuration under the **Materials Management** node deals with settings specific to the following goods movements:

- Goods issues
- Transfer postings
- Other movements excluding goods receipts (e.g., inventory difference postings)

Create Storage Location Automatically

In this configuration step, you can specify whether the automatic creation of storage location data is allowed for goods issues and transfer postings.

You have to allow automatic creation per plant first of all. You then have to explicitly allow the creation of storage location data for each movement type. The storage location data is only created if the quantity is posted to standard storage location stock. It's not created for receipts into a special stock (e.g., into sales order stock).

The configuration menu path for auto-creation of storage locations is **SPRO • SAP Reference IMG • Materials Management • Inventory Management and Physical Inventory • Goods Issue/Transfer Postings • Create Storage Location Automatically** (Transactions OMB2/OMB3). Figure 6.9 shows plant and movement settings for auto-creation of storage locations for material.

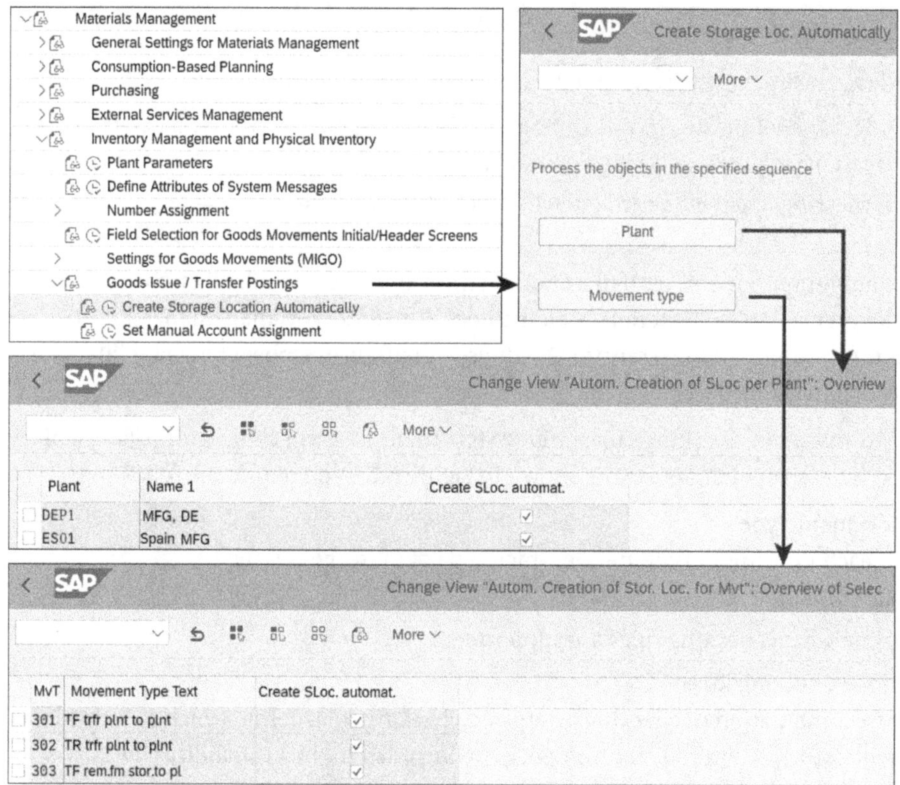

Figure 6.9 Setting for Automatic Storage Location Creation for Goods Issue/Transfer Postings

> **Note**
>
> Basically, this configuration step is not creating a storage location for a plant, it's creating storage location data for the material used in the transaction. If the material isn't extended to the storage location involved in the transaction, it won't result in a transaction error, but the goods movement posting will automatically create storage location data for the material. As a result, an auto entry is created in table MARD.

Configure the Scope of the Availability Check

In this configuration step, specific settings are maintained in Customizing regarding the available-to-promise (ATP) check in materials management. In inventory management, two types of availability check are carried out:

- Availability check during the entry of reservations
- Availability check during the entry of goods movements

The system checks whether the material to be withdrawn is available and issues a message, if necessary.

> **Prerequisite**
>
> The availability check is only carried out if the following are true:
>
> - The material master record contains a checking group for the availability check (material requirements planning [MRP] data).
> - A checking rule is assigned to the transaction in Customizing.

The configuration menu path maintaining the scope of the availability check is **SPRO • SAP Reference IMG • Materials Management • Inventory Management and Physical Inventory • Goods Issue/Transfer Postings • Configure Scope of Availability Check** (Transaction OMCP).

How to maintain checking rules and assign them to a checking group (availability check) is detailed in Chapter 5. As shown in Figure 6.10, you define the following:

- **Movement Type**
  For each movement type, define which type of message is issued.
- **Checking rule**
  Define which checking rules are supported.
- **Define Checking Rule**
  For a combination of checking group and checking rule, define which objects (documents, stocks, material requirements) are included in the availability check.
- **Transaction Code**
  For inventory management transactions, define which checking rule applies.

Standard SAP recommends message type **A** (warning message only in the case of nonavailability) for all relevant movement types. For inventory management, checking rule (**ChR**) **03** is provided.

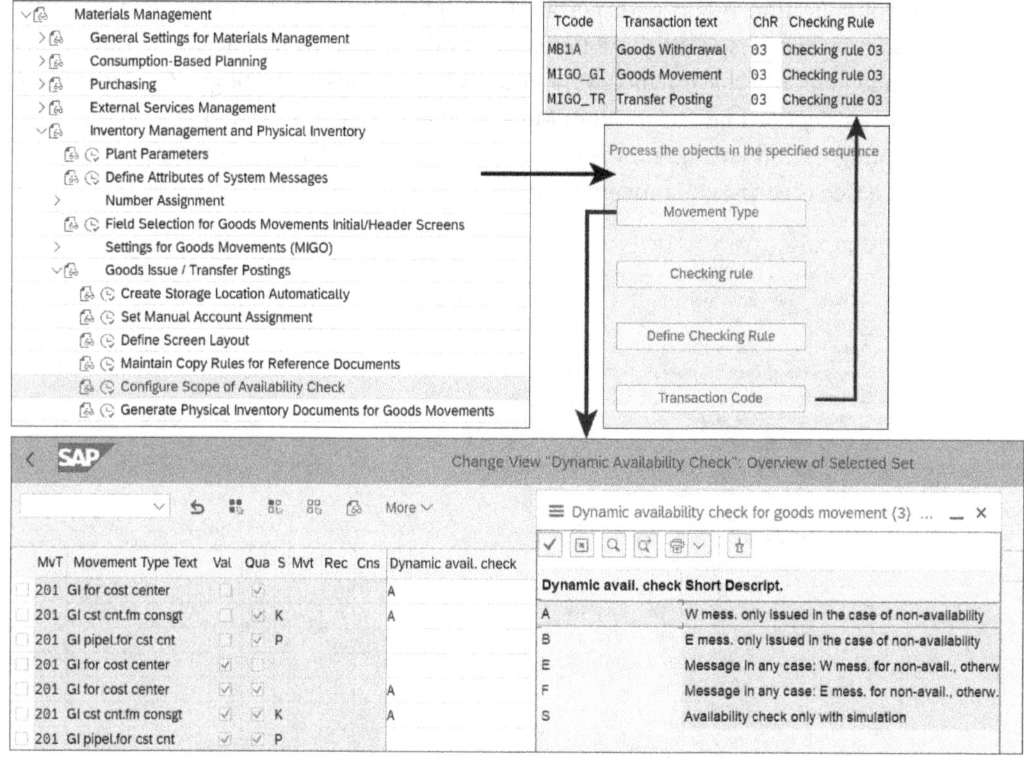

Figure 6.10 Configuration for Scope of ATP Check during Goods Issue/Transfer Postings

Generate Physical Inventory Documents for Goods Movements

SAP provides the option to create a physical inventory document in addition to the material document at the time of a goods movement. This function is important when inventory differences are posted from the warehouse management system (WMS). These inventory differences are thereby documented in inventory management with a physical inventory document.

To create an inventory document in addition to a material document for a goods movement, the following conditions must be met:

- In the plant, you must define the transaction/event type as **WV** for automatically generated physical inventory documents. You must define the document number assignment for this transaction/event type.
- Automatic generation of physical inventory documents must be allowed for the movement type.

In this configuration step, you define for each plant and movement type whether a physical inventory document is automatically generated when posting a goods movement.

The configuration menu path pertaining to generation of inventory document for goods movement is **SPRO • SAP Reference IMG • Materials Management • Inventory Management and Physical Inventory • Goods Issue/Transfer Postings • Generate Physical Inventory Documents for Goods Movements** (Transaction OMCC). Figure 6.11 shows the configuration settings for the generation of inventory documents for goods movement based on plant and movement type.

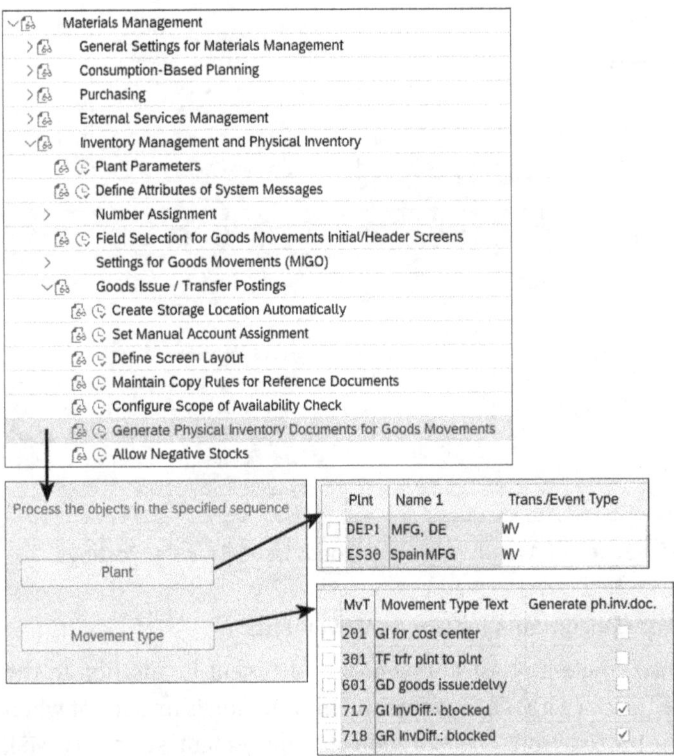

Figure 6.11 Settings for Generation of Physical Inventory Documents along with the Material Document

In the standard system, automatic generation of physical inventory documents is allowed for movement types 711 to 718.

6.1.4 Goods Receipt

This section of configuration under the **Materials Management** node deals with settings specific to goods receipt transactions.

Set Tolerance Limits

While processing a goods receipt, the system checks each item to determine whether the goods receipt varies from the purchase order or the material master data. The different types of variances are defined by tolerance keys. For each tolerance key, tolerance limits can be set per company code.

The following variances are checked in inventory management:

- **Variance in the purchase order price quantity**
  For this variance, two tolerance keys are provided in the standard:
  - Tolerance key B1 (error message): This key is used to define the maximum percentage variance. If the variance is greater than the maximum variance, the system creates an error message. The goods receipt can't be posted.
  - Tolerance key B2 (warning message): This key defines the percentage variance from which a warning message is to be issued. It's possible to post the goods receipt.
- **Variance for the moving average price of the material**
  SAP standard tolerance key VP can be used to define the percentage variance from which a warning message is issued. This warning message indicates a price change.

The configuration menu path to set tolerance limits is **SPRO • SAP Reference IMG • Materials Management • Inventory Management and Physical Inventory • Goods Receipt • Set Tolerance Limits** (Transaction OMCO), as shown in Figure 6.12.

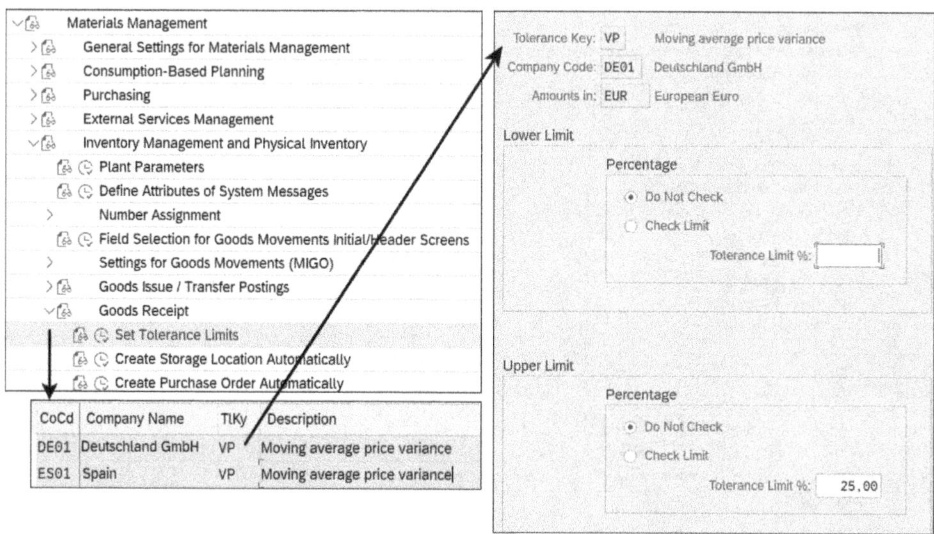

Figure 6.12 Settings for Tolerance Key for Goods Receipt Transaction

Create Storage Location Automatically

In this configuration step, you can specify whether the automatic creation of storage location data is allowed for goods receipt postings. The configurations settings are similar to the settings described for goods issue/transfer postings. In addition, as explained previously, the automatic function is creating storage location data for the material used in the transaction.

The configuration menu path for auto-creation of storage location is **SPRO • SAP Reference IMG • Materials Management • Inventory Management and Physical Inventory • Goods Receipt • Create Storage Location Automatically** (Transaction OMB3).

Set Dynamic Availability Check

This configuration setting is similar to that explained in Section 6.1.3. The only difference is the movement types because it also covers inbound movement–specific movement codes such as goods receipt, vendor return, and so on.

For goods receipt postings, you can also set up an ATP check. In this case, you'll receive a warning message when posting goods receipt if the material is a missing part. To receive the message, you have to set up the ATP check, the missing part check, and the periods in the scope of availability check (Transaction OVZ9).

The configuration for the missing part check is through menu path **SPRO • SAP Reference IMG • Materials Management • Inventory Management and Physical Inventory • Goods Receipt • Set Missing Parts Check** (Transaction OMBC). Figure 6.13 shows the configuration menu path and setup for the missing parts check.

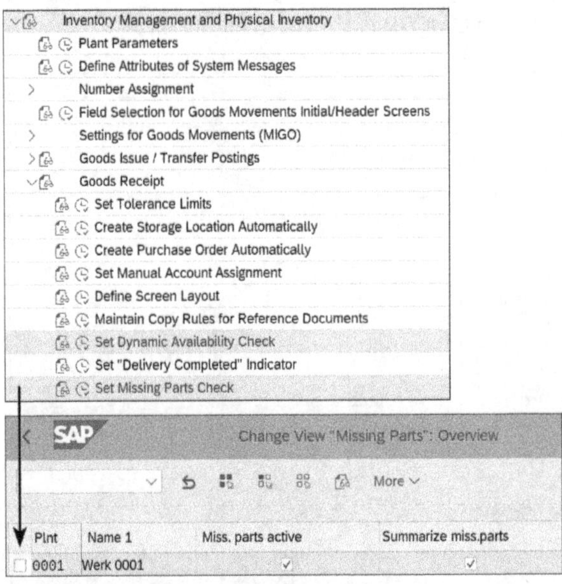

Figure 6.13 Set the Missing Parts Check

Set the "Delivery Completed" Indicator

In this step, you can specify that the system automatically suggests the inward delivery completed indicator for goods receipts if the delivered quantity lies within the under/overdelivery tolerance.

The "inward delivery completed" indicator has the following effects:

- The item is regarded as closed, even if the total quantity wasn't delivered.

- A further delivery isn't expected, may, however, arrive. In the next goods receipt, the item appears on the selection list but has no selection indicator.
- The item can be reorganized even if the total quantity hasn't been delivered.

Here, we specify for each plant whether the system proposes the "delivery completed" indicator automatically.

> **Note**
>
> Even if the "delivery completed" indicator hasn't been set, an item is considered closed once the total quantity has been delivered. This means that the "delivery completed" indicator isn't necessary in this case.

The configuration menu path for setting up default delivery completed indicator at plant level is **SPRO • SAP Reference IMG • Materials Management • Inventory Management and Physical Inventory • Goods Receipt • Set Delivery Completed Indicator.**

Figure 6.14 shows the configuration menu path and setting **Del. Compl. Default** as the default at the plant level.

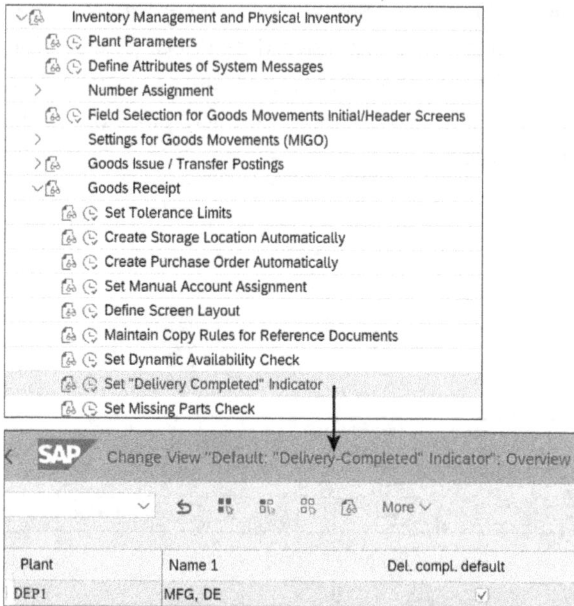

Figure 6.14 Set "Delivery Completed" Indicator for Goods Receipt

Set Expiration Date Check

In this configuration step, you maintain the material shelf-life expiration date check for goods receipts for each plant and movement type. If the shelf-life check expiration date check is active, the shelf-life expiration date or the production date of the material must be entered at the time of goods receipt.

At the time of goods receipt, the system checks whether the remaining shelf life is sufficient. If this isn't the case, a warning or error message is issued, depending on the system settings. When the goods receipt is posted, the shelf-life expiration date is recorded in the material document. For materials handled in batches, the date is also stored in the batch master record.

Prerequisite for the Shelf-Life Expiration Date check

The shelf-life expiration date of a material can only be entered if the following are true:

- The minimum remaining shelf life is maintained in the material master record. The minimum remaining shelf life is the time the material must still be usable, so that the goods receipt is accepted by the system.
- You can define the time unit the system uses in the storage data of the material master record in the **Period ind.** field.
- The shelf-life expiration date check is active in the plant.
- The shelf-life expiration date check is active for the movement type.

The configuration settings for the shelf-life expiration date check can be done using menu path **SPRO • SAP Reference IMG • Materials Management • Inventory Management and Physical Inventory • Goods Receipt • Set Expiration Date Check** (Transaction OMJ5). Figure 6.15 shows the configuration menu path, the shelf-life expiration date check at the plant level as **BBD/ProdDt**, and selection options for activating the shelf-life expiration date check for movement types.

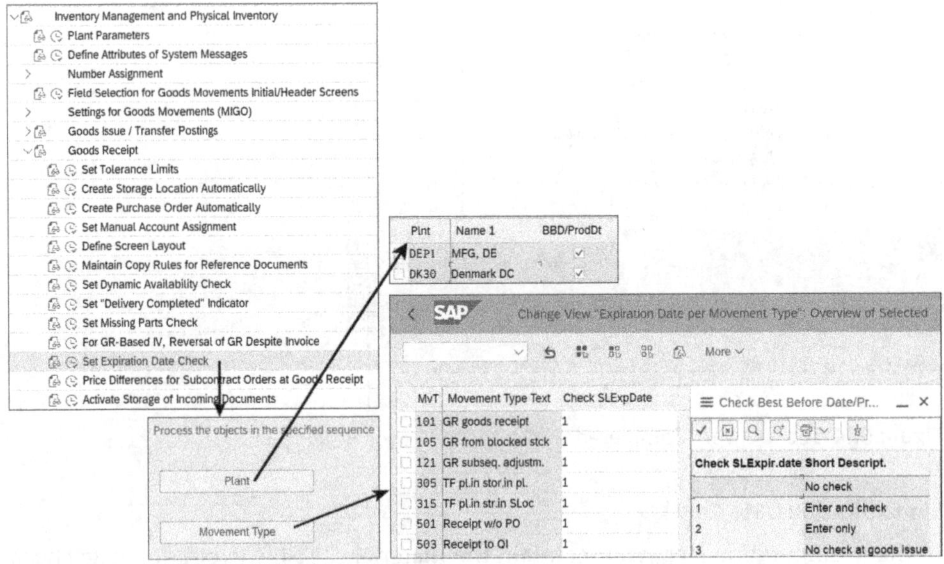

Figure 6.15 Setting Up the Shelf-Life Expiration Date Check for Plant and Movement Type

6.1.5 Automatic Movements

This section deals with system settings for goods movements that are posted automatically. Some examples are as follows:

- Goods issue for a delivery via shipping
- Goods movements for repetitive manufacturing
- Inventory adjustment postings in inventory management
- Status changes of batches

As we'll discuss in this section, you can only set those movements that can't be entered manually in inventory management.

Create Storage Location Automatically

This configuration setting specifies whether the automatic creation of storage location data is allowed for automatic goods movements. This is recommended for goods movements via shipping or when you post physical inventory differences, for example.

You have to allow automatic creation per plant first. You then have to explicitly allow the creation of storage location data for each movement type. The storage location data is only created if the quantity is posted to standard storage location stock; that is, it's not created for receipt into a special stock (e.g., into sales order stock).

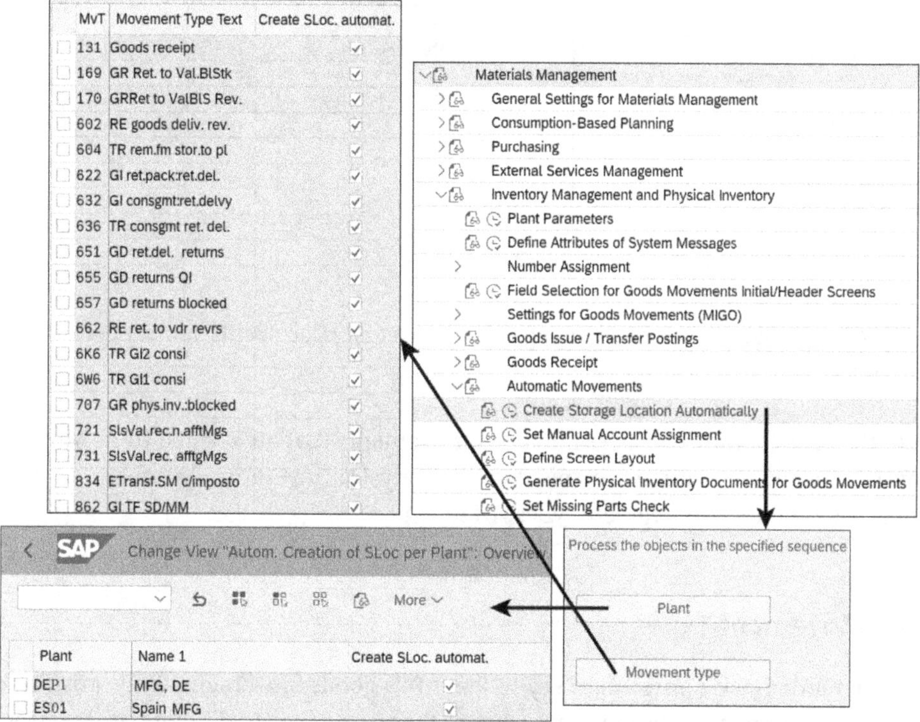

Figure 6.16 Automatic Creation of Storage Locations for Automatic Movements

The menu path for automatic creation of storage locations is **SPRO • SAP Reference IMG • Materials Management • Inventory Management and Physical Inventory • Automatic Movements • Create Storage Location Automatically** (Transaction OMJ8). Figure 6.16 shows specific configurations for automatic storge location creation for plants and movement types.

Generate Physical Inventory Documents for Goods Movements

This configuration step specifies for each plant and movement type whether a physical inventory document is created automatically when a goods movement is posted.

The configuration menu path is **SPRO • SAP Reference IMG • Materials Management • Inventory Management and Physical Inventory • Automatic Movements • Generate Physical Inventory Documents for Goods Movements** (Transaction OMJC). Figure 6.17 shows selection or activation at the plant and movement type levels, related to generation of physical inventory documents for goods movement

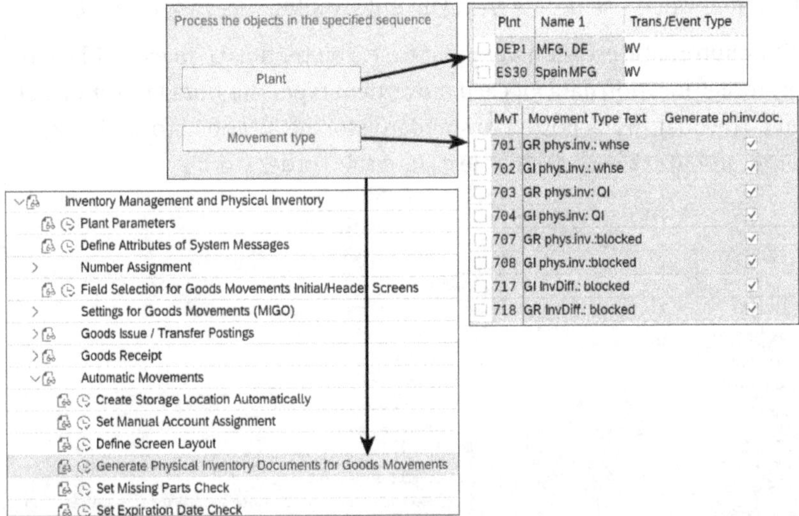

Figure 6.17 Configuration for Generating Physical Inventory Documents during Goods Movement

Note that this indicator isn't set for goods movements that are created from within a physical inventory exercise. In this case, the physical inventory document already exists. This applies to movement types 701 to 708.

6.2 Movement Types

A movement type is a three-digit key to identify a goods movement. While posting a goods movement in the system, a movement type is required to differentiate between the various goods movements.

Movement types indicate the purpose of the goods movement (e.g., from storage location to a storage location, receipt from purchase order, issue for the delivery, receipt from production). SAP movement types control several functionalities in SAP S/4HANA, such as their primary role in account determination, Transaction MIGO, and many others, as you've just learned in previous configuration activities.

The movement type has important control functions in inventory management:

- Updating of quantity fields
- Updating of stock and consumption accounts
- Determining which fields are displayed during entry of a document in the system
- Defining the screen layout
- Checking shelf-life expiration dates, dynamic availability, and so on
- Printing goods receipt/issue slips

As examples of the role of movement types in account determination, movement type 101 used for goods receipt is configured to post accounting documents on the stock account as debits and GR/IR accounts as credits, whereas movement type 201 used for goods issue is configured to post accounting documents on the stock account as credits and the cost center account as debits.

Movement types can be categorized into the following:

- Movement types for goods receipts
- Movement types for goods issue
- Movement types for transfer posting
- Special movement types
- Movement types for physical inventory
- Reversal movement types

Next, we'll discuss the configuration settings pertaining to movement types. In this configuration step, you can modify the standard, predelivered movement types or create a new one per business requirements.

The menu path for changing standard movement types or creating a new custom one is **SPRO • SAP Reference IMG • Materials Management • Inventory Management and Physical Inventory • Movement Types • Copy, Change Movement Types** (Transaction OMJJ).

To set up this configuration, select a standard movement type that is a close match to one that needs to be created. Execute Transaction OMJJ, and select the **Movement type** checkbox in the resulting popup screen. In the next screen, enter the movement types (both the standard one and the custom one that you intend to create), as shown in Figure 6.18.

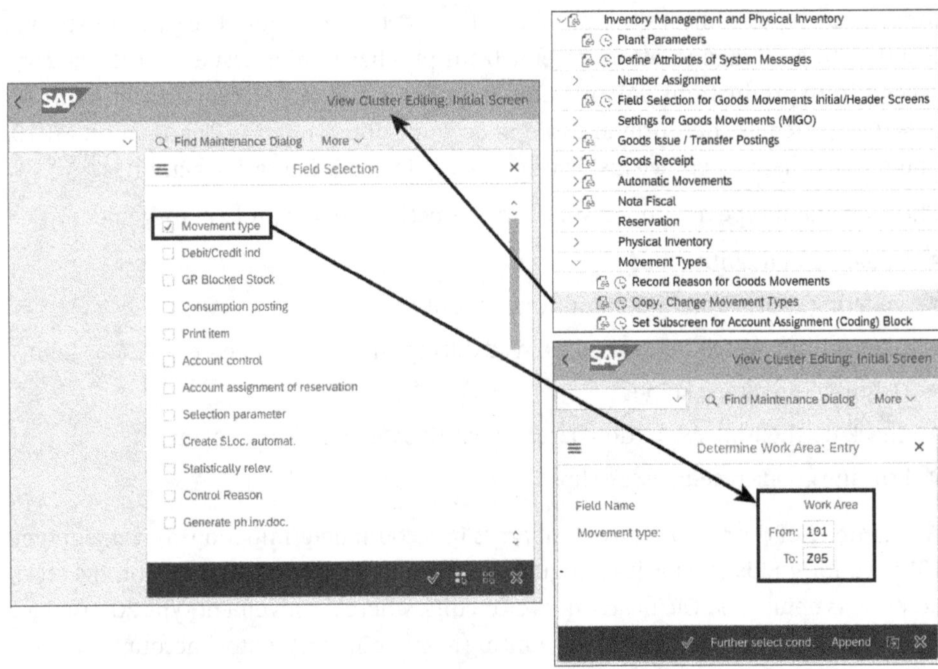

Figure 6.18 Selection Criteria for Movement Type Configuration

Next, select the standard movement type, and execute the **Copy as** function that brings in all dependent entries to the new or custom movement types, as shown in Figure 6.19.

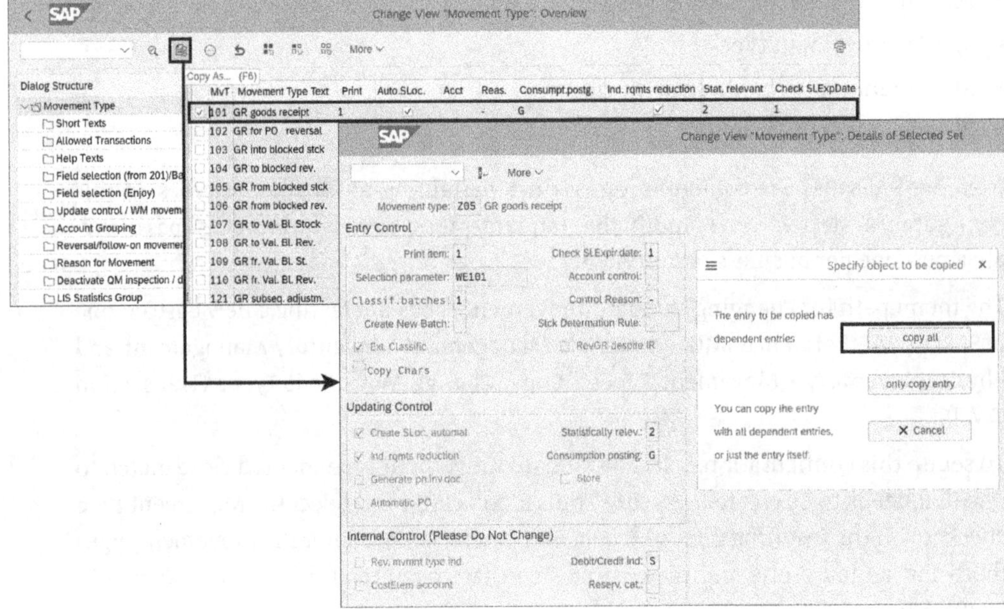

Figure 6.19 Creating a Custom Movement Type Using a Standard Movement Type as Reference

Once all the entries are available in the custom movement type, these can be fine-tuned per business requirements.

We have to limit our scope to this point, as configuring the movement type requires good in-depth knowledge of the concepts and processes in the inventory management component of materials management. As movement type is closely linked to account determination, this kind of configuration needs to be discussed jointly with SAP finance experts for accounting requirements and any other possible impacts.

6.3 Output Determination

In inventory management, output is created at the time of goods movements. Output documents can be of the following types:

- Goods receipt/issue slips (e.g., GR/GI slips, pallet slips)
- Labels
- Kanban cards
- Mail messages for communicating with other departments (e.g., purchasing or materials planning), or communication with third-party logistics partners
- cXML messages for communicating goods receipt information to the Ariba Network
- IDoc output (special function) for communication with third-party logistics partners

In this section, the output control features available with SAP S/4HANA will be explained along with relevant configuration details for output determination in inventory management.

6.3.1 Output Control in SAP S/4HANA

Table 6.1 provides an overview of existing features and functions of the new output control.

Supported Channels	▪ Printer: Using the spool ▪ Email: Allows flexible configuration of sender and recipient(s) as well as use of email templates ▪ XML: Integration with the Ariba Network ▪ IDoc: Limited support only
Supported UI Technologies	▪ SAP Fiori ▪ SAP GUI and SAP GUI for HTML

Table 6.1 Features and Functions of Output Control in SAP S/4HANA

Supported Form Technologies	▪ PDF-based print forms with fragments (default) ▪ PDF-based print forms ▪ Smart forms ▪ SAPscript
Master Form Templates	A master form template defines a universally valid layout and features reusable parts of a form, such as header, footer, or logo (only available for PDF-based print forms with fragments). This allows you to maintain general content and application-specific content separately. If you make changes in the general content (master form), the application forms will be updated accordingly.
Determination of Master Form Templates	Master form templates can be determined based on preconfigured parameters, such as company, sender country, or channel. This makes it easy to define separate logos or footer texts per company.
Output Parameter Determination via BRFplus	All output parameters can be automatically determined via a configuration activity based on BRFplus. The benefits are as follows: ▪ Multiple messages can be sent to multiple recipients using multiple channels at the same time. ▪ Easily configurable without the need for ABAP exits. ▪ Each output determination step can be based on application-specific fields that can be extended. ▪ Predelivered content/determination rules can be used to run output control out of the box. ▪ Extensibility via Core Data Services (CDS) is supported.
Dispatch Time	Immediately or scheduled via batch job (application-specific).
Application Log	The system creates one application log per output request instance. In addition, you can filter messages by output request item. You can display the log in the business application or via Transaction SLG1 once an output item has been processed.
Used Frameworks/Services	▪ Business Object Processing Framework (BOPF) ▪ BRFplus ▪ Knowledge Provider (KPro) ▪ SAP Gateway Services ▪ SAP Fiori ▪ Web Dynpro ▪ SAP GUI

Table 6.1 Features and Functions of Output Control in SAP S/4HANA (Cont.)

> **Note**
> SAP Note 2228611, Output Management in SAP S/4HANA, provides information regarding usage of output management and output control in the latest versions of SAP S/4HANA. You can also look at the output configuration details in Chapter 5, Section 5.4, for how to configure output for delivery documents.

Configuration steps are similar to what is explained in Chapter 5. You must check or maintain the following objects so that the system creates an output document whenever a goods movement is entered:

- Condition table
- Access sequence
- Output types
- Output determination procedure
- Condition records
- Smart forms for the output type

> **Note**
> For communication to third-party logistics partners of the stock status notifications for any stock adjustments in SAP S/4HANA, the output for goods movement can be used.

6.3.2 Maintain Condition Tables

Condition tables consist of a combination of key fields to identify an individual condition record. A condition record is defined as how the system stores the specific conditions. Figure 6.20 shows the following:

❶ Selected fields for a condition table

❷ Field catalog for goods movement (structure KDMB)

The configuration menu path for defining a condition table is **SPRO • SAP Reference IMG • Materials Management • Inventory Management and Physical Inventory • Output Determination • Maintain Condition Table • Output: Create Conditions Table** (Transaction M703).

You may find that the certain fields aren't available in the **Field Catalog** for the condition key and output determination (structure KOMB). This can be appended with the corresponding field. Alternatively, SAP Note 2367296, Movement Type in Field Catalog for Inventory Management, explains that the movement type field is available in the SAP standard with the Discrete Industries and Mill Products (DIMP) business function. So, you should have the movement type available if you activate DIMP or if you're on SAP S/4HANA where DIMP is always on.

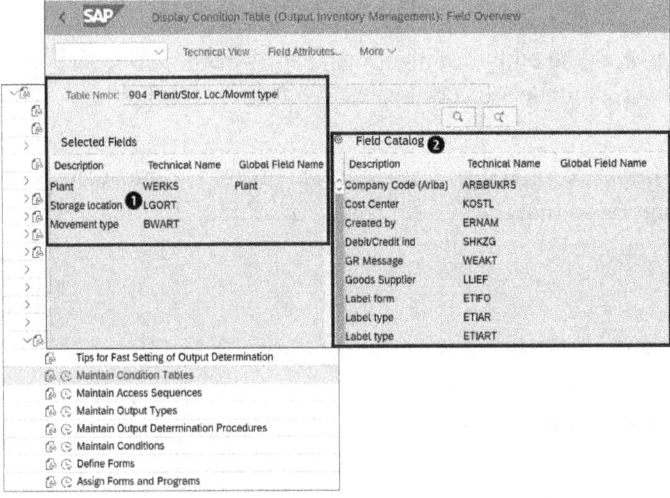

Figure 6.20 Maintain Condition Table for Goods Movement

6.3.3 Maintain Access Sequences

An access sequence needs to be defined for every condition table because it sequentially accesses condition table fields. The access sequence can be maintained using configuration menu path **SPRO • SAP Reference IMG • Materials Management • Inventory Management and Physical Inventory • Output Determination • Maintain Access Sequences** (Transaction M710). Figure 6.21 shows the step-by-step configuration of the access sequence for inventory management output.

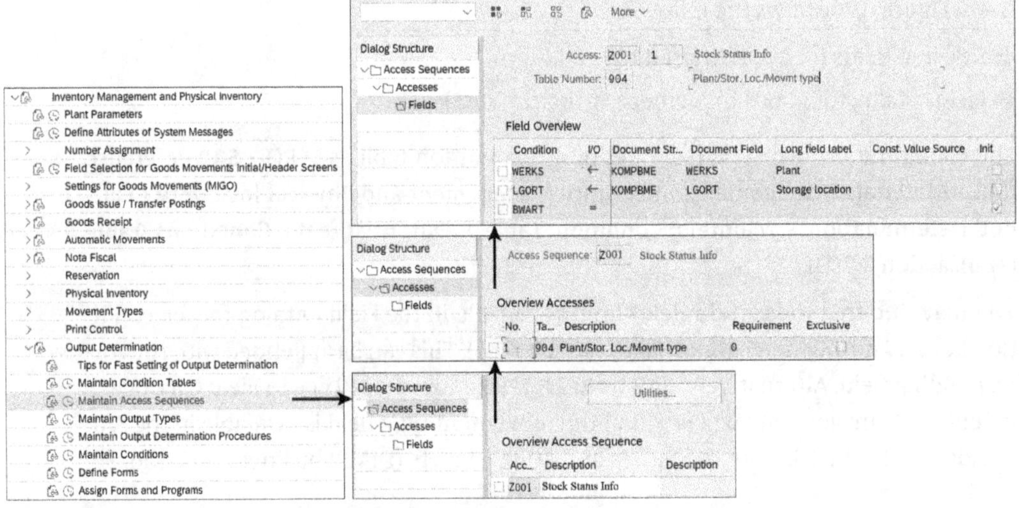

Figure 6.21 Maintain Access Sequence for Goods Movement Output

6.3.4 Maintain Output Types

In this step, individual output types are defined. For each output type, maintain the following:

- Relevant access sequence (and thus the linked condition table)
- Output parameters (e.g., printout, mail message, time of output, etc.)
- Print parameters for printer determination
- Processing routines if special function or IDoc is used
- Partner functions

Output type configuration is maintained using menu path **SPRO • SAP Reference IMG • Materials Management • Inventory Management and Physical Inventory • Output Determination • Maintain Output Types** (Transaction M706). Figure 6.22 shows the configuration settings for output type in inventory management.

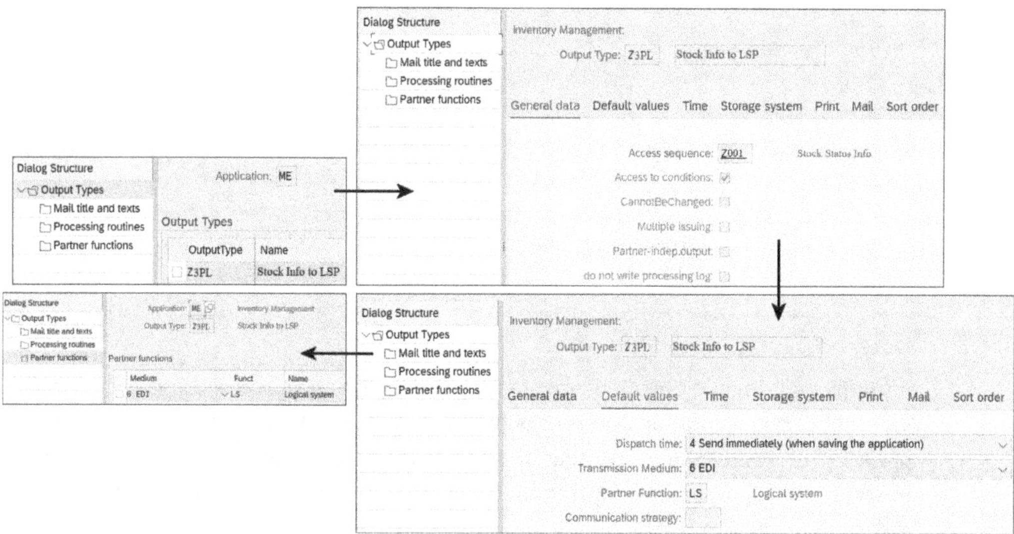

Figure 6.22 Maintain Output Types

For processing routines for IDoc messages, standard driver program RSNASTED along with form routine `EDI_PROCESSING` can be used. Alternatively, if the logistics service provider isn't EDI enabled, a mail output can also be configured with similar setups.

6.3.5 Maintain Output Determination Procedure

An output determination procedure groups together all output types of an application. For inventory management, output determination procedure `ME0001` is predefined. In the output determination procedure, you can enter a requirement for individual output types that must be fulfilled for the output to be created.

Configuration of output determination procedure for inventory management documents can be done using menu path **SPRO • SAP Reference IMG • Materials Management • Inventory Management and Physical Inventory • Output Determination • Maintain Output Determination Procedures** (Transaction M708).

Figure 6.23 shows configuration related to output determination procedures and assignment of relevant output types (for the purpose of communicating material document information to third-party logistics partners).

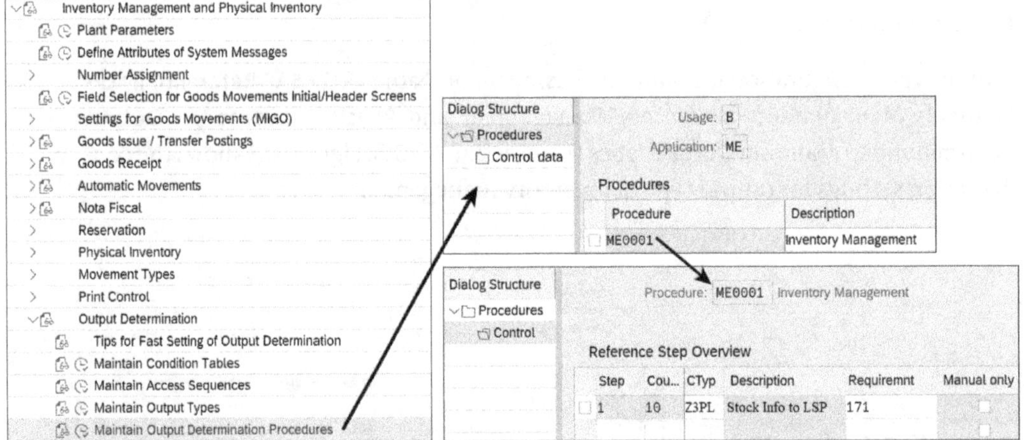

Figure 6.23 Maintain Output Determination Procedure for Goods Movement

Note

It's important to note one disclaimer here. The nomenclature and settings used in figures are for representation purposes only. Consultants working on third-party logistics projects may need to judiciously decide the best output mechanism based on a client's needs, technical capabilities of third-party logistics, and change management associated with the design.

6.4 Inventory Reconciliation

The third-party logistics service provider is typically a separate business with its own WMS. As a result, the ERP system of the customer isn't updated with the most recent data from the third-party logistics warehouse system. The ERP system will therefore be unable to execute plans effectively. This leads to the demand for inventory synchronization on a real-time basis. This sort of integration requires communication between two different business partners, and there may be hundreds and thousands of inventory transactions happening during a single day. It may be difficult to achieve a real-time sync or update, and, for all practical purposes, most of the clients rely on daily, twice-a-day, or—at the most—hourly updates from logistics service providers.

When an ERP runs its MRP, it may have the current inventory located at third-party logistics service provider warehouses, and such inventory synchronization or up-to-date data on inventory will help the planners to plan accurately.

A couple of the EDI messages related to inventory transaction are listed here:

- **EDI X12 846 - Inventory Snapshot**
  This message can be sent by the third-party logistics warehouse to the client on a daily basis to inform the client of the current inventory position at the third-party logistics location.
- **EDI X12 947 - Inventory Movements (Adjustment)**
  There will be daily inventory adjustments within the third-party logistics warehouse. This may include moving the inventory from sellable to nonsellable due to quality checks, or may even include scrapping or destroying the inventory.

Figure 6.24 shows the EDI/IDoc communication between third-party logistics and host systems through middleware SAP Process Integration/SAP Process Orchestration.

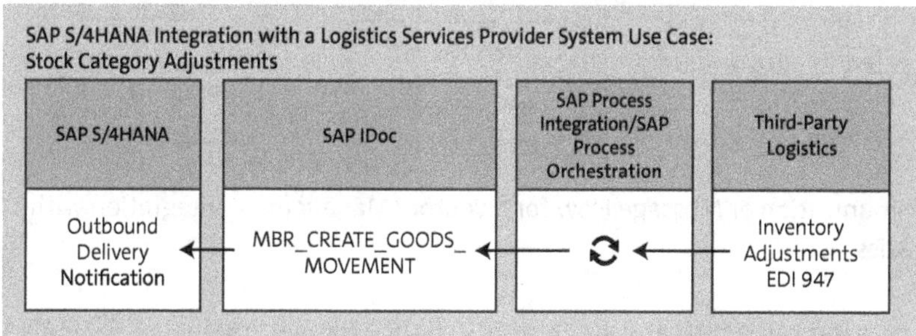

Figure 6.24 Inventory Management Integration with Third-Party Logistics

Whenever an internal movement takes place in a logistics service provider, the same movement needs to be passed on to SAP S/4HANA. Internal movements can refer to transfer postings, goods issues, and inventory adjustments based on the inventory cycle count. Based on the file sent by the logistics service provider, middleware will create an IDoc in SAP S/4HANA against which a batch program will post transactions in SAP S/4HANA for that material, plant, batch, and storage location combination as the case may be.

The following are examples of inventory adjustments that third-party logistics service providers can make and communicate to SAP S/4HANA:

- Scrapping samples taken during receipt
- Warehouse scrap with reason for scrapping (preconfigured scrap code)

- Physical inventory adjustments resulting from physical inventory in their warehouse
- Stock category changes, for example, logistics service provider changing stock category (e.g., blocked to unrestricted, quality to unrestricted, unrestricted to blocked, etc.) in their system and communicating to SAP S/4HANA
- Transfer from storage location to storage location
- Plant-to-plant transfer

Figure 6.25 indicates the communication chain between the organization and its third-party logistics partner system for inventory integration.

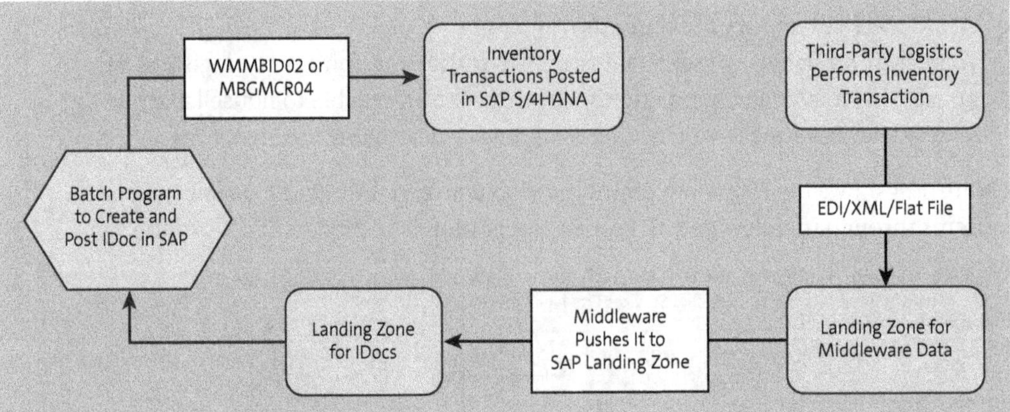

Figure 6.25 Communication or Message Flow for Inventory Management Integration with Third-Party Logistics

Note that information related to any stock status change happening in SAP S/4HANA will be sent to the logistics service provider through middleware or directly through a point-to-point connection. The output type configuration for triggering IDocs has been described in detail in Section 6.3.

While designing such an integration, some aspects related to conversion of data needs to be considered in detail:

- **Plant and storage location conversion**
  Logistics service providers may have their own naming convention in their system for their plants, warehouses, and storage locations. These need to be mapped to the SAP S/4HANA system.
- **Movement types**
  The various types of inventory movements in SAP are classified by movement types. These movement types are unique to a process in SAP S/4HANA. For logistics service providers to identify the different inventory movements, the movement types should be mapped with the logistics service provider system.

- **Unit of measure conversion**
  Logistics service providers may use their own units of measures and may not match exactly with what is configured in SAP S/4HANA.
- **Stock type/stock status identification**
  Stock type and stock status indicators (naming) may be significantly different at the logistics service providers. They will need mapping with SAP S/4HANA.
- **Master data**
  Logistics service providers should have material master, customer master, batch master, and cost center data as in SAP S/4HANA.

All these mappings and rules for conversion can be maintained easily using BRFplus applications. Moreover, BRFplus enables superusers or core user groups to easily maintain rule logic with minimal or no technical intervention.

6.5 Summary

Inventory management integration with third-party logistics addresses the key requirements of up-to-date inventory visibility and inventory tracking. This is a two-way communication between the company and third-party logistics partners for inventory transactions involving stock adjustments, miscellaneous receipts and issues, and transfers, as well as receiving inventory snapshots from third-party logistics on a regular interval for stock analysis and comparison reports in SAP S/4HANA. This includes WMS transactions for physical count and reconciliations. A robust integration and synchronizations address issues such as frequent stock-outs, inaccurate MRP, inventory carrying costs for nonmovable or out-of-date stocks, and so on.

Chapter 7
Warehousing

Warehouse management is the essential link within the supply chain and, if designed and managed well, leads to efficient and organized processes thereby enhancing productivity. Warehousing involves tasks such as receiving, storage, cross-docking, shipping, and monitoring activities. Warehouse operations may also need handling equipment, labor, maintenance, and other overheads. Therefore, operating warehouses require significant involvement in planning storage structures, inventory management, resource planning, and so on. The modern warehouse offers not just storage but many other options such as sophisticated ecommerce operations, high-velocity retail distribution, vendor-managed inventory, and a variety of other value-added services. Any company or business will find it more cost-effective to outsource warehousing operations to a third-party logistics service provider, rather than developing the facilities itself.

Businesses with a need for high-quality and specialized supply chain requirements can take advantage of third-party logistics warehouse operations. The following warehousing requirements may explain the benefits of using third-party logistics resources and facilities:

- Expansive or specialized storage facilities
- Fluctuations in demand
- Exploration of new routes of delivery or customer base
- Proximity to customers or better responsiveness
- Training the workforce and awareness of latest trends in supply chain practices
- Innovation and technology adoption capabilities in areas of robotics, analytics, Internet of Things (IoT), 3D printing, and so on
- Maximizing performance and long-term cost savings

There are two types of warehouses offered by third-party logistics: *contract warehousing* and *public warehousing*. Contract warehousing is mostly suited for large businesses with predictable order volumes. It's more expensive options and dedicated warehouse storage at third-party logistics can be rented for business needs. Third-party logistics service providers may run the operation and provide the labor, equipment, and systems needed. In many cases, contract warehousing may provide better value-added services along with enhanced responses to business needs.

With public warehousing, share or multitenancy models allow businesses to share space along with other "tenants." It's a cheaper option than contract warehousing because the business only needs to pay for the required storage and services. Third-party logistics providers can adjust the space and services per your requirements thus supporting scalability and seasonality of demand. For multitenant warehousing, third-party logistics must be able to differentiate between each client's material and support the same purchase order numbers, sales order numbers, item numbers, and batch numbers of different clients in the same warehouse.

The beginning of this chapter focuses on the deployment options third-party logistics partners may have to opt for best-of-breed WMS solution to manage their warehouse operations. These are either non-SAP or SAP solutions with coverage of WMS functions and analytics, as well as integration requirements with other enterprise resource planning (ERP) solutions. Then, we'll discuss system landscapes for embedded and extended warehouse management (EWM) with an SAP S/4HANA system. Once the scenarios and deployment options are understood, it's imperative to discuss in detail the integration mechanism between SAP S/4HANA and EWM systems and various master data needed for the integration.

7.1 Deployment Scenarios for Third-Party Logistics

SAP S/4HANA supports multiple scenarios of deployment. On one hand, it uses the embedded EWM features for its own warehousing needs, and, on the other, it can seamlessly integrate with any third-party WMS applications. Let's take a look at these integration options.

7.1.1 Warehouse Management Systems

There are many non-SAP, best-of-breed warehouse management systems (WMSs) available in the market with connectivity options through electronic data interchange (EDI), XML, or application programming interfaces (APIs) and thus integrate seamlessly with SAP systems. In this section, we'll walk through system integration options and process flows for third-party logistics with WMSs.

Integration Options

The features available in most of the WMSs are listed here:

- **Inventory management**
  Recording and managing stocks per key characteristics such as warehouse location, available quantity, pallet ID, batch number, lot number, expiration date, and so on.

- **Track and trace**
Real-time tracking of orders, delivery, and shipping status, as well as the capability to trace by SKU, UPC, serial number, batch, lot, package, and part/component.
- **Control inbound and outbound logistics**
Helping provide effective picking patterns that minimize labor costs by reducing overlaps and eliminating repeat runs.

  The systems for third-party logistics must prioritize essential functionality above all else, much like any conventional WMS. This calls for strong cross-docking assistance, the ability to set up dock door appointments, performance evaluation for carriers and suppliers, flow-through monitoring, and the returns processing.

  A robust third-party logistics WMS will incorporate auditing and quality inspections at every stage and help generate audit and quality assurance (QA) documents automatically. WMSs should make your audits easy to view, analyze, and share so you can improve your processes and keep clients informed.
- **Multiple device support**
Support for devices should go beyond desktops and laptops to include all scanners and tracking devices, as well as modern smartphones and tablets. Smartphones are always present on the work floor thanks to bring your own device (BYOD) regulations, and they may now be just as crucial to maintaining control as radio frequency identification (RFID) reader or barcode scanner assistance.
- **Slotting capabilities**
Typically, a single third-party logistics warehouse keeps stocks owned by multiple clients—this increases chances for mis-picks and other mistakes. Therefore, some of the clients of third-party logistics are very specific about the requirement of a slotting module to better manage the warehouse's physical layout according to individual clients. Slotting modules factor in metrics such as pick velocity (how often an item is picked) that helps determine the best layout for warehouse space.
- **Advance billing features**
When offering third-party logistics services to numerous clients, billing becomes a challenging process. The process of manually keeping track of transactions and compiling them into invoices can take days. Such an undertaking leads to wasted time and strained customer relations. Invoicing is prone to errors when there are many transactions to process. Billing errors are a crucial oversight and may impact a client's trust adversely.
- **Integration with the ERP system**
The WMS can integrate and communicate with other platforms via EDI, XML, or APIs. These may also include transport management systems; eCommerce platforms such as Amazon, Shopify, and Magento; and carrier portals (e.g., FedEx, USPS, and UPS) for real-time shipping updates.

- **Labor management**
  A WMS can examine and monitor crucial performance measures, and its implementation could be beneficial for labor management.
- **Reporting and analytics**
  An efficient WMS provides comprehensive and accurate insights into warehouse activities and assists in efficiently tracking inventory in and out. Therefore, it helps warehouse managers make informed decisions to significantly enhance their business practices through real-time analytics and reporting.

These features are also shown in Figure 7.1.

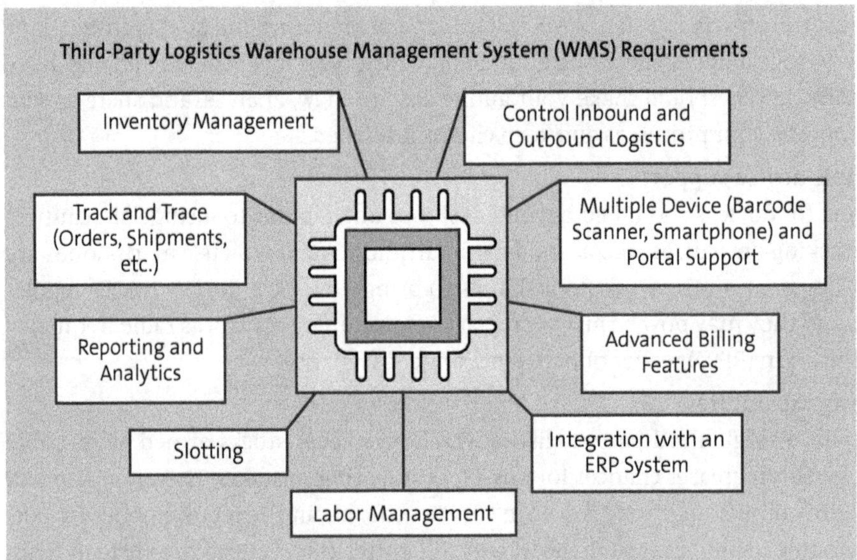

Figure 7.1 Features of WMS Applications

Process Flows

SAP S/4HANA integrates with third-party logistics WMSs using IDocs and APIs in the following ways:

- **Procurement with inbound delivery**
  In this scenario, third-party logistics WMS is notified about impending inbound shipments through outbound IDoc `SHP_IBDLV_SAVE_REPLICA` triggered at the time of inbound delivery creation in the SAP S/4HANA system. The logistics service partner arranges for physical pick-up from the supplier and delivery at the customer location and confirms the same in their WMS. This in turn sends the goods receipt details through IDoc `SHP_IBDLV_CONFIRM_DECENTRAL` to the SAP S/4HANA system. Figure 7.2 shows the end-to-end process flow for integration of the SAP S/4HANA system with the non-SAP WMS.

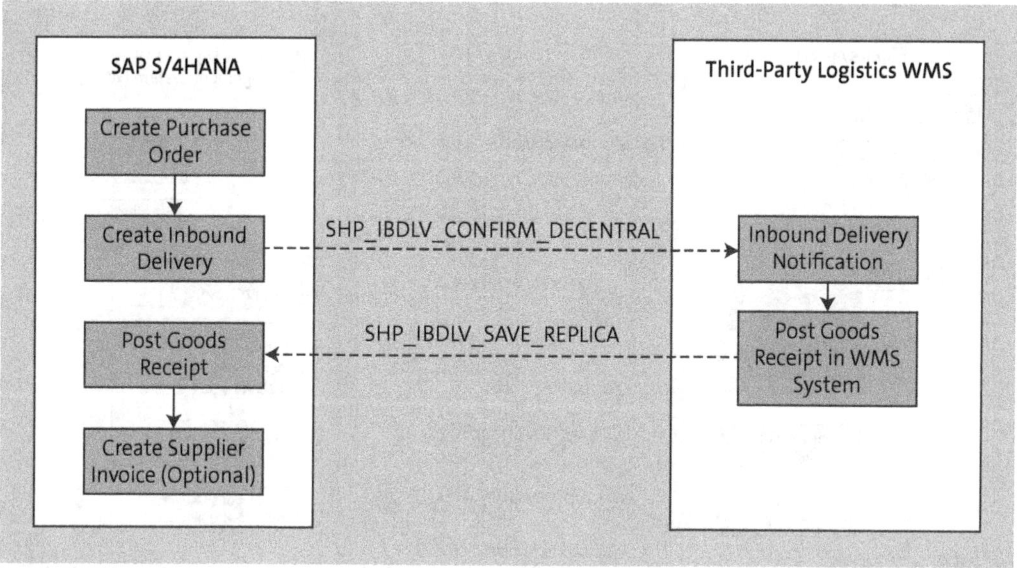

Figure 7.2 Inbound Logistics (Procurement) Scenario Integration Using IDoc Message Types

- **Sales with outbound delivery (API integration)**
  The scenario represents outbound logistics integration with third-party logistics using APIs. This includes a point-to-point connection with the third-party logistics partner system wherein an outbound API is triggered upon creation of the outbound delivery for the sales order. Once received in the third-party logistics WMS, it creates an outbound delivery request followed by a warehouse task for picking. If there is a quantity and batch split in delivery, it sends the information back to the source system. In SAP S/4HANA, the delivery document is changed, and a new additional delivery is created. The information pertaining to the changed and new delivery is communicated to the WMS. Upon shipment confirmation by the third-party logistics in their system, post goods issue information is sent back to the SAP S/4HANA system for confirming the delivery in SAP S/4HANA. The Simple Object Access Protocol (SOAP) API message types are shown in Figure 7.3. SOAP is a standard messaging protocol that operating systems use to communicate via HTTP and XML.
- **Sales with outbound delivery (IDoc integration)**
  This scenario is similar to the outbound shipment scenario with API integration just explained. As shown in Figure 7.4, the difference is in message types as IDocs (instead of APIs). In addition, the split in delivery is shown as optional and hence not represented in detail. The treatment of split delivery will be similar to what is explained in the API integration scenario.

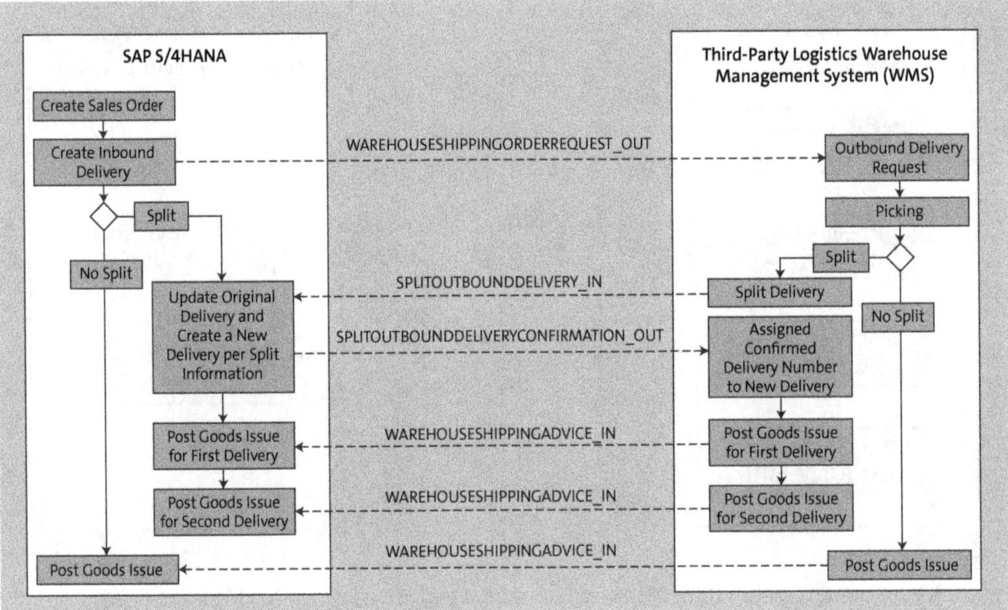

Figure 7.3 Outbound Logistics (Sales) Scenario Integration Using SOAP APIs

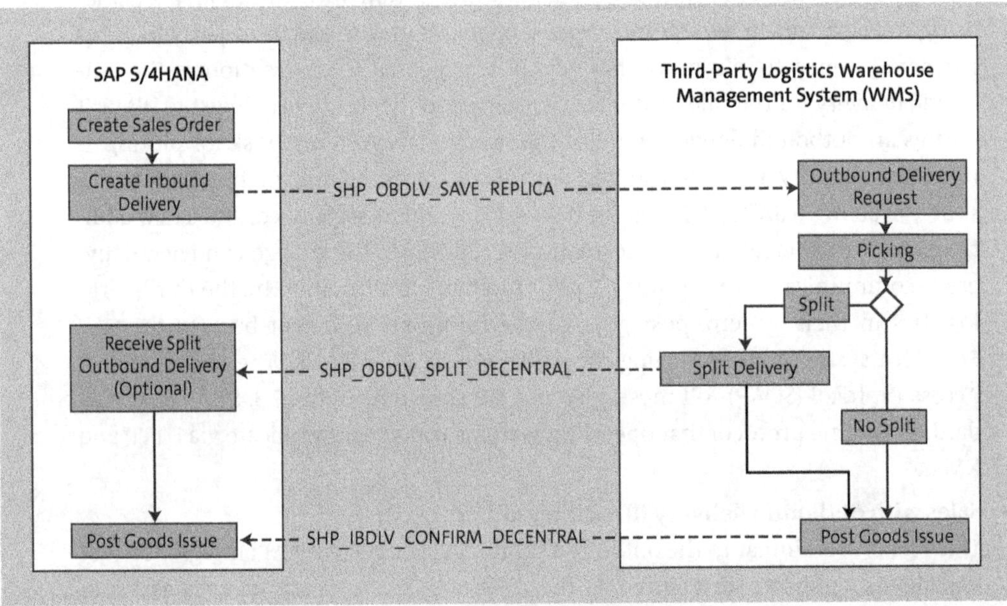

Figure 7.4 Outbound Logistics (Sales) Scenario Integration Using IDocs

- **Two-step stock transfer (IDoc integration)**
  A stock transfer order (STO) involves both outbound and inbound movements. For outbound movement of material from the supplying plant, an outbound delivery is created for the STO. The delivery information, when received at the third-party logistics system, creates an outbound delivery request. As a follow-up warehouse task, picking and post goods issue are performed in the third-party logistics WMS at the time of goods shipment.

  In this scenario, as shown in Figure 7.5, IDoc `SHP_OBDLV_CONFIRM_DECENTRAL` brings in the post goods issue information from the WMS to SAP S/4HANA and processes the confirmation of relevant outbound deliveries. In the stock transfer scenario, post goods issue or confirmation of outbound delivery should automatically create the inbound delivery. In this process, the batch and packing information from the outbound delivery is automatically replicated in the corresponding inbound delivery.

  For cases involving inventory management, the post goods receipt can be done SAP S/4HANA itself. However, for warehouse-managed inventory (managed by the third-party logistics WMS), the inbound delivery is again communicated to the WMS. Upon receipt of material at the destination location or plant, the third-party logistics partner confirms the receipt in the WMS, and the same is communicated to SAP S/4HANA through the IDoc. The rest of the process is self-explanatory.

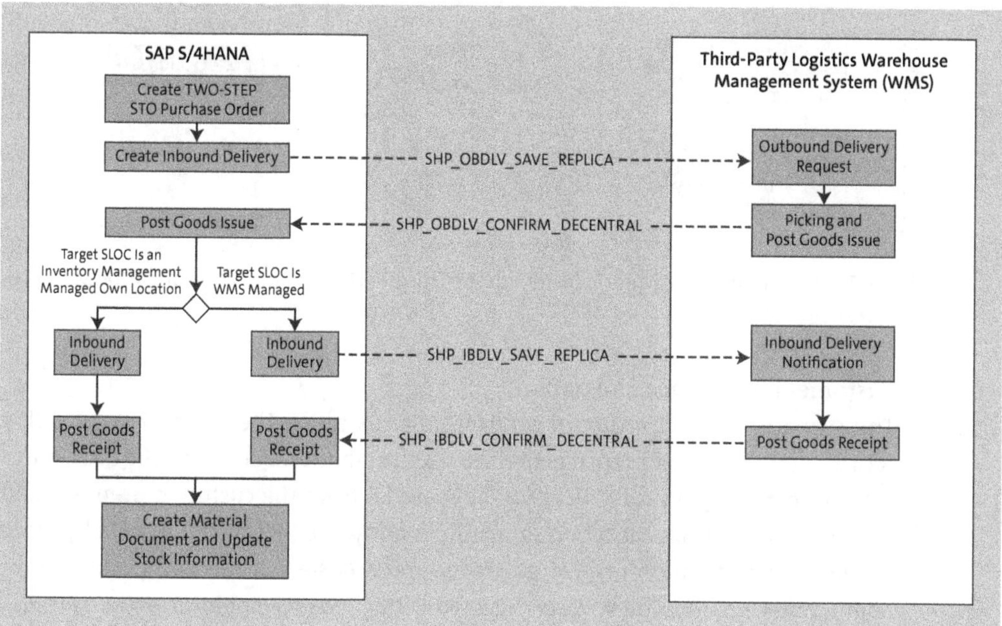

Figure 7.5 Two-Step Stock Transfer Scenario Integration with a Third-Party Logistics WMS Using IDocs

- **Vendor return (IDoc integration)**
  The return to supplier scenario involves creating a return purchase order and then a return delivery against the purchase order. Similar to the scenarios for outbound logistics, the information pertaining to return delivery (delivery back to the supplier) is communicated to the third-party logistics WMS through IDoc SHP_OBDLV_SAVE_REPLICA. This, in turn, creates an outbound delivery request, and the follow-up warehouse task picks and ships to the supplier. Upon actual shipment, the goods issue or post goods issue is posted in WMS for this return delivery, and all the relevant details are communicated back to SAP S/4HANA for auto-confirmation of inbound delivery in SAP S/4HANA. Figure 7.6 shows the end-to-end process flow involving IDoc message types.

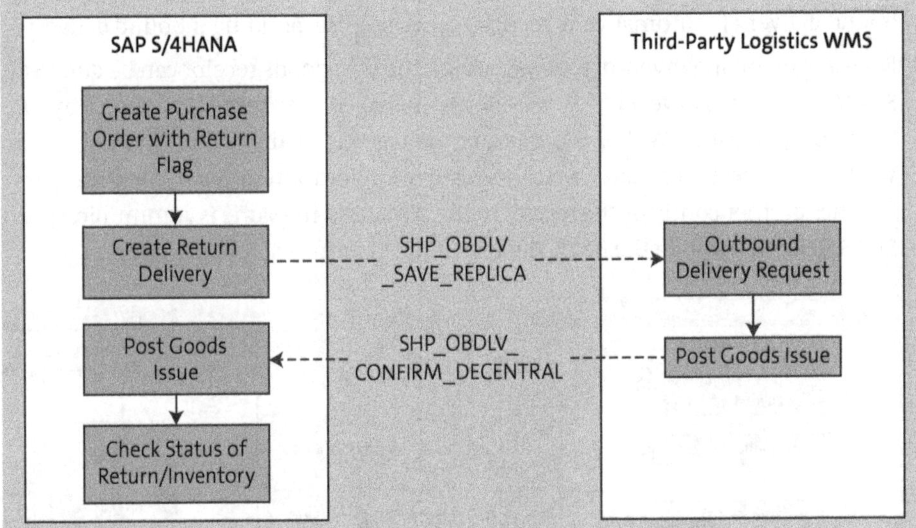

Figure 7.6 Vendor Return Scenario Involving the Third-Party Logistics WMS and IDoc Message Types

- **Customer return (IDoc integration)**
  This scenario involves return of goods from a customer. The scenario starts with a return sales order that in turn generates a return inbound delivery. The third-party logistics partner physically picks up the material from the customer premises and delivers it to the source location or manufacturing site (per the order details). Once delivered, confirmation or post goods receipt is posted in the WMS, and an outbound message from the WMS is triggered with goods receipt information. Receipt of the IDoc message in SAP S/4HANA helps in auto-posting of the goods receipt. Figure 7.7 shows the end-to-end process flow involving IDoc message types.

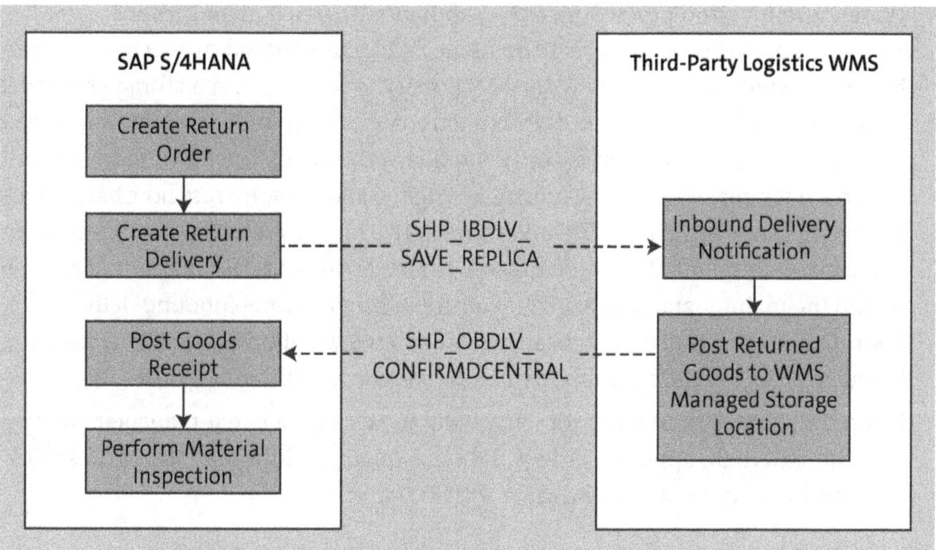

Figure 7.7 Customer Return Scenario Involving Third-Party Logistics WMS and IDoc Message Types

7.1.2 Extended Warehouse Management with SAP S/4HANA

In this section, we'll discuss the integration options and process flows for inbound and outbound movements when using EWM with SAP S/4HANA.

Integration Options

Businesses using their own SAP S/4HANA system may opt to use one of the following on-premise deployment options:

- Embedded EWM on SAP S/4HANA (version 1610 onward)
- Decentralized EWM in SAP S/4HANA (version 1809 FPS 02 onward)

Note

SAP Note 1606493, SAP EWM Deployment Options Best Practices, covers the deployment options for classic EWM and SAP S/4HANA EWM systems.

Process Flows

SAP best practices for EWM recommend the following process flows as standards for inbound and outbound movements:

- **Warehouse inbound processing from supplier with batch management**
  The scope of this process covers the usage of batch-managed products in a warehouse inbound process. In SAP S/4HANA, stock is managed on a storage bin level using warehouse management (WM) functionality. When ordering batch-managed items for procurement from external suppliers, the goods use batch management with the example of a batch—country of origin—and demonstrate how batch characteristics can be used to create and characterize batches received from a supplier. The batch management–specific process characteristics are the creation of a batch ID and the maintenance of a batch's country of origin during inbound delivery creation. Once characterized, batches are stored in the warehouse, and batch selection based on characteristics is supported in the warehouse outbound processing.

  Figure 7.8 shows inbound logistics involving the receipt of batch-managed material from an external supplier to the warehouse-managed location. The process flow involves integration of SAP S/4HANA with EWM.

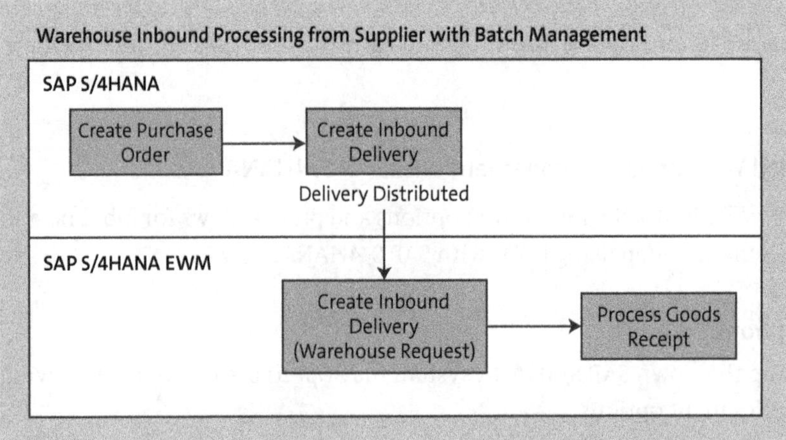

Figure 7.8 Process Flow for Receipt from External Vendors/Suppliers Involving Batch Management

- **Basic warehouse inbound processing from supplier with quality management**
  This process allows basic inbound processing using the WM functionality seamlessly integrated with quality management in SAP S/4HANA. The process starts with the creation of a purchase order and an EWM inbound delivery. Upon arrival of the purchased goods, they are unloaded and packed, and the goods receipt is posted. Upon goods receipt posting, the system creates an inspection lot for the inbound delivery items if they are inspection relevant.

  Upon creation of the inspection lot, the system updates the EWM inbound delivery by adding the inspection lot number to the document flow, and the stock is posted to stock type quality inspection stock. The warehouse agent triggers the creation of warehouse tasks to move the goods to the quality inspection center for a physical inspection.

The inspection lot is now available in the quality inspection worklist, and upon inspection of the delivered goods, inspection results are recorded. The quality department makes a usage decision based on the result recorded, and stock is either fully or partially accepted or rejected.

After the decision, handling units (HUs) with either goods to be stored or goods to be moved to the blocked stock area are created in the system. The creation of warehouse tasks to move the HUs to the corresponding area in the warehouse is triggered automatically. The process ends with the confirmation of the warehouse tasks.

The basic inbound warehouse process with quality management integration is shown in Figure 7.9. The process involves receipt of material into the warehouse, quality inspection, and movement of accepted and rejected stocks to the appropriate storage locations.

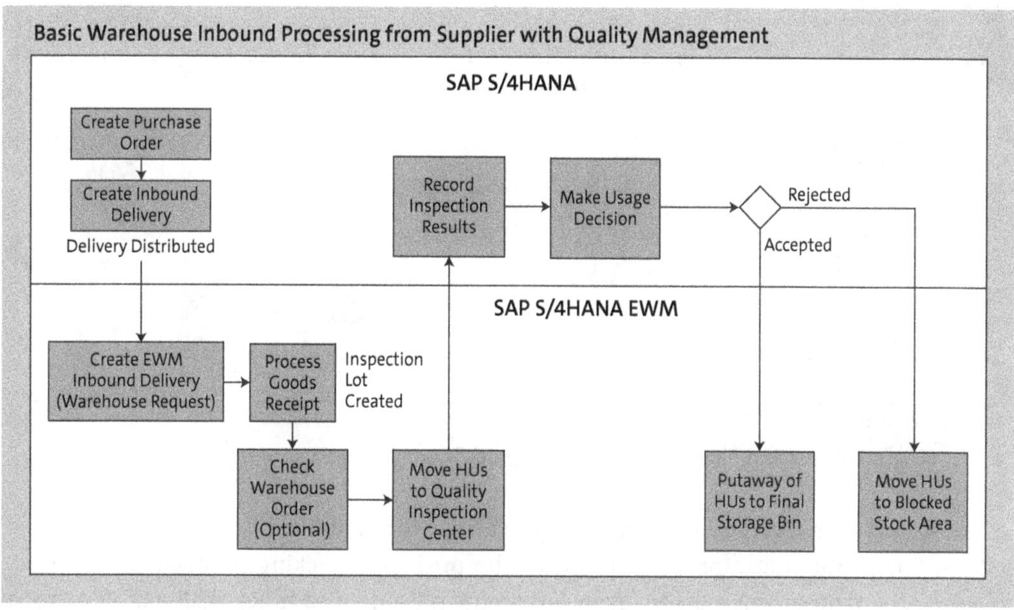

Figure 7.9 Process Flow for Receipt from External Vendors/Suppliers Involving Quality Management

SAP S/4HANA EWM supports other inbound delivery scenarios, including returns from the customer, goods from production, and goods received from expected goods receipt processing.

- **Basic warehouse outbound processing to customer**
  The scope of this process covers shipping ordered goods to external customers. This process usually starts with the creation of a sales order and an outbound delivery,

which is later used to trigger the outbound processing in the warehouse. During the creation of warehouse tasks, the system determines the location for picking the stock and allocates the stock to the warehouse task. At the same time, the system automatically makes sure that all goods for a customer are routed to the same place at the packing work center, where shipping HU labels and content lists are printed. After goods are loaded, the goods issue is posted. At the end of the process, follow-up actions (e.g., the creation of the billing document) are executed.

Figure 7.10 shows basic outbound logistics operations for shipment to customers. The end-to-end process integrates operations across SAP S/4HANA and SAP S/4HANA EWM platforms.

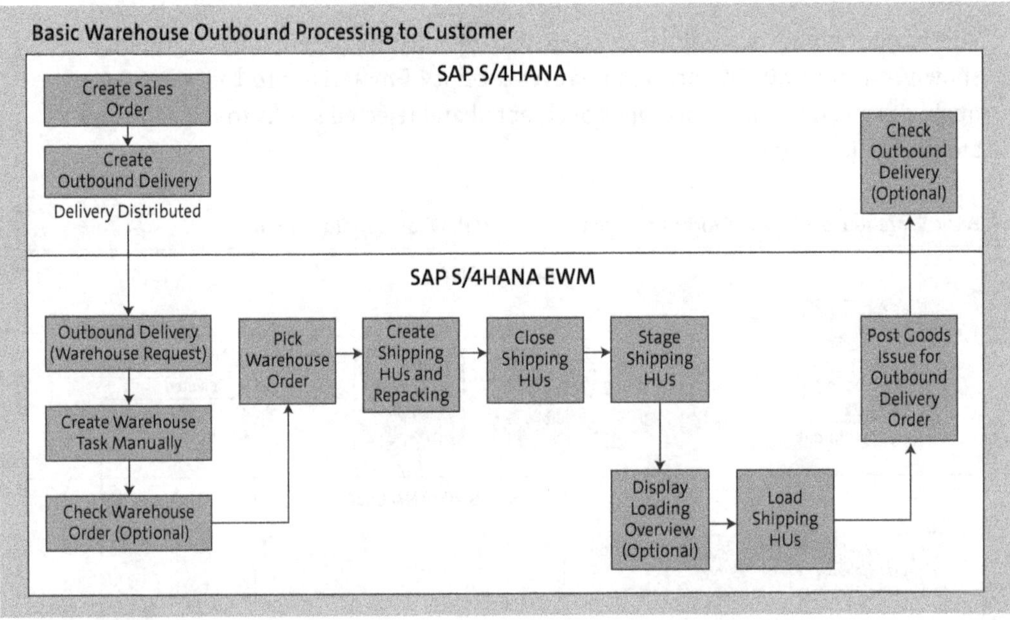

Figure 7.10 Process Flow for Shipping Ordered Goods to the Customer

- **Advanced warehouse outbound processing to customer**
  This warehouse process optimizes the process of picking products and sending them to external customers by leveraging the extended functionality of WM in SAP S/4HANA. It extends the basic warehouse inbound process with wave management and transportation unit (TU) handling.

  The process usually starts with the creation of a sales order and an outbound delivery, which is later used to trigger the outbound processing in the warehouse. Based on defined criteria, the system automatically combines items from the outbound warehouse requests in waves. The warehouse clerk (EWM) releases the picking waves by using the warehouse monitor. The system determines the location for picking the stock and allocates the stock to the warehouse task. At the same time, the system automatically makes sure that all goods for a customer are routed to the same place at the packing work center, where shipping HU labels and content lists are printed. The Shipping Cockpit is used to create TUs to check them in and to assign the TUs to a door. After goods are loaded, the system posts goods issue for all stock that was loaded on the TUs. At the end of the process, follow-up actions (e.g., the creation of the billing document) are executed.

Figure 7.11 shows EWM integration with SAP S/4HANA and the usage of advanced EWM functions such as pick waves and TUs.

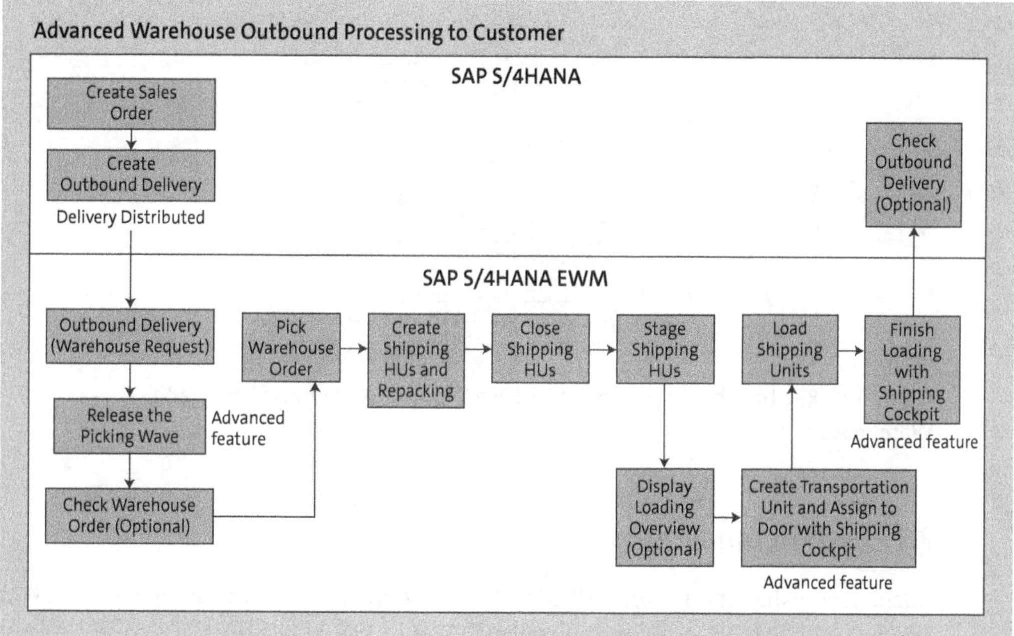

Figure 7.11 Advanced Outbound Warehouse Functions with Pick Waves and TUs

- **Warehouse request**
  The inbound or outbound delivery document is also called a warehouse request, and all subsequent actions in the warehouse are based on that document. Warehouse activities include picking, putaway, posting changes, stock transfers (within the warehouse), and scrapping. These are performed using warehouse tasks, and these documents are created with reference to the respective warehouse request. Using the warehouse management monitor, you can display and supervise all documents that are based on a warehouse request.
- **Warehouse task**
  Warehouse tasks are used to execute good movements in the warehouse, and these movements may be physical or just stock changes. Warehouse tasks are needed for picking, putaway, internal movements, posting changes, good receipt postings, and

goods issue postings. For each warehouse request item, the warehouse task is created either manually or automatically by the Post Processing Framework (PPF). When the warehouse task is executed, the task must be confirmed upon completion.

- **Warehouse order**
  Warehouse tasks are grouped (or bundled) in a warehouse order. The warehouse order represents an executable work package that a warehouse employee should complete within a specific time frame. It contains one or more warehouse tasks or physical inventory items. Figure 7.12 depicts how the warehouse request, warehouse task, and warehouse order are interrelated.

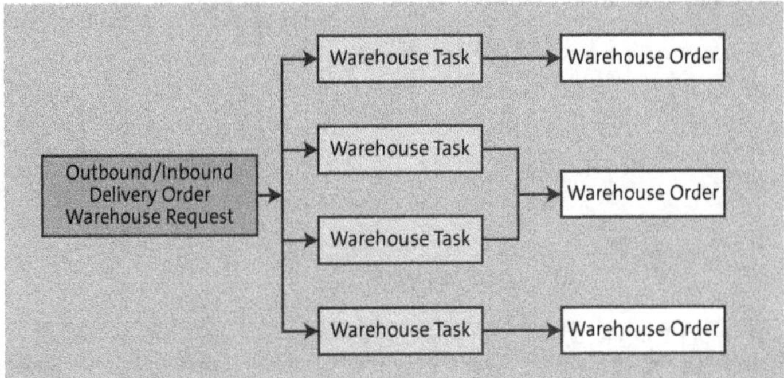

Figure 7.12 Relationship between Warehouse Request, Warehouse Tasks, and Warehouse Orders

7.2 System Landscapes

Businesses using their own SAP S/4HANA system may opt to either embed EWM within SAP S/4HANA or use decentralized EWM in SAP S/4HANA. We'll cover the system landscape of each option in the following sections.

7.2.1 Embedded Extended Warehouse Management in SAP S/4HANA

Extended warehouse management was integrated into SAP S/4HANA Enterprise Management (on-premise edition 1610) in 2016. Extended warehouse management, also referred to as SAP S/4HANA EWM, officially is part of SAP S/4HANA Supply Chain.

Embedded EWM is ideal for small- and medium-scale warehouses. The solution is best used for warehouses located in the same country or on-premise locations of SAP S/4HANA. Warehouses that operate in many continents or countries might not fit the bill because such operations require faster response times and handling large data volumes.

Embedded EWM comes with two license versions. Customers have the choice to select the one based on their application requirements. The options available with each license are as follows:

- **Basic warehouse management**
  This is included in the SAP S/4HANA ERP license. It covers basic warehousing processes and functions: inventory management, inbound and outbound processing, internal warehouse movements, physical inventory, and reporting.
- **Extended warehouse management**
  This covers more advanced and additional warehousing functions and provides process optimization features, for example, slotting, wave management, TU handling, transportation management (TM) integration, dock appointment scheduling, labor management, value-added services, kitting, cross-docking, warehouse billing, cationization planning, distribution equipment, just-in-time (JIT), and material flow system (MFS) integration.

Some of the key benefits with embedded EWM in SAP S/4HANA include the following:

- EWM is easy to deploy and reduces total cost of ownership (TCO).
- Data is simplified, and data redundancy and master data duplications are eliminated. For example, business partner data, material master data, and batch master data can be directly accessed from SAP S/4HANA, thus eliminating the need for data replication.
- Redundancy in Customizing and transactions is avoided.
- Built-in integration with the ERP system results in transactional data, such as purchase order data or production order data, can be read directly.
- The use of additional documents is eliminated, such as outbound delivery documents.
- SAP S/4HANA and SAP S/4HANA EWM run on the same landscape.

There are a couple of limitations as well:

- Embedded EWM can only integrate to local ERP storage locations in the same system and can't be integrated with storage locations coming from another ERP system.
- Embedded EWM can't run if the SAP S/4HANA system is down.

7.2.2 Decentralized Extended Warehouse Management in SAP S/4HANA

The decentralized EWM deployment option was introduced with SAP S/4HANA version 1809, FPS 01. EWM is deployed on a separate server and can be connected to SAP S/4HANA or any other ERP system. Decentralized EWM is highly scalable, and it can support high data volume and transactional loads without impacting the system performance of other processes. EWM system upgrades can be planned without any dependency on the ERP system. The EWM system can remain operational even if the ERP system is down.

Transfer of master data (material, batches, business partners) from the ERP system to decentralized EWM in SAP S/4HANA is carried out using IDocs or web services. This is a change from when the Core Interface (CIF) was used to transfer master data from the ERP to EWM systems based on SAP NetWeaver or SAP Supply Chain Management (SAP SCM).

Decentralized EWM can be activated using menu path **SPRO • SAP Reference IMG • SCM Extended Warehouse Management • Extended Warehouse Management • Enable Decentralized EWM**, as shown in Figure 7.13.

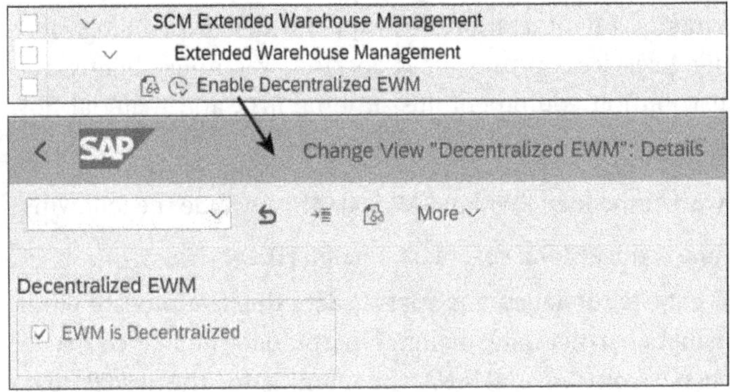

Figure 7.13 Enabling Decentralized EWM in SAP S/4HANA

> **Note**
>
> SAP has provided guidelines for decentralized EWM in SAP Note 2840129, Release Information and Restrictions of Decentralized EWM in SAP S/4HANA 1909.
>
> Following are other relevant SAP Notes:
>
> - 2806070 - SAP S/4HANA 1909: Release Information and Restrictions for EWM in SAP S/4HANA
> - 2911449 - Updated Maintenance Timelines for SAP EWM Product Versions
> - 2782080 - Overview of Guides or Best Practices Documents for Extended Warehouse Management

Some of the key benefits with extended EWM in SAP S/4HANA include the following:

- Highly scalable and supports high data volume and transactions.
- Can connect to multiple ERP systems
- Can be upgraded independently to an ERP system
- Can remain operational even if the ERP system is down

A limitation for extended EWM is that it has a higher TCO.

Figure 7.14 is a schematic representation of the difference between embedded and extended EWM systems.

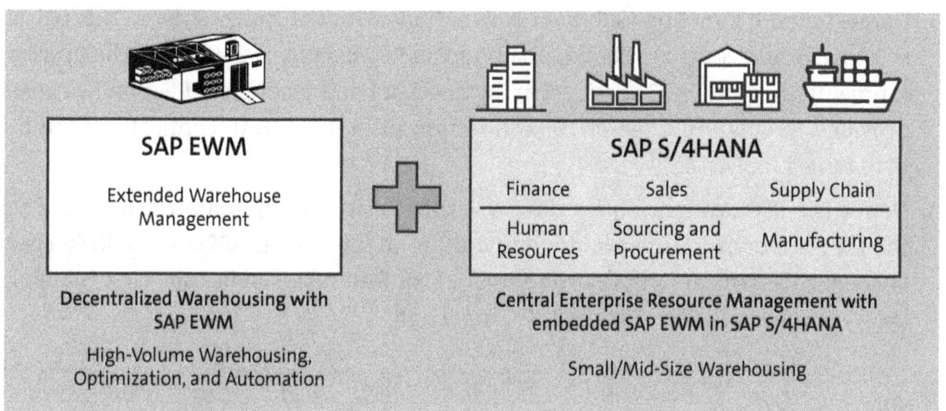

Figure 7.14 Extended EWM versus Embedded EWM in SAP S/4HANA

7.3 Integration between SAP S/4HANA and SAP Extended Warehouse Management

In this section, we'll cover the essentials of integration between SAP S/4HANA and EWM, which can be broadly classified into three categories: system landscape, organizational structure, and process-specific integration. Most of the settings used for decentralized EWM are common for embedded EWM as well.

7.3.1 System Landscape Integration between SAP S/4HANA and SAP Extended Warehouse Management

Basic integration settings are essential to establish a connection between the SAP S/4HANA system and the EWM, as we'll cover in the following sections. A key step in linking an SAP S/4HANA system and an EWM system is configuring the necessary Application Link Enabling (ALE) settings. However, most of these steps aren't really EWM specific; they are required in general for the communication between two SAP systems.

Settings in SAP S/4HANA Systems

Follow these steps to configure the settings in your SAP S/4HANA system:

1. Define logical systems for both SAP S/4HANA and EWM systems in SAP S/4HANA systems using configuration menu path **SPRO • SAP Reference IMG • ABAP Platform • Application Server • IDoc Interface/Application Link Enabling (ALE) • Basic Settings • Logical Systems • Define Logical Systems.**
2. Assign the logistics system just created to the client using IMG menu path **SPRO • SAP Reference IMG • ABAP Platform • Application Server • IDoc Interface/Application Link Enabling (ALE) • Basic Settings • Logical Systems • Assign Logical System to Client.**

3. Define Remote Function Call (RFC) destinations with the same name as the logical systems. In addition, you need to maintain an RFC user in each system. The process for maintaining the RFC destination for the client and testing connection has been provided in detail in Chapter 10. Connection of type 3 (ABAP connection) can be maintained for EWM.
4. Define the RFC destination for the EWM client using configuration maintained in the previous step. The relevant configuration menu path is **SPRO • SAP Reference IMG • ABAP Platform • Application Server • IDoc Interface/Application Link Enabling (ALE) • Communication • Create RFC Connection.**

Note

Keep the following points in mind during these steps:

- Define RFC destinations with the same name as the logical systems.
- Define an RFC user in each target system/client before this step.
- RFC connection details can be viewed in table RFCDES.

Figure 7.15 presents screenshots of the following configurations, along with configuration menu paths in a single frame.

Maintaining logical systems ❶ includes the following:

1. Assigning the logical system to the client ❷
2. Adding the RFC destination ❸ for the EWM system

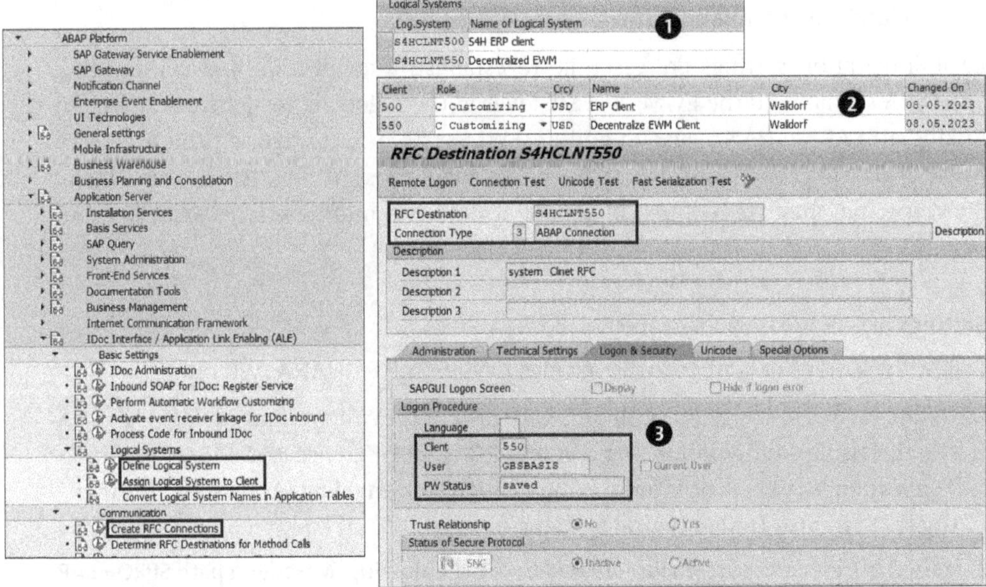

Figure 7.15 Assigning the Logical System to the Client, and Adding the RFC Destination for EWM

Settings in Extended Warehouse Management Systems

Follow these steps to configure the settings in your EWM system, as shown in Figure 7.16:

❶ Define a logical system in EWM using configuration menu path **SPRO • SAP Reference IMG • Integration with Other SAP Components • Extended Warehouse Management • Basic Settings for Setting Up System Landscape • Name Logical System.**

❷ Assign the logical system for SAP S/4HANA and EWM systems to the client being used in each server via menu path **SPRO • SAP Reference IMG • Integration with Other SAP Components • Extended Warehouse Management • Basic Settings for Setting Up System Landscape • Assign Logical System to a Client.**

❸ Define **RFC Destination** with the same name as the logical systems (in our example, "S4HCLNT500" and "S4HCLNT550").

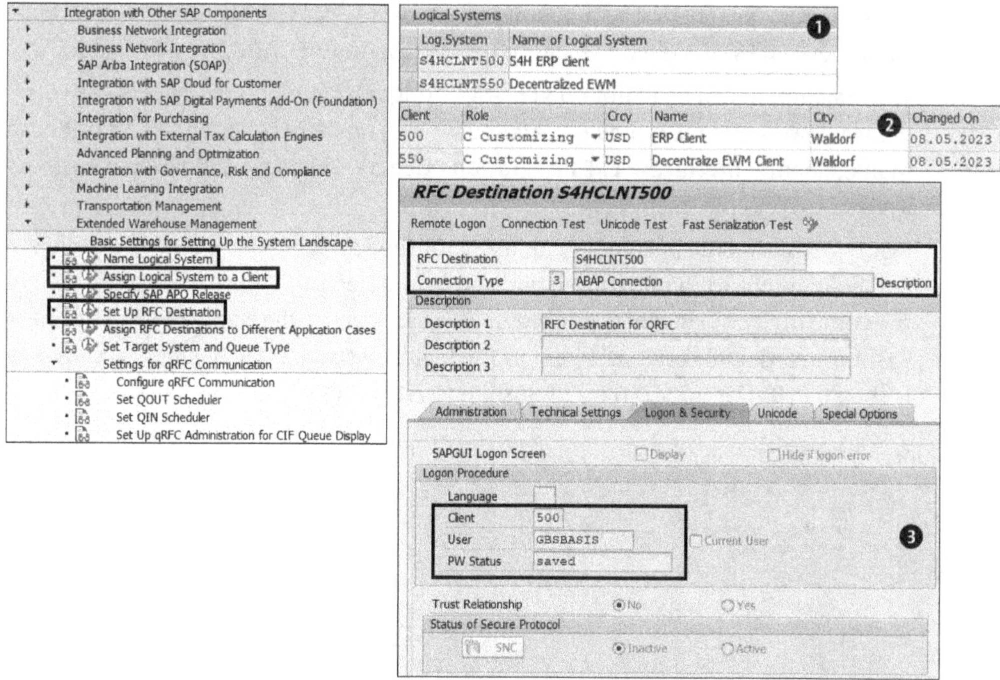

Figure 7.16 Defining the Logical System, Assigning to the Client in EWM, and Defining the RFC Destination for the ERP System

Next, we'll focus on organization-level integration between SAP S/4HANA and EWM. This involves logistics organizational entities such as plants, storage locations, and warehouses within each system. Follow these steps:

1. **Maintain business systems and assignments.**
   Business systems for EWM (company's own business system) and for the ERP systems are maintained in this step. The EWM system stores the business system in the

document flow tables of the cross-system and cross-partner document flow. The business system is a system key that you define in the system landscape directory (SLD).

In addition, business systems need to be mapped to their corresponding logical systems. This step is important for the communication of the SAP S/4HANA and EWM because the delivery documents sent from the SAP S/4HANA system to the EWM system contain only the logical system number of the SAP S/4HANA system. Logical systems need mapping based on business systems to differentiate between deliveries originating from different ERP clients or systems. Therefore, it provides a connection between logical systems and business systems.

In addition, while sending confirmations from EWM to the SAP S/4HANA system, EWM uses the logical system that originally contained the delivery. EWM therefore maps the business system onto the logical system. This means that only one logical system must be assigned to a business system, so that the assignment is unique. Only then can EWM send messages to the corresponding SAP S/4HANA system.

The assignment can be maintained using configuration menu path **SPRO • SAP Reference IMG • SCM Extended Warehouse Management • Extended Warehouse Management • Interfaces • ERP Integration • General Settings • Define Own Business System**.

Figure 7.17 shows the business system key for the SAP S/4HANA-based EWM system and assignment to the logical system, as maintained in EWM. The maintenance of the company's own business system in EWM appears in the **Name of Own Business System** area ❶, and the assignment of the logical system to the business system key is shown in **Buffer for SLD Data of Business Systems** ❷.

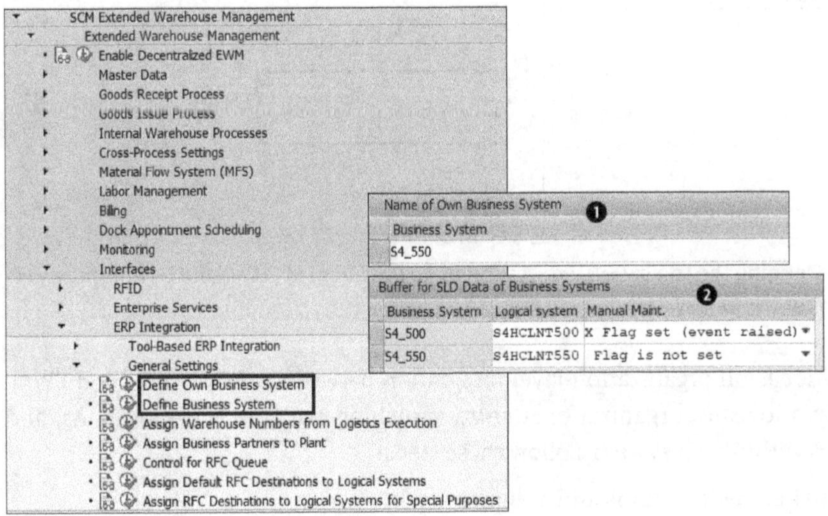

Figure 7.17 Defining Business Systems for EWM, and Assigning Logical Systems to Business System Keys

EWM stores the business system in the warehouse request document. Mapping of the logical system to the business system helps to determine the business system based on the logical system so that the warehouse request document in EWM can be updated with the corresponding business system. Figure 7.18 shows the business system information in the warehouse request document.

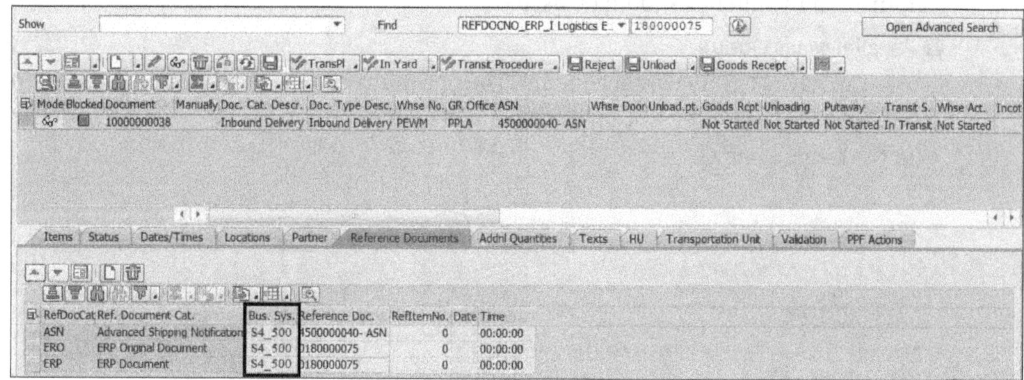

Figure 7.18 Business System Key in a Warehouse Request

2. **Define control for the RFC queue.**

   Queued RFC (qRFC) is used as the communication technology because the delivery-related messages must be serialized. Strict serialization is needed to ensure that the messages are sent to SAP S/4HANA in the correct order. This means that data inconsistencies can be avoided. The following types of messages are sent from EWM to SAP S/4HANA:

   Messages with reference to a delivery in SAP S/4HANA (delivery interface):

   - Reporting delivery changes (e.g., quantity changes)
   - Reporting goods movements for the delivery (goods receipt or goods issue)

   Messages without reference to a delivery in SAP S/4HANA (goods movement interface):

   - Reporting stock postings without reference to a delivery
   - Posting changes without reference to a delivery
   - Reporting inventory differences
   - Scrapping without a delivery in SAP S/4HANA

In this Customizing activity, you define the control of the RFC message queues sent from EWM to SAP S/4HANA: EWM uses RFC-enabled function modules to report data to SAP S/4HANA. The configuration menu path for defining control for the RFC queue is **SPRO • SAP Reference IMG • SCM Extended Warehouse Management • Extended Warehouse Management • Interfaces • ERP Integration • General Settings • Control for RFC Queue.** The following points are shown in Figure 7.19:

❶ **Business System**
Define a business system name for each target system.

❷ **Queue Type**
Define the queues that are used for the inbound/outbound communication between SAP S/4HANA and EWM. For the communication to an SAP S/4HANA system, inbound queues should be used.

❸ **Parallel Mass Queues**
You can have 1 to 999 parallel queues. The optimum number depends on the system resources of the target system. Each parallel queue can occupy its own work process in the target system.

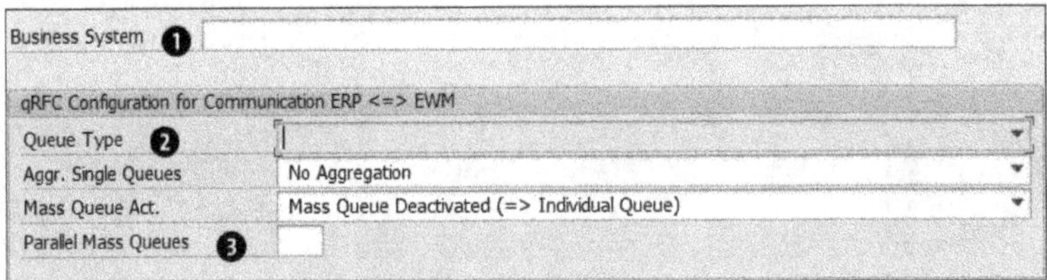

Figure 7.19 Control Parameters for qRFC

Figure 7.20 is a schematic representation of document information exchange between SAP S/4HANA and EWM using qRFC communication.

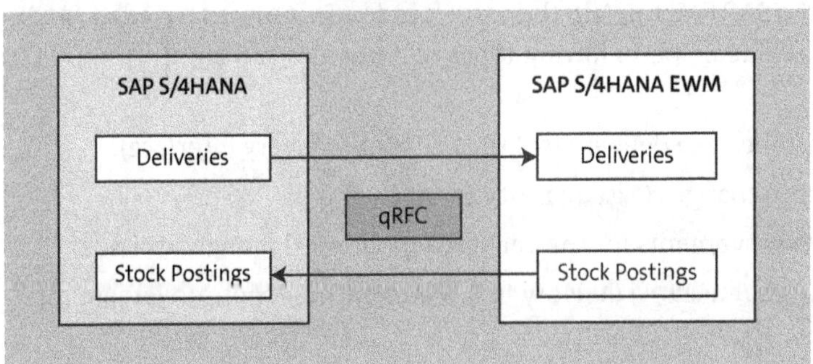

Figure 7.20 Replication of Documents between SAP S/4HANA and SAP S/4HANA EWM Systems

7.3.2 Organizational Structure Integration

Organizational structure integration between SAP S/4HANA and EWM involves configuring logical systems on both sides and assigning them to respective clients. Figure 7.21 is a snapshot of configuration activities on the SAP S/4HANA and EWM sides.

We'll walk through these activities in the following sections.

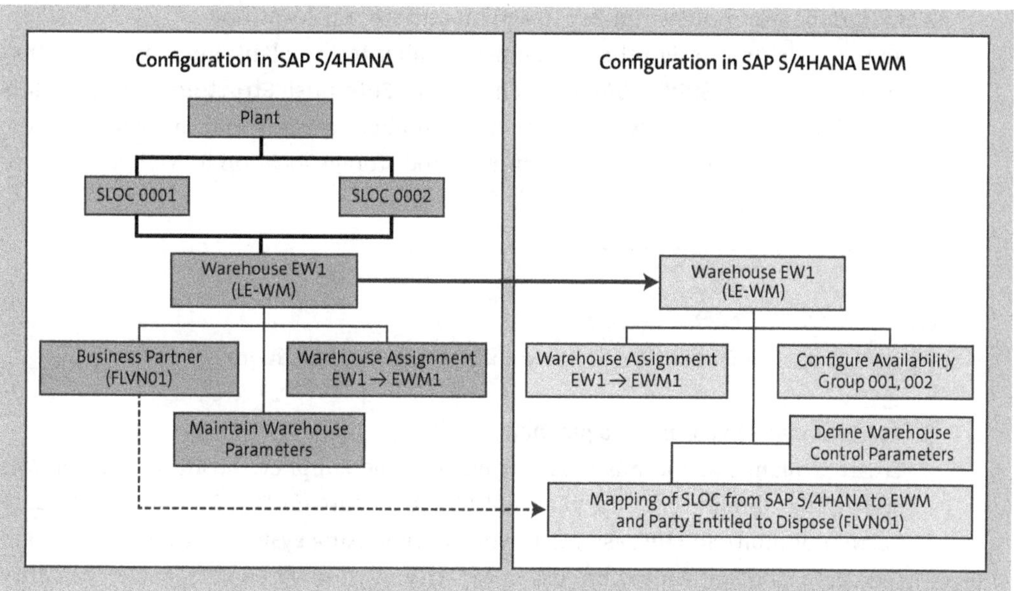

Figure 7.21 Configuration Activities in SAP S/4HANA and SAP S/4HANA EWM

Configuration Settings in SAP S/4HANA

In SAP S/4HANA, the traditional logistics execution/inventory management/WM settings are required for linkage with EWM. This section covers the configuration setup in SAP S/4HANA. An intermediate warehouse number is created and assigned to the combination of plant and storage locations for which EWM is required. This warehouse number is the logical representation of the EWM warehouse in the SAP S/4HANA system.

Follow these configuration steps:

1. **Define the warehouse number in logistics execution.**
   The warehouse number in logistics execution is created using menu path **SPRO • SAP Reference IMG • Enterprise Structure • Definition • Logistics Execution • Define Warehouse Number**. Figure 7.22 shows a warehouse number and its description in WM under logistics execution.

W...	Warehouse No. Description
EW1	EWM Warehouse

Figure 7.22 Defining the Warehouse Number in Logistics Execution

> **Note**
> Don't create an SAP S/4HANA warehouse by copying the template warehouse, as it doesn't require all the structural elements of an SAP S/4HANA warehouse.

2. **Assign the warehouse number to a plant and storage location.**
   As the next step, assign the warehouse number to the plant and storage location using menu path **SPRO • SAP Reference IMG • Enterprise Structure • Assignment • Logistics Execution • Assign Warehouse Number to Plant and Storage Location.** Figure 7.23 shows the assignment of plant and storage location to warehouse number.

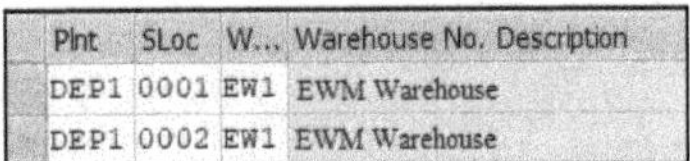

Plnt	SLoc	W...	Warehouse No. Description
DEP1	0001	EW1	EWM Warehouse
DEP1	0002	EW1	EWM Warehouse

Figure 7.23 Assignment of Warehouse Number in Logistics Execution

3. **Assign a plant to a business partner.**
   Create a plant-specific business partner of type supplier (business partner role FLVN01) that should act as the party entitled to dispose (PETD). EWM allows the connection of multiple ERP systems to your warehousing system without the need to segregate stocks, meaning no loss of capacity in your warehouse. The PETD entity allows stock from one system to sit in a bin next to stock from another system. PETD also drives picking of the correct stock to ensure there are no mistakes.

   Creation of PETD is done via Transaction BP and business partner role FLVN01. This is a master data setup for a vendor with a plant assignment. Figure 7.24 shows the **Vendor: General Data** tab reflecting plant assignment.

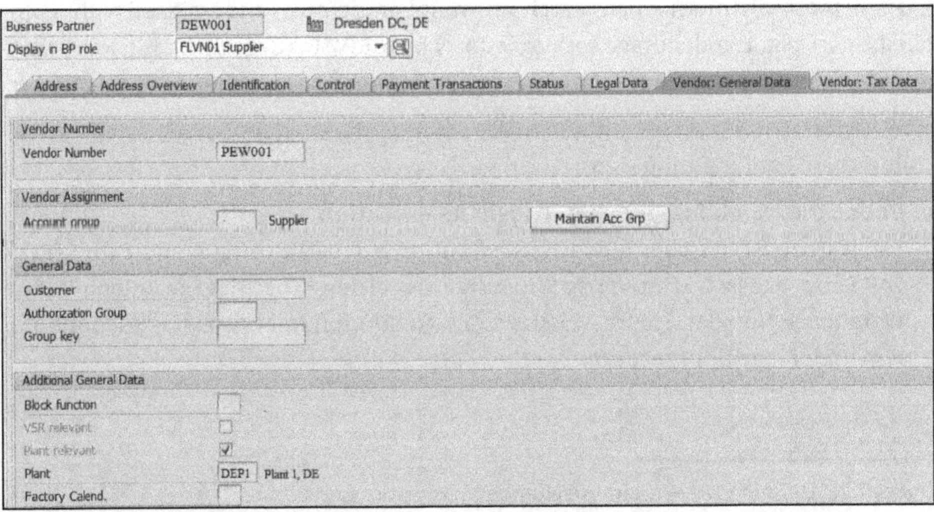

Figure 7.24 Creating a Business Partner and Assigning a Plant to a Business Partner

4. **Maintaining EWM-specific parameters for an SAP S/4HANA warehouse.**
   The WMS in SAP S/4HANA is activated by assigning the plant/storage location to the warehouse number. If this warehouse number is set to SAP S/4HANA EWM, this ensures that the system manages the warehouse stock with an SAP S/4HANA EWM

system. In addition, you can define the kind of communication with the EWM system and activate the distribution of unchecked deliveries to the warehouse. The parameters for an EWM warehouse can be set using configuration menu path **SPRO • SAP Reference IMG • Integration with Other SAP Components • Extended Warehouse Management • Basic Settings for EWM Linkage • Configure SAP EWM-Specific Parameters.**

As shown in Figure 7.25, during the activation of the SAP S/4HANA warehouse as the EWM warehouse, the following settings can be chosen:

- **Dist. Mode** (distribution mode): The setting controls if a delivery is distributed immediately at creation (default entry) or not (stop distribution). If the distribution isn't done immediately, you trigger the distribution in the outbound delivery monitor (for distribution) or with Transaction VL06D.
- **UD** (unchecked deliveries): This setting controls if unchecked deliveries are to be distributed. An unchecked delivery is a delivery with a reduced scope of check. This reduced scope of check refers mainly to the availability checks but can also affect other checks. Customer relationship management (CRM) can create unchecked deliveries in SAP S/4HANA.
- **BatchDetEW** (batch determination in EWM via batch attribute replication): This indicator controls the behavior of the automatic batch determination in SAP S/4HANA outbound deliveries.
- **GR fr. EWM** (goods receipt from EWM only): This indicator enables you to control whether the process of posting the goods receipt for a manufacturing order into an EWM-managed warehouse can be started from both SAP S/4HANA and EWM or from EWM only.
- **Comm. WM**: The correct entry for EWM is **Queued and Serialized Asynchronous RFC**.

SN Dec. WM (allow serial numbers in decentralized WMS): Once this indicator is set, it's not changeable anymore. It means that you can work with serial numbers in EWM, and the details are controlled by the serial number profiles.

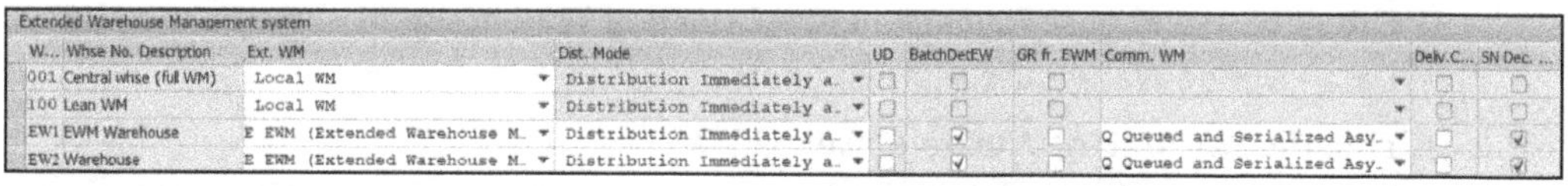

Extended Warehouse Management system

W...	Whse No. Description	Ext. WM	Dist. Mode	UD	BatchDetEW	GR fr. EWM	Comm. WM	Delv.C...	SN Dec. ...
001	Central whse (full WM)	Local WM	Distribution Immediately a...	☐	☐	☐		☐	☐
100	Lean WM	Local WM	Distribution Immediately a...	☐	☐	☐		☐	☐
EW1	EWM Warehouse	E EWM (Extended Warehouse M...	Distribution Immediately a...	☐	☑	☐	Q Queued and Serialized Asy...	☐	☑
EW2	Warehouse	E EWM (Extended Warehouse M...	Distribution Immediately a...	☐	☑	☐	Q Queued and Serialized Asy...	☐	☑

Figure 7.25 Maintaining EWM-Specific Parameters for ERP Warehouse Number

5. **Link the SAP S/4HANA warehouse to the EWM warehouse.**
   In this configuration activity, the warehouse number of the decentralized EWM system is assigned to the warehouse number of the SAP S/4HANA system. This assignment is necessary to select the materials required in EWM during distribution of the material master data. The previously mentioned assignment can be done using con-

figuration menu path **SPRO • SAP Reference IMG • Integration with Other SAP Components • Extended Warehouse Management • Assign Warehouse Number to EWM Warehouse Number**. As shown in Figure 7.26, the warehouse number from WM in SAP S/4HANA is assigned to the decentralized EWM warehouse number.

Assign Warehouse No. to Warehouse No. of Decentr. SCM Sys

W...	Whse No. Description	WHN.DecSCM
EW1	EWM Warehouse	EWM1

Figure 7.26 Linking a Warehouse Management Warehouse Number to the EWM Warehouse

Configuration Settings in Extended Warehouse Management

The following settings in SAP S/4HANA deal with defining the EWM warehouse, establishing linkage with the SAP S/4HANA warehouse and settings related to the availability group, and so on:

1. **Define the EWM warehouse number.**
   The EWM warehouse is maintained using configuration menu path **SPRO • SAP Reference IMG • SCM Extended Warehouse Management • Extended Warehouse Management • Master Data • Define Warehouse Number**. The resulting screen where you define the number is shown in Figure 7.27.

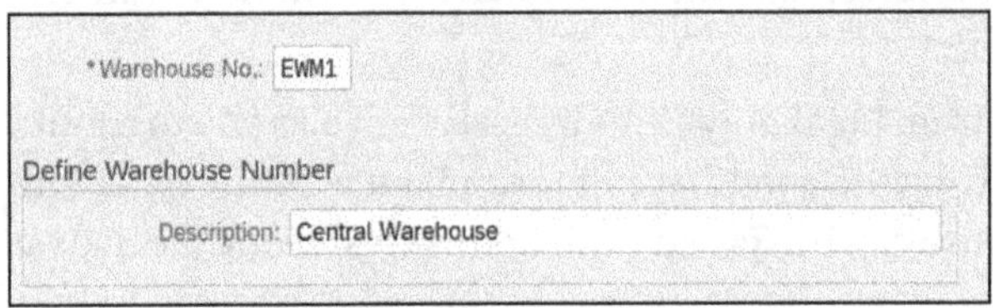

Figure 7.27 Defining the EWM Warehouse Number

2. **Define the warehouse control parameter.**
   In this configuration step, the basic data items for a warehouse number—the **Weight Unit, Volume Unit, Time Unit,** and **Unit of Length**—are defined, as shown in Figure 7.28. To define these settings, use menu path **SPRO • SAP Reference IMG • SCM Extended Warehouse Management • Extended Warehouse Management • Master Data • Define Warehouse Number Control.**
3. **Link the SAP S/4HANA warehouse to EWM.**
   In this Customizing activity, the warehouse numbers from SAP S/4HANA are mapped to the warehouse numbers in EWM. You can assign more than one SAP S/4HANA warehouse number to each warehouse number in EWM. While using decentralized EWM based on SAP S/4HANA, you can also enter the business system. Entry of a business system is needed for exception cases, for example, where an SAP S/4HANA warehouse number maps to a different EWM warehouse number if the data from SAP S/4HANA arrives from a particular business system.

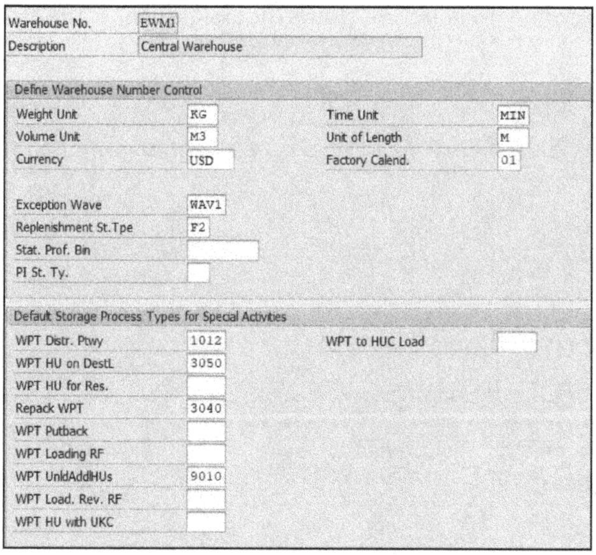

Warehouse No.	EWM1		
Description	Central Warehouse		
Define Warehouse Number Control			
Weight Unit	KG	Time Unit	MIN
Volume Unit	M3	Unit of Length	M
Currency	USD	Factory Calend.	01
Exception Wave	WAV1		
Replenishment St.Tpe	F2		
Stat. Prof. Bin			
PI St. Ty.			
Default Storage Process Types for Special Activities			
WPT Distr. Ptwy	1012	WPT to HUC Load	
WPT HU on DestL	3050		
WPT HU for Res.			
Repack WPT	3040		
WPT Putback			
WPT Loading RF			
WPT UnldAddlHUs	9010		
WPT Load. Rev. RF			
WPT HU with UKC			

Figure 7.28 Basic Settings for the EWM Warehouse Number

Prerequisites

The prerequisites for this configuration step are as follows:

- Warehouse numbers are maintained for the EWM system.
- Business system is defined for the SAP S/4HANA system.

The required mapping is done using configuration menu path **SPRO • SAP Reference IMG • SCM Extended Warehouse Management • Extended Warehouse Management • Interfaces • ERP Integration • General Settings • Assign Warehouse Number from Logistics Execution**. Figure 7.29 shows the assignment of the SAP S/4HANA **Business System** and warehouse number to the EWM warehouse number.

Mapping for Warehouse Number

Business System	Whse No. ERP	Warehouse Number
S4H_010	EW1	EWM1

Figure 7.29 Linking the SAP S/4HANA Warehouse to the EWM Warehouse

4. **Assign the PETD to the SAP S/4HANA plant.**
   In this Customizing activity, you maintain the link between a PETD (disposal party) and a plant. This links the plant to a business partner that is only used to manage stock in the warehouse. This business partner can differ from the business partner in the vendor role or the customer role that is assigned to a plant and used as the ship-to party or ship-from party in the stock transfer order processes. It can also differ from the business partner used as the default ship-to party of the warehouse in direct outbound delivery orders. For embedded EWM in SAP S/4HANA, assignment

of a disposal party to a plant is optional. Figure 7.30 shows the assignment of the disposal party to the plant.

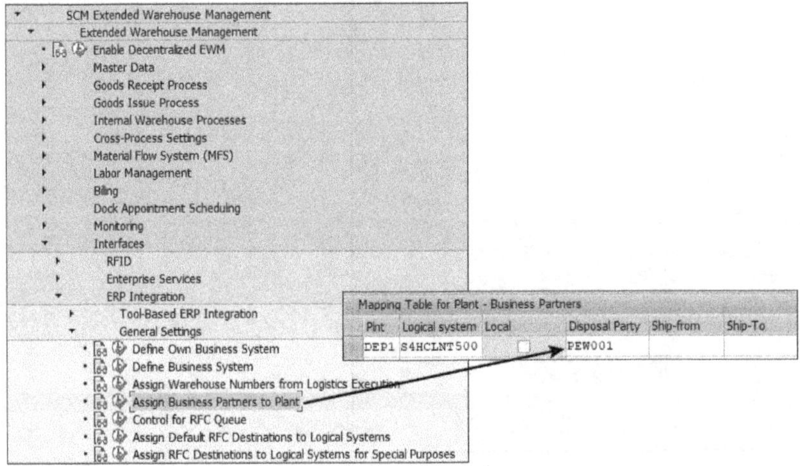

Figure 7.30 Assignment of the Disposal Party to the Plant in SAP S/4HANA

Prerequisites

For embedded EWM in SAP S/4HANA, you must have assigned a logical system to the client. For decentralized EWM, you must have defined the logical system for the SAP S/4HANA system with active inventory management and stock management. It's the system in which the plant exists. For both deployment options, you must have defined the business partner.

Note

Create an entry for each plant in the productive system. You can't transport the entries.

5. **Map a storage location from SAP S/4HANA to EWM.**
   In this Customizing activity, you map storage locations onto the warehouse numbers in EWM and assign an availability group to each storage location. Create an entry for each plant and storage location assigned to the warehouse. If a business partner has already been assigned to the plant, the system displays a PETD.

   From an organizational structure perspective, storage locations don't exist in the EWM system. An availability group is used to map the SAP S/4HANA storage locations, which in turn is linked to the EWM stock type and SAP S/4HANA stock type (location-independent stock type). This is maintained in the availability group settings.

   The configuration menu path for mapping the SAP S/4HANA storage location to EWM is **SPRO • SAP Reference IMG • SCM Extended Warehouse Management •**

Extended Warehouse Management • Interfaces • ERP Integration • Goods Movement • Map Storage Locations from ERP System to EWM. Figure 7.31 shows the assignment of the plant and storage location combination to the EWM warehouse and availability group.

Customizing Mapping Table for ERP Plant Storage Location

Plnt	SLoc	Logical system	Warehou...	Avail.Grp	Description	Disposal	Desc. Party Ent. to Dispose
DEP1	0001	ERPCLNT010	EWM1	001	Goods in Putaway	PEW001	Ontario DG DC, CA
DEP1	0002	ERPCLNT010	EWM1	002	Goods Completely Available	PEW001	Ontario DG DC, CA

Figure 7.31 SAP S/4HANA Storage Location Mapping to EWM with the Availability Group Assignment

6. **Configure the availability group for putaway.**
   This Customizing step is used to define the availability groups of the warehouse and their relationship to the stock types and, optionally, to the storage types. The availability group is an attribute of a stock type and represents a storage location linked to a warehouse.

Let's consider an example: Assume there is a need to keep stock in the goods receipt zone, separate from stock in the final bins in terms of ATP availability. This can be done by maintaining two availability groups, each representing one storage location. Each availability group needs to be assigned to a set of stock types. In this way, the goods receipt zone will handle stock in stock types of one availability group. In final storage types, stock is managed in stock types of the other availability group. Upon confirmation of putaway warehouse tasks, the system automatically posts the stock from the stock type of one availability group to the corresponding stock type of the other availability group.

The configuration menu path for this step is **SPRO • SAP Reference IMG • SCM Extended Warehouse Management • Extended Warehouse Management • Goods Receipt Process • Configure Availability Group for Putaway.**

Prerequisites

The prerequisites for configuring this step are as follows:

- In Customizing for the Logistics Inventory Management Engine (LIME), the stock type is already defined. This can be done using Transaction /N/LIME/86000068.
- Assignment of availability groups to storage locations in Customizing for EWM appears under **Map Storage Locations from ERP System to EWM.**

Sequential configuration steps are indicated in Figure 7.32. These configuration activities include the following:

❶ Under **Define Availability Group**, define an availability group for each storage location that needs to be linked to the warehouse.

❷ Under **Define Non-Location-Dependent Stock Type**, create the following predefined, location-independent stock types corresponding to the stock types in inventory management:

- **BB Blocked Stock**
- **FF Unrestricted-Use Stock**
- **QQ Stock in Quality Inspection**
- **RR Returns Blocked Stock** (if you use this stock type in your warehouse)

❸ Under **Configure Stock Type**, create a set of stock types for each availability group. Assign a location-independent stock type and an availability group to each stock type.

❹ Under **Assign Availability Group to Storage Type**, assign an availability group to a storage type and specify whether the availability group is mandatory in the storage type.

Figure 7.32 Configure the Availability Group for Putaway

Figure 7.33 depicts a typical EWM warehouse structure. Businesses may have the flexibility to use either WM in SAP S/4HANA or EWM (embedded or decentralized) in SAP S/4HANA. Depending on the system selected, appropriate customization may need to be adopted. For clients using WM in SAP S/4HANA, WM-specific setups will be required.

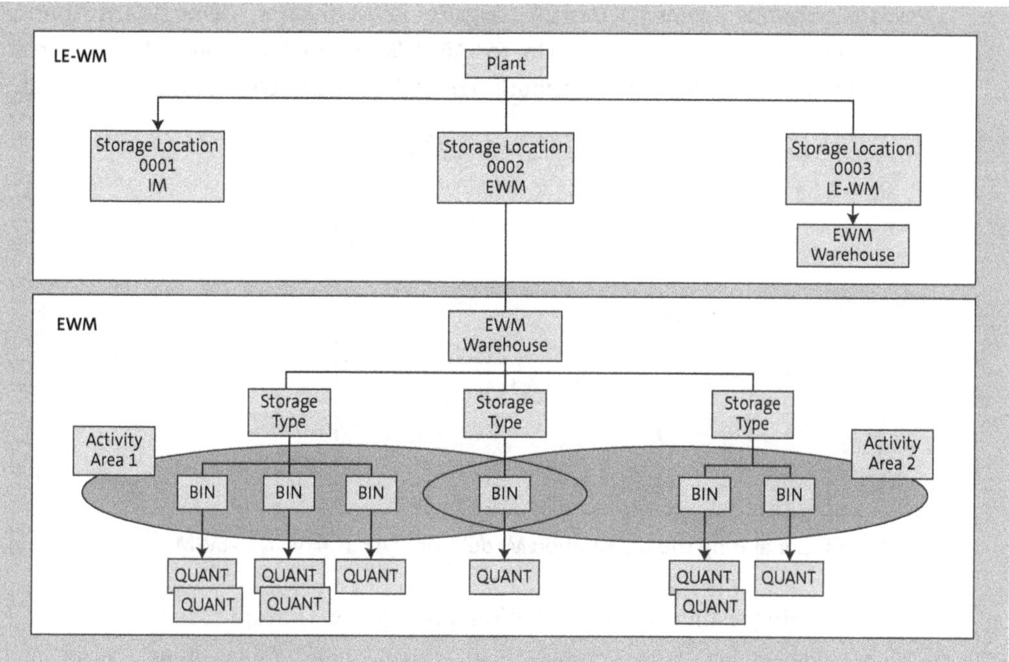

Figure 7.33 EWM Warehouse Organizational Structure

7.3.3 Business Process Integration

So far, we've looked at how integration occurs between SAP S/4HANA and EWM systems, how organizational structures are set up and linked, and how parameters specific to organizational structures and control are maintained. Next comes the business process integration through linkages between deliveries in SAP S/4HANA and EWM, which can be configured using the steps laid out in the following sections.

Configuration Settings in SAP S/4HANA

Take the following steps to configure business process integration settings in your SAP S/4HANA system:

1. **Generate a distribution model for transaction data transfer to EWM.**
   In this Customizing activity, you generate the distribution model that determines how objects are transferred from a central SAP S/4HANA system to one or more EWM systems by qRFC. These objects include inbound deliveries, outbound deliveries, and production material requests. Business processes in SAP S/4HANA are integrated with EWM by means of delivery documents. The business objects for integration include inbound delivery, outbound delivery, and production material request.

   Generation of the distribution model is carried out using configuration menu path **SPRO • SAP Reference IMG • Integration with Other SAP Components • Extended**

Warehouse Management • Basic Settings for EWM Linkage • Generate Distribution Model for Transaction Data Transfer to SAP EWM. Figure 7.34 shows selection screen inputs while generating the distribution model using this configuration menu path.

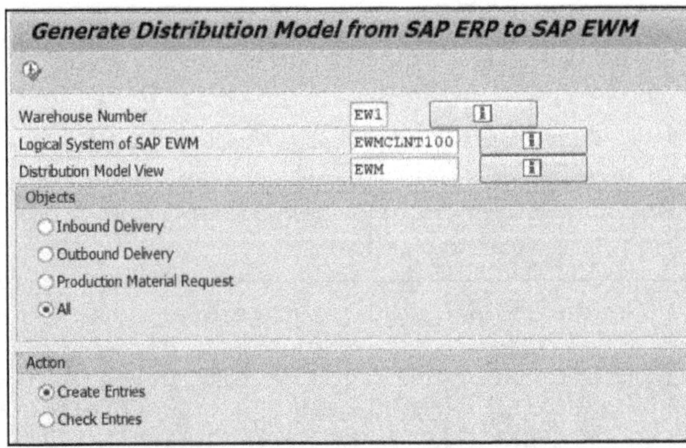

Figure 7.34 Generating the Distribution Model from SAP S/4HANA to EWM

The following results of the distribution model generation are possible:

- Selecting the **Outbound Delivery** option to generate the distribution model will create `OutboundDeliverySPE` objects.
- Selecting the **Inbound Delivery** option to generate the distribution model will create `InboundDelivery` objects.
- Selecting the **Production Material Request** option to generate the distribution model will create `ManufacturingOrder` objects.

These objects are shown in Table 7.1, along with the corresponding methods.

Model View	Sender	Object Type	Method	Receiver	Warehouse Number
EWM	ERPCLNT010	BUS2015	SAVEREPLICA	EWMCLNT100	EW1
EWM	ERPCLNT010	BUS2015	DELIVERYCHANGE	EWMCLNT100	EW1
EWM	ERPCLNT010	/SPE/LIKP	SAVEREPLICA	EWMCLNT100	EW1
EWM	ERPCLNT010	/SPE/LIKP	DELIVERYCHANGE	EWMCLNT100	EW1
EWM	ERPCLNT010	/SPE/LIKP	REQUESTCHANGES	EWMCLNT100	EW1
EWM	ERPCLNT010	BUS2048	STAGEEWMPARTS	EWMCLNT100	EW1

Table 7.1 Customizing Distribution Model SAP S/4HANA to EWM

2. **Linking an SAP S/4HANA delivery to an EWM document (inbound delivery).**
   In this IMG activity, you can set various attributes depending on the inbound delivery type.

The configuration menu path for setting up the attributes of the inbound delivery and assigning to EWM delivery types is **SPRO • SAP Reference IMG • Logistics Execution • Extended Warehouse Management Integration • Extended Inbound Delivery Processing • Settings for Inbound Delivery Processing** and **Determine Delivery Type for Manufacturing-Related Inbound Deliveries**, as shown in Figure 7.35.

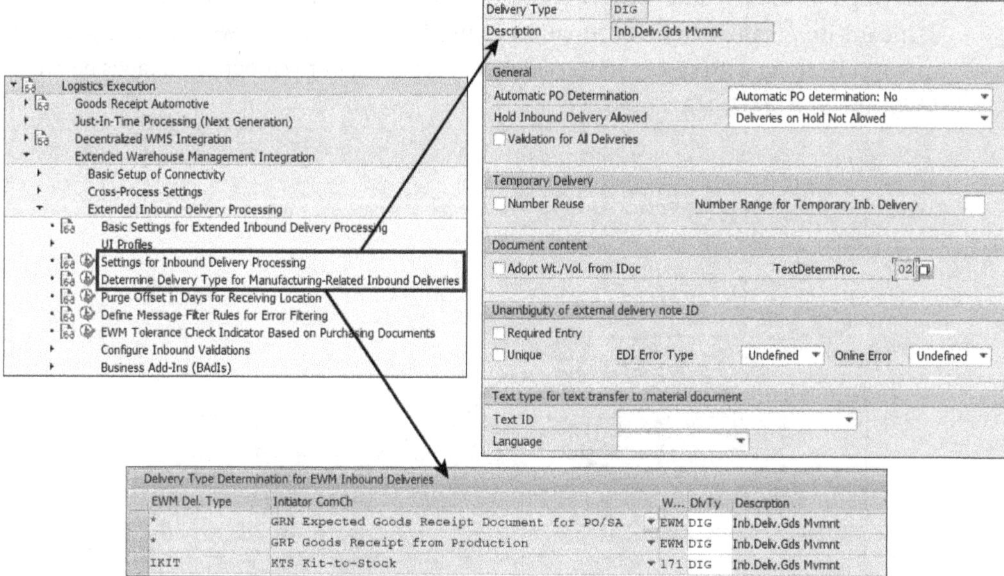

Figure 7.35 Linking an SAP S/4HANA Delivery to an EWM Document (Inbound Delivery)

3. **Linking an SAP S/4HANA delivery to an EWM outbound delivery.**
   In this IMG activity, you define which delivery types are used for direct outbound deliveries in SAP S/4HANA. The delivery type is mandatory for creating the direct outbound delivery in SAP S/4HANA if the delivery is directed to an external customer. For stock transfer deliveries, this activity is optional. The relevant configuration menu path is **SPRO • SAP Reference IMG • Logistics Execution • Extended Warehouse Management Integration • Outbound Process • Direct Outbound Deliveries • Determine Delivery Types for Direct Outbound Deliveries.**

Additionally, delivery EWM item categories can also be linked to SAP S/4HANA item categories using menu path **SPRO • SAP Reference IMG • Logistics Execution • Extended Warehouse Management Integration • Outbound Process • Direct Outbound Deliveries • Determine Item Categories for Direct Outbound Deliveries**. In this IMG activity, you define the item categories that must be used for the items of a direct outbound delivery. By means of the item category, you can, for example, control which movement type is used for the related inventory posting. The item category determination is also used for additional packaging items that are added to replenishment deliveries or to outbound deliveries to customers.

Usually, the SAP S/4HANA item category is determined based on the delivery type, the item category group from the material, and the item usage (see the standard item category determination). The current IMG activity provides the possibility to overrule the standard item category determination and to determine the item category dependent on the business processes for which direct outbound deliveries are used.

Carrying out the current activity is optional. If you don't make any entries here, the standard item category determination is used. Figure 7.36 shows the configuration menu path along with screenshots of delivery type and item category linkages between EWM and SAP S/4HANA systems.

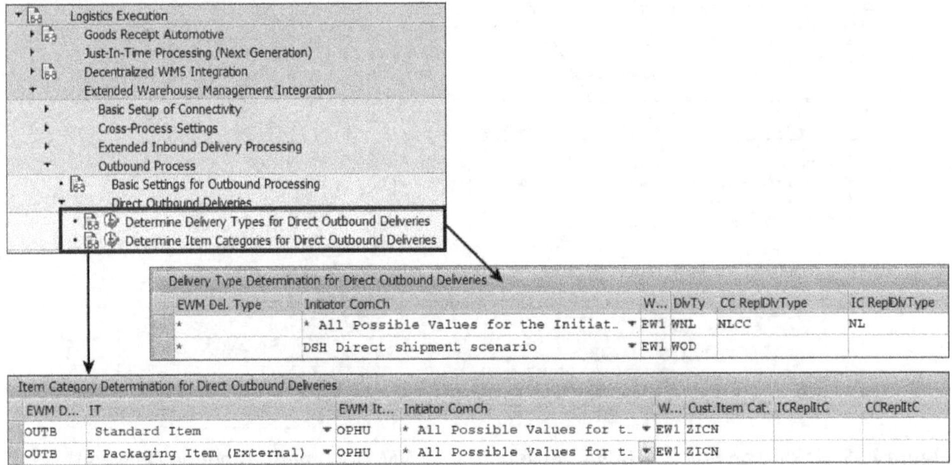

Figure 7.36 Delivery Type (Outbound Delivery) Determination in EWM

Configuration Settings in Extended Warehouse Management

Let's now look at few configuration settings in EWM pertaining to mapping of delivery documents:

1. **Map document types from the SAP S/4HANA system to EWM.**
   In this IMG activity, you map the document types from the SAP S/4HANA to the document types for the warehouse request. This enables the following:
   - Matching of the keys for SAP S/4HANA document types with the keys for the document types for the warehouse request
   - Assignment of an SAP S/4HANA document type to various document types for the warehouse request (e.g., to have more differentiated control over your processes in the warehouse)
   - Assignment of more than one SAP S/4HANA document type to a document type for the warehouse request
   - Mapping of document types from the SAP S/4HANA system to EWM, as well as in the other direction

The mapping logic for EWM delivery types is shown in Figure 7.37. The business system plays an important role if more than one SAP S/4HANA system is connected to decentralized EWM. The code for initiator of a communication chain is an additional key, which helps to distinguish different processes.

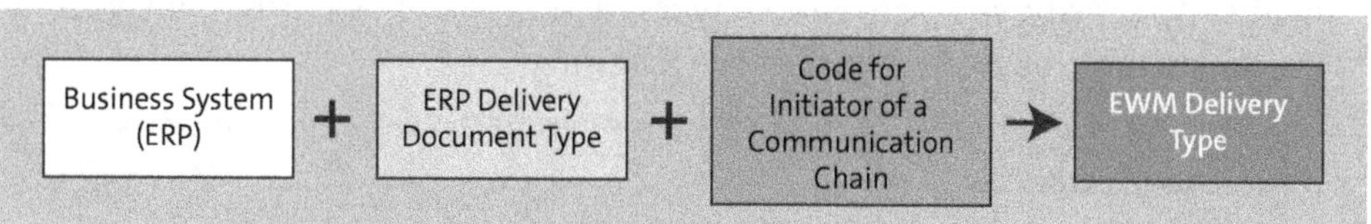

Figure 7.37 Mapping Logic for EWM Delivery Types

The configuration menu path for mapping EWM delivery types is **SPRO • SAP Reference IMG • SCM Extended Warehouse Management • Interfaces • ERP Integration • Delivery Processing • Map Document Types from ERP System to EWM**. As shown in Figure 7.38, the **Cde Init. C** (code for initiator of a communication chain) column acts as a differentiator to bring in further granularity in mapping.

- SCM Extended Warehouse Management
  - Extended Warehouse Management
    - Enable Decentralized EWM
    - Master Data
    - Goods Receipt Process
    - Goods Issue Process
    - Internal Warehouse Processes
    - Cross-Process Settings
    - Material Flow System (MFS)
    - Labor Management
    - Billing
    - Dock Appointment Scheduling
    - Monitoring
    - Interfaces
      - RFID
      - Enterprise Services
      - ERP Integration
        - Tool-Based ERP Integration
        - General Settings
        - Delivery Processing
          - Define Number Ranges for LE Deliveries
          - Define ERP Document Types for Differentiation Attribute
          - Map Document Types from ERP System to EWM
          - Map Item Types from ERP System to EWM
          - Map Date Types from ERP System to EWM

Mapping Delivery Type - Document Type

Business System	DocTypeERP	Cde Init. C...	Doc. Type
	DIG		DIG2
	DIG	GRP	INBI
	DIG	PMD	DIG5
	DIG	RNP	INBM
	DOG		DOG3
	DOG	PMD	DOG5
	DOG	PSD	DOG1
	DOG	TPI	DOG2

Figure 7.38 Configuration for Mapping an EWM Delivery Document Type to an SAP S/4HANA Delivery Type

In a decentralized EWM system, the EWM delivery document type would be an inbound delivery notification, an outbound delivery request, or a posting order request.

2. **Map item types from the SAP S/4HANA system to EWM**
   In this IMG activity, you map the item types from SAP S/4HANA to the item types for the warehouse request. This enables the following:
   - Matching of keys for SAP S/4HANA item types to the keys for item types for the warehouse order.
   - Assignment of an SAP S/4HANA item type to various item types for the warehouse order (e.g., to control process in your warehouse in a more differentiated manner).
   - Assignment of more than one SAP S/4HANA item type to an item type for the warehouse request.
   - Mapping of item types from the SAP S/4HANA system to EWM, as well as in the other direction.
   - Mapping logic for delivery item types is shown in Figure 7.39.

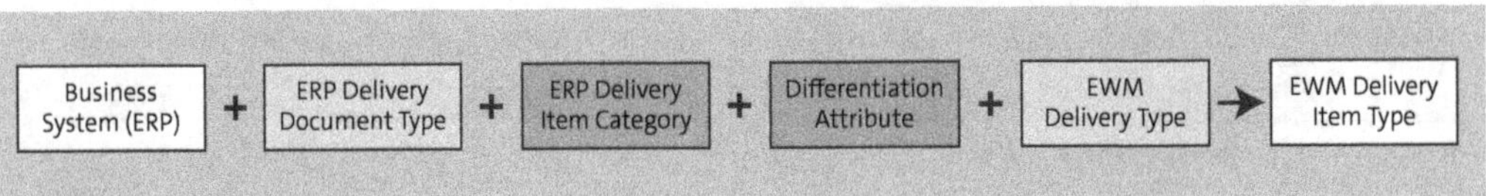

Figure 7.39 Mapping Logic for EWM Delivery Types

The differentiation attribute is an additional key that helps to distinguish different processes. The relevant configuration menu path is **SPRO • SAP Reference IMG • SCM Extended Warehouse Management • Interfaces • ERP Integration • Delivery Processing • Map Item Types from ERP System to EWM**. Figure 7.40 shows the configuration for the delivery item type linkages between SAP S/4HANA and EWM systems.

We've seen some of the key integration aspects in EWM. There are many more as we integrate the organizational structure, inbound deliveries, and outbound deliveries, as well as automate control of transfer order creation (WM) or warehouse tasks. However, we need to limit the scope of this chapter as most of these configurations are specific to operations or processes within WM and may not be directly related to the third-party logistics integration topic.

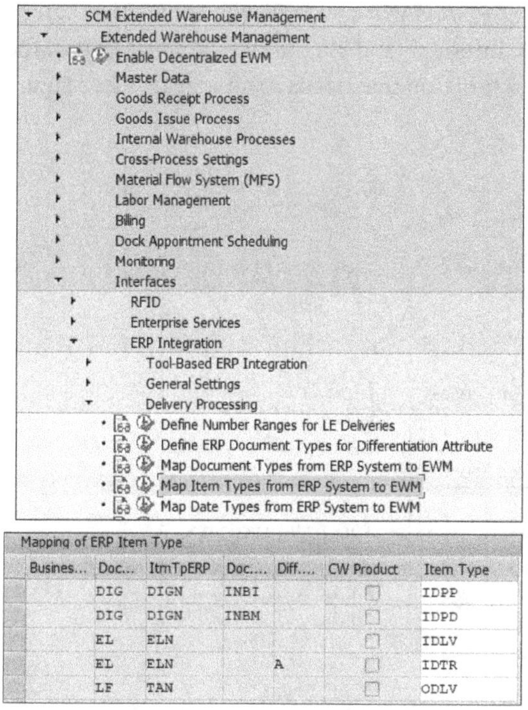

Figure 7.40 Configuration for Mapping Delivery Item Type Linkage between EWM and SAP S/4HANA

7.4 Master Data in Extended Warehouse Management

Table 7.2 is a snapshot of how master data can be replicated from SAP S/4HANA to EWM. Transaction DRFIMG provides a customization menu for the Data Replication Framework (DRF) that helps in setting up the parameters for master data replication using the methods mentioned in the table.

ERP system	EWM version	Method
SAP ERP	Decentralized EWM in SAP S/4HANA, version 1809 onward	ALE, DRF (enterprise services, IDoc, file, RFC)
SAP S/4HANA	Embedded EWM in SAP S/4HANA, version 1610 onward	Master data replication not necessary
SAP S/4HANA	Decentralized EWM in SAP S/4HANA, version 1809 onward	ALE, DRF (enterprise services, IDoc, file, RFC)

Table 7.2 Standard Methods of Replication between ERP Systems and EWM Systems with Different Versions

When using decentralized EWM in an SAP S/4HANA platform, the connection can be established using the familiar ALE/IDoc interface, and via the DRF, which is part of the master data governance (MDG) business function that needs to be activated (see Figure 7.41).

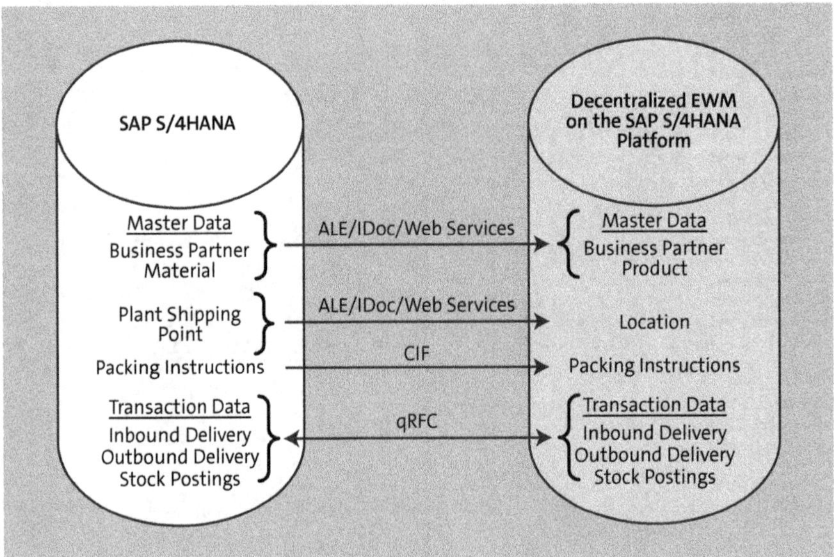

Figure 7.41 Replication of Master and Transactional Data between SAP S/4HANA and EWM

Within the DRF, different communication channels are available, and, in most cases, using IDocs is the preferred option, as it offers the advantage of using existing message types. DRF has the flexibility that you can switch to replication via services, file, or RFC easily with a few changes. You can use Transaction DRFIMG to open settings for the data replication model, as shown in Figure 7.42.

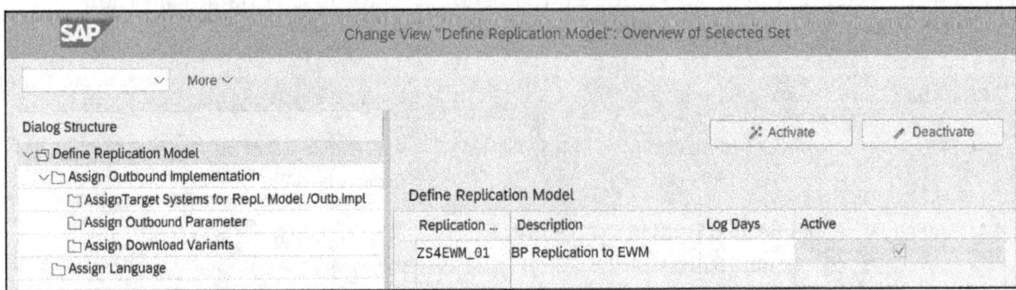

Figure 7.42 Setting Up the Data Replication Model Using Transaction DRFIMG

To create master data replication from SAP S/4HANA to decentralized EWM, you must first set up the reduced message type in SAP S/4HANA and activate the segments.

Basic types for master data are supplied in the standard SAP system. If you want to reduce the basic types, you can select the segments and fields that you want to distribute. To do this, you activate the segments and fields that you require and generate a new message type.

Reduction is carried out for logical message types associated with an IDoc type. Mandatory segments and mandatory fields can't be deactivated.

Creating custom reduced message types for only selected segments and fields can be maintained using configuration menu path **ABAP Platform • Application Server • IDoc Interface/Application Link Enabling (ALE) • Modeling and Implementing Business Processes • Master Data Distribution • Scope of Data for Distribution • Message Reduction • Create Reduced Message Type** (Transaction BD53).

Figure 7.43 shows Transaction BD53 along with custom reduced message type **Z_MATMAS**, which can contain only selected segments and fields for transfer. As shown in the figure, you can select/deselect the fields for each specific segment.

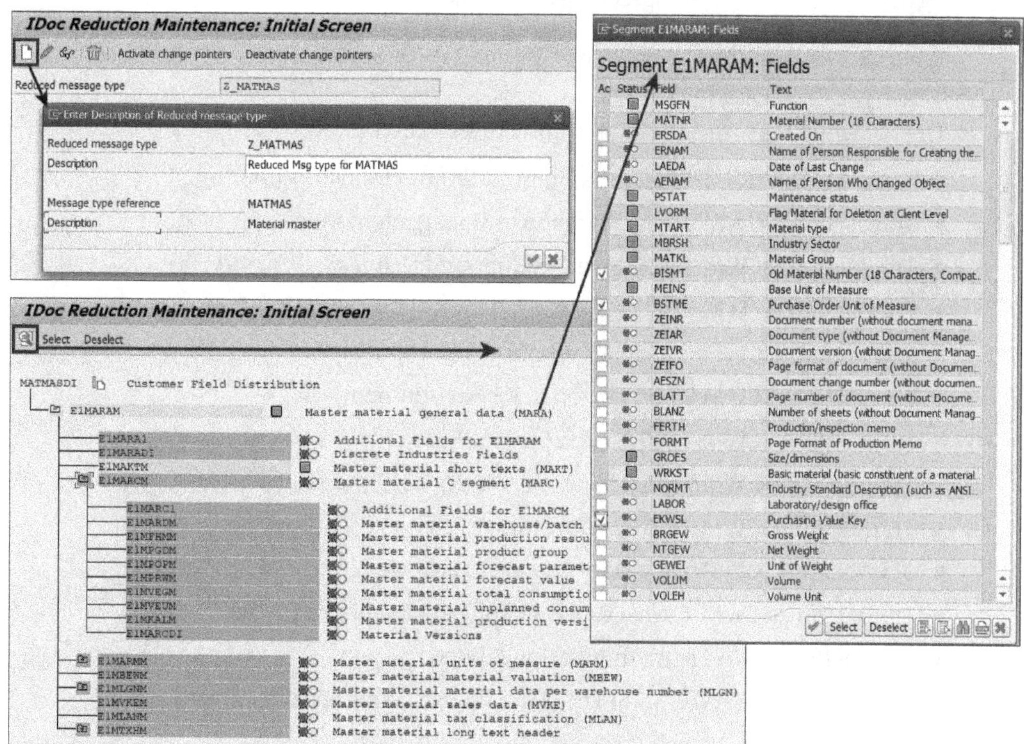

Figure 7.43 Creation of Reduced Message Type for IDoc MATMAS

Table 7.3 shows the reference IDoc types that can be used for creating reduced message types for distribution of master data.

Reference IDoc Type	Data Object
Master Data	
DEBMAS	Customer master
CREMAS	Vendor master
MATMAS	Material master
BATMS	Batch master
CHRMAS	Characteristics master
CLSMAS	Class master
CLFMAS	Classification master
ADRMAS	Addresses
MATQM	Inspection setup

Table 7.3 IDoc Types for Distribution of Master Data through Reduced Message Type

The following are some common transactions in EWM:

- Transaction /SCWM/PROD: Maintain Outbound Delivery Order
- Transaction /SCWM/MON: Warehouse Management Monitor
- Transaction /SCWM/PRDI: Maintain Inbound Delivery
- Transaction /SCWM/TODLV_O: Stock Removal for Outbound Delivery Order
- Transaction /SCWM/TO_CONF: Confirm Warehouse Task
- Transaction /SCWM/RFUI: Log on to RF Environment
- Transaction /SCWM/RFUI: Create Storage Bin
- Transaction /SCWM/MAT1: Maintain Warehouse Product
- Transaction /SCWM/LS01: Create Storage Bin
- Transaction /SCWM/LS11: Mass Change to Storage Bins
- Transaction /SCWM/GWL: Display Graphical Warehouse Layout
- Transaction CFM1: Create Integration Model
- Transaction /SCWM/PACKSPEC: Maintain Packaging Specifications

Figure 7.44 through Figure 7.47 capture various document-specific details of warehouse tasks generated from replication of inbound delivery documents from SAP S/4HANA to EWM. Figure 7.44 shows an inbound delivery notification containing header and item details such as warehouse number, delivery document details, item details, item category, and item type determined from configuration.

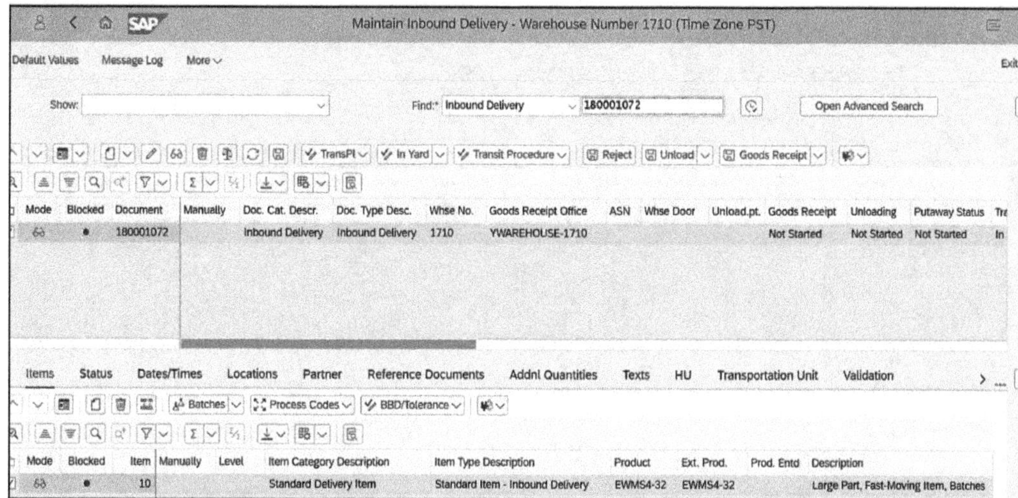

Figure 7.44 Inbound Delivery in EWM

Figure 7.45 shows the receiving **Stock Type** as **Stock in QI** (stock in quality inspection), along with batch number and staging bin.

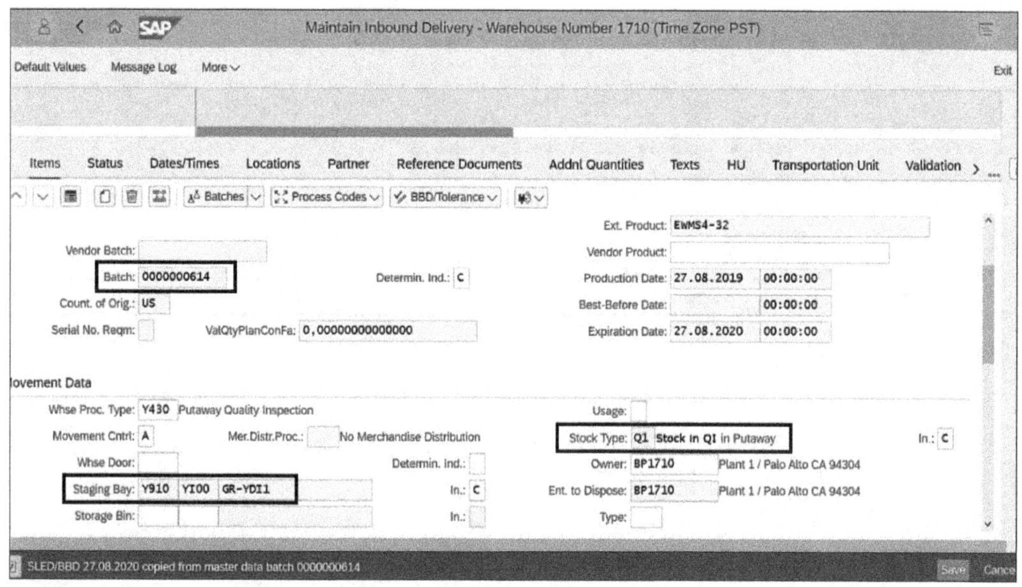

Figure 7.45 Inbound Delivery in EWM with Batch Number Created and Stock Type as Stock in Quality Inspection

Figure 7.46 shows the **Warehouse Management Monitor** screen displaying the status of the physical stock after goods receipt. Scrolling to the right on this screen may show the stock type, batch number, inspection lot number, and expiration date, as per your column settings.

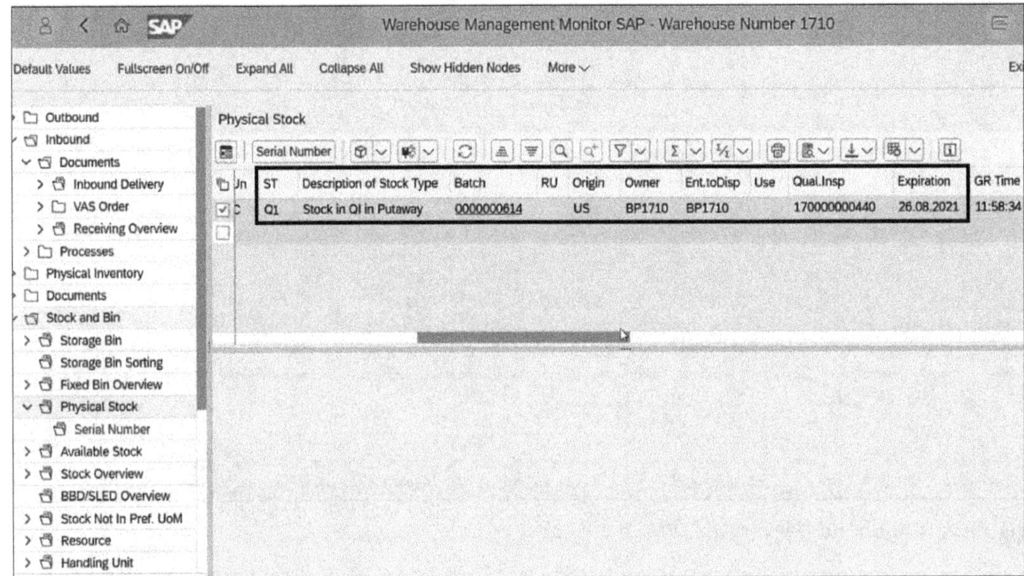

Figure 7.46 Warehouse Management Monitor Showing Physical Stock with Batch Number and Stock Type Details

Figure 7.47 reflects the physical stock at the staging bin.

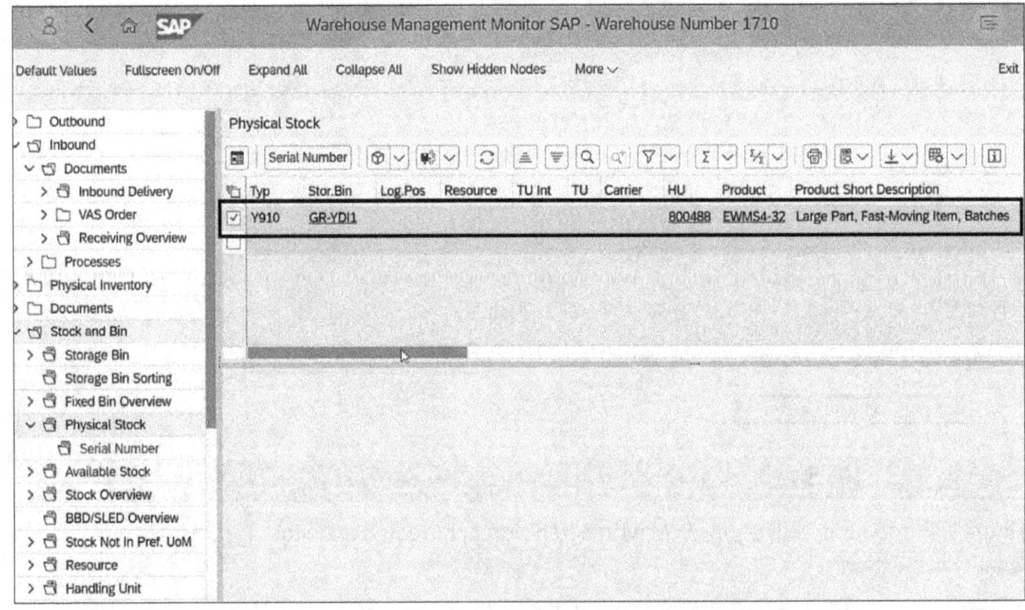

Figure 7.47 Warehouse Management Monitor Showing Stock in Staging Bin

7.5 Summary

The chapter dealt with key WM aspects that are most relevant to setting up integration with third-party logistics partners. It started with coverage of various deployment scenarios involving both non-SAP and SAP solutions for WM and the possible integration options and communication mechanisms. It was followed by a detailed look at all configurational aspects involving system integration, setting up and linking relevant organizational entities between SAP S/4HANA and WMSs, and process integration areas covering both inbound and outbound processes. Options and settings for master data replication from SAP S/4HANA to EWM was discussed at length with a few insights into technical aspects of setting up the communication channel.

SAP S/4HANA EWM is a kind of all-inclusive WM solution that can enable efficient management of warehouse operations. It offers a number of features that can help in enhancing inventory accuracy by streamlining the warehouse processes that, in turn, enable effective execution of warehouse operations.

EWM is based on the robust in-memory computing capabilities of the SAP HANA platform. This enables EWM to process warehouse data quickly and precisely, as well as quickly access and examine warehouse data for use in reporting and decision-making.

There are many non-SAP WMS applications available in the market today that cater to most of the warehouse needs of third-party logistics providers, and some of them are well matched with EWM and equally cost-effective. So, it becomes imperative for consultants or architects to understand the design perspective for both situations, so we covered the integration aspects for both SAP and non-SAP solutions.

Chapter 8
Transportation

An efficient transportation management system provides an overall, end-to-end solution for transportation planning, execution, track and trace, and visibility of movement of goods from supplier to customer or consignee. This results in major time savings for logistics service providers through process automation, and this precious time can be invested elsewhere for more productive work. It also enables both shippers and logistics service providers to interact through system integration via secure channels such as electronic data interchange (EDI) and application programming interfaces (APIs) to share rates and quotes, transportation requirements, and freight information on a real-time basis.

The beginning of this chapter focuses on SAP Transportation Management (SAP TM) deployment options, as well as subcomponents of SAP TM and their applicability in different scenarios. Some screenshots were created using the SAP training system, and some were taken from best practice client systems. We'll discuss the organizational structure and master data needed for system setup and how SAP TM interacts with SAP S/4HANA. Later in the chapter, we'll cover the functional aspects of shipper and logistics service provider scenarios.

8.1 Third-Party Logistics Providers and Transportation

A third-party logistics provider is a firm that provides end-to-end logistics services either fully or in part. This enables the provider's customers to outsource its logistics and supply chain management functions, as well as scale it up or down per their business requirement. Third-party logistics providers typically specialize in integrated operations, warehousing, and transportation services that can be scaled and adapted to the customer's needs. They are based on market conditions and the demands and delivery service requirements for their products and materials.

Figure 8.1 provides a pictorial representation of the services typically offered by third-party logistics partners.

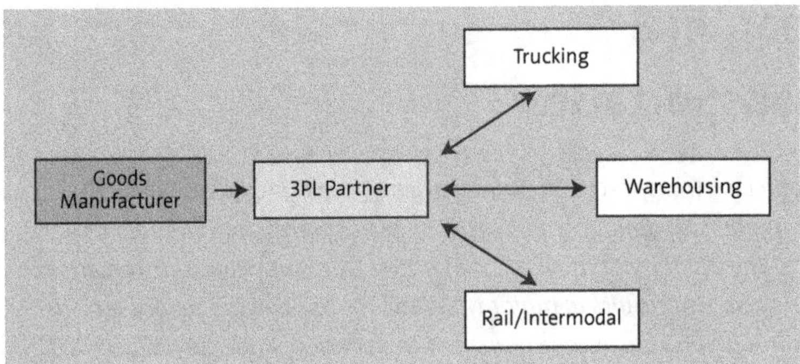

Figure 8.1 Services Offered by Third-Party Logistics Partners (Logistics Service Providers/ Freight Forwarders)

SAP Transportation Management focuses on end-to-end business scenarios involving outbound and inbound transportations (or shipments) and their nuances. It's highly user-friendly, scalable, and built on a solid architectural framework. Standard SAP S/4HANA solutions and licenses provide embedded SAP TM solutions that support basic shipping and some additional basic features such as manual planning and charge settlements.

Figure 8.2 and Table 8.1 illustrate and compare the different features of basic (embedded) and advanced SAP TM solutions.

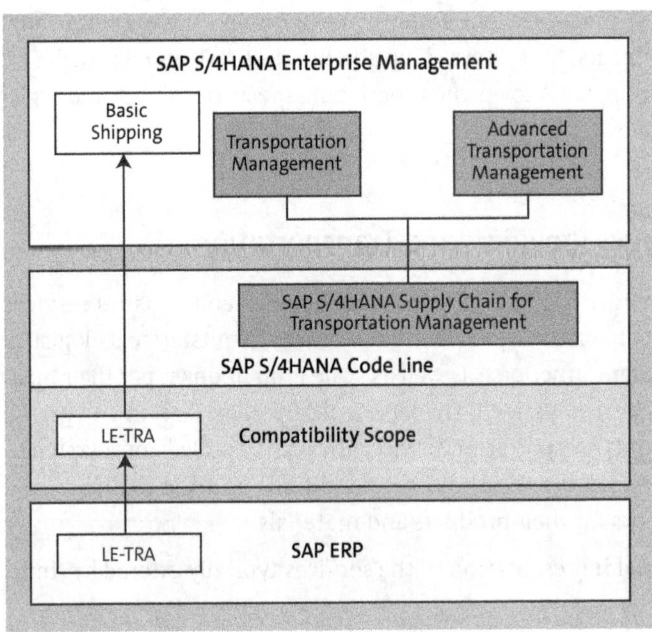

Figure 8.2 Advanced Transportation Management Compared to Basic Shipping

Basic Shipping/Transportation Management	Advanced Transportation Management
■ Shipping industries only ■ Delivery-based planning ■ Create freight orders with grouped inbound and outbound deliveries ■ Manual and rule-based planning ■ Monitor freight orders without event management ■ Basic charge calculation and freight settlement	■ Shipping and freight forwarding industries ■ Holistic end-to-end transportation management ■ Early planning based on orders ■ Plan combined inbound and outbound deliveries/orders ■ Optimized transportation planning ■ Peer-to-peer/broadcast tendering ■ Advanced charge calculation and freight settlement ■ Dispute management ■ Support real-time data-driven decision-making with core data services analytics ■ Strategic freight management

Table 8.1 Basic Shipping and Transportation Compared to Advanced Transportation Management

SAP Note 3065464 provides a comprehensive overview on which objects and functions are part of the basic transportation management scope and require an additional license (advanced transportation management).

8.2 Deployment Scenarios for Third-Party Logistics

To understand the deployment scenarios involving embedded SAP TM functions, it's essential to discuss the role of different business partners and their responsibilities. We'll walk through these first and then dive into the integration options available.

8.2.1 Business Partners and Their Roles

The following list provides an overview of the business partners and their responsibilities during transportation:

- **Shipper**
  A shipper can be a manufacturer or seller (called a consignor) that requires goods to be transported between two locations in the supply chain. The shipper, or exporter, is the party responsible for packing and preparing all the goods being sent, as well as handling all documents and paperwork needed.
- **Customer**
  A customer (called a consignee) is the receiver of the shipment and is usually the owner of the goods.

- **Third-party logistics or logistics service provider**
  Logistics service providers or third-party logistics is the party that arranges for transportation of goods per customer requirements. Many logistics service providers (asset-based) have global operations and manage their own fleet of carriers and warehousing facilities. However, some business entities such as freight forwarders (nonasset-based) use carrier companies to arrange for transport.
- **Carrier**
  They sell transportation services to their customers and have the obligation to execute the transportation with their own assets and transportation resources.

Figure 8.3 shows the responsibilities or functions of various business entities in the context of transportation.

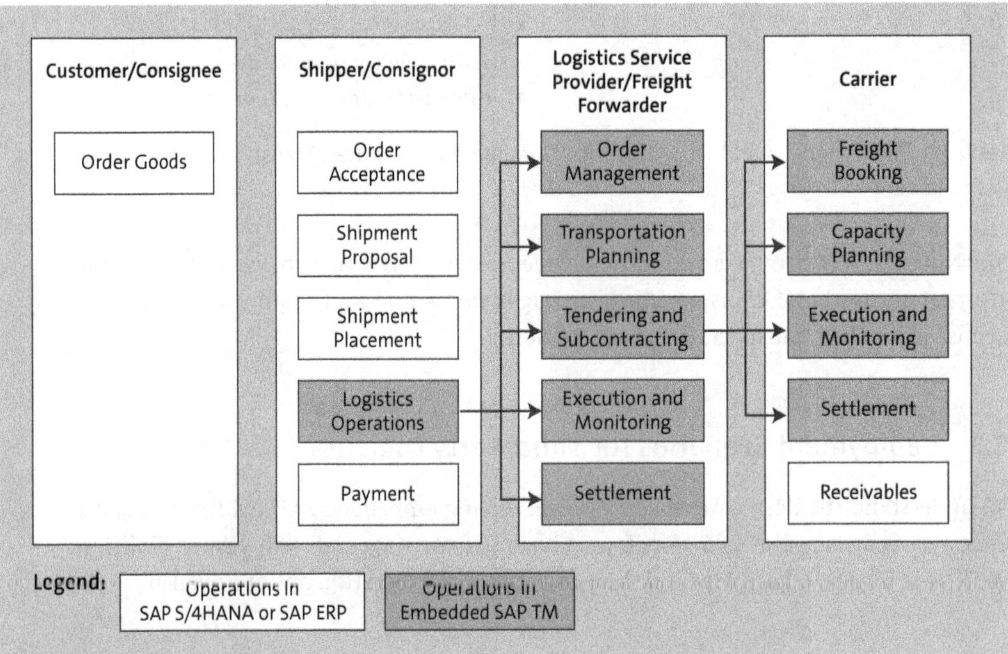

Figure 8.3 Business Entities and Their Functions in a Transportation Scenario

Depending on the nature of the role of the organization and business scenario, the deployment options can be one of the following:

- A shipper/manufacturing organization using embedded SAP TM or the advanced transportation management license can use both order-based (sales/purchase order) and delivery-based (outbound/inbound delivery) transportation requirements creation in SAP TM.
- A logistics service provider/forwarder using embedded SAP TM may create FWOs in SAP TM directly (or based on FWQs).

Figure 8.4 provides an illustration of the sequence or flow of transaction documents in embedded SAP TM. This represents the shipper scenario where the shipper is having SAP TM implemented with the associated SAP S/4HANA system.

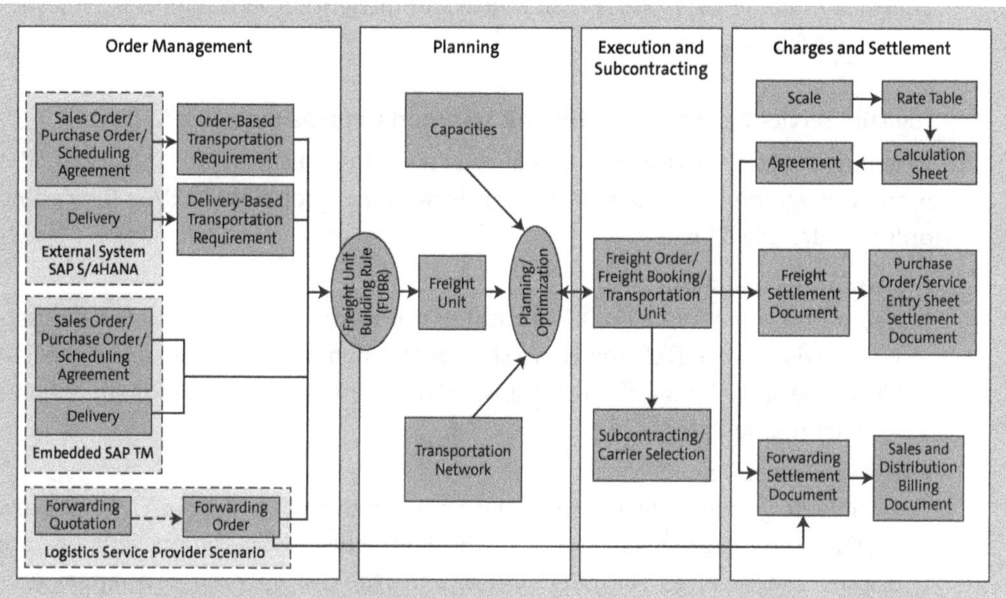

Figure 8.4 Document Flow in Embedded SAP TM

Figure 8.5 shows the basic flow in embedded SAP TM.

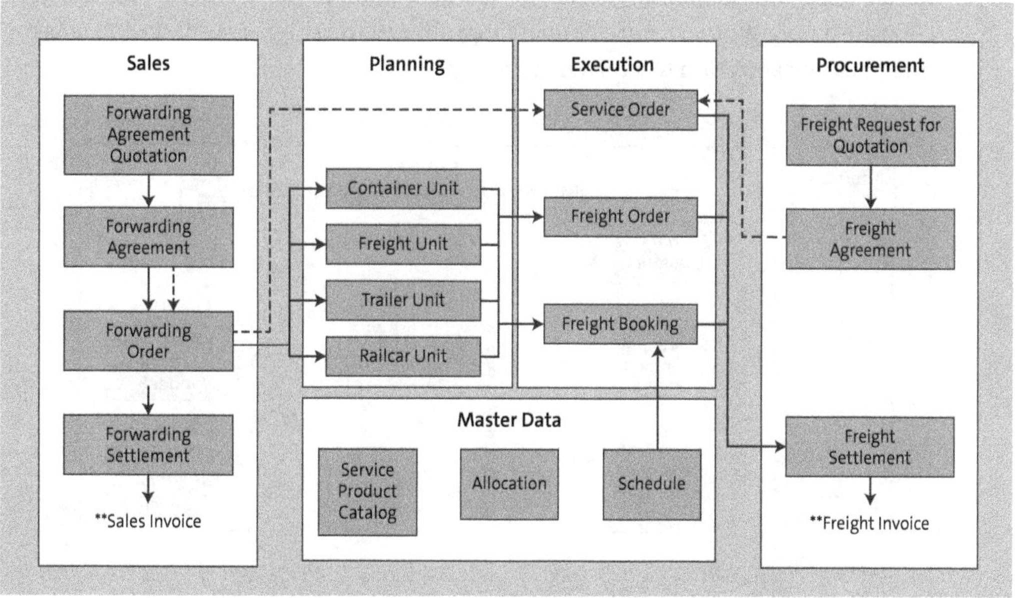

Figure 8.5 Basic Flows in Embedded SAP TM

8.2.2 Integration Options

In this section, we'll look into the details of various integration options between embedded SAP TM and traditional SAP S/4HANA procurement and sales functions. There are multiple ways SAP TM can be used for integration with shippers or logistics service provider systems.

Logistics Service Providers with the SAP Transportation Management System

Assuming logistics service providers already have an embedded SAP TM system, the process integration will happen through forwarding quotations and/or forwarding orders, as described here:

- **Forwarding quotation**
  A forwarding quotation (FWQ) is a business document that a carrier or logistics service provider proactively sends to a shipper (consignor) or customer (consignee) to bid for the provisioning of future transportation services in one or more trade lanes for a defined period.
- **Forwarding order**
  A forwarding order (FWO) in SAP TM is an order given by a shipper/customer to a logistics service provider. Based on the FWO, the transportation activities are planned and executed. The FWO is also used for the customer settlement process.

In most business scenarios, usually the customer requests the logistics service provider/freight forwarder for transportation of goods. Typically, a logistics service provider scenario begins with a FWO. Based on transportation requirements, freight units can be created automatically from FWOs. In addition, logistics service providers can generate a freight order (or freight booking) alike to the shipper scenario and subcontract the transportation to a carrier company.

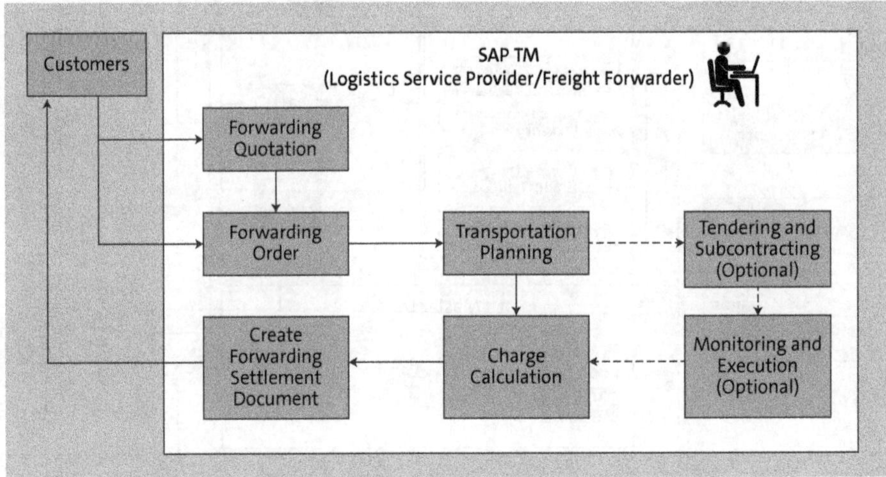

Figure 8.6 Integration Scenario with Logistics Service Provider Having SAP TM Implemented

Shown in Figure 8.6 is an end-to-end transportation process flow, including carrier hiring and charge settlements between the logistics service provider or freight forwarder and goods manufacturer (for shipment of goods to the end customer).

Third-party integration between SAP TM and legacy systems is possible either as a point-to-point connection (IDoc/API) or using middleware such as SAP Process Orchestration (Figure 8.7).

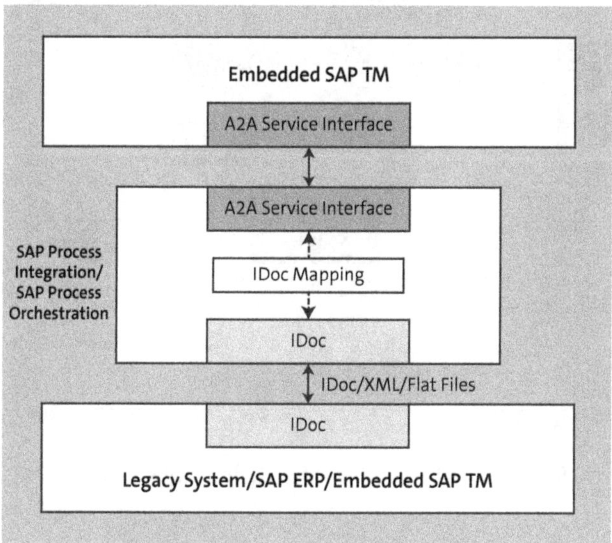

Figure 8.7 Third-Party Integration for Document Exchange between SAP TM and Legacy Systems

The SAP TM component called Forwarding Order Management offers fast entry screens for logging the request details of the customer quickly with additional enablers as previously saved templates, previous booking references, and so on.

The logistics service provider scenario generally starts with a FWO. Freight units can be created either manually or automatically from FWO based on the transportation requirement. The logistics service provider also can subcontract the transportation to another company and can create a freight order. If the logistics service provider is a freight forwarder and hires a carrier for transportation, it may result in two different settlements, as described here:

- **Forwarding settlement**
  Freight charges are calculated based on a forwarding agreement (FWA) between the logistics service provider and its customers. This will create the billing document to the customer of logistics service provider.
- **Freight settlement**
  Freight charges are calculated on the freight order (or freight booking) based on the

freight agreement (FA) of the logistics service provider with their subcontractors or carriers. This will create a purchase order and service entry sheet.

Figure 8.8 depicts an end-to-end flow from order initiation to settlement for a scenario involving a logistics service provider using embedded SAP TM.

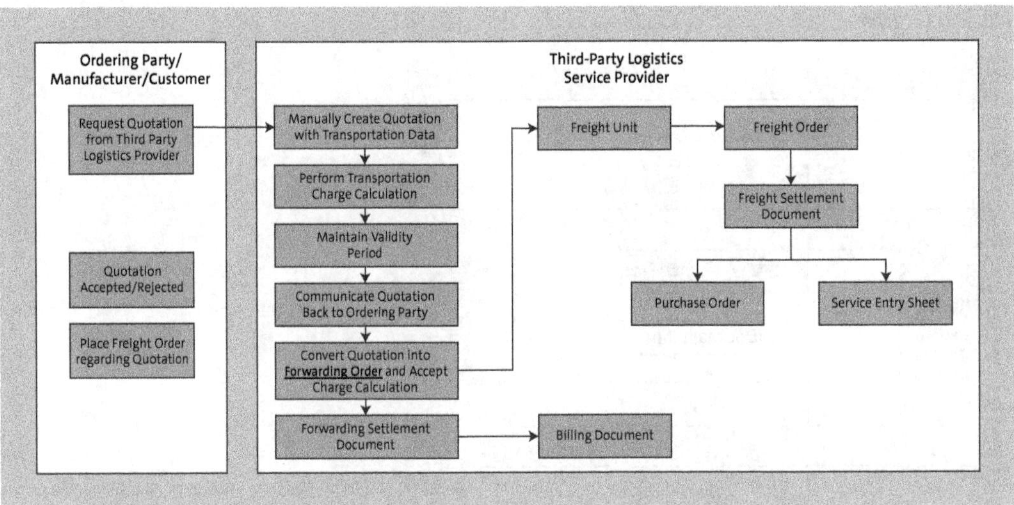

Figure 8.8 End-to-End Process Cycle for Logistics Service Providers Using Embedded SAP TM

A document for a typical FWO is shown in Figure 8.9.

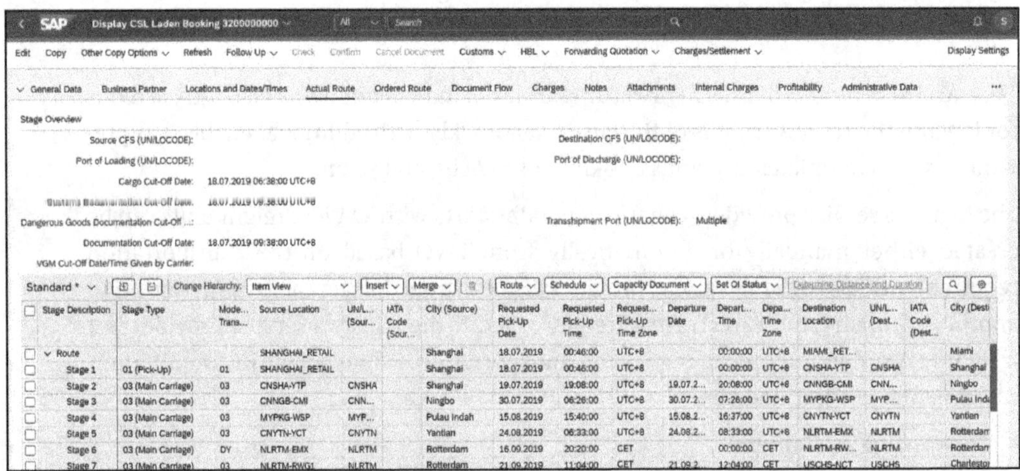

Figure 8.9 Example of a Forwarding Order

Key field attributes in a typical FWO include the following:

- **Organizational Unit**
  This is the same as sales organization and is mandatory for creating freight units.

- **Settlement Terms**
  This represents an incoterm. It's a free text field for logistics service providers because it can't be mapped to a location. In a shipper scenario, it can be mapped to a location.
- **Order Date**
  The order date is important for charge calculation.
- **House Bill of Lading**
  This is used as a tracking reference or in event management.
- **Transportation Mode/Traffic Direction/Shipping Type**
  This can be configured to be the default when defining the FWO document type.
- **Goods Value for Customs/Goods Value**
  This is used for custom duty declarations.

Figure 8.10 shows a snapshot of a FWO in SAP Fiori containing some of these key attributes.

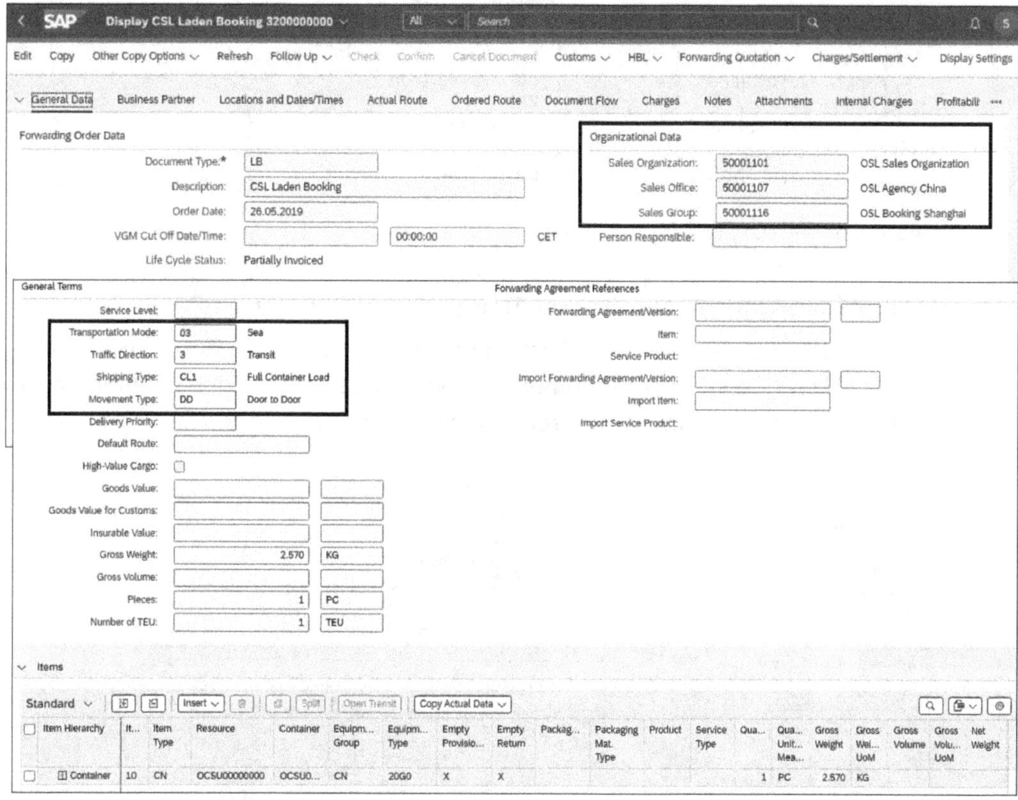

Figure 8.10 Snapshot of a Forwarding Order

Logistics service providers may have a number of key clients with recurring transports over a time period. In such a scenario, FWO templates can be created (with as much information as necessary) and r-used because the details are almost the same or identical in every transport. FWO templates are based on FWO types and appear the same as FWOs but without fields for dates as these need to be filled while creating the FWO using this template.

When a customer orders a transport for a specific date on one of the recurring transports, the logistics service provider only must create a FWO from the template. All the data is copied from the template into the FWO.

Figure 8.11 shows the utility of template FWO for recurring transport requirements of a similar nature.

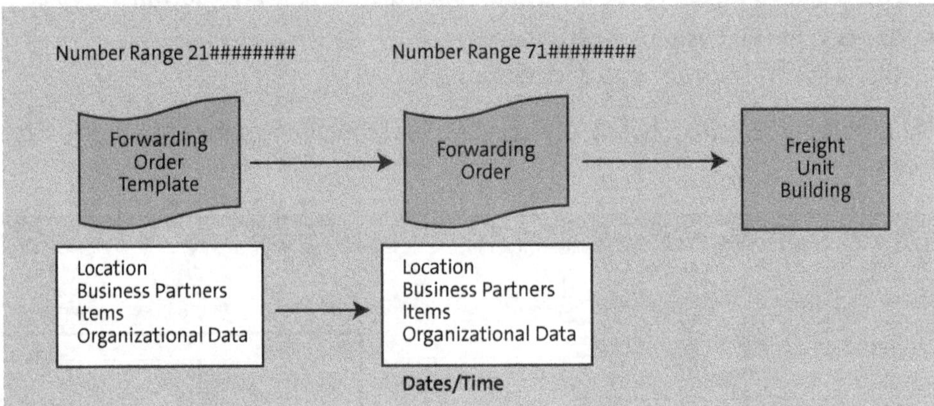

Figure 8.11 Template Forwarding Order

Shipper Scenario with Embedded SAP Transportation Management

This scenario represents cases where the shipper or goods manufacturer is using embedded SAP TM and a group company handles the transportation requirements as a freight forwarder or carrier. The process flow is depicted in Figure 8.12.

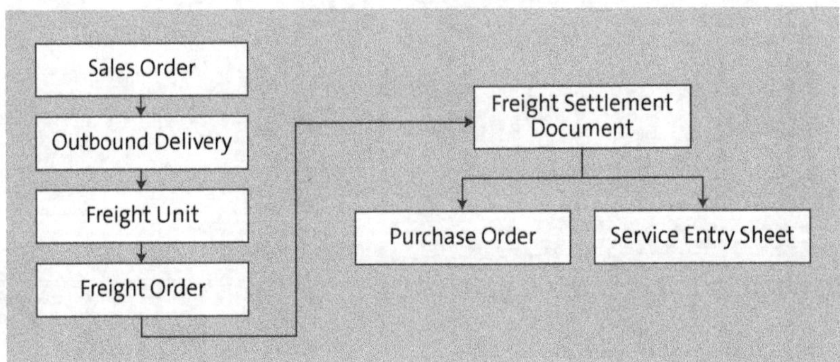

Figure 8.12 Shipper Scenario with Embedded SAP TM

Following are the main partners in the shipping process:

- **Shipper**
  The shipper has the requirement for shipment of goods.
- **Logistics service provider**
  The logistics service provider takes care of the shipment.
- **Consignee**
  The consignee is the recipient of the shipment.

There could be multiple scenarios, such as the following:

- The shipper or its group company uses its own vehicle fleet to carry out the delivery; in this case, the shipper acts as logistics service provider. This resembles a shipper scenario.
- The shipper outsources the task to a third-party logistics partner or logistics service provider, which could either be asset-based or non-asset-based such as a freight forwarding service. This resembles a true logistics service provider scenario.

8.3 Master Data and Organizational Structure

In this section, we'll look into some of the key organizational data and master data in embedded SAP TM. This will be followed by details about a few key configurational aspects.

8.3.1 Organizational Units in Embedded SAP Transportation Management

In SAP TM, the organizational structure consists of the following organizational units:

- **Company organization**
  This organization unit corresponds to the SAP S/4HANA company code containing the local currency. The company organization unit is used by transportation charge management (TCM) for invoicing and settlement.
- **Sales organization, office, and group**
  This is relevant for logistics service provider/third-party logistics and FWO settlement.
- **Purchase organization and group**
  This is relevant for freight booking and freight settlements.
- **Planning and execution organization and group**
  This is relevant for transportation planning and execution.

Figure 8.13 shows the organizational unit hierarchy in SAP TM.

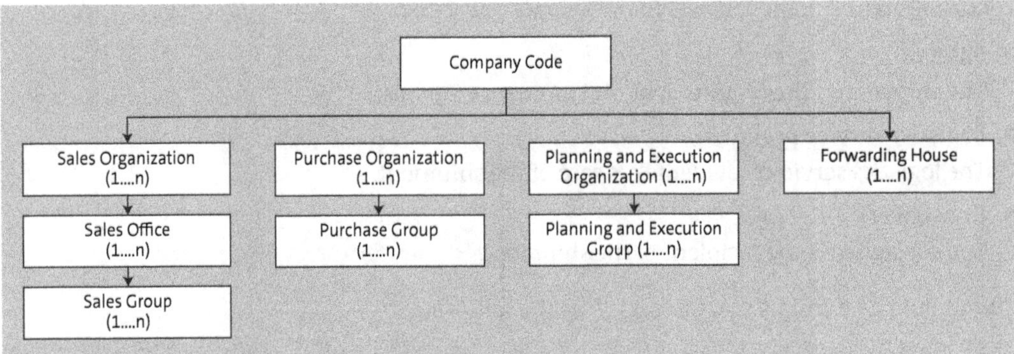

Figure 8.13 Organizational Structure in Embedded SAP TM

Purchase, sales, and planning organizations in SAP TM are linked to the materials management/sales and distribution organizational units in SAP S/4HANA. Figure 8.14 shows how you link the two using Transaction PPOMA in embedded SAP TM:

❶ SAP TM purchasing organization

❷ Materials management purchasing organization

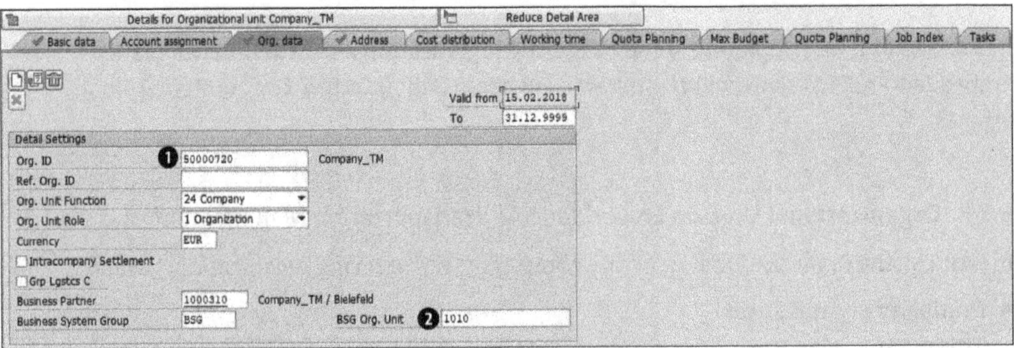

Figure 8.14 Linking SAP S/4HANA Materials Management Purchase Organization to SAP TM Purchase Organization

Mapping the purchasing organization is required for a shipper using both SAP TM and sales and distribution/materials management that purchases transportation services from third parties. The same is relevant for logistic service providers that are subcontracting freight to external carriers. The initial determination point is during the creation of the freight order/booking. The same purchasing organization needs to be defined in the freight agreement, which is a long-term contract that represents the contractual relationship with a carrier from whom you're buying transportation services.

SAP TM generates a freight settlement document (FSD) when a freight order is settled. The FSD is posted to materials management. When the invoice is received, the invoice verification process from materials management is used. If you perform evaluated receipt settlement, the automatically generated invoice is based on the data from the FSD.

Figure 8.15 shows the end-to-end freight booking and settlement process and the role of the purchase organization in it.

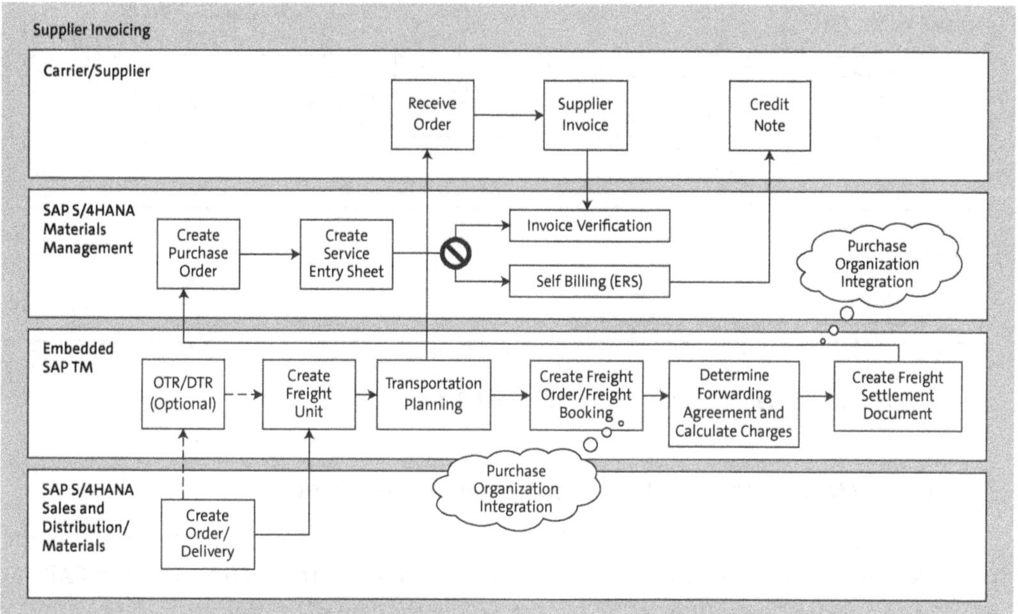

Figure 8.15 Freight Booking and Settlement Process

As part of the supported integration with SAP TM, the integration of the sales organization is essential if an SAP S/4HANA backend is present. In the case of a logistics service provider using both SAP TM and SAP S/4HANA, the sales organization is integrated at two points. The first integration point is the FWO in SAP TM. To calculate charges and settle the FWO, a FWA for the same sales organization is required. A FWA is a long-term contract that represents the contractual relationship with a customer to whom you're selling transportation services.

In SAP TM, you can perform forwarding settlement with your ordering parties by creating a forwarding settlement document (FWSD) and sending it to sales and distribution for billing and subsequent posting to financials.

Figure 8.16 shows the end-to-end freight booking and settlement process and the role of the purchase organization in it.

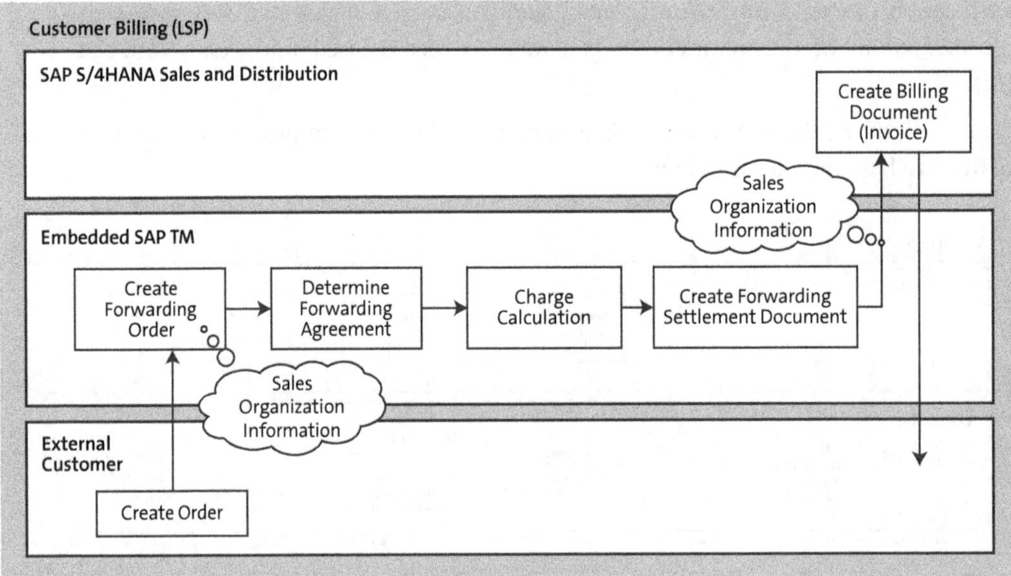

Figure 8.16 Forwarding Order and Settlement Process

8.3.2 Master Data in SAP Transportation Management

The SAP TM system can perform its functions, to some extent, independently of the existing product, location, and partner master data. The transactional data in SAP TM can be created without existing master data in the case of third-party logistics providers. Planning and execution can be performed in SAP TM with only a minimal master data set that can be created and maintained exclusively in SAP TM. As an additional scenario, SAP TM can work with master data, customer, and materials to support an end-to-end scenario for a manufacturer or distributor.

Master data can be distributed between SAP S/4HANA and SAP TM using standard communication mechanisms (Figure 8.17).

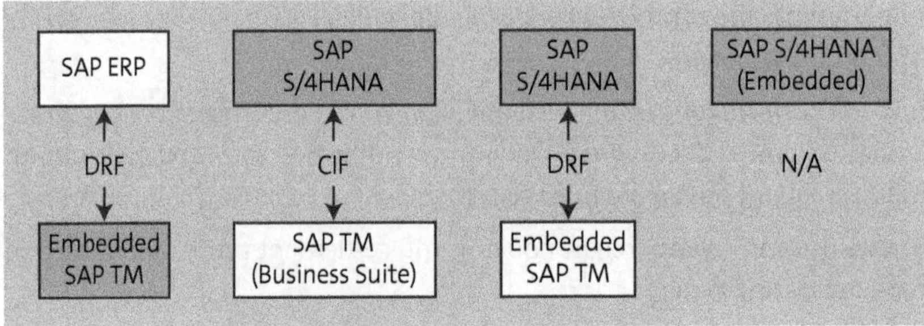

Figure 8.17 Master Data Integration

SAP TM can be integrated into sales and distribution and materials management processes and can use corresponding master data (customer, vendor, material). The master data elements shown in Figure 8.18 are used in SAP TM to support transactions.

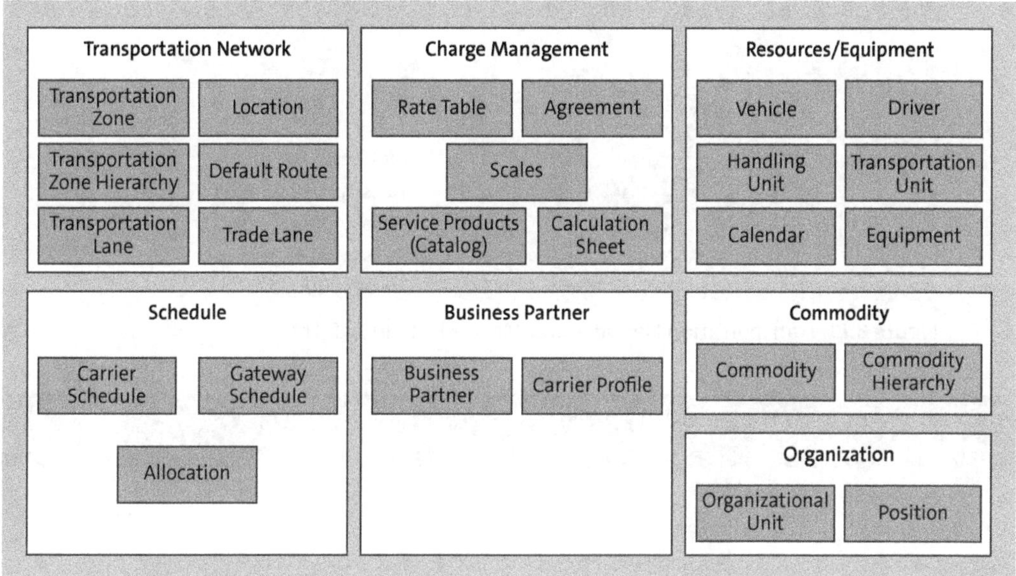

Figure 8.18 Standard Master Data in SAP TM

Let's look at few master data elements used in SAP TM:

- **Business partner**
  We've already seen the details of business partners and their roles in Chapter 3. For SAP TM, it may be essential to maintain a business partner role for carrier (CRM010) as carrier-specific fields are available in this role that flow to SAP TM.
- **Transportation network**
  In SAP TM, the transportation network is a combination of the following master data objects:
  - Location
  - Transportation zone
  - Transportation lane
  - Schedule

  In conjunction with the means of transport, the transportation network defines the direct accessibility of transportation zones and locations for a particular vehicle resource. Figure 8.19 shows a transportation network and Figure 8.20 shows a snapshot of SAP Fiori tiles for creation of master data.

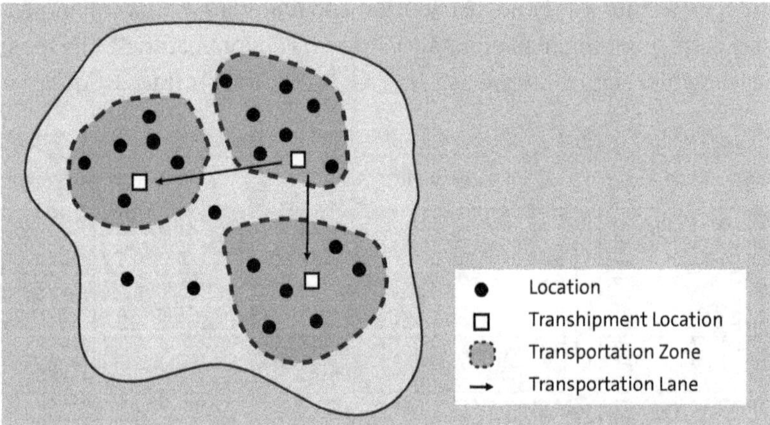

Figure 8.19 Transportation Network and Master Data in SAP TM

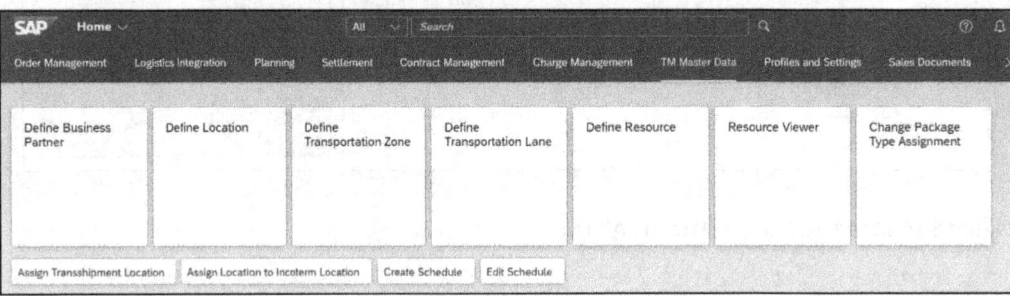

Figure 8.20 SAP Fiori Tiles – Master Data Creation

- **Location**
  In the SAP S/4HANA system, each inventory-relevant location can be assigned one or more shipping points. A shipping point is an independent organizational unit at a fixed location that processes and monitors inbound and outbound deliveries. The physical address of a shipping point is represented as a location in master data. Location in SAP TM has the following attributes:
  - Can act as a transshipment location
  - Can be assigned to a transportation zone
  - Can be specific to transactional data (one-time address)
  - Is a node of the transportation network

  Locations have the following purposes:
  - Freight is moved between locations.
  - A freight unit always has both a source location and a destination location.
  - A freight unit may travel through a number of locations.
  - Figure 8.21 shows the various ways to create locations in SAP TM, for example, via reports or business add-ins (BAdIs), or, in some cases, automatically.

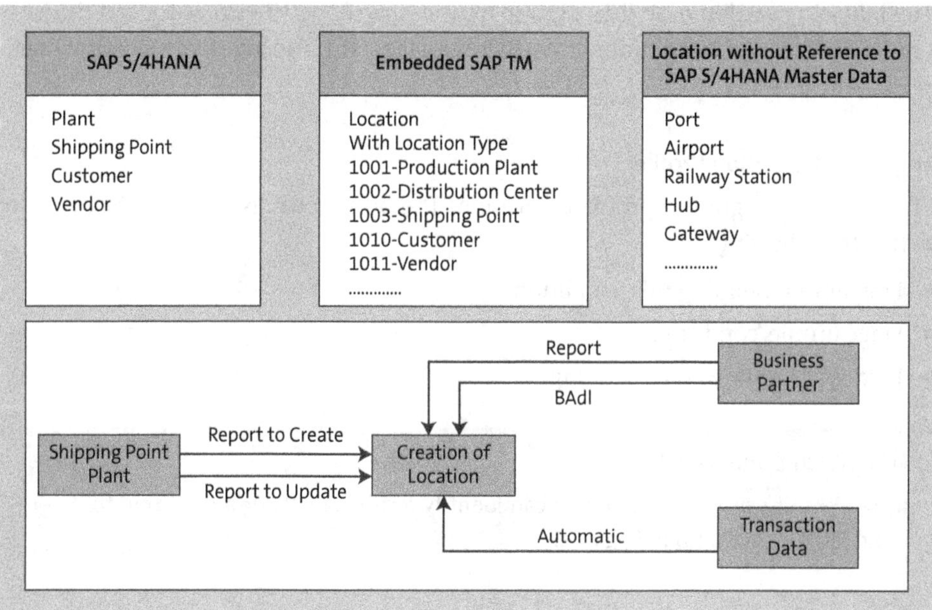

Figure 8.21 Creation of Locations in Embedded SAP TM

We won't discuss in detail each master data item in SAP TM as they aren't all relevant for the topic of this book.

8.4 Basic Configuration Setups

In this section, we'll look into some key configurational aspects that may impact the process flow for scenarios involving logistics service providers and shippers. It may be pertinent to emphasize that the shipment process needs to be understood in detail as the processes may differ when the logistics service provider handles the shipment as opposed to when shippers manage the transportation.

8.4.1 Logistics Integration in SAP Transportation Management

The initial event that triggers the transportation management process is a transportation demand. In the basic shipping scenario, an outbound or inbound delivery can act as a transportation demand. In advanced transportation scenarios, order-based planning is also possible, that is, sales orders, purchase orders, stock transfer orders (STOs), or materials management scheduling agreements can act as transportation requirements.

SAP TM can be integrated with SAP S/4HANA internal components or with an external system. For external integration, the embedded SAP TM component acts as the target system for an external source system such as SAP S/4HANA or another ERP system.

The following section describes a group of user-specific settings that control the integration of sales and distribution, materials management, and logistics execution documents into SAP TM.

Logistics Integration Profile

The logistics integration profile determines the basic settings for SAP TM. Here we define the following:

- Freight unit building rule/condition
- Stage profile/condition
- Incoterm location stage building

> **Freight Unit Building Rules**
>
> If freight units are to be built, you can identify a rule directly or a condition to determine the freight unit building rule.

Figure 8.22 shows the **Freight Unit Building Rule** in an SAP Fiori application

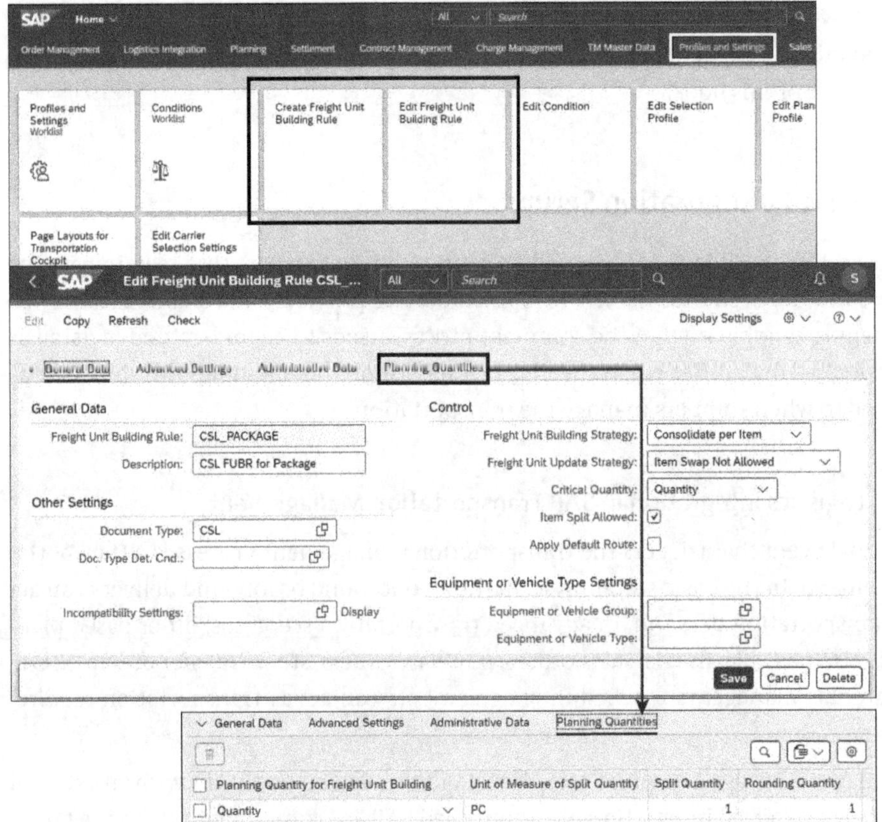

Figure 8.22 Freight Unit Building Rule

The logistic integration profile can be defined using configuration menu path **SPRO • SAP Reference IMG • Transportation Management • Integration • Logistics Integration • Internal TM Component Integration • Define Logistics Integration Profile.** Figure 8.23 shows the configuration parameters of **Logistics Integration Profile.**

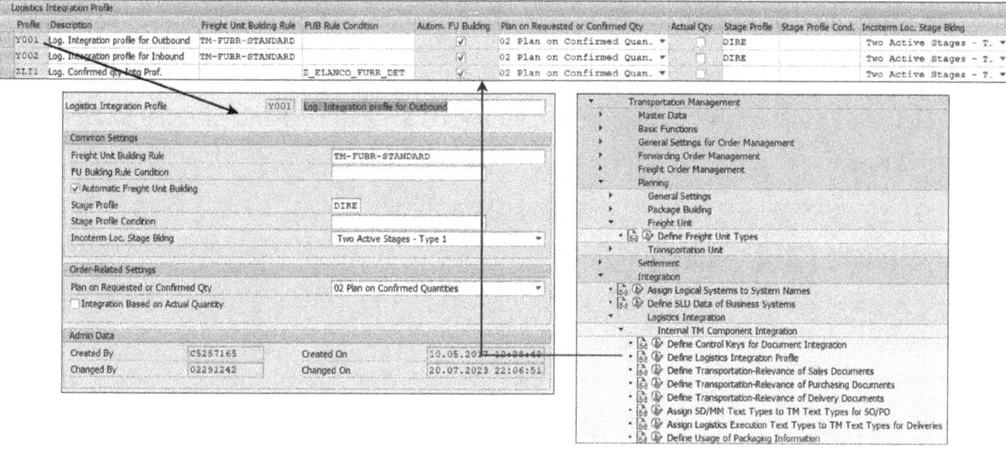

Figure 8.23 Logistics Integration Profile

Control Keys for Document Integration

Control keys are used for document integration with the internal SAP TM component. Control keys can be maintained using configuration menu path **SPRO • SAP Reference IMG • Transportation Management • Integration • Logistics Integration • Internal TM Component Integration • Define Control Keys for Document Integration.**

Here, the integration mode needs to be specified or maintained. For SAP S/4HANA, you can select the internal SAP TM component or choose to work with an external SAP TM system. If an external SAP TM system is selected, then the following documents can be integrated:

- Sales orders
- Sales and distribution scheduling agreements
- Purchase orders
- Materials management scheduling agreements
- STOs
- Returns orders (e.g., sales returns, vendor returns, and store returns)
- Outbound deliveries
- Inbound deliveries

If an internal SAP TM component is selected, the following documents can be integrated:

- Sales orders
- Purchase orders
- STOs
- Returns orders (e.g., sales returns, vendor returns, and store returns)
- Outbound deliveries
- Inbound deliveries

Figure 8.24 shows the control keys maintained for external and internal SAP TM systems.

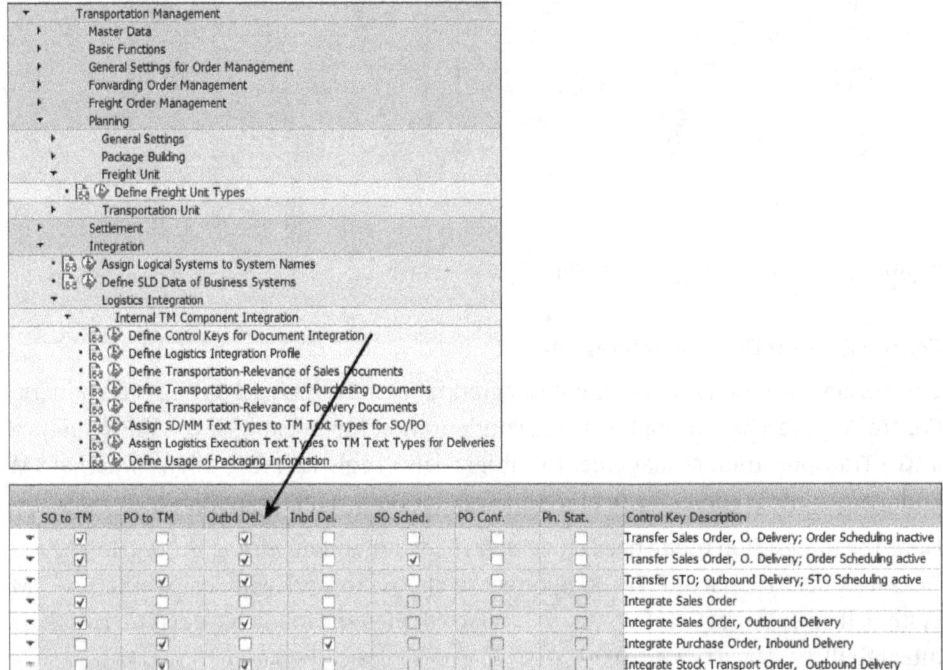

Figure 8.24 Control Keys for Logistics Integration

Defining Transportation Relevance of Documents in SAP S/4HANA

While defining different document integration scenarios, the control key is assigned to a shipping point, delivery type, and shipping condition. It's also assigned to the logistics integration profile that will be used.

When the delivery is relevant for document integration, the control key is copied to the delivery document. The logistics integration profile can be used to differentiate the freight unit building logic if multiple logistics integration profiles are used.

For *order integration*, the configuration menu path is **SPRO • SAP Reference IMG • Transportation Management • Integration • Logistics Integration • Internal TM Component**

Integration • Define Transportation-Relevance of Sales Documents/Define Transportation-Relevance of Purchasing Documents or Define Transportation-Relevance of Delivery Documents.

For *delivery integration*, the configuration menu path is **SPRO • SAP Reference IMG • Transportation Management • Integration • Logistics Integration • Internal TM Component Integration • Define Transportation-Relevance of Delivery Documents.**

Figure 8.25 shows the linkage of the control key and logistics integration profile to the document type and responsible organization entities. The top screen shows order integration while the bottom screen shows delivery integration.

Sales Document Integration

Sales org.	Distr. Chl	Division	Sales Doc. Type	Shp.Cond.	Control Key	TM No.	Log. Int. Profile	Control Key Description
DE01	40	10	ZEXP		0052		Y001	Integrate Sales Order, Outbound Delivery

Delivery Document Integration

Ship Point	Del. type	Shp.Cond.	Control Key	TM No.	Log. Int. Profile	Control Key Description
DEM5	LB	01	0053		Y002	Integrate Outbound Delivery

Figure 8.25 Setting Up Transportation Relevance of Documents

Forwarding Order Configuration (Logistics Service Provider)

So far, we've seen the master data and organizational structure for SAP TM and key aspects of integration with SAP S/4HANA transactional documents. There are many other configurational aspects in SAP TM such as freight order type configuration, FWO-specific setups, and so on. As we're focusing on the third-party logistics scenarios, it's worth taking a look at only the relevant configurational aspects that may impact such operations

FWO-specific configuration can be done using menu path **SPRO • SAP Reference IMG • Transportation Management • Forwarding Order Management** (Figure 8.26). Relevant configuration details pertaining to nodes numbered ❶ to ⓫ in this figure are described in the following sections in the sequence of the actual configuration steps.

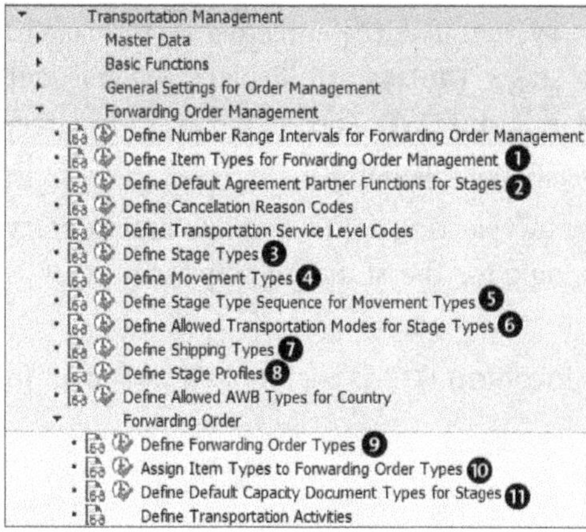

Figure 8.26 Configurations for Forwarding Orders

Define Item Types for Forwarding Order Management

In this Customizing activity, item types are defined for items in forwarding documents and FWQs. Parameters for an item such as the item category, equipment type, or default unit of measure (UoM) can be specified here.

For vehicle resources, such as trailers and railcars, you must select the item category **Passive Vehicle Resource**, and specify a **Resource Class**, as shown in Figure 8.27.

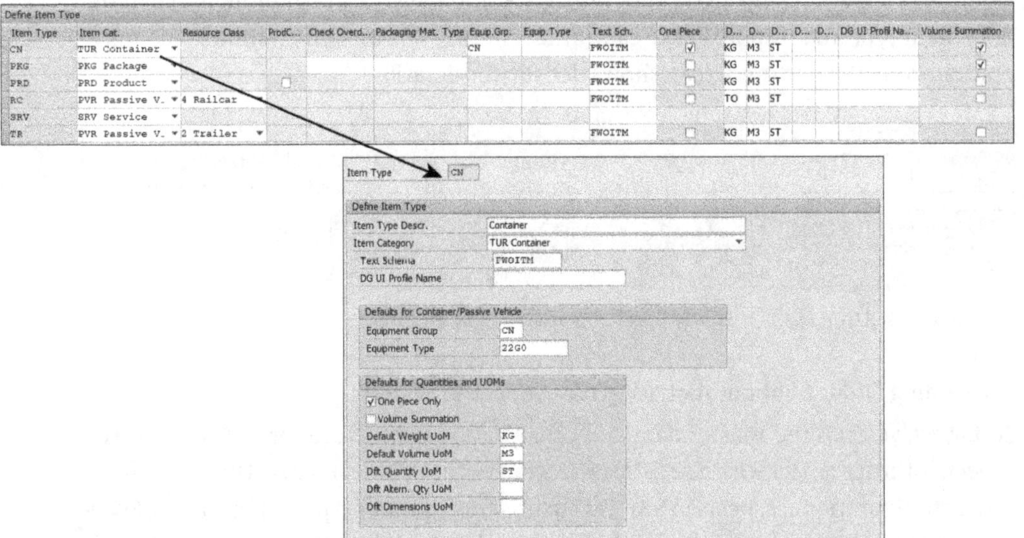

Figure 8.27 Define Item Types and Specifications for Forwarding Order/Forwarding Quotation

Define Default Agreement Partner Functions for Stages

In this configuration step, the agreement party role for a stage type and incoterm combination can be maintained. You can specify either **Prepaid Agreement Party** or **Collect Agreement Party** as the agreement party role.

When an incoterm is specified in a FWO, the system uses the agreement party role to identify the default agreement party for a stage. The system displays the relevant agreement party in the **Stages** tab page in the relevant FWO. The system uses the agreement party to determine the contract during charge calculation.

You can change the default agreement party by selecting a different agreement party role in the **Stages** tab page. Configuration logic for the standard shipper process is explained in the following steps:

1. Assume that the configuration table for incoterm FOB is maintained as shown in Table 8.2.

Incoterm	Stage Type	Agreement Party Role
FOB	Pickup	Prepaid agreement party
FOB	Pre-carriage	Prepaid agreement party
FOB	Main carriage	Collect agreement party
FOB	On-carriage	Collect agreement party
FOB	Delivery	Collect agreement party

Table 8.2 Defining Agreement Party Role for Each Stage Type per Incoterm

2. Business partners can be maintained for each party role, as shown in Table 8.3.

Business Partner	Party Role
BP1	Prepaid agreement party
BP2	Collect agreement party

Table 8.3 Business Partner for Each Party Role

3. Per the Customizing logic, if a FWO uses incoterm FOB, the system assigns two stages by default in the FWO shown in Table 8.4, based on the settings shown earlier in Table 8.2 and Table 8.3.

Stage Type	Agreement Party Role	Agreement Party
Pre-carriage	Prepaid agreement party	BP1
Main carriage	Collect agreement party	BP2

Table 8.4 Determination of Agreement Party in the FWO Based on Incoterm and Stage Types

Figure 8.28 provides a snapshot of the configuration for stage type definitions and assignment of the agreement party role to the info record and stage type combination:

❶ **Define Default Agreement Party Types for Stages**
Each stage of incoterm **FOB** gets an agreement party role.

❷ **Define Stage Types**
A stage type specifies whether a stage is main carriage, pre-carriage, or on-carriage.

Define Stage Types

In this step, stage types are defined by assigning stage categories. A stage type specifies whether a stage is main carriage, pre-carriage, or on-carriage. Standard settings have stage types 01 to 05. However, user-defined stage types can also be maintained.

While entering a stage in a FWO or FWQ, a stage type must be maintained. A stage type allows follow-up processes, such as invoicing or charge calculation, to identify what type of stage exists in the business document (e.g., pre-carriage or main carriage).

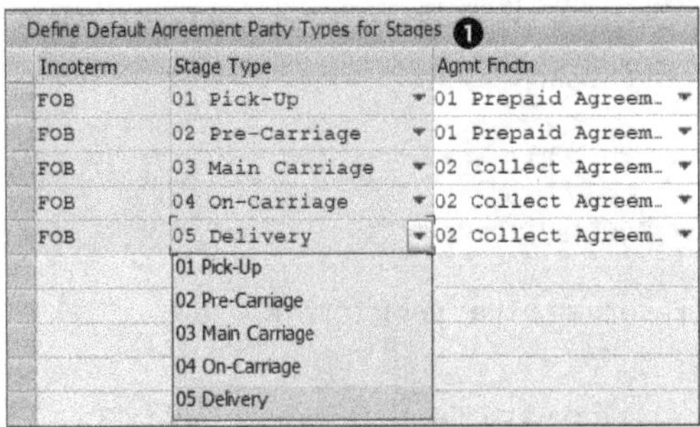
Define Default Agreement Party Types for Stages ❶

Incoterm	Stage Type	Agmt Fnctn
FOB	01 Pick-Up	01 Prepaid Agreem...
FOB	02 Pre-Carriage	01 Prepaid Agreem...
FOB	03 Main Carriage	02 Collect Agreem...
FOB	04 On-Carriage	02 Collect Agreem...
FOB	05 Delivery	02 Collect Agreem...

Define Stage Types ❷

Stage Type	Description	Stage Cat.	Ins. Set
01	Pick-Up	P Pre-Carriage	
02	Pre-Carriage	P Pre-Carriage	
03	Main Carriage	M Main Carriage	
04	On-Carriage	O On-Carriage	
05	Delivery	O On-Carriage	

Figure 8.28 Stage Types and Assignment of Agreement Party Role to the Combination of Incoterm and Stage Types

Define Movement Types

A movement type specifies for which stages the logistics service provider is in charge of organizing the transportation of the cargo. Furthermore, it defines which stage types are valid for the business document.

Movement type is maintained in the **General Data** settings of a FWO or FWQ. This is mandatory for the follow-on transactions such as transportation planning and the creation of freight orders or freight bookings.

In other words, movement type determines the routing, that is, source and destination locations. The system automatically sets the locations as a default value (location of the shipper/location of the consignee) based on the business partners maintained in a FWO or FWQ.

Define Stage Type Sequence for Movement Types

This configuration step handles the assignment of allowed stage types and their sequence to a specific movement type. The defined stages are then automatically entered in the actual route on the FWO or FWQ documents.

Stage types that aren't defined for a movement type can't be used in FWOs or FWQs with the movement type. Stage types that are defined for a movement type must be part of the actual route of the FWO or FWQ.

Figure 8.29 shows the configuration details for movement types, allowed stage types, and sequence per movement type:

❶ **Define Movement Types**
The movement type determines the routing, that is, the source and destination locations. Furthermore, it defines which stage types are valid for the business document.

❷ **Define Stage Type Sequence for Movement Type**
Checked for mandatory stage types that must occur when the respective movement type is selected. Stage types 02, 03, and 05 are mandatory, and 02 and 04 can be elective.

Define Movement Types ❶

Mov. Type	Description	Src. Loc.	Dest. Loc.
CC	CFS to CFS	☐	☐
CD	CFS to Door	☐	☑
CP	CFS to Port	☐	☐
DC	Door to CFS	☑	☐
DD	Door to Door	☑	☑
DP	Door to Port	☑	☐
PC	Port to CFS	☐	☐
PD	Port to Door	☐	☑
PP	Port to Port	☐	☐

Define Stage Type Sequence for Movement Type ❷

Mov. Type	Seq. No.	Stage Type	StgeTpeOcc	StageProp.	Det. Rule	Set. Rule
CC	1	02	CN Stage type can occur in ...	☐	Not Pl.-Rel. if Ex. and Sa...	Not Relevant for Internal ...
CC	2	03	N Stage type must occur at ...	☑	Not Pl.-Rel. if Ex. and Sa...	Not Relevant for Internal ...
CC	3	04	CN Stage type can occur in ...	☐	Not Pl.-Rel. if Ex. and Sa...	Not Relevant for Internal ...
CD	1	02	CN Stage type can occur in ...	☐	Not Pl.-Rel. if Ex. and Sa...	Not Relevant for Internal ...
CD	2	03	N Stage type must occur at ...	☑	Not Pl.-Rel. if Ex. and Sa...	Not Relevant for Internal ...
CD	3	04	CN Stage type can occur in ...	☐	Not Pl.-Rel. if Ex. and Sa...	Not Relevant for Internal ...
CD	4	05	N Stage type must occur at ...	☑	Not Pl.-Rel. if Ex. and Sa...	Not Relevant for Internal ...

Figure 8.29 Movement Types and Allowed Stage Types per Movement Type

Define Allowed Transportation Modes for Stage Types

In this Customizing activity, transportation modes are assigned to a stage type. This allows the system to determine which transportation modes can be used for a stage type. You can also define a default transportation mode for a stage type. If this transportation mode complies with the transportation mode that has been entered for the entire FWO or FWQ, it's entered automatically for a stage of this type when the stage is created on the FWO or FWQ document.

If no transportation modes have been defined for a particular stage type, all transportation modes can be used for a specific stage type. In such a case, the system takes the transportation mode from the general data.

> **Prerequisite**
> Transportation modes are maintained in Customizing.

Figure 8.30 shows **Define Allowed Transportation Modes for Stage Type** area with **Stage Type 01** (Pickup), which can use transportation mode 01 and 02 only, even though transportation mode 01 is set as default for **Stage Type 01**.

Define Allowed Transportation Mode for Stage Type

Stage Type	Transpo...	Dft TrM
01	01	☑
01	02	☐
02	01	☑
02	02	☐
02	04	☐
03	01	☐
03	02	☐
03	03	☐
03	04	☐
03	05	☐
04	01	☑
04	02	☐
04	04	☐
05	01	☑
05	02	☐

Transportation Mode Code

Restrictions

MT	Description
01	Road
02	Rail
03	Sea
04	Inland Waterway
05	Air
06	Postal Service
07	Intermodal
51	Parcel Service

Figure 8.30 Maintaining Allowed Transportation Mode(s) per Stage Type for Forwarding Order/Forwarding Quotation

Define Shipping Types

In this Customizing activity, you define shipping types and assign them to transportation modes. You can define shipping types and assign them to transportation modes, as shown in Figure 8.31. Specify whether cargo items such as packages and products are to be assigned to equipment items.

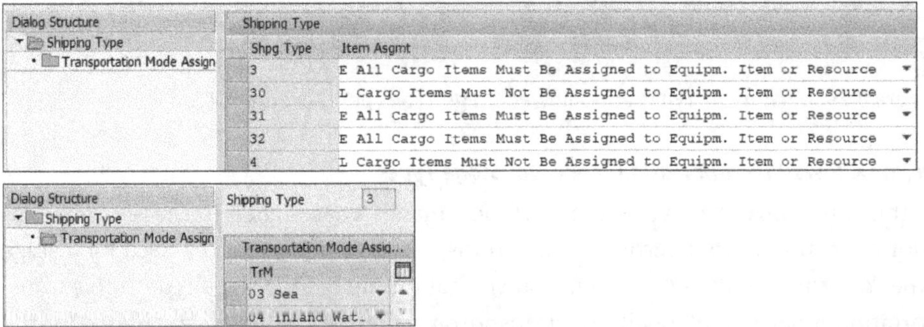

Dialog Structure
- Shipping Type
  - Transportation Mode Assign

Shipping Type

Shpg Type	Item Asgmt
3	E All Cargo Items Must Be Assigned to Equipm. Item or Resource
30	L Cargo Items Must Not Be Assigned to Equipm. Item or Resource
31	E All Cargo Items Must Be Assigned to Equipm. Item or Resource
32	E All Cargo Items Must Be Assigned to Equipm. Item or Resource
4	L Cargo Items Must Not Be Assigned to Equipm. Item or Resource

Dialog Structure
- Shipping Type
  - Transportation Mode Assign

Shipping Type 3

Transportation Mode Assig...

TrM
03 Sea
04 Inland Wat...

Figure 8.31 Maintaining Shipping Types and Their Assignment to Transportation Modes

Define Stage Profiles

A stage profile specifies which stage types are to be created for freight units in a FWO or a FWQ, the sequence of the stage types, and certain properties of the stages. While creating a stage profile, the system checks that the sequence of the stage types is correct, as determined by the corresponding stage category. The following sequence is allowed:

1. Customer self-delivery
2. Pre-carriage
3. Main carriage
4. On-carriage
5. Customer pick-up

A stage profile can be assigned to the business document type in the following ways:

- Assign a stage profile directly to the business document type.
- Assign a stage profile to a condition and then assign the condition to the business document type. Condition type stage profile (/SCMTMS/TRQ_STG_PRF) must be used. This condition enables you to define the allowed stages for the movement type, transportation mode, traffic direction, or a combination of these values.

Note

If both options are used to assign a stage profile to a business document type, the condition always takes priority over the directly assigned stage profile.

When the system creates the stages for the freight units, it automatically enters the stages according to the settings in the relevant stage profile. The stages are entered in the actual route on the **Actual Route** tab page of the FWO or FWQ. Stage types that aren't specified in the stage profile can't be used in business documents of this type. Stage types that are specified in a stage profile must be part of the actual route of business documents of this type.

Allowed stage types can also be specified for a specific movement type in the Customizing activity **Define Stage Type Sequence for Movement Types** (Figure 8.29).

Note

While defining the business document type (e.g., FWO type), it must be defined whether stage types are to be determined based on the movement type or the stage profile.

Figure 8.32 show a typical stage profile:

❶ The **StageProp.** column indicates that a stage proposal is automatically generated.

❷ The **StgeTpeOcc** column indicates the number of times that a particular type of a transportation stage occurs in a sequence of stages. For the **Stage type must occur at least once in this field** option, the **Stage Prop.** checkbox must be selected.

❸ The **Det. Rule** checkbox indicates the rule that the system applies to determine whether a stage is relevant for planning.

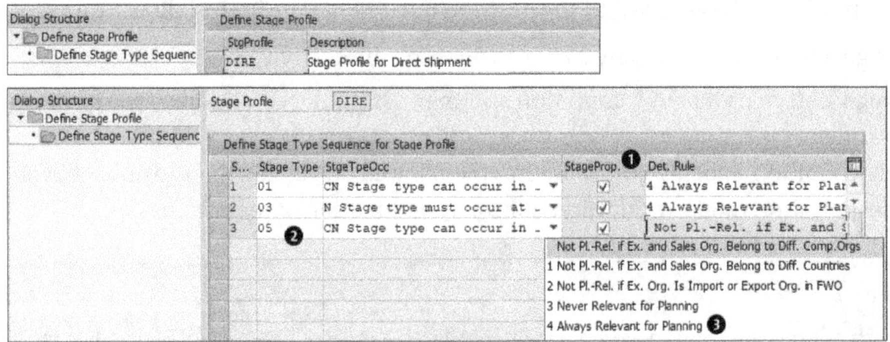

Figure 8.32 Maintaining the Stage Profile

Define Forwarding Order Types

Due to the requirement to operate globally, third-party logistics use different means of transport such as road, rail, sea, or air freights. Each transportation mode requires its own specific information to be maintained in FWO. In most cases, the mode of transportation is predefined by the customer based on the locations involved in the transportation. Different processing methods may be needed for FWOs based on the customer, region, or product being handled. Figure 8.33 illustrates the different components of a FWO and their purpose.

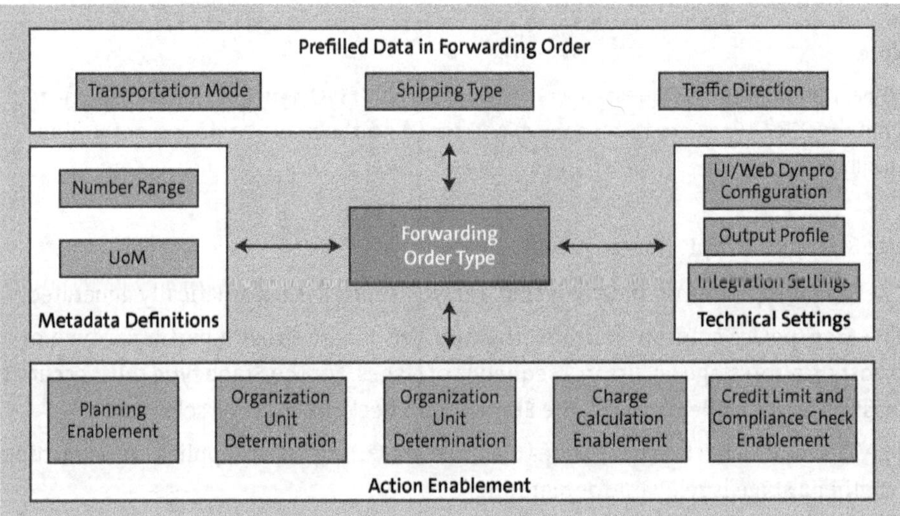

Figure 8.33 Forwarding Order Types

In this Customizing activity, types of FWOs can be maintained along with parameters for each section as depicted in Figure 8.33. These parameters influence how the system processes this business document; for example, the FWO type determines the number range interval that is used to generate a number for a FWO, or it defines whether freight units are created automatically when you save the FWO.

Furthermore, you can assign planning profiles and dangerous goods profiles that determine how follow-up steps such as planning and dangerous goods checks are performed.

Organizational unit determination parameters define how the system determines the sales organization. The system response happens in the following sequence:

1. The system checks if **Condition** is maintained for **Default Sales Organization**.
2. If no such condition is maintained or if the condition doesn't find a sales organization, the system checks if the **Consider User Assignment for Org. Unit Determination** checkbox is selected, and if the user is assigned to a sales organization.
3. If the system doesn't find anything, it uses the sales organization that is maintained in the **Sales Organization** field in the FWO.

Figure 8.34 provides a snapshot of FWO type configuration, showing how to define FWO types (**Default Units of Measure**).

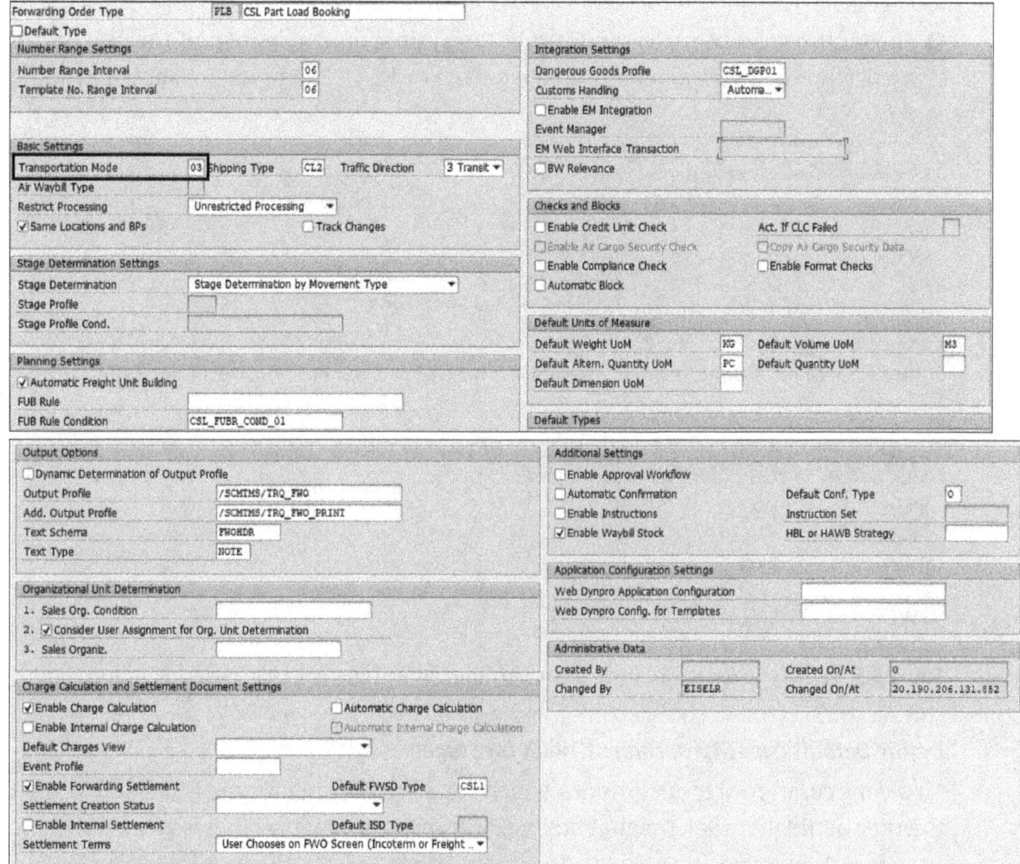

Figure 8.34 Configuration of Forwarding Order Types

Assign Item Types to Forwarding Order Types

This configuration step specifies valid item types and item categories for a FWO type. This specification limits the creation of items of the item type and category (only as specified in configuration) while creating a FWO of the corresponding FWO type.

In addition, you can define the following:

- Specify a default item type for a combination of item category and FWO type. This means that you can predefine a default item type for items of a certain item category (e.g., a product) that are created for a FWO of a certain FWO type. It's used only in cases in which FWO data is transferred using EDI.
- Define that items of a certain type are to be entered automatically as initial items when you manually create a FWO.

Figure 8.35 shows the **Assign Item Types to Forwarding Order Types** screen:

❶ The **DefItm Type** column indicates that this item type is the default type for items in this category. Used only in cases when FWO data is transferred using EDI.

❷ The **InitItm** column indicates whether the system enters an item of this type automatically when you manually create a FWO or FWQ. Examples are shown in Table 8.5.

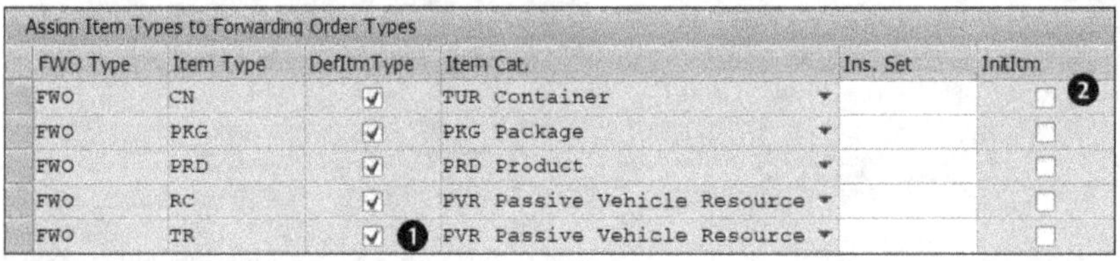

Assign Item Types to Forwarding Order Types

FWO Type	Item Type	DefItmType	Item Cat.	Ins. Set	InitItm
FWO	CN	☑	TUR Container		☐ ❷
FWO	PKG	☑	PKG Package		☐
FWO	PRD	☑	PRD Product		☐
FWO	RC	☑	PVR Passive Vehicle Resource		☐
FWO	TR	☑ ❶	PVR Passive Vehicle Resource		☐

Figure 8.35 Maintaining Item Types and Default Options per Forwarding Order Type

FWO Type	Item Type	InitItm
FWO	CN	X
FWO	PKG	X
FWO	SRV	

Table 8.5 Item Type Assignment to a Forwarding Order Type with the Default Option

Define Default Capacity Document Types for Stages

This configuration step determines which capacity document type is created against the type of freight order, freight booking, or transportation unit that is created specific to a stage. While creating capacity documents directly from the user interface of a forwarding order or its successor documents (such as freight units or transportation units), the relevant capacity document type is determined based on this setting. You can also specify which actions are allowed when you create or delete (or assign and unassign) capacity documents directly from the forwarding order, freight unit, or transportation unit.

The following combination of system objects are used to define this setting:

- Forwarding order type
- Shipping type
- Stage type
- Transportation mode
- Sales organization

If you leave one or more of these fields empty, the specified parameters apply to all possible values of these fields, similar to a wildcard.

Prerequisites

You must have first defined the forwarding order types, stage types, and transportation modes in the following Customizing activities:

- Define Forwarding Order Types
- Define Stage Types
- Define Transportation Mode

An example of parameter setting is shown in Table 8.6. In this example, a stage is maintained for a forwarding order with shipping type LCL. The same stage type (Main Carriage) is maintained with transportation mode Sea. For this stage, both entries from the example Customizing are valid. However, the system can consider only one of these entries and so one needs to define an access sequence for them. Figure 8.36 represents parameter settings for capacity document types.

Forwarding Order Type	Shipping Type	Stage Type	Transportation Mode	Sales Organization	Document Type
FWO1	LCL	Main Carriage	–	–	FRO1
FWO1	–	Main Carriage	Sea	–	FRO2

Table 8.6 Capacity Document Types Determination Based on Forward Order Type, Stage Type, and Transportation Mode

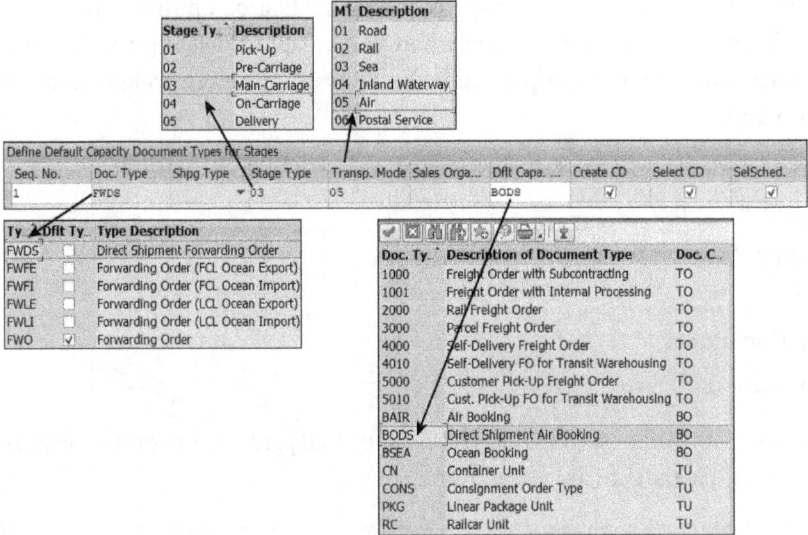

Figure 8.36 Parameter Settings for Capacity Document Type

8.4.2 Financial Processes in SAP Transportation Management

When shipping products to customers via a third party, an organization needs to account for and recognize the costs involved in moving goods. SAP TM provides a broad and flexible calculation process. With the integration into materials management, integration to the financial accounting and controlling applications allows an organization to perform payment and execute billing transactions, as well as trigger the settlement process to accrue the proper transportation charges to financial accounts.

In the following sections, we'll look into the basic principles of charge calculation in SAP TM, its role in freight and FWAs, and all relevant configurational aspects. This is followed by a few examples of charge calculation in freight and forwarding settlements.

Charge Management Master Data

The system calculates transportation charges based on the following master data in charge management (see Figure 8.37):

- **Scale template**
  This template is used to easily create the scale.
- **Scale**
  This represents the dimensions of a rate table.
- **Rate table**
  A rate is a price for a specific transportation service that is valid for the period defined by its validity. A rate is listed in a rate table, where the freight rates are maintained based on the scale specifications.

- **Charge types**
  The charge type describes a specific charge, for example, HAUL (haulage), BAF (bunker adjustment factor), or FSC (fuel surcharge).
- **Calculation sheet**
  This is a hierarchical table used to calculate transportation charges. It combines the charge types permitted for a document and the sequence in which the system takes these charge types into account during the calculation.
- **Forwarding (or freight) agreement**
  A FWA describes the contractual relationship between the customer and the purchasing organizational unit. It contains payment terms, limits, validity period, and valid calculation sheets. Information such as organization unit, contracting parties, and conditions can be specified.

Agreements are the basis for calculating transportation charges and can involve one or more parties. Use FWAs to calculate transportation charges billable to the customer. An agreement includes contractual data, as follows:

- Organizational unit, for example, the sales or purchasing organization
- Involved parties, for example, the ordering party or carrier
- Terms of payment
- Validity dates

The system determines the agreement based on the organization, business partner, validity period, and preconditions. Agreement determination is based mainly on the partner relationship.

Figure 8.37 is a schematic representation of components of charge calculation and how they are created and linked with each other.

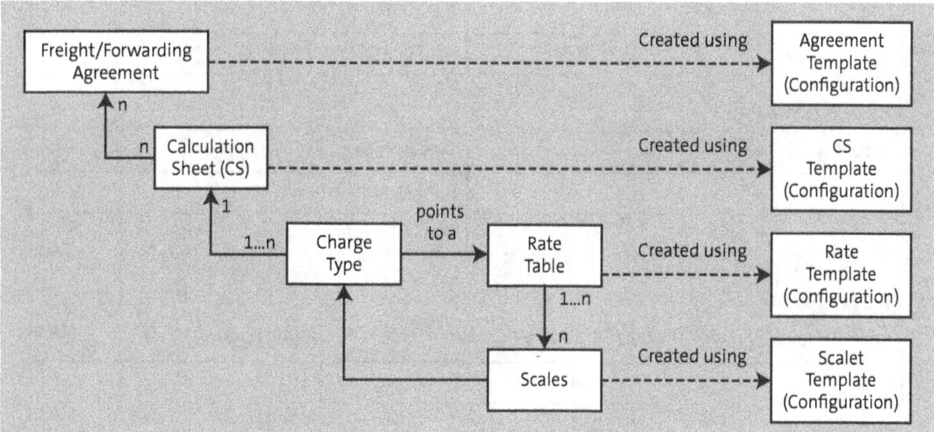

Figure 8.37 Transportation Charge Management Master Data

Calculation Sheet

Calculation sheet maintenance is a step in the process of setting up master data in charge management, and it's integrated with agreements, rate tables, and scales. You can use a template to create a calculation sheet and calculation sheet line items.

The calculation sheet is a hierarchical table used to calculate transportation charges. It combines the charge types permitted for a document and the sequence in which the system takes these charge types into account during the calculation. The system uses the calculation sheet to specify which transportation charges to calculate and how to calculate them. This helps to pay suppliers or carriers for subcontracted transportation services. The calculation sheet points to charge items, and each charge item can result in a calculated amount. Calculation sheets can consist of the following items:

- Charge type
- Rate and rate table
- Scales
- Scale base and calculation base

We'll discuss each of these in more detail in the following sections.

Charge Type

A charge type is the classification of a charge line that determines how the system calculates the transportation charges for the charge line, for example, a base rate, surcharge, or discount. A charge type can result in a positive or negative value for a charge line. Charge types can be specified as an amount or a percentage value.

Charge Category

Category	Description
004	Basic Freight
011	Accessorial Charges

Charge Subcategory

Subcategory	Description
100000	Bunker Adjustment Factor (BAF)
101021	Ocean Freight

Charge Types

Charge Type	Category	Subcategory
HAUL	004	101021
ZS01	011	100000

Figure 8.38 Charge Types, Charge Categories, and Subcategories

Charge types can be grouped into charge categories. Charge subcategories of charge types are used to group and categorize charge types in a more granular way than charge categories. Figure 8.38 depicts the relationship between charge type, charge category, and charge subcategory.

For charge category Customizing, use menu path **SPRO • SAP Reference IMG • Transportation Management • Basic Functions • Charge Calculation • Basic Settings • Define Charge Categories**. Charge subcategories for charge types can be maintained using menu path **SPRO • SAP Reference IMG • Transportation Management • Basic Functions • Charge Calculation • Basic Settings • Define Charge Subcategories**. Charge categories and subcategories can be assigned in the charge type Customizing activity, and they can be used to determine the service material during settlement.

Charge types can be defined using menu path **SPRO • SAP Reference IMG • Transportation Management • Basic Functions • Charge Calculation • Basic Settings • Define Charge Types**. As shown in Figure 8.39, you can do the following:

❶ Specify whether a charge type can result in a positive or negative value.

❷ Specify whether a charge type can be an absolute value or a percentage value.

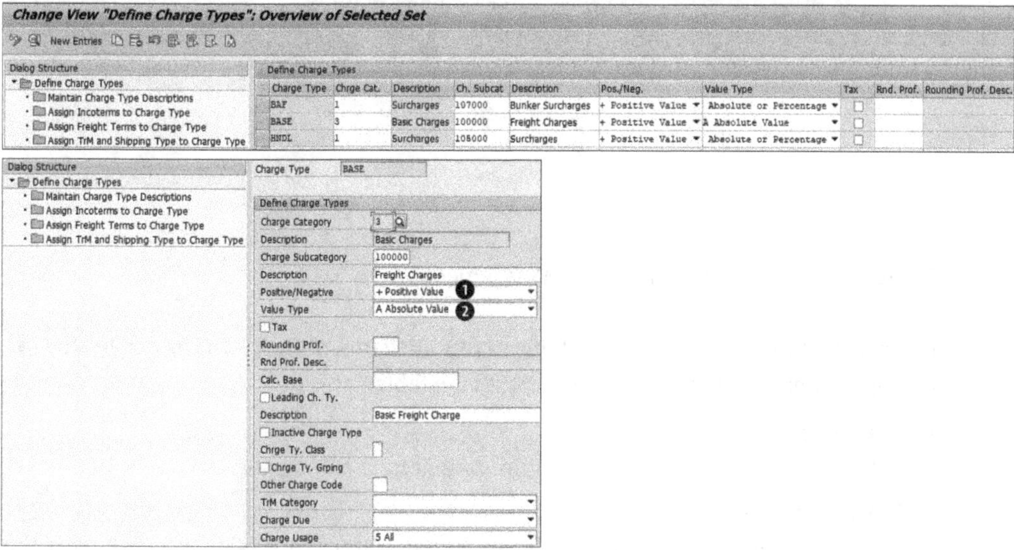

Figure 8.39 Charge Type Configuration

Rate and Rate Table

A rate is a price for a certain transportation service that applies only during its validity period. In SAP TM, rates are listed in a rate table. Figure 8.40 provides a snapshot of the rate table construct along with validity and status.

Validity Period: 01.01.2024 - 12.31.2024 | Status: RELEASED

Destination Zone	Origin = Plant Chicago				
	⇐ 100 kg	⇐ 500 kg	⇐ 1,000 kg	⇐ 2,000 kg	⇐5,000 kg
TX	$60.00	$51.00	$45.90	$43.61	$30.52
IN	$30.68	$26.08	$23.47	$22.30	$15.61
KY	$35.68	$30.33	$27.30	$25.93	$18.15
CA	$85.00	$72.25	$65.03	$61.77	$43.24
FL	$55.90	$47.52	$42.76	$40.63	$28.44
LA	$57.00	$48.45	$43.61	$41.42	$29.00
CO	$45.00	$38.25	$34.43	$32.70	$22.89
GA	$55.90	$47.52	$42.76	$40.63	$28.44
WA	$55.90	$47.52	$42.76	$40.63	$28.44

Figure 8.40 Overview of a Rate Table

Figure 8.41 shows the SAP Fiori tiles used for charge management master data creation.

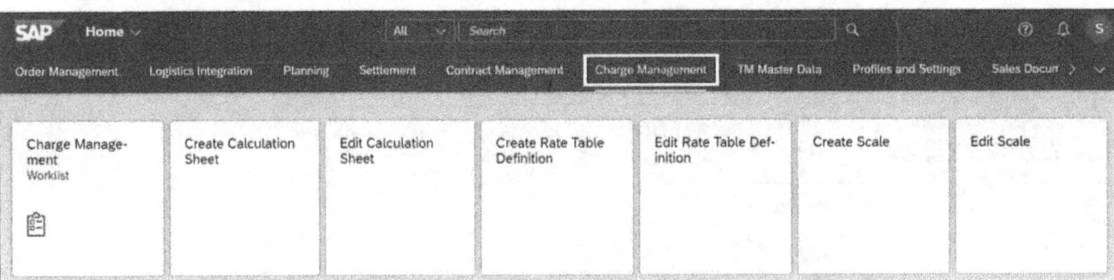

Figure 8.41 SAP Fiori Tiles for Charge Management

Figure 8.42 shows an example rate table maintained in SAP Fiori. Among others, the following elements can be maintained in the rate table:

- **Charge Usage**
  Define the charge usage as relevant for payment to the carrier.
- **Charge Type**
  The system only assigns rate tables to a charge item in a calculation sheet if the charge types match. You can use this field to search for a rate table.
- **Dimensions**
  You add each dimension on which you want your rate to be defined and looked up. For example, if you want the rate to be defined based on destination location and weight, add a dimension for destination location and for weight. The number of dimensions that you can use in one rate table is restricted to three for the (basic) transportation management scope.

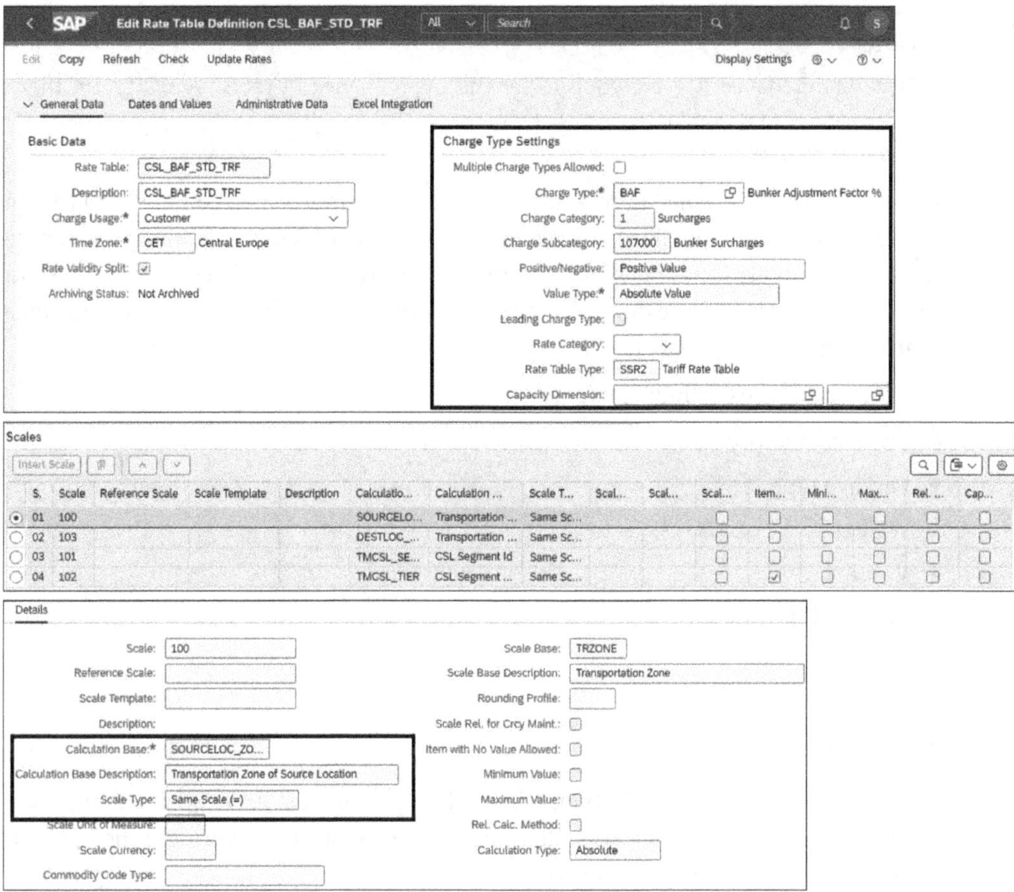

Figure 8.42 Different Sections of a Rate Table (Part 1)

- **Calculation Base (for each scale)**
  You can define the calculation base for the scale. The calculation base is the actual base or the factor on which the rate retrieval is performed.
- **Relevance for Calculation Method**
  If you select this checkbox, you define whether you want the system to apply a special calculation method, such as break weight or clipping, to the rate table on the level of a charge item.

Scales

If a rate is a price for a certain transportation service that applies only during its validity period, a scale represents a dimension of the rate. In effect, a scale is used to define a parameter that then defines a rate. For example, if a rate depends upon distance and weight, you must define a separate scale for distance and for weight. The distance and weight scales are then used to define the rate.

Note

Scale maintenance is a prerequisite for the maintenance of rates, as scales are the dimensions of rate tables.

A scale can be maintained independently of the agreement and the rates, and can be reused in multiple rate tables. A scale has a scale base, which is essentially a database field tied to an attribute of the freight order. The scale base defines how the system interprets the scale. For example, a rate based on distance has to have the distance maintained.

Possible scale bases include the following:

- Product quantity
- Package type
- Weight
- Volume
- Means of transport
- Location
- Distance
- Business partner

Each scale is based on a scale type. The scale type controls how the scale levels are defined. It indicates whether the scale values that were specified represent the upper or lower boundaries for the scale levels.

Depending on the scale base, a scale UoM can be defined. This is a mandatory field if the UoM is relevant for the scale base. Weight, for example, requires a UoM.

In relation to calculation type, you can choose between absolute and relative. For non-numeric scales, you can only choose absolute. The system uses the calculation type on the scale level as the proposed value for rate table maintenance.

Figure 8.43 shows different sales with scale bases of weight, distance, and country. While the scale types for weight and distance are numeric (to scale), country has a scale type as absolute (same scale).

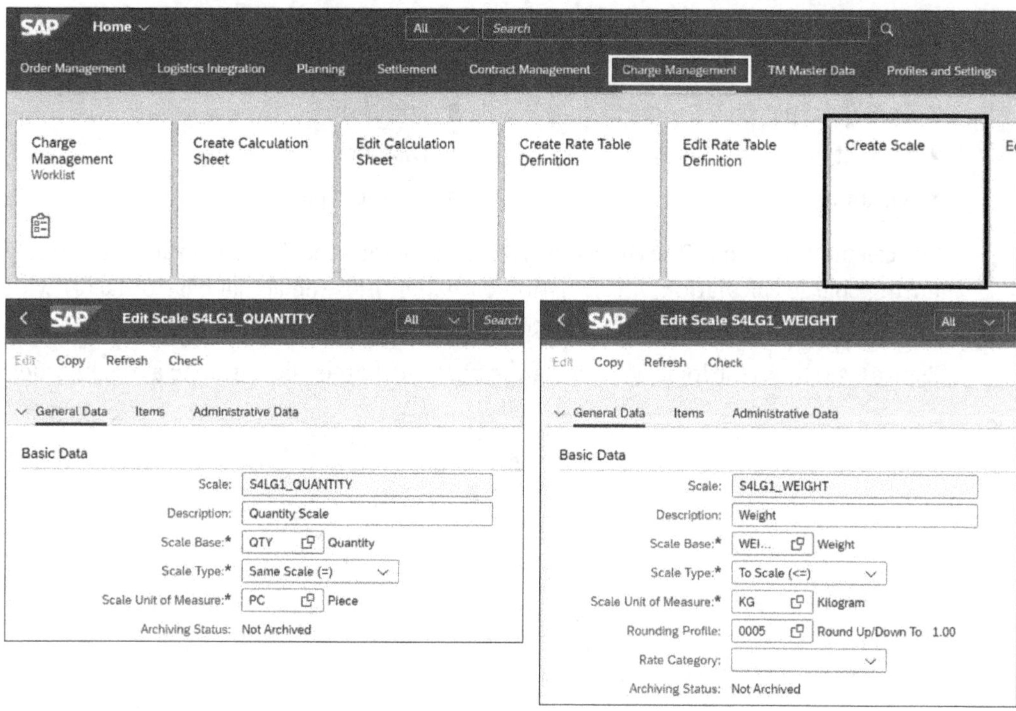

Figure 8.43 Scales with Different Scale Bases

Scale Base and Calculation Base

A scale base is a parameter or a factor that can influence transportation charges (see Figure 8.44).

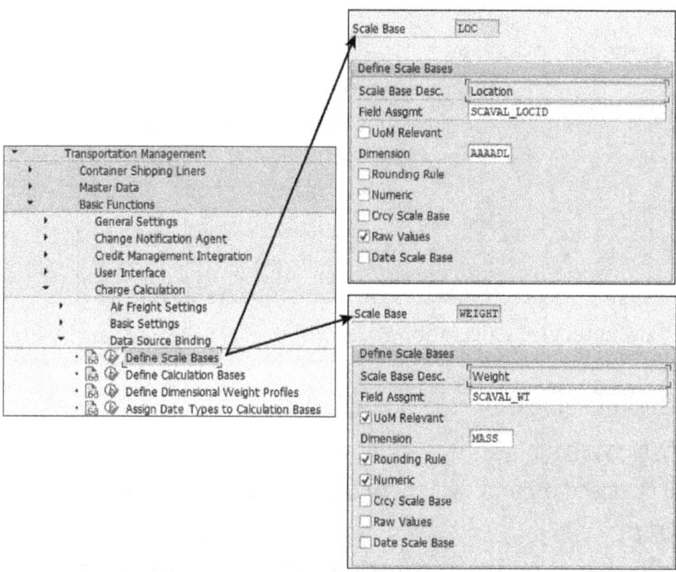

Figure 8.44 Scale Base for WEIGHT and LOCATION

It helps the system interpret the scale. Possible scale bases include the following:

- Product
- Package type
- Weight
- Volume
- Means of transport
- Location
- Distance
- Business partner

The calculation base defines how the system will determine the actual value of the scale during charge calculation. A calculation base identifies the actual base or factor with which the system calculates the charges for the scale.

Figure 8.45 shows the logic of a calculation base for charge calculation in a freight order.

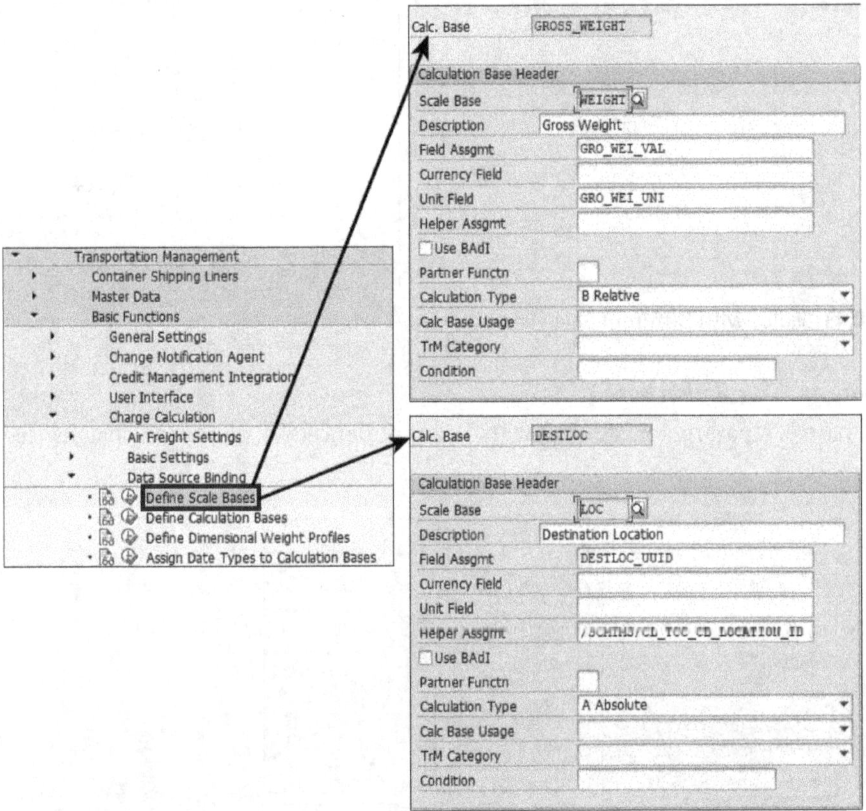

Figure 8.45 Configuration Parameters for Calculation Base

Figure 8.46 provides a snapshot of logic for calculation bases:

❶ **Calculation base: GROSS_WEIGHT**
Charge calculation will be based on total weight of freight order/booking.

❷ **Calculation base: DESTLOC**
Charge calculation will be based on the destination location of each stage of the freight order/booking.

Helper Assignment

BAF/Container Dry			
	Newark	Portland	Miami ❷
Up to 10 tons ❶	$42.00	$48.00	$55.00
Up to 10 tons	$36.00	$39.60	$48.00
Up to 10 tons	$30.00	$33.65	$42.00

Figure 8.46 Calculation Base

A calculation base is defined for every scale base. This calculation base is the final element in the charge determination. This means that the calculation base is used to determine the charges.

In Figure 8.46, the charges are dependent on weight and the destination location. The scale has the scale base location. This scale base defines that the scale uses locations only. The scale base can be used for both the source and destination location. The calculation base DESTLOC determines that the destination location of the freight order is used. Technically, in the Customizing of the calculation base, you define which field is to be used and which program or class the system will call to determine the field value.

If the resolution base in the calculation sheet points to the freight order stage, the system will derive the values from the specific stage and use it in the calculation.

Standard Charge Calculation

In this process, the system calculates the charges for transporting goods between locations based on the rates maintained in a rate table. The system uses master data, such as agreements, calculation sheets, and rate tables, when calculating transportation charges.

Charge calculation can be triggered from the following SAP TM components:

- Forwarding order management
- Freight order management
- Planning
- Forwarding settlement
- Freight settlement

Note

The system allows manual editing of the transportation charges even after charge calculation.

Figure 8.47 shows the logic and process of charge calculation in SAP TM.

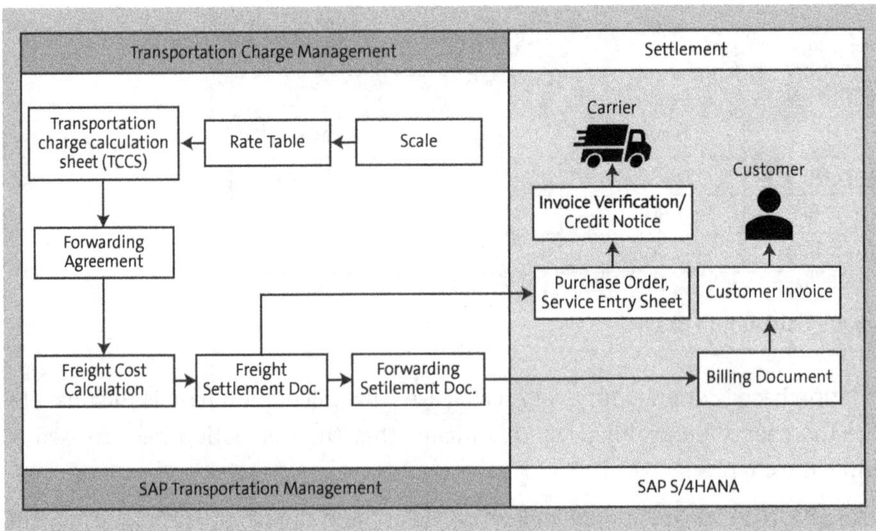

Figure 8.47 Charge Calculation and Settlement Process

Figure 8.48 provides a snapshot of freight agreements, including the **Purchasing Organization**, **Carrier**, **Valid-From Date**, **Valid-To Date**, **Document Currency**, **Calculation Sheet**, and **Rate Table**.

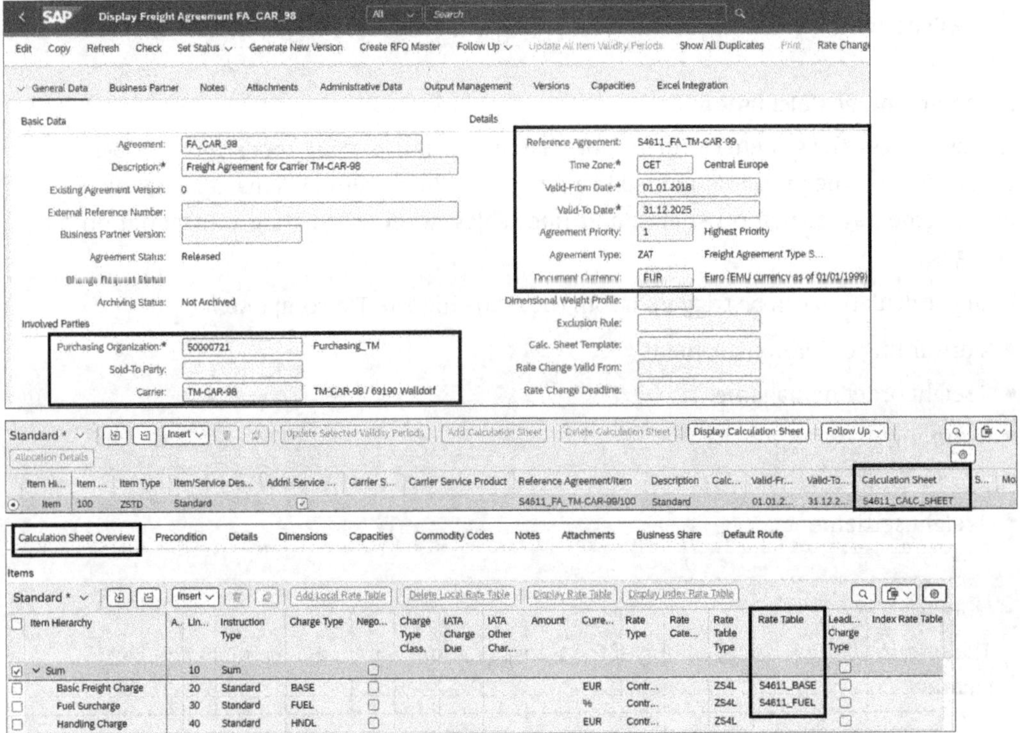

Figure 8.48 Freight Booking Document

Figure 8.49 gives an overview of FWAs, including the **Valid-From Date**, **Valid-To Date**, **Document Currency**, **Sales Organization**, and **Calculation Sheet**.

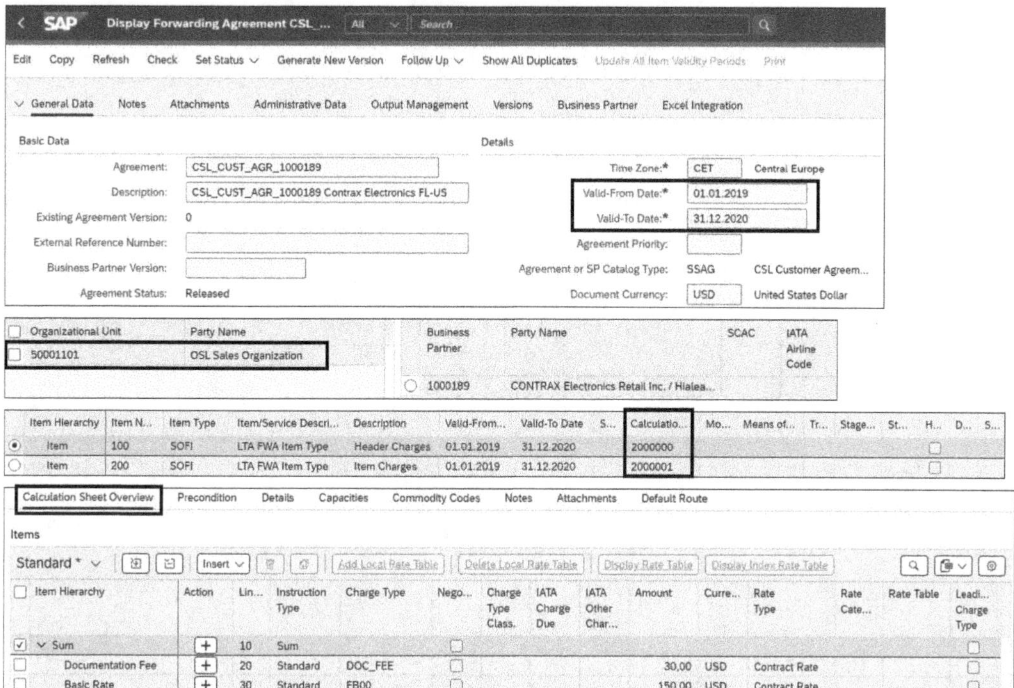

Figure 8.49 Forwarding Agreement Document

8.5 Summary

An efficient transportation management system that is properly integrated with the overall supply chain network enables businesses to manage all aspects of shipping materials, components, and finished goods using external freight carriers or an internal fleet. For global organizations that have complex logistics processes or logistics service providers handling transportation requirements of multiple clients, a transportation management system has become a necessity for effective planning and efficient execution of transportation functions.

State-of-the-art transportation management systems provide direct communication and integration with freight carriers, and some cloud-based systems give shippers access to carrier networks with thousands of prescreened carriers, easing the selection and management of multiple cargo carriers.

For third-party logistics partners, a transportation management system helps logistics service providers efficiently manage service-based rate sheets aligned with agreed-upon terms and conditions with vendors/customers.

Chapter 9
Application Link Enabling and IDocs

The SAP interfaces enable business process integration and asynchronous data communication between two or more SAP systems or between SAP and external systems. Electronic data interchange (EDI) is a system or method for exchanging business documents with trading partners and uses a highly secure, machine-readable, standardized format. This chapter focuses on the widely used EDI IDoc method that can seamlessly integrate with any SAP or non-SAP systems. You'll learn how to configure and set up Application Link Enabling (ALE) and IDocs in SAP S/4HANA.

The IDoc interface is basically used for exchanging data between SAP S/4HANA and various external trading partner systems. To exchange data between SAP S/4HANA systems and non-SAP systems, we normally use ALE. The IDoc format is used for exchanging data between systems that have a common data structure format.

When configuring ALE in SAP S/4HANA, the core task is to create what is called a distribution model. There are many predefined distribution models available as standard in an SAP S/4HANA system, but you can also create a customizable distribution model of your own exit.

In this chapter, we'll discuss ALE and IDoc technologies in details, including what you need to integrate third-party logistics systems with SAP S/4HANA systems.

9.1 IDoc Interface and Application Link Enabling

IDoc is an SAP proprietary format for storing business data that needs to be communicated across systems. It uses a standard data structure, along with processing logic for this data structure. The IDoc interface is used in the following scenarios:

- With EDI
- When connecting other business application systems (e.g., PC applications, external workflow tools) by IDoc
- With ALE

ALE is a technology to create and run distributed applications. ALE is used for distributing large volumes of similar data between different logical systems.

This section will mainly focus on system readiness and configuration steps for ALE.

9.1.1 Logical Systems and Assignments

A logical system is a unique ID within an SAP S/4HANA system that identifies a partner communication system. This partner system can be another SAP system or SAP S/4HANA system, or it can be an external system.

To avoid any confusion, logical systems in a distributed landscape should use a unique ID (name). This name should be assigned to one of the clients in an SAP or SAP S/4HANA system. If you want to set up two clients for the purpose, you must notify them which logical systems exist in the distribution landscape and provide a description of the client.

When you execute a transaction from SAP S/4HANA that triggers the output type as external and sends an IDoc from a source logical system to a destination logical system (e.g., sending and receiving partner), the transaction is implemented by sending the IDocs from one client to the other. Generally, both the source and destination end of the transaction must be configured, but in our environment, the source logical system and client are already configured for us. We're only interested in configuring a destination to send IDocs.

In the following subsections, we'll follow these steps to use ALE, IDoc, and Remote Function Calls (RFCs).

Define Logical System

The menu path to create logical systems in SAP S/4HANA is **SPRO • Reference IMG • Application Server • IDoc Interface/Application Link Enabling (ALE) • Logical Systems • Define Logical System** (Transaction SALE). Choose **New Entries**, and in the subsequent screen, enter a name and its description for the logical system you need to create, and click on the **Save** button (Figure 9.1). Per the naming standard, we normally follow this convention: "system name" + "CLNT" + client number (numeric).

Assign Logical System to Client

Next, you need to map the logical system you just created to the client. To assign the logical system to the client in SAP S/4HANA, you can use menu path **SPRO • Reference IMG • Application Server • IDoc Interface/Application Link Enabling (ALE) • Logical Systems • Assign Logical System to Client** (Transaction SALE).

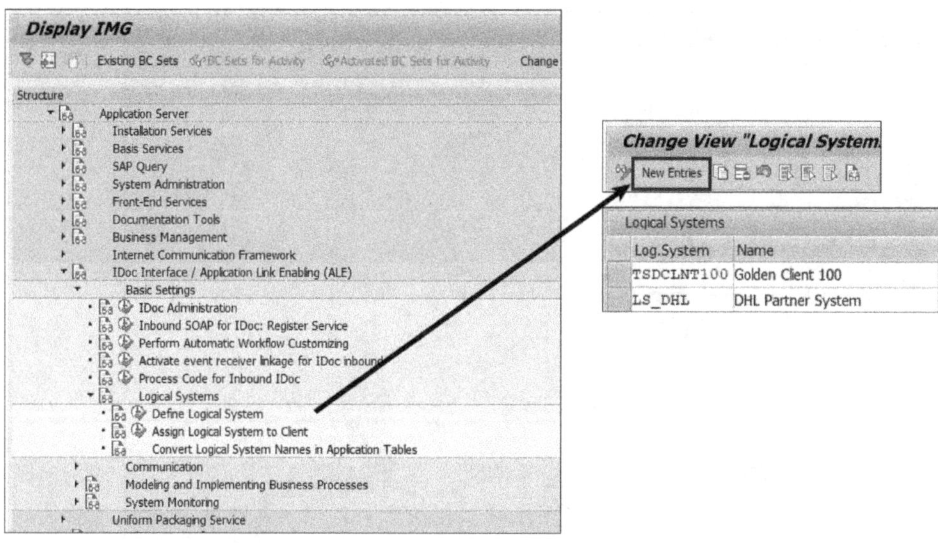

Figure 9.1 Navigation Path for Creation of Logical Systems

Click the **New Entries** button, and enter a client number not already defined in the system, as shown in Figure 9.2. In addition, you need to enter a city and the logical system name created in the previous section. Select the name of a currency code from the list (e.g., **USD** = US dollars), and choose a client role from the list. The most important selection is the **Logical system**. Finally, save your entry. These entries are summarized in Table 9.1.

Entries	Value
Client number	"100"
City	<Name of the City>
Logical system	"TSDCLNT100"
Std currency	"USD"
Client role	**Development**, **Test**, or **Production**

Table 9.1 Assign the Logical System to the Client

> **Note**
>
> You might not have all the authorizations in the system, as some of the transactions require Basis and security teams to look into.

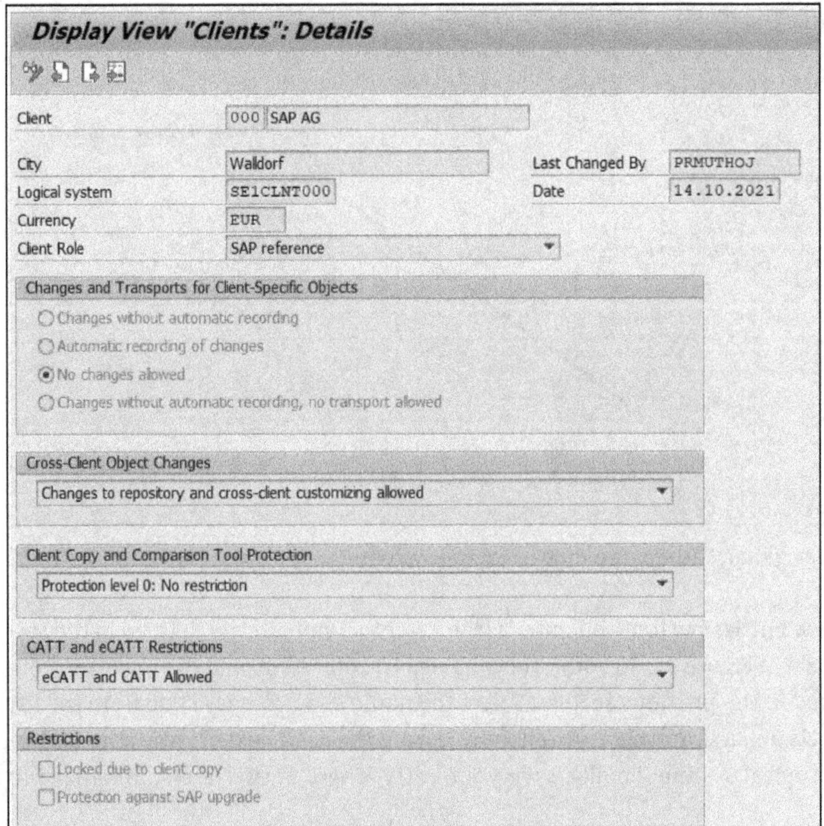

Figure 9.2 Assignment of the Logical System to the Client

9.1.2 Target Systems for Remote Function Calls

You must know how to connect two logical systems to communicate with each other by creating an RFC destination. Because we're dealing with an external system via SAP Process Integration/SAP Process Orchestration, we've configured a transactional RFC (tRFC) connection via TCP/IP to send/receive IDocs from the sending SAP S/4HANA system and to send a status update back that indicates we've received the IDoc.

To create RFC connections in SAP S/4HANA, you can use configuration path **SPRO • SAP Reference IMG • SAP NetWeaver • Application Server • IDoc Interface/Application Link Enabling (ALE) • Communication • Create RFC Connections** (see Figure 9.3). You can directly create RFC connections by executing Transaction SM59.

Select **TCP/IP Connections** of type **T**, and click on the **Create** button (Figure 9.4).

> **Note**
>
> If the RFC destination/connection is already there, then don't create it again.

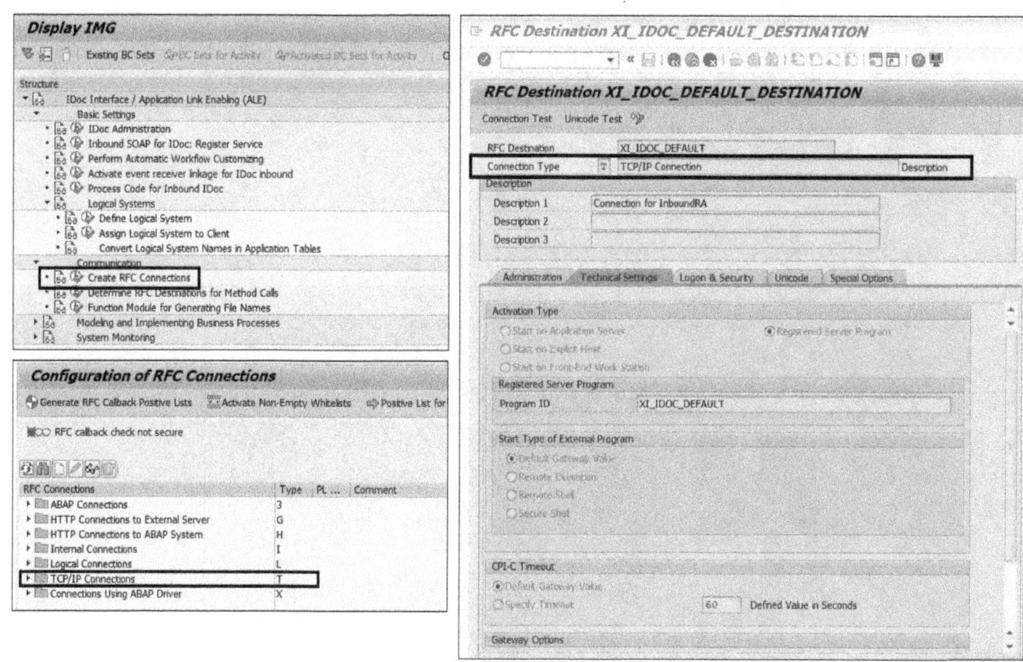

Figure 9.3 Transaction SPRO Navigation Path for RFC Connection and Overview

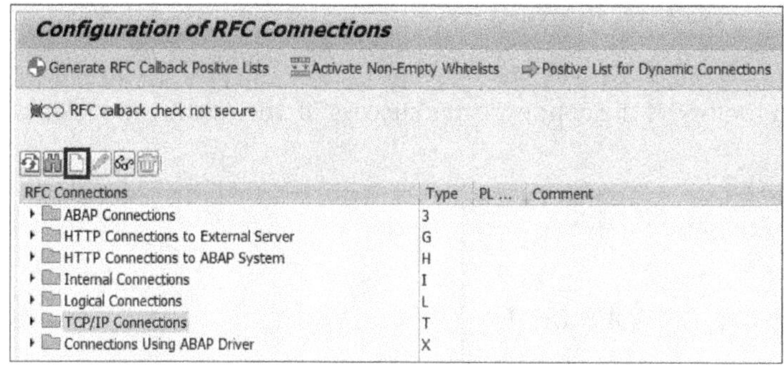

Figure 9.4 Transaction SM59: RFC Connection Overview

Enter the **RFC Destination** name "<System_ID>CLNT<Client>"; you can use a name of your choice. Add a description in one of the fields under **Description** (e.g., Connection RFC 1). In the **Technical Settings** tab under **Activation Type** and **Registered Server Program**, enter the program ID under which the dispatcher is registered on the SAP Gateway in the **Program ID** field (Figure 9.5).

Note

To registering a program on SAP Gateway, follow the relevant SAP Note 353597, Registration of RFC Server Programs.

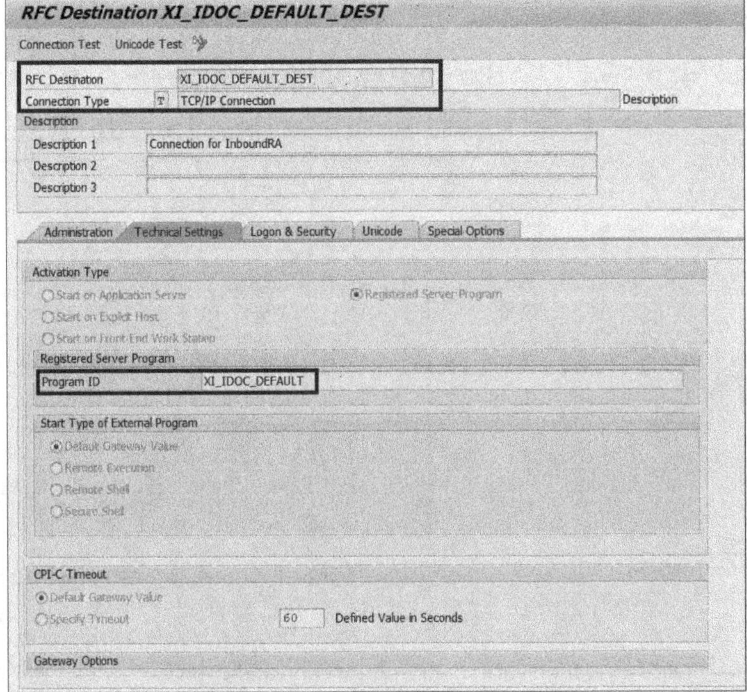

Figure 9.5 Attributes of the RFC Connection

Once the RFC connection is saved, you can test the connectivity by clicking **Connection Test** in the application toolbar. If the connection is successful, the result will look like Figure 9.6.

RFC - Connection Test

Connection Test XI_IDOC_DEFAULT_DEST

Connection Type TCP/IP Connection

Action	Result
Logon	2 msec
Transfer of 0 KB	1 msec
Transfer of 10 KB	1 msec
Transfer of 20 KB	1 msec
Transfer of 30 KB	1 msec

Figure 9.6 RFC Connection Connectivity Test Results

9.1.3 Remote Function Call Destinations for Method Calls

RFC destinations are assigned to logical systems using a tree graphic that can display all the logical systems of the whole system. RFC destinations are used for synchronous business Application Programming Interface (BAPI) calls.

Use menu path **SPRO • SAP Reference IMG • SAP NetWeaver • Application Server • IDoc Interface/Application Link Enabling (ALE) • Communication • Determine RFC Destinations**

for Method Calls (see Figure 9.7). In addition, you can create the RFC connection directly by executing Transaction BD97.

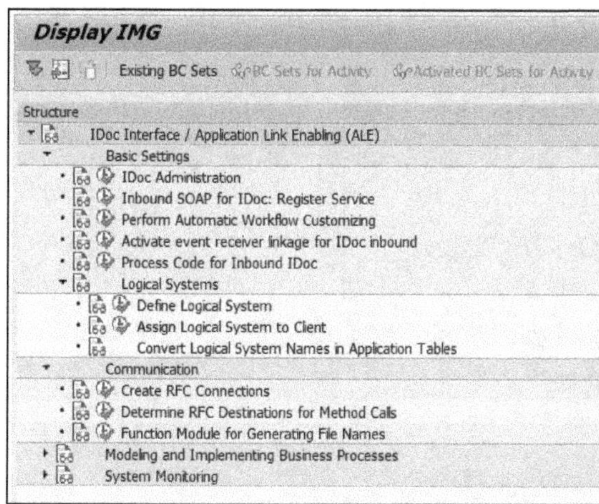

Figure 9.7 Navigation Path for RFC Destination Method Call

Note

RFC destinations can't be assigned for your own logical system because your own logical system is always called locally. For this reason, your own system doesn't appear in the list.

These settings aren't transported in the Change and Transport System (CTS). You must make the settings in each system.

In the **Assign RFC Destinations for Synchronous Method Calls** screen, select the logical system node you've created (Figure 9.8).

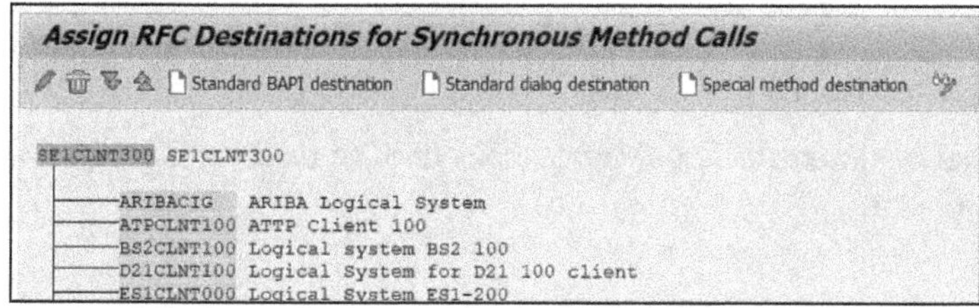

Figure 9.8 Assigning the RFC Destination

Choose **Standard BAPI Destination** from the toolbar (you can use value help). In the **Assign RFC Destinations for Synchronous Method Calls** dialog box, in the **RFC destination for BAPI calls** field, enter the RFC destination of the SAP S/4HANA <system> (in capital letters or use the value help), as shown in Figure 9.9. Click on the **Continue** (or press Enter). Click on the **Save** button.

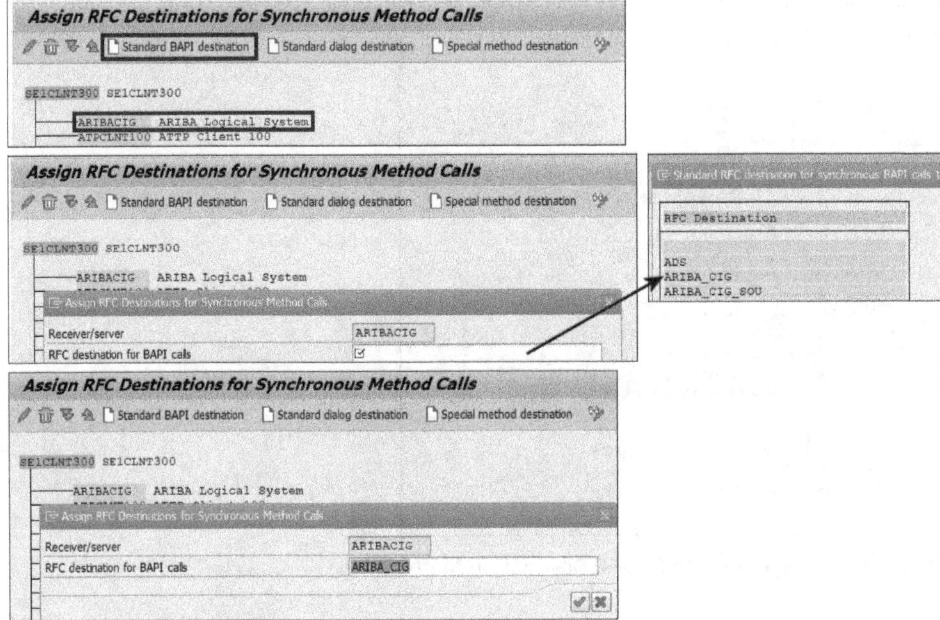

Figure 9.9 Assign a BAPI in the RFC Destination

9.1.4 Distribution Model and Distributing Views

A distribution model defines the sender logical system, the receiver logical system, and the message type (logical message type, not the IDoc type). Normally, the distribution model is configured in the sending system because the sending system knows which messages need to be distributed to which system and which messages are coming from which system. IDoc distribution is based on and directly controlled by the distribution model. The distribution model mainly is used for distributing the master data, user transaction data, or company code data.

> **Note**
>
> Sender and receiver logical systems and RFC connections need to be created prior to setting up ALE distribution.

In this section, you'll see how to create a distribution model, add a message type to the model, and configure the partner profile.

Creation of a Distribution Model

To create a distribution model, use Transaction SPRO and menu path **SAP Reference IMG • SAP NetWeaver • Application Server • IDoc Interface/Application Link Enabling**

(ALE) • Modeling and Implementing Business Processes • Maintain Distribution Model and Distribute Views (see Figure 9.10). Alternatively, you can use Transaction BD64.

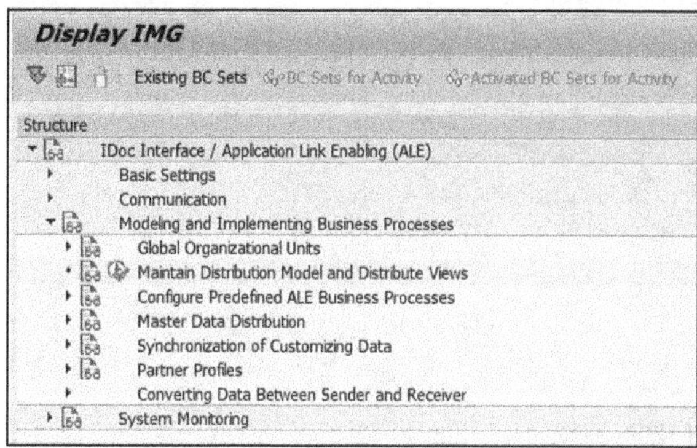

Figure 9.10 Navigation Path for the Distribution Model

Go to change mode by clicking **the Change Mode** icon, and click on **Create Model View** (Figure 9.11).

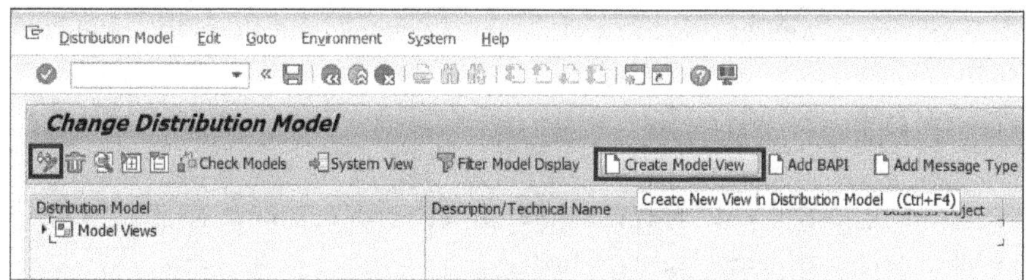

Figure 9.11 Creation of the Distribution Model

In the **Create Model View** dialog box (Figure 9.12), enter a short description of the model view in the **Short text** field and a technical name in the **Technical Name** field. Choose **Continue** or press [Enter].

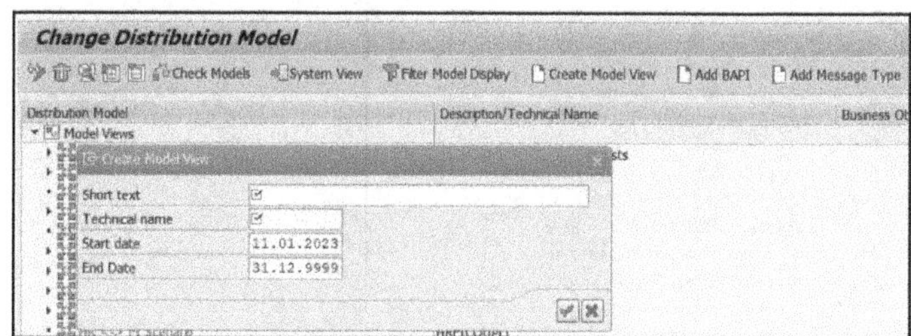

Figure 9.12 Creation of the Model View

Add Message Type to the Model View

After generating the distribution model, click on **Add Message Type**, and the model view will automatically prompt for which type you want to add (Figure 9.13).

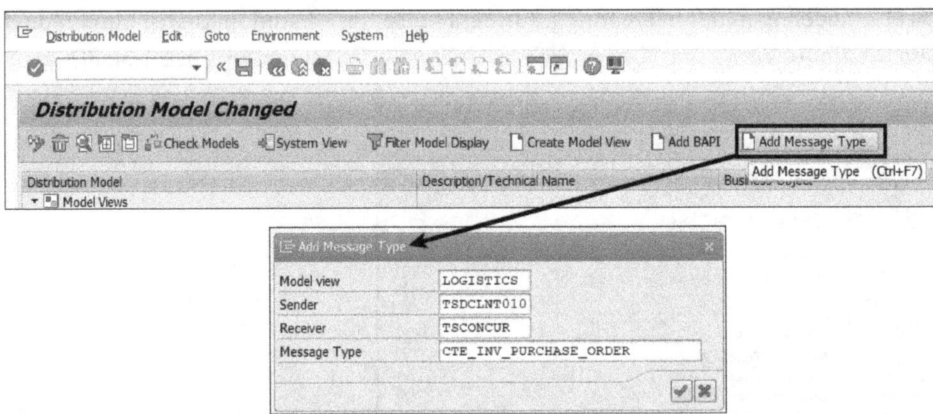

Figure 9.13 Add Message Type to the Model View

You need to enter the sender system in the **Sender** field. In SAP S/4HANA, there can be multiple clients, depending on your business requirements. You need to configure the sender in each client. The **Receiver** field is the system which is supposed to receive the messages. Configure the **Message Type** by using value help or entering it in all caps (e.g., "MATMAS"). After configuring the appropriate message type, press the **Save** icon to save the new distribution model.

Note

The standard message type provided in SAP S/4HANA can be seen in Transaction WE81 (Figure 9.14); if not, create a message type first.

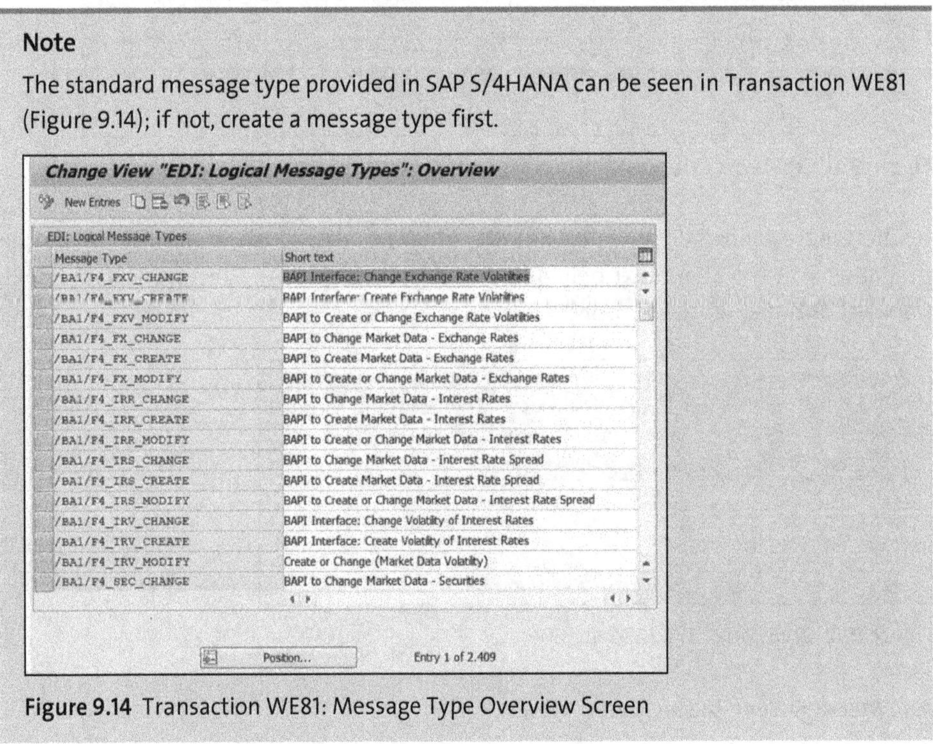

Figure 9.14 Transaction WE81: Message Type Overview Screen

> **Note**
>
> You must complete the setup described in Sections Section 9.1.1 to Section 9.1.4 before moving on to further steps.

Generation of Partner Profiles

After generation of the distribution model, it needs to be distributed to the target logical system. Select the model, and choose **Environment • Generate Partner Profiles** (Figure 9.15).

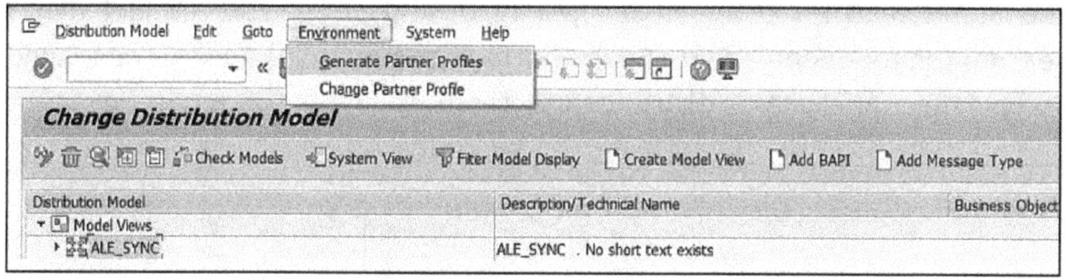

Figure 9.15 Generate Partner Profile (1)

Enter the **Model View** name for which you want to generate the partner profile (Figure 9.16). Select the **Pass IDoc immediately** and **Trigger immediately** radio buttons. Click **Execute** (F8).

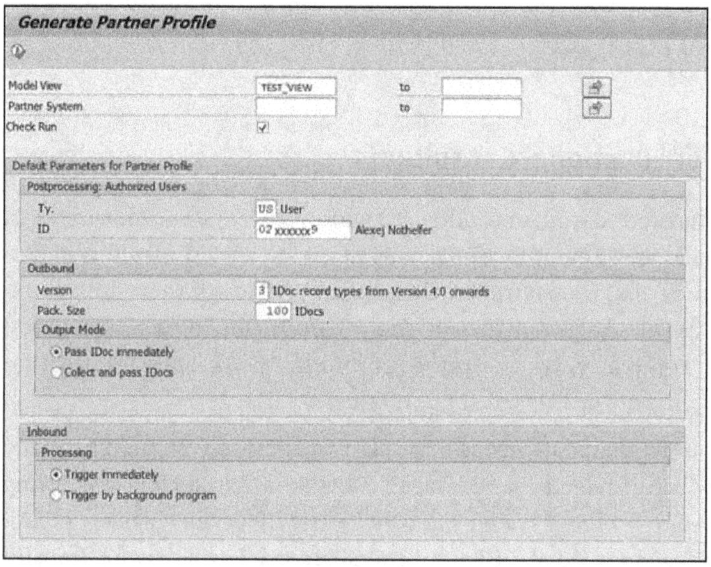

Figure 9.16 Generate Partner Profile (2)

You should get a list and status in green, which means it executed successfully (Figure 9.17).

Generate Partner Profile

Log for Partner Profile Generation

Task	System	Status	Result	Port Number	RFC Destination	Message Type	Message Var.	Message Funct.	Basic Type	Enh. Type	Process Code
Check Run											
Partner	System		System already exists as partner								
	System CLNT010		System CLNT010 already exists as partner								
Port	System		Port A00000000 with RFC destination CLNT010 already exists	A0000000	CLNT010						
Outbound			Outbound parameters for message type SYNCH SYNCHRON already exist			SYNCH			SYNCHRON		
Inbound			Inbound parameters for message type HRMD_A already exist			HRMD_A					

Figure 9.17 List Displaying the Generated Partner Profile

Back to the main screen, select the model view again, and choose **Edit • Model View • Distribute** from the top menu (Figure 9.18). Continue by pressing `Enter`. You should get a list saying the model view is distributed successfully.

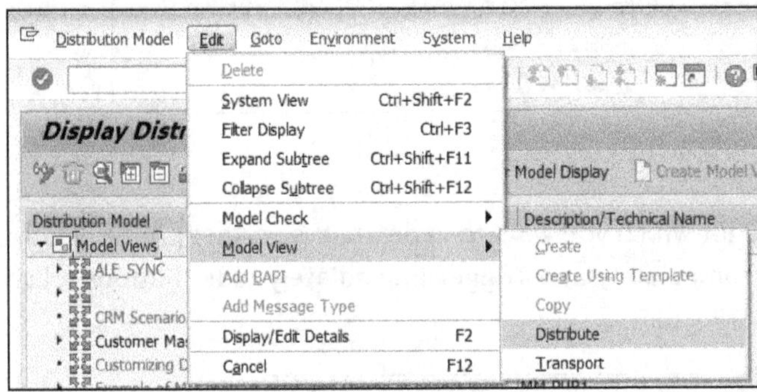

Figure 9.18 Distribute the Model View

9.2 IDoc Basics for Functional Consultants

Today, messages are transferred from one SAP system to another system and vice versa in the form of IDoc messages. For a functional consultant, it's very difficult to understand how messages are being transferred across systems. Although there are a number of documents available freely on the internet, it's difficult for a nontechnical person to understand. This is an effort to capture the minimum necessary information that you might need to handle projects or support issues on IDocs.

When a business's transactional data needs to be sent to another system, it's usually transferred in IDocs, which are electronic messages. IDocs are also used for transferring business data from SAP to non-SAP systems and vice versa. If a transfer from an SAP system to a non-SAP system is done, EDI subsystems are used, whereas for transfer between two SAP systems, ALE is used.

In this section, we'll discuss the various concepts of IDoc types and segments, how IDocs can be extended, the IDoc structure, and messages types, including how those message types can be assigned to partner profiles.

9.2.1 Overview

IDocs can be generate in an SAP system or an EDI subsystem. Depending on the direction in which the IDocs are sent, it's either called an inbound IDoc or an outbound IDoc. For an outbound flow, an IDoc is triggered in SAP S/4HANA through document message control, which is then sent to an EDI subsystem. EDI converts the data from an IDoc into XML or an equivalent format and then sends the data to a partner system via the internet. For an inbound flow, EDI converts the partner data, and the IDoc is created in SAP S/4HANA. After successful processing of this IDoc, an application document is posted in SAP S/4HANA (see Figure 9.19).

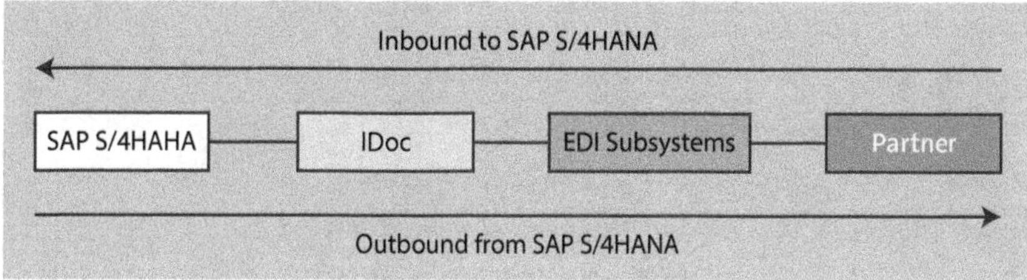

Figure 9.19 High-Level Flow Diagram

EDI is the electronic exchange of business document between the computer systems of business partners using a standard format over a communication network. For transmission of information electronically, two widely used standards are ANSI ASC X12 and EDIFACT. ANSI is a committee formed by representatives of major organizations, government bodies, and EDI software companies that defines standards and guidelines for information interchange over EDI. UN/EDIFACT stands for United Nations EDI for Administration, Commerce and Transport and was formed in 1985 using ANSI X12 and United Nations Trade Data Interchange (UNTDI) as base standards.

ANSI X12 describes business documents as transactions, and each transaction is represented by a three-digit number (e.g., 850 for Purchase Order, 855 for Purchase Order Acknowledgement, etc.). EDIFACT describes business documents as messages, represented by standard names (e.g., ORDERS for purchase orders).

9.2.2 IDoc Types and Segments

In SAP, IDoc types are based on the EDI standard and mostly based in EDIFACT standards. An IDoc type (basic) defines the structure of an IDoc, which consists of IDoc segments, data fields, type, and size. You can create, change, or display IDoc types in Transaction WE30. IDoc segments contain the actual business data that is sent to or received from a partner (see Figure 9.20):

❶ IDoc type

❷ Parent segment

❸ IDoc segment

❹ Child segment

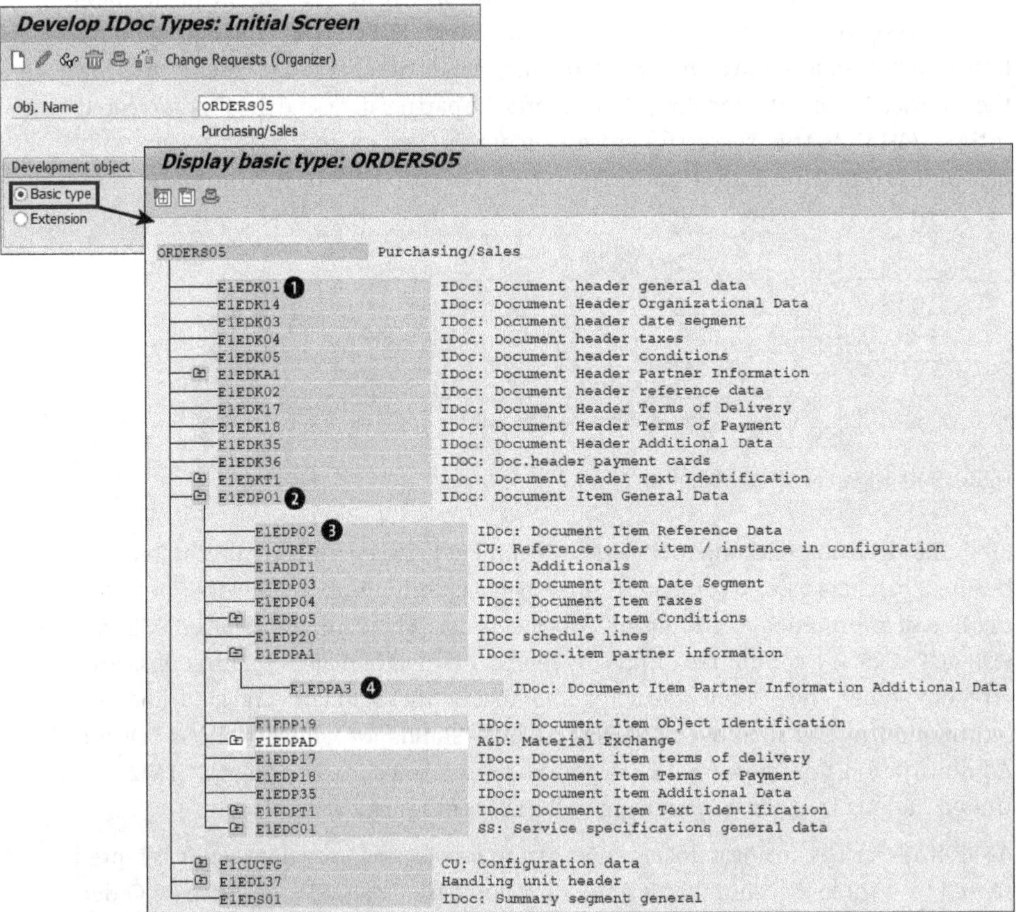

Figure 9.20 IDoc Basic Type and IDoc Segments

In IDoc, the **Minimum number** of the segment is 1, and the **Maximum number** represents the maximum number of segments that can be present in an IDoc structure. The **Parent segment** is the used if the segment contains its own segment, and the dependent segment will be called the child segment (see Figure 9.20). If you click on **Segment editor**, it will navigate you to the details screen where you can see the fields the segment contains and its data type (Figure 9.21).

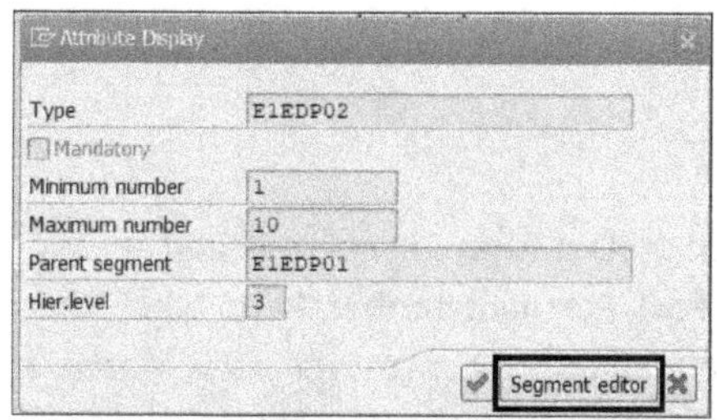

Development segments: Display segment definition E2EDP02002

Segment type attributes

Segment type: E1EDP02 ☑ Qualified segment

Short Description: IDoc: Document Item Reference Data

Definition: E2EDP02002 ☑ Released

☐ Primary segment

Last Changed By: SAP

Fields in Segment

Pos...	Field Name	Data element	ISO c...	Exp...
1	QUALF	EDI_QUALFR	☐	3
2	BELNR	EDI_BELNR	☐	35
3	ZEILE	EDI1082_A	☐	6
4	DATUM	EDIDAT8	☐	8
5	UZEIT	EDITIM6	☐	6
6	BSARK	EDI_ORGID	☐	35
7	IHREZ	EDI3413LA	☐	30
8	RMA	LABNR	☐	20
9	REASON	AUGRU	☐	3
10	RSNTXT	BEZEI40	☐	40

Figure 9.21 IDoc Segment Details

9.2.3 IDoc Extension

IDoc extensions describe how basic IDoc structures and procedures can be modified or improved to meet particular business needs. If necessary, the IDoc structure may need to have new segments created, old segments modified, or even special data added. If you need to send additional data that isn't available as part of the standard structure field, then you can use the extension feature of the basic type (see Figure 9.22).

Develop IDoc Types: Initial Screen

Change Requests (Organizer)

Obj. Name: ORDERS05

Development object

○ Basic type

◉ Extension

Figure 9.22 IDoc Extension

9.2.4 IDoc Structure

The IDoc structure consists of three parts:

- **Control record**
  This record contains administrative data such as IDoc type, logical message description, recipient and sender information (e.g., **Port**, **Partner Number**, **Part. Type**, **Function**, and **Partner Role**), and so on. All data related to the control record is stored in database table EDIDC. **Direction** determines the direction of the IDoc flow: **1** means an outbound IDoc, and **2** means an inbound IDoc (Figure 9.23).

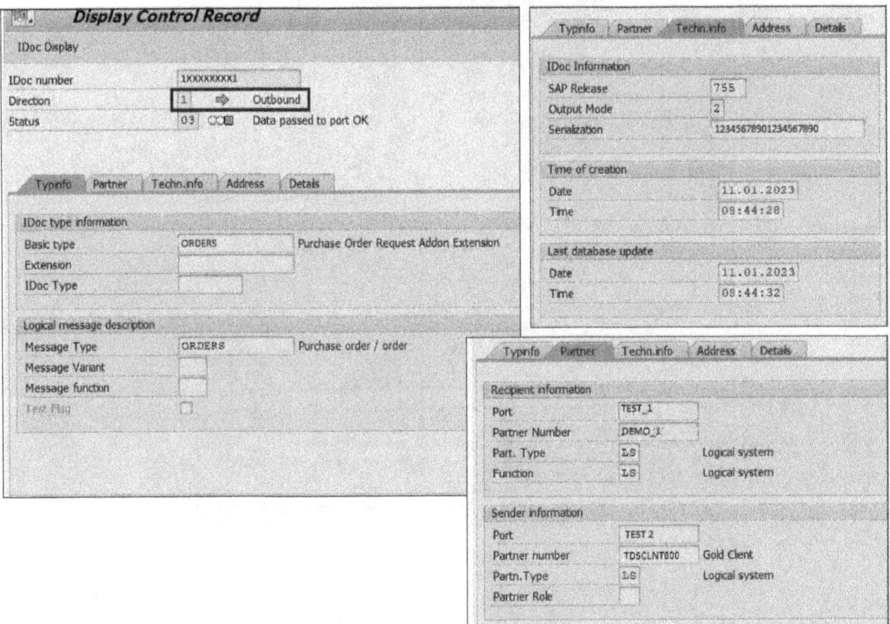

Figure 9.23 IDoc Control Record

- **Data record**
  This record contains the actual data to send or receive in the SAP S/4HANA system, and it can be a single record or multiple records (Figure 9.24).

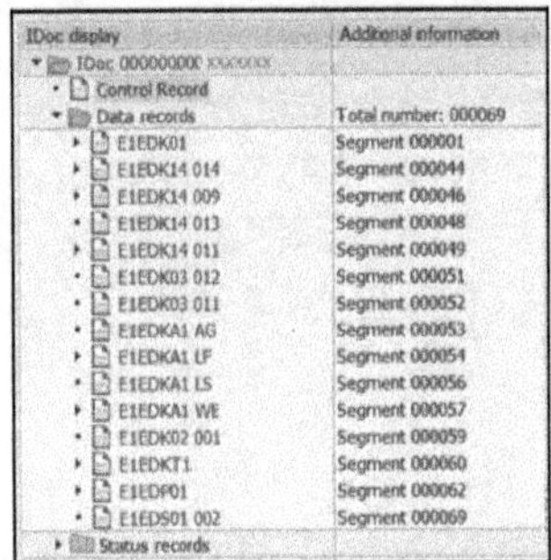

Figure 9.24 IDoc Data Records Overview

All the data related to the data record is stored in tables EDID2 to EDID4. In table EDIDD, you can see the record's components. Table EDIDD-SDATA contains the actual data, which is of data type **LCHAR** with a **Length** of **1000**, as shown in Figure 9.25.

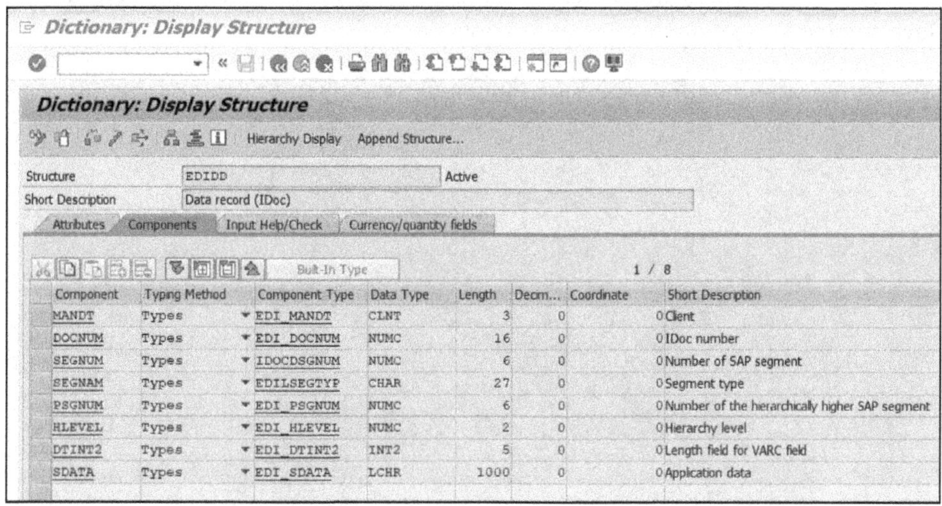

Component	Typing Method	Component Type	Data Type	Length	Decim...	Coordinate	Short Description
MANDT	Types	EDI_MANDT	CLNT	3	0	0	Client
DOCNUM	Types	EDI_DOCNUM	NUMC	16	0	0	IDoc number
SEGNUM	Types	IDOCDSGNUM	NUMC	6	0	0	Number of SAP segment
SEGNAM	Types	EDILSEGTYP	CHAR	27	0	0	Segment type
PSGNUM	Types	EDI_PSGNUM	NUMC	6	0	0	Number of the hierarchically higher SAP segment
HLEVEL	Types	EDI_HLEVEL	NUMC	2	0	0	Hierarchy level
DTINT2	Types	EDI_DTINT2	INT2	5	0	0	Length field for VARC field
SDATA	Types	EDI_SDATA	LCHR	1000	0	0	Application data

Figure 9.25 Table EDIDD Structure View

- **Status record**

  Status records gives you the information about the various stages an IDoc has been passed through (Figure 9.26). All the data related to the status record is stored in table EDIDS. SAP has provided standard status numbers as well: 01–49 for outbound processes and 50–75 for inbound processes.

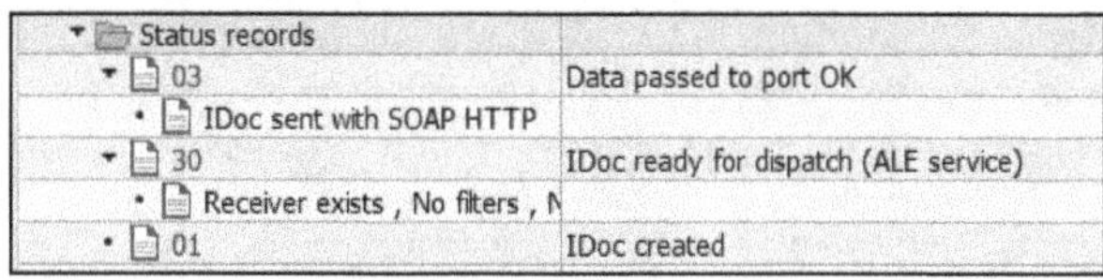

Figure 9.26 Status Record Details

9.2.5 Message Type

In SAP S/4HANA, documents are exchanged between two partners as IDoc messages, and each document is of a specific type—order, good receipt, invoice, notification, and so on. Normally, message types are associated with different basic IDoc types and vice versa. The system-available message types in Transaction WE8 are shown in Figure 9.27.

The linkage between the IDoc message type, function module, and basic types can be seen using Transaction WE57. As shown in Figure 9.28, the basic type (**BasicType**) for

IDoc message **IDOC_INPUT_DESADV** is **DELVRY01–DELVRY07**. However, SAP recommends using the latest basic type (e.g., **DELVRY07**) because it may contain the latest additions regarding segments and fields. Older versions may continue to be used because previous client implementations may have included or used them during development.

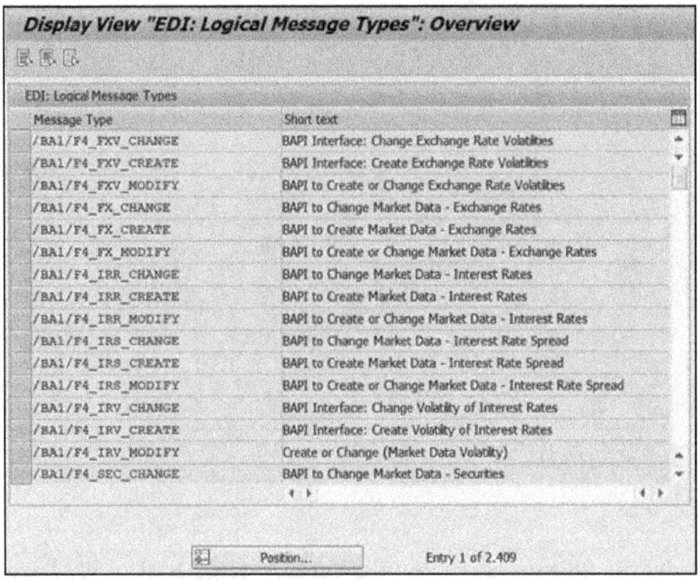

Display View "EDI: Logical Message Types": Overview

EDI: Logical Message Types

Message Type	Short text
/BA1/F4_FXV_CHANGE	BAPI Interface: Change Exchange Rate Volatilities
/BA1/F4_FXV_CREATE	BAPI Interface: Create Exchange Rate Volatilities
/BA1/F4_FXV_MODIFY	BAPI to Create or Change Exchange Rate Volatilities
/BA1/F4_FX_CHANGE	BAPI to Change Market Data - Exchange Rates
/BA1/F4_FX_CREATE	BAPI to Create Market Data - Exchange Rates
/BA1/F4_FX_MODIFY	BAPI to Create or Change Market Data - Exchange Rates
/BA1/F4_IRR_CHANGE	BAPI to Change Market Data - Interest Rates
/BA1/F4_IRR_CREATE	BAPI to Create Market Data - Interest Rates
/BA1/F4_IRR_MODIFY	BAPI to Create or Change Market Data - Interest Rates
/BA1/F4_IRS_CHANGE	BAPI to Change Market Data - Interest Rate Spread
/BA1/F4_IRS_CREATE	BAPI to Create Market Data - Interest Rate Spread
/BA1/F4_IRS_MODIFY	BAPI to Create or Change Market Data - Interest Rate Spread
/BA1/F4_IRV_CHANGE	BAPI Interface: Change Volatility of Interest Rates
/BA1/F4_IRV_CREATE	BAPI Interface: Create Volatility of Interest Rates
/BA1/F4_IRV_MODIFY	Create or Change (Market Data Volatility)
/BA1/F4_SEC_CHANGE	BAPI to Change Market Data - Securities

Position... Entry 1 of 2.409

Figure 9.27 Message Types Displayed in Transaction WE81

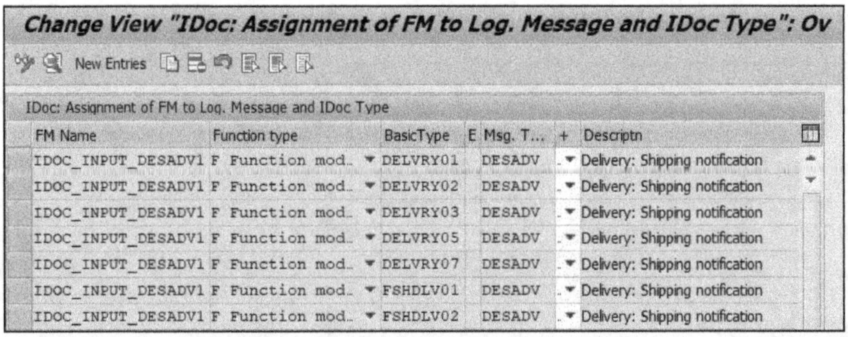

Change View "IDoc: Assignment of FM to Log. Message and IDoc Type": Ov

New Entries

IDoc: Assignment of FM to Log. Message and IDoc Type

FM Name	Function type	BasicType	E	Msg. T...	+	Descriptn
IDOC_INPUT_DESADV1	F Function mod...	DELVRY01		DESADV		Delivery: Shipping notification
IDOC_INPUT_DESADV1	F Function mod...	DELVRY02		DESADV		Delivery: Shipping notification
IDOC_INPUT_DESADV1	F Function mod...	DELVRY03		DESADV		Delivery: Shipping notification
IDOC_INPUT_DESADV1	F Function mod...	DELVRY05		DESADV		Delivery: Shipping notification
IDOC_INPUT_DESADV1	F Function mod...	DELVRY07		DESADV		Delivery: Shipping notification
IDOC_INPUT_DESADV1	F Function mod...	FSHDLV01		DESADV		Delivery: Shipping notification
IDOC_INPUT_DESADV1	F Function mod...	FSHDLV02		DESADV		Delivery: Shipping notification

Figure 9.28 Link between IDoc Message Type, Basic Type, and Function Module

Detailed documentation on IDoc basic types with explanation of each field can be viewed through Transaction WE60 (Figure 9.29).

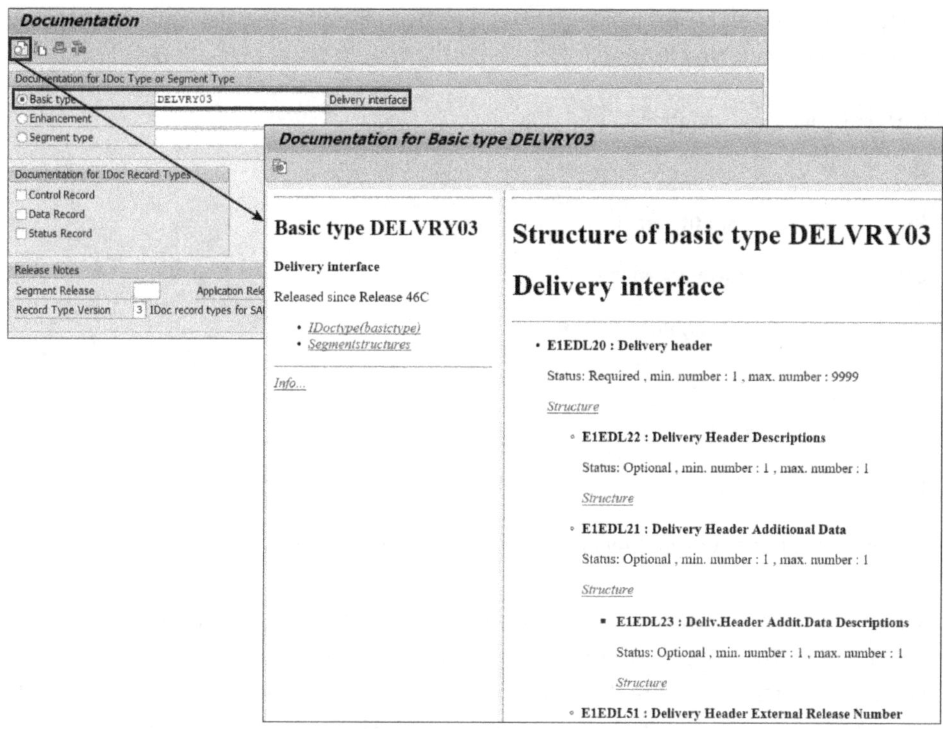

Figure 9.29 Detailed Documentation on IDoc Basic Types

9.2.6 Partner Profiles

To send or receive data, you must maintain all the business partners with whom you want to communicate, so a partner profile must be maintained in the SAP S/4HANA system. To create a partner profile, use Transaction WE20, as shown in Figure 9.30.

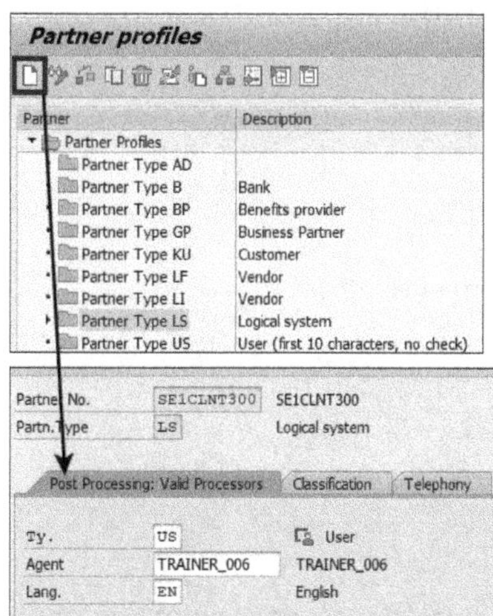

Figure 9.30 Creation of a Partner Profile

You need to select the appropriate partner type from the available list. Table 9.2 lists the partner types available as standard.

Partner Type	Description
B	Bank
BP	Benefits provider
GP	Business partner
KU	Customer
LI	Vendor
LS	Logical system
US	User (first 10 characters, no check)

Table 9.2 List of Partner Types Available in SAP S/4HANA Systems

Each of the types is suitable for a different usage. For example, **Partner Type LS** is used for ALE and **Partners Type LI** and **KU** are used for EDI. You need to select the partner type (e.g., **LS**) and click on the **Create** button to create a new partner (refer to Figure 9.30). Enter the **Partner No.** and **Partn. Type** for **Logical System** here (you can use value help F4). If you enter a partner number that wasn't created earlier in Section 9.1.1, the system will throw an error (Figure 9.31).

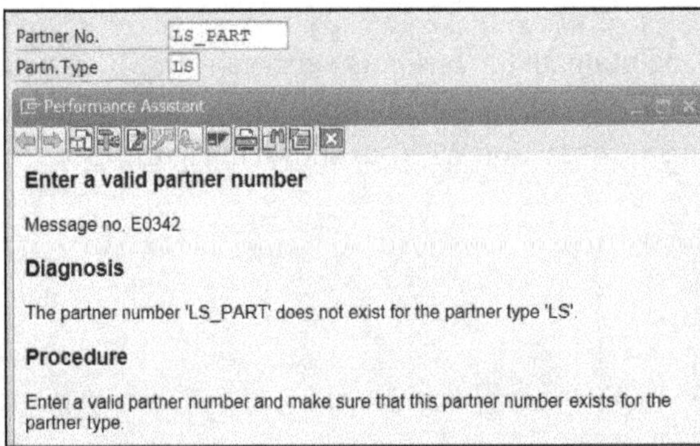

Figure 9.31 System Error due to Selecting the Wrong Partner Number

> **Note**
>
> The logical system must be created already; refer to Sections Section 9.1.1 and Section 9.1.2.

Figure 9.32 shows how the partner profile of type LS (logical system) is maintained with the appropriate message types:

❶ Object type can be work center, organizational unit, job, person, position, or user.

❷ Use the **Create**, **Copy**, **Delete**, or **Display** buttons with the inbound/outbound message type for any partner.

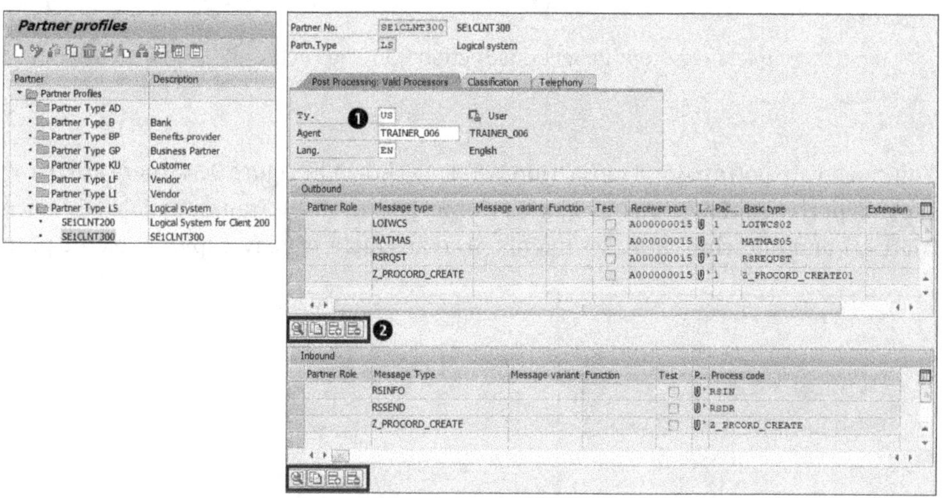

Figure 9.32 Maintaining the Partner Profile of Type LS

While adding a new outbound message type, the **Receiver port** and **Process code** need to be added (Figure 9.33). The process code can be standard or custom developed per the business logic.

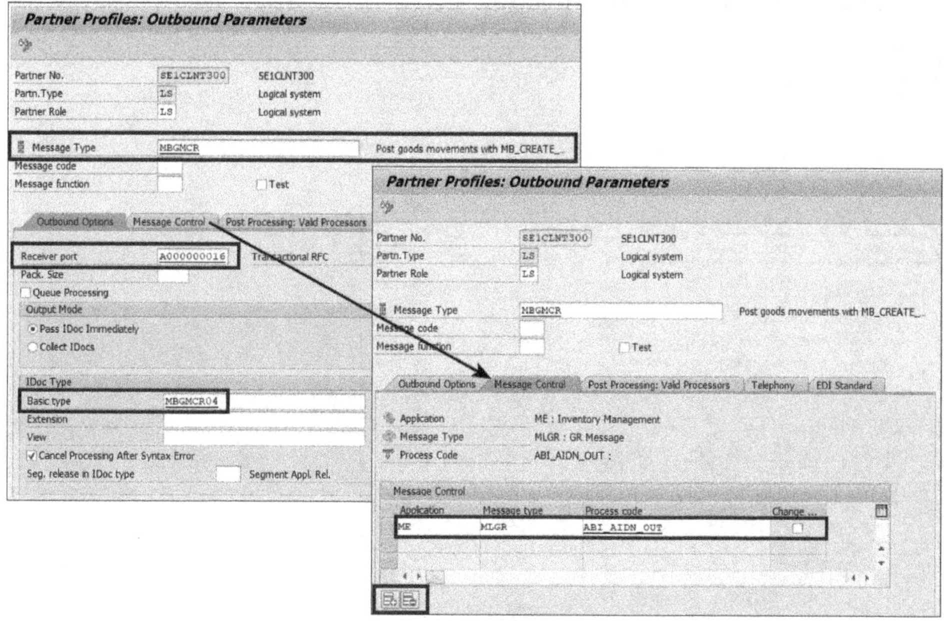

Figure 9.33 Adding an Outbound Message Type to a Partner

9.2.7 Ports

Ports play a key role in communication with different SAP or non-SAP systems.

> **Note**
>
> You must complete the steps in Sections Section 9.1.1 and Section 9.1.2 prior to creating ports.

You can create ports using Transaction WE21. As shown in Figure 9.34, there are many types of ports in SAP S/4HANA, but most commonly used are **Transactional RFC**, which ALE uses, and **File**, which EDI uses mainly. You can create, display, copy, or delete a port.

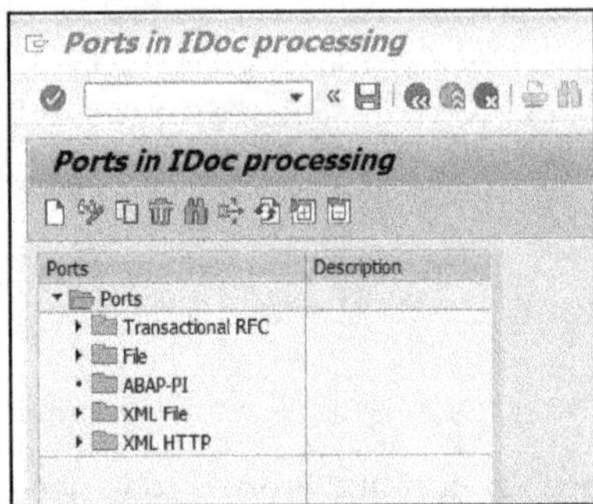

Figure 9.34 SAP S/4HANA Ports

Figure 9.35 displays a tRFC port attribute.

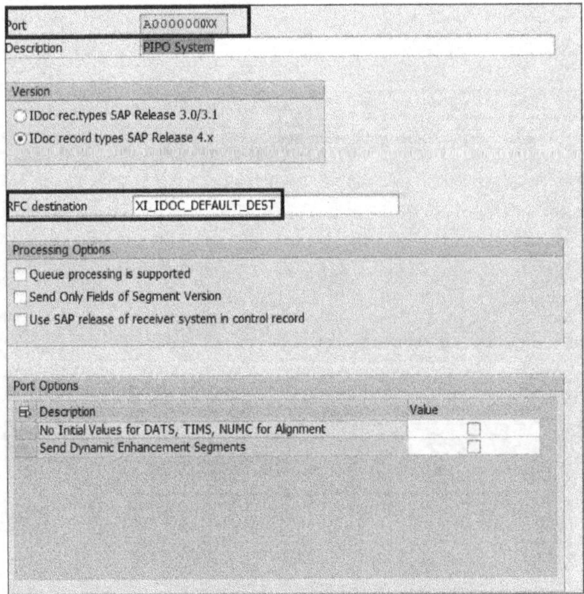

Figure 9.35 SAP S/4HANA tRFC Port Creation

For tRFC ports, you have to give the name of the logical destination created in Section 9.1.2 using Transaction SM59 or Transaction SALE. When using file ports, you can specify the **Directory** as "/usr/sap/<system id>/SYS/<dir>" where the IDoc file should be placed. The other system or the middleware will pick up the file from here. **Function module** can be used to generate a file name for the IDoc (Figure 9.36).

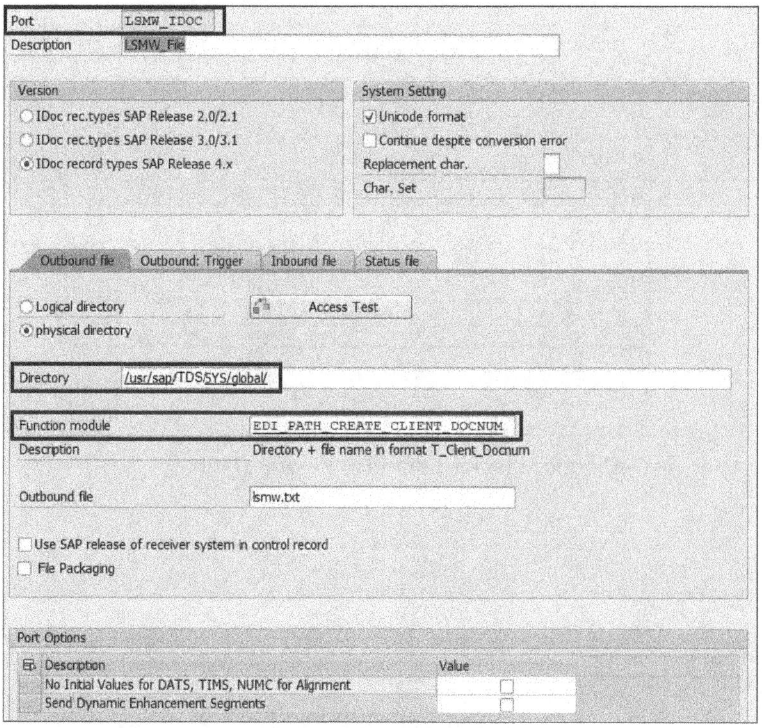

Figure 9.36 SAP S/4HANA File Port Creation

9.2.8 Inbound and Outbound Processing

During inbound (client) processing, IDocs are transferred to the interface and stored in the SAP S/4HANA system. Outbound processing in SAP S/4HANA involves event handling. An event in SAP S/4HANA is defined as an occurrence of a status change in an object. Events are created when the relevant status change occurs. Triggering outbound IDocs through output is dealt with in Chapter 6, Section 6.3.

The process codes for handling inbound and outbound messages are defined during partner profile creation. The upcoming figures represent how the process code is maintained in a partner profile. A process code is needed for processing IDoc messages. The process code contains the details of the function modules that are used for IDoc processing. The IDoc message type can be linked to the process code.

In Figure 9.37, you can see how **Process code** for outbound messages can be assigned from your partner profile, and the same is shown in Figure 9.38 for inbound messages.

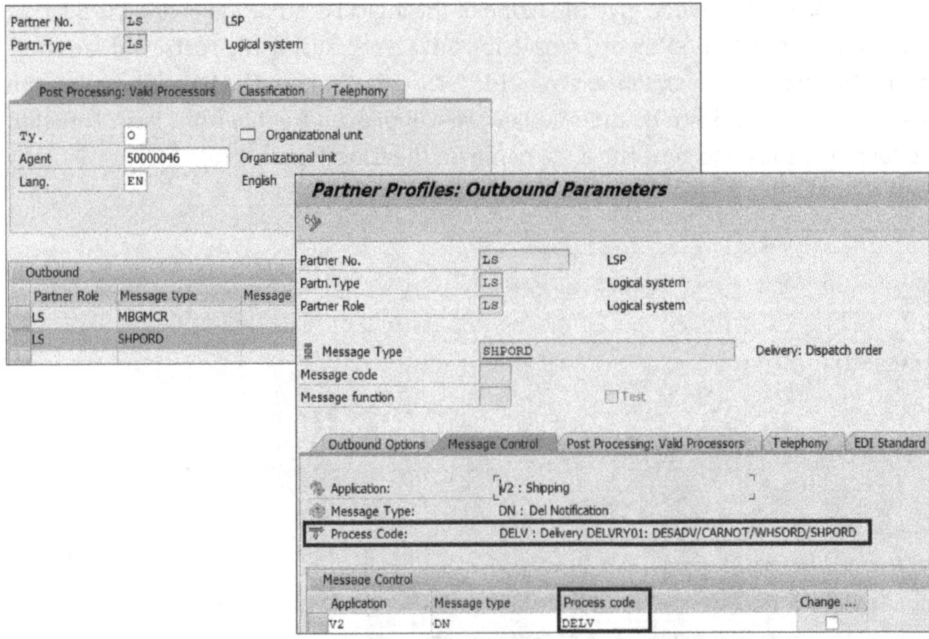

Figure 9.37 Process Code for Outbound Message Control in a Partner Profile

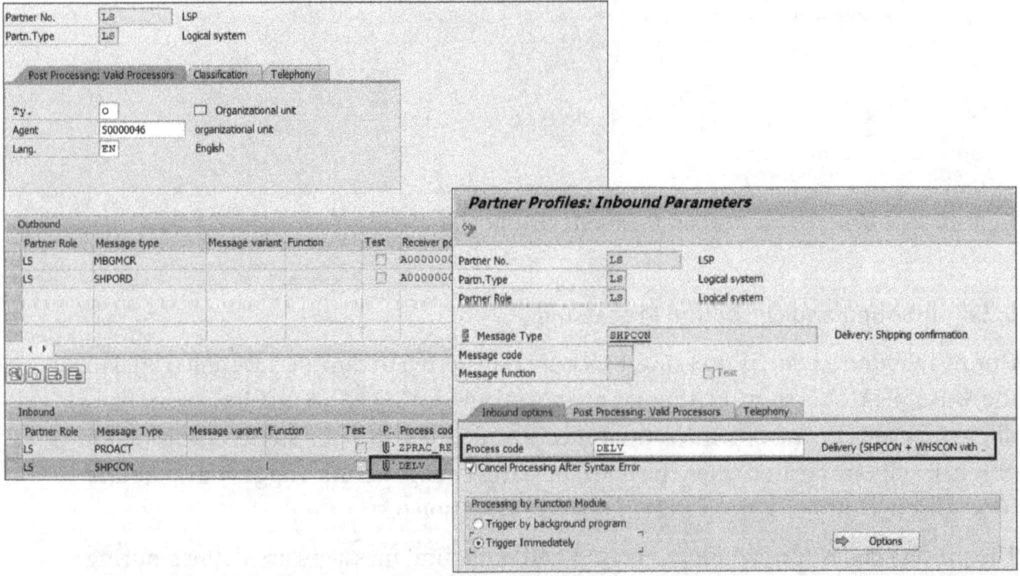

Figure 9.38 Process Code for Inbound Message Control in a Partner Profile

Transaction WE41 shows the related **Function module** and linked IDoc message types for an outbound process code (see Figure 9.39).

Transaction WE42 shows the related function module and linked IDoc message types for an inbound process code (see Figure 9.40).

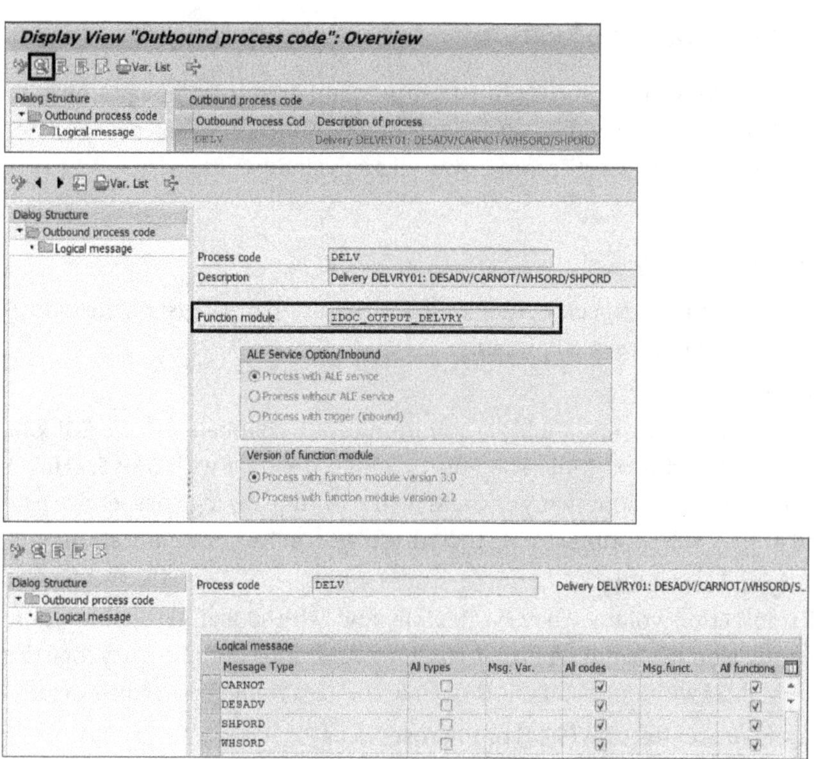

Figure 9.39 Process Code Details for Outbound Message Types

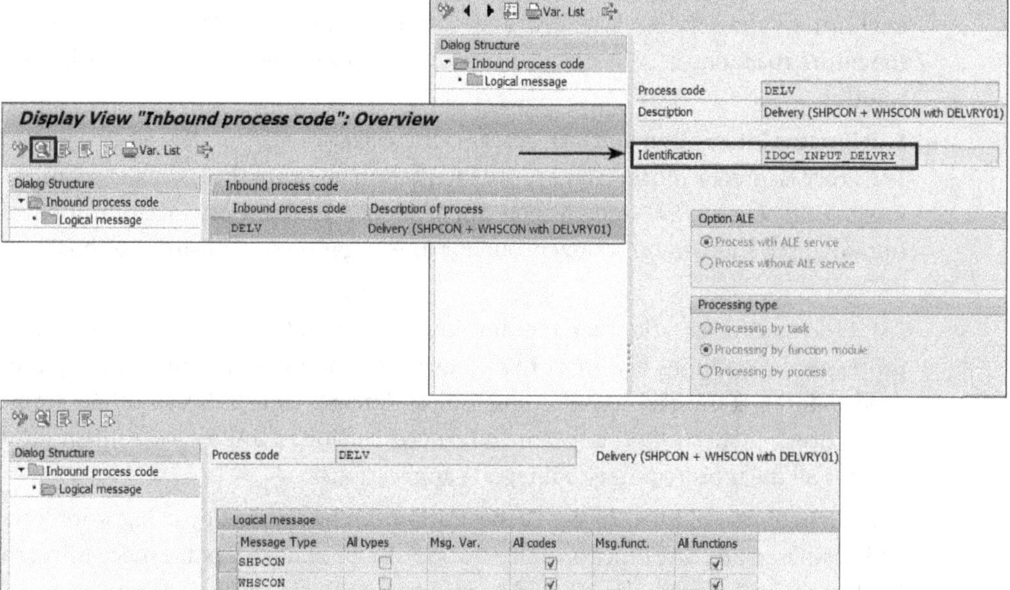

Figure 9.40 Process Code Details for Inbound Message Types

9.3 Technical Integration Features

In this section, we'll look into some essential features and drivers for integration with third-party logistic service providers, means of integration, and technical aspects of standard integration modes most commonly used across industries.

9.3.1 Integration Strategy

While considering technical aspects of integration with third-party logistics, the following factors may need to be assessed properly:

- **Data transformation**
  Logistics service providers use multiple communication channels such as EDI x.12, flat file, EDIFACT, and IDoc/XML. The effort to map the data with SAP S/4HANA should be considered without many changes on the third-party logistics service provider's end as they will continue to use the same platform for communication.
- **Extensibility and scalability**
  Data and transaction volume may be flexible, and with global expansion, there should be provision for handling bigger volumes. Additions of third-party logistics across geographies without major effort or code changes would be an advantage.
- **Service-oriented architecture (SOA) or microservices**
  SOA and microservices are both approaches to architectural design that break applications into components. In microservices, applications are decoupled, and each service can have independent data storage—unlike SOA. Both use APIs to connect applications and services. With SAP's recommendation to use APIs, it's going to be the future road map as APIs efficient framework accelerates the time to market the solution.
- **SAP Process Integration and SAP Process Orchestration**
  SAP Process Integration is an enterprise integration platform that provides seamless integration between SAP and non-SAP applications within the organization application-to-application (A2A) or even outside the organization business-to-business (B2B).

  SAP Process Orchestration is a tool for automation and optimization of business processes. It combines features of SAP Business Process Management (BPM), SAP Process Integration, and SAP Business Rules Management (BRM). In other words, SAP Process Orchestration is a more advanced version of SAP Process Integration and has all the tools required to integrate applications.

  The full potential of SAP Process Orchestration can be realized by using workflows and business rules. SAP Process Orchestration also provides a flexible mechanism to implement and maintain rule-based transformation logic and action triggers.

- **Enterprise application integration (EAI)**
  EAI uses technologies and methodologies to integrate business applications, data, private and public clouds, processes, and devices across the entire IT landscape.
  The enterprise service bus (ESB) is the latest in integration solutions with guaranteed reliability. These are fully abstracted from the application layer, follow a consistent pattern, and can be designed and configured with minimal additional code and with no modification to the systems that need to be integrated.
  EAI provides several advantages:
  - Transparency through central configuration of endpoints, and the consumer application need not know about the source in order to receive the message
  - Acceptability for all major protocols used by third-party systems and easy protocol conversion
  - Efficient routing based on preconfigured rules
  - Ability to retrieve missing data in incoming messages, based on the existing message data, and append it to the message before delivery to its final destination
  - Retaining messages in offline mode
  - Handling messages in a fully secure manner, and negotiating between the security assurance systems used by each of the systems that will be integrated
  - Easy method of monitoring the performance of the system, and a simple means of managing the flow of messages between systems

With SAP S/4HANA, SAP has been very explicit in recommending API as the preferred approach for integration. Use of SAP Process Orchestration provides extensive message and workflow management capabilities along with the use of published SAP S/4HANA APIs to retrieve information from or push information to SAP S/4HANA.

APIs from the SAP Business Accelerator Hub can be called by SAP Process Orchestration and dynamically routed to proven EDI channels for connecting external business partners. This provides the option to use APIs as an alternative to IDocs. However, whether the new API structures will match the current complete IDoc capabilities is yet to be seen. Version 1809 of the SAP Business Accelerator Hub offers the first outbound APIs for an on-premises installation.

Figure 9.41 provides a schematic representation of a third-party logistics integration scenario using SAP S/4HANA, SAP Solution Manager with AIF, and SAP Process Orchestration.

Third-party logistics solutions can be designed in many ways, and each solution may have some uniqueness in one way or other. This can be attributed to various factors such as the nature of the client's business, geographical scope, third-party logistics systems, cost of onboarding third-party logistics, technical changes required for accommodating the solution at the third-party logistics end, and so on. There can't be one solution for all situations, and consultants should be aware of the pros and cons of the technology and platforms used during integration. Overall, this requires a bit of techno-functional flair for someone to work on such a topic or assignment.

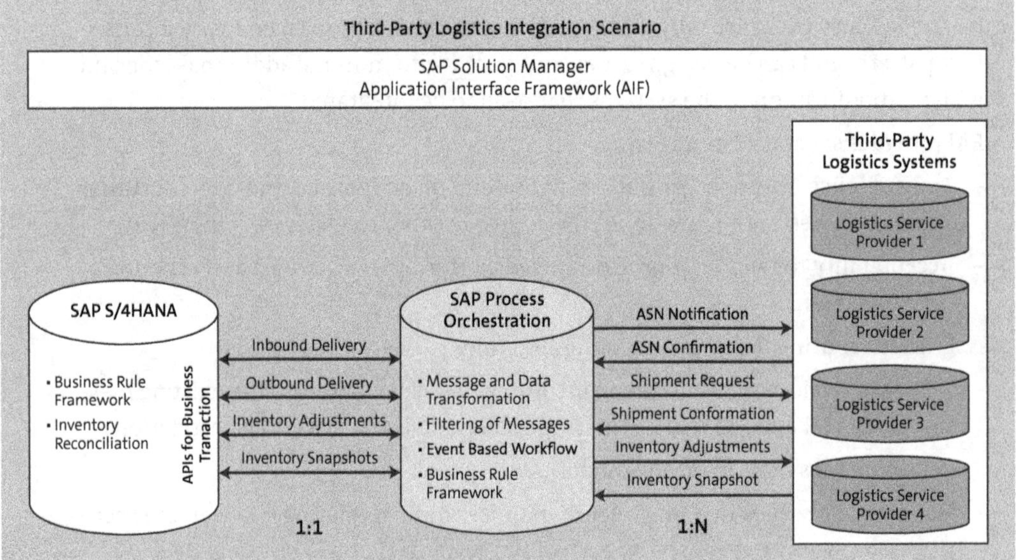

Figure 9.41 Third-Party Logistics Integration Architecture

Table 9.3 shows the IDoc message types that are used commonly in third-party logistics scenarios.

IDoc Message Type	Description
Master Data Replication with Third-Party Logistics Systems	
MATMAS	Material Master
DEBMAS	Customer Master Data Distribution
CREMAS	Vendor Master Data Distribution
Order Processing	
ORDERS	Purchase Order/Order
ORDRSP	Purchase Order/Order Confirmation
Inbound/Outbound Delivery Processing	
SHPORD	Delivery: Dispatch Order
WHSORD	Delivery: Stock Order
DESADV	Delivery: Shipping Notification

Table 9.3 Common List of IDoc Message Types Used in Third-Party Logistics Scenarios

IDoc Message Type	Description
SHPCON	Delivery: Shipping Confirmation
WHSCON	Delivery: Stock Confirmation
Stock Adjustments/Movements	
MBGMCR	Post Goods Movements with MB_CREATE_GOODS_MOVEMENT
WMMBXY	IDoc Report Goods Movements in Inventory Management
Inventory Reports	
INVRPT	Inventory Advice
PROACT	Stock and Sales Data
Invoicing/Billing	
INVOIC	Invoice/Billing Document

Table 9.3 Common List of IDoc Message Types Used in Third-Party Logistics Scenarios (Cont.)

You may also have to look into setting up the initial master data in sync between SAP S/4HANA and third-party logistics systems. This is essential because, otherwise, the transaction messages sent to third-party logistics systems are bound to fail due to unrecognizable master data.

9.3.2 Electronic Data Interchange

As mentioned previously, EDI is the intercompany communication of business documents in a standard format. EDI allows one company to send information to another company electronically rather than with paper. Business entities conducting business electronically are called trading partners.

In this section, we'll look at the various standard formats of EDI, the transmission methods that can be used in third-party logistics, and the file structure and printer.

Standard Format

The standard format is at the core of EDI conversations. Several EDI standards are in use, and the most common EDI file formats are listed here:

- cXML
- xCBL
- ebXML
- CSV
- ANSI X12
- EDIFACT
- EDIFICE (Information Technology)
- EDITRANS (Transportation)
- ETIS (Telecommunications)

For each standard, there are many different versions (e.g., ANSI 5010 or EDIFACT version D12, Release A, etc.). For two businesses to decide and agree on the exchange of EDI documents, they must agree on the specific EDI standard and version.

Transmission through Electronic Data Interchange

In general, there are two basic types of EDI transmission:

- **Point-to-point or direct connections**
  Systems connect with no middleware or intermediary over the internet, generally with secure protocols.
- **Value-added network (VAN)**
  A third-party network manages data transmission.

EDI internet transmission protocols include Secure File Transfer Protocol (SFTP), Applicability Statement 2 (AS2), an HTTPS-based protocol, Simple Object Access Protocol (SOAP), and others.

Metalanguages such as Extensible Markup Language (XML) or JavaScript Object Notation (JSON) complement rather than replace EDI.

EDI exchange typically involves the following processes:

- **Outbound process**
  Involves generating and sending EDI files with in the following step sequence:
  - Obtain data from the existing system.
  - Generate the EDI file.
  - Validate the EDI file.
  - Send the EDI file to the trading partner. Some common methods of transmission are FTP, HTTP, and AS2.
- **Inbound process**
  Involves receiving and translating the EDI files in the following step sequence:
  - Get the EDI file.
  - Validate and acknowledge the EDI file by sending back a TA1, 997, or CONTRL EDI file.
  - Translate the EDI file.
  - Save data to a database.

Electronic Data Interchange File Structure: X12 Primer

An EDI file is a binary computer file that contains data arranged in units called data elements. Data elements are separated by element terminators, and a group of data elements make up a data segment. Data segments are separated by segment terminators.

All EDI transactions are controlled by EDI message standards. It's essential to have proper governance processes for data quality. The EDI document may not be processed correctly if information is missing or in the wrong place.

The X12 message structure has primary levels that are hierarchical. From highest to the lowest, this structure contains the following:

- Interchange Envelope (identified by pairing of ISA/IEA segments)
- Functional Group (identified by pairing of GS/GE segments)
- Transaction Set (identified by pairing of ST/SE segments)

The rules for structure ensure the integrity of the data and the efficiency of the information exchange. A schematic structure of X12 envelopes is shown in Figure 9.42. The levels are explained in detail in the following sections.

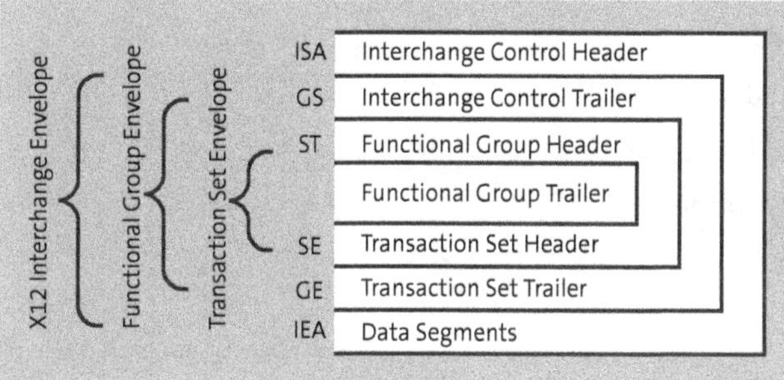

Figure 9.42 ASC X12 File Structure

Interchange Control Envelope (ISA/IEA)

The interchange envelope acts as wrapper for all the data to be sent in one transmission. It can contain one or multiple functional groups, which means the data of one or multiple transactions can be included in the interchange envelope, with each type of transaction stored in a separate functional group.

The interchange envelope is identified by the header and trailer. The beginning of an EDI file starts with the interchange control header (designated ISA), and the interchange control trailer (designated IEA) appears at the end.

Within the ISA and IEA segments, you can find the following:

- Data element separators and data segment terminator
- Sender and receiver identification
- Control information (to verify the message is correctly received)
- Authorization and security information, if applicable

The EDI file transmits the following information in sequence:

1. ISA
2. Optional interchange-related control segments

3. Actual message information, grouped by transaction type into functional groups
4. IEA

Figure 9.43 shows the elements and their definition in ISA and IEA segments of EDI:

❶ Authorization information qualifier
❷ Authorization information
❸ Security information qualifier
❹ Security information
❺ Interchange ID qualifier
❻ Interchange sender ID
❼ Interchange ID qualifier
❽ Interchange receiver ID
❾ Date the envelope was created
❿ Time the envelope was created
⓫ Interchange control standard identifier
⓬ Interchange control version number
⓭ Interchange control number
⓮ Usage indicator
⓯ Sub element separator
⓰ Number of included functional groups
⓱ Interchange control number

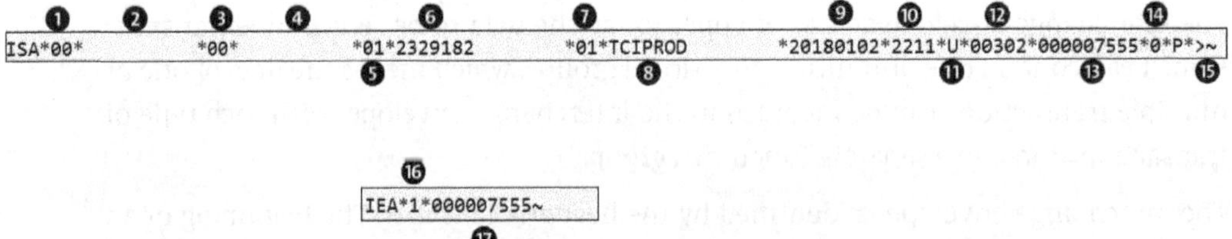

Figure 9.43 Elements and Their Definition in ISA and IEA Segments of EDI

Functional Groups Envelopes (GS/GE)

Functional group contains information of one or more transaction sets. One functional group envelope must include transactions of the same type, which can be batched together into one transmission.

The functional group is defined by the header and trailer segments. The functional group header (designated GS) segment appears at the beginning, and the functional group trailer (designated GE) segment appears at the end. Many transaction sets of the same type can be included in the functional group.

Each transaction set is assigned a functional identifier code, which is the first data element of the header segment. The transaction sets that constitute a specific functional group are identified by this functional ID code.

Figure 9.44 shows the definition for each element for GE and GS segments belonging to a functional group:

❶ Functional ID code (two-character transaction code, e.g., PO = Purchase Order, IN = Invoice, etc.)

❷ Sender ID code

❸ Receiver ID code

❹ Date the GS envelope was created

❺ Time the GS envelope was created

❻ Functional group control number

❼ Responsibility agency code (X is the qualifier for X12)

❽ Version release identifier code is the EDI version

❾ Sub element separator

❿ Number of transaction sets included (e.g., ST/SE combinations)

⓫ Functional group control number (originated and maintained by the sender)

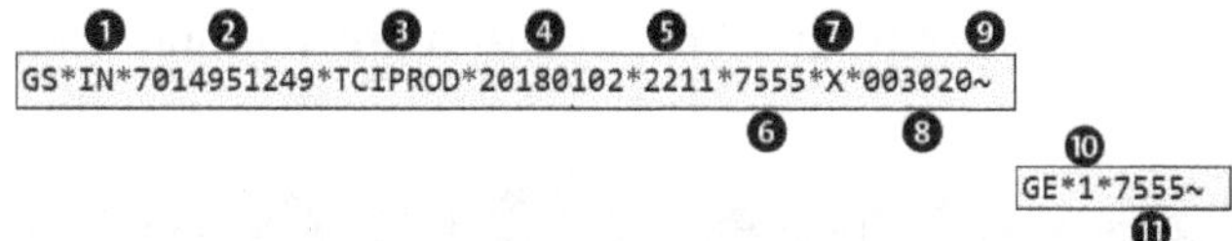

Figure 9.44 Definition for Each Element for Segments Belonging to Functional Groups

Transaction Set Envelopes (ST/SE)

Each transaction set envelope contains the following sections:

- Transaction set header (designated ST)
- Transaction set trailer (designated SE)
- Single message, enveloped within the header and footer

A transaction set may contain multiple segments. For example, the address segment might be used repeatedly to convey multiple sets of address information. The X12 standard defines the sequence of segments in the transaction set and also the sequence of elements within each segment. The relationship between segments and elements can be compared to the relationship between records and fields in a database environment.

Figure 9.45 shows the definition of elements in the transaction set envelope and sample EDI 810 file:

❶ Transaction set identifier code (three-character code to represent transaction EDI type, e.g., 850 = Purchase Order, 810 = Invoice, etc.)

❷ Transaction set control number

❸ Number of included segments in the transaction (includes ST and SE segments)

❹ Transaction set control number

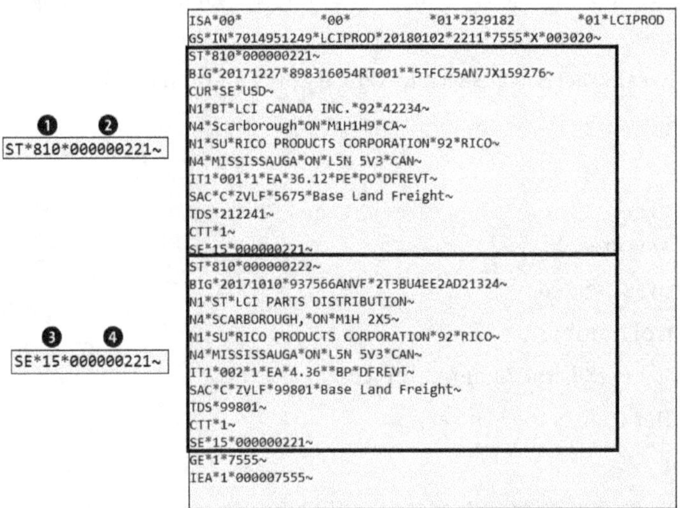

Figure 9.45 Definition of Elements for the Transaction Set and Sample EDI 810 File

From a logistics and third-party logistics perspective, Table 9.4 lists the commonly used X12 and EDIFACT messages.

EDI X12 Transaction Number	EDI Transaction Name/ Document Type	Description
810	Invoice	Provides invoice information for the billing of goods and services provided.
846	Inventory Inquiry/ Advice	The transaction set can be used in the following ways: (1) for a seller of goods and services to provide inventory information to a prospective purchaser, with no obligation to the purchaser to acquire these goods or services; (2) for a representative of a seller of goods and services to supply inventory information to that seller; (3) for one location to supply another location with inventory information; and (4) for an inquiry as to the availability of inventory with no obligation on the seller of goods and services to reserve that inventory.

Table 9.4 Commonly Used X12 and EDIFACT Messages

EDI X12 Transaction Number	EDI Transaction Name/ Document Type	Description
850	Purchase Order	Provides for the placement of purchase orders for goods and services.
855	Purchase Order Acknowledgment	The transaction set can be used to provide for customary and established business and industry practice relative to a seller's acknowledgment of a buyer's purchase order. This transaction set can also be used as notification of a vendor-generated order.
856	Ship Notice/Manifest	The transaction set can be used to list the contents of a shipment of goods as well as additional information relating to the shipment, such as order information, product description, physical characteristics, type of packaging, marking, carrier information, and configuration of goods within the transportation equipment.
860	Purchase Order Change Request – Buyer Initiated	The transaction set can be used to provide the information related to buyer-initiated changes to a previously submitted purchase order.
940	Warehouse Shipping Order	This transaction set can be used to enable the depositor or other business party to advise a warehouse to make a shipment, confirm a shipment, or modify or cancel a previously transmitted shipping order.
945	Warehouse Shipping Advice	The transaction set can be used by the warehouse to advise the depositor or other business party that shipment was made. It's used to reconcile order quantities with shipment quantities.
947	Warehouse Inventory Adjustment Advice	The transaction set can be used to inform a warehouse/depositor or other business party of a quantity or status change to inventory records. This transaction set provides detailed information concerning the internal adjustments that occur between a warehouse and a depositor or other business party.

Table 9.4 Commonly Used X12 and EDIFACT Messages (Cont.)

9

EDI X12 Transaction Number	EDI Transaction Name/ Document Type	Description
888	Item Maintenance	This transaction set can be used to provide information about new products or changes in existing product specifications. Price changes won't be sent with this transaction set but can be sent with the Price Change Transaction Set (879).
879	Price Information	This transaction set can be used to enable a manufacturer, supplier, broker, or agent to provide a trading partner with pricing information.
214	Transportation Carrier Shipment Status Message	This transaction set can be used by a transportation carrier to provide shippers, consignees, and their agents with the status of shipments in terms of dates, times, locations, route, identifying numbers, and conveyance.

Table 9.4 Commonly Used X12 and EDIFACT Messages (Cont.)

Note

You can find the exhaustive list of EDI transaction sets in the parent company's website at *https://x12.org/products/transaction-sets*.

We've discussed the purpose of EDI communication, its structure, and commonly used EDI messages in the logistics/supply chain domain. EDI isn't the only communication method but it's very common and widely used in industries. You may come across other modes of communication such as XML, flat files, or other standards of EDI messages as explained in previous sections.

Figure 9.46 shows a schematic representation of EDI message flows between the customer, third-party logistics, and shipper in the third-party logistics integration context. In this case, the shipper may invoice the customer for the services rendered because customers often ask the third-party logistics providers for shipment of goods. However, it may be shipper initiated as well.

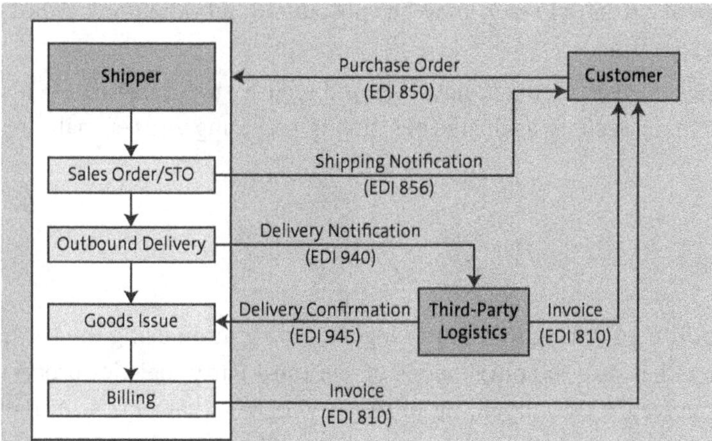

Figure 9.46 EDI Message Flow in a Third-Party Logistics Scenario

Advantages and Disadvantages of Electronic Data Interchange

Some advantages of EDI include the following:

- Provides savings in terms of cost and effort from EDI automating processes that were previously manually executed with paper documents
- Improves productivity and efficiency as more business documents are shared and processed in less time with greater accuracy
- Reduces errors through rigid standardization and governance, which helps to ensure information and data are correctly formatted before they enter business processes or applications
- Improves traceability and reporting because electronic exchange of documents can be tracked for data collection, visibility, and analysis.
- Provides reliability and positive customer experiences for reliable product and service delivery

Limitations of EDI include the following:

- Requires a heavy investment to launch and maintain the technology. Small companies might find it too expensive, considering implementation and running costs.
- It may limit your business partners as large companies running EDI may find it difficult to integrate with small non-EDI partners.
- Too many standard formats to consider for integration (e.g., ANSI ASC X12, GS1 EDI, HL7, TRADACOMS, and UN/EDIFACT). In addition, one organization may be using the latest version of the standard while its trading partners may still be using older versions.
- As a system, EDI isn't easy to learn, use, and implement.

- Time and effort spent on initial setup may be substantial and may prevent fast-tracking of the EDI solution.
- EDI requires significant investments in networks and security mechanisms. Such a system would need to be secured against cyber threats, including viruses, malware, and hacking.

9.4 Summary

Common issues encountered during third-party logistics integration using the IDoc/EDI combination are either due to complexities of the third-party logistics process itself or challenges involved with onboarding third-party logistics partners as it may require negotiating technical changes in their system as well. Integration challenges exist with both new and existing trading partners. As third-party logistics providers cater to a huge number of customers, sometimes it may be quite frustrating for implementation partners to negotiate timeline and change efforts from the logistics provider's end, leading to undue extension of project deadlines. Nevertheless, a successful third-party logistics integration not only makes the process immensely efficient but also dramatically improves customer service leading to a delightful customer experience.

In this chapter, we focused on the concept of ALE and IDoc and also looked at how to configure and make your system ready from the logical system creations to creating your RFC destination for your third-party logistics vendor system.

As the IDoc topic itself requires quite a bit of learning, we covered the basics of IDocs for nontechnical consultants as well, starting from the port creation to setting up of IDoc messages and assigning partner profiles. In addition, we looked into some essential features and drivers for integration with third-party logistic service providers, means of integration, and technical aspects of standard integration modes most commonly used across industries.

Chapter 10
Direct Connectivity and Middleware

Integrating SAP with other internal or external applications is one of the most ubiquitous challenges for companies running on SAP. Organizations can't fully automate and optimize their business operations without adequate integration between SAP and these systems. Discover some of the most typical problems related to integration, a quick rundown of SAP interfaces and interaction points, and many different approaches for integrating SAP systems.

The use of SAP, one of the most well-liked integrated enterprise resource planning (ERP) solutions available, has helped businesses optimize several of their most crucial business processes, including finance, human capital management, enterprise performance management, logistics, and many other areas. However, connecting SAP with other systems inside or outside their organization is one of the most common issues for SAP businesses to address. Organizations are unable to completely automate and optimize their business operations without adequate connectivity between SAP systems and other systems.

The challenge that comes with typical businesses using IT resources to support and manage their everyday activities is that the IT infrastructure has evolved over the past few decades into a complex and diverse environment. Regardless of the underlying technology, platform, or operating system (OS) of these parts, steadily emerging company requirements were the primary forces behind the ongoing introduction of new components into the IT infrastructure. All these various systems will be able to interface with one another in a flexible and effective way, thanks to the establishment of a middleware layer.

SAP can potentially interact with a wide range of other corporate software. The on-premise SAP S/4HANA or SAP S/4HANA Cloud will replace SAP ERP, which will be retired in 2027. In this chapter, we'll examine the many communication capabilities that SAP S/4HANA's middleware responsibilities in creating connections with various systems offer. We'll also analyze additional techniques, such as the usage of application programming interfaces (APIs) in electronic data interchange (EDI) solution connection creation in business-to-business (B2B) situations.

10.1 Direct Connectivity Options

Software applications need to be able to interface with a broad variety of diverse system settings, which necessitates a complicated network of various technologies. To ensure smooth communication in both directions inside the SAP environment, it's crucial to connect traditional technology written in ABAP with non-ABAP structures.

In this section, you'll learn about basic communication capabilities and different connectors used to link different systems, how you can use RFC and web service calls as an alternative in interacting across various systems, and how SAP Business Technology Platform (SAP BTP) can be used in establishing connections as an alternative.

10.1.1 Synchronous versus Asynchronous Communication

Let's understand the functional characteristics of different communication technologies. The two main forms of communication between two systems are synchronous and asynchronous communication. Both methods of communication have unique benefits and drawbacks that might have an impact on either system or business applications.

One prerequisite for synchronous communication is that both the sender and the receiver's systems must be operational to transmit or receive messages. In typical synchronous communication, shown in Figure 10.1, a single function call is used, and the transmitting system waits for the receiving system's answer.

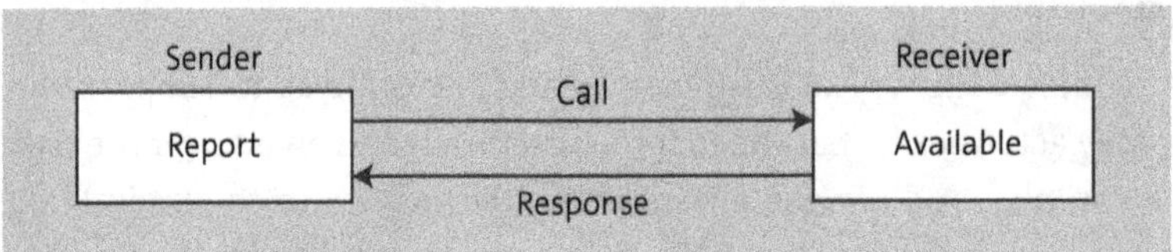

Figure 10.1 Typical Synchronous Communication

A business scenario where typically synchronous communications are seen is performing an availability check (available-to-promise [ATP]) for material during sales order creation or any sort of inquiry performed from other systems or from SAP Customer Relationship Management (SAP CRM) or SAP Supplier Relationship Management (SAP SRM).

It's not necessary for the receiving system to be online when a function call is sent from the sending system in the case of asynchronous communication. The call can be received and handled by the receiving system at a later time, as shown in Figure 10.2.

The function call is kept in the transmitting system's outbound queue if the receiving system is unavailable, and it's replayed periodically there until the receiving system can handle it.

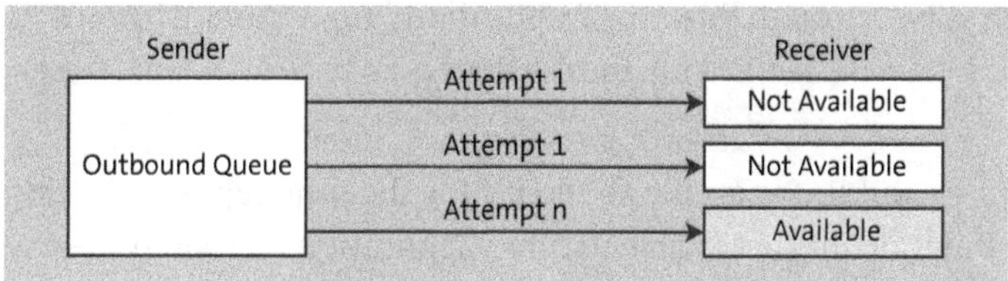

Figure 10.2 Typical Asynchronous Communication

The business scenarios where typically asynchronous communications are seen include material management and logistics scenarios such as third-party logistics and master data replication, among others.

10.1.2 Connectors

SAP also offers numbers of standalone components to connect with SAP systems and non-SAP systems. We'll walk through five different examples of typical connectors in the following sections.

SAP Business Connector

SAP Business Connector connects many applications and technologies to SAP systems using open and nonproprietary standards. It communicates over the internet (via HTTP and XML) and allows you to greatly automate your company procedures. SAP Business Connector has mapping and routing techniques that allow you to, for example, convert IDocs to XML messages and send them through the internet to any recipient system. Thus, SAP Business Connector seamlessly integrates different IT architectures with SAP systems, as shown in Figure 10.3.

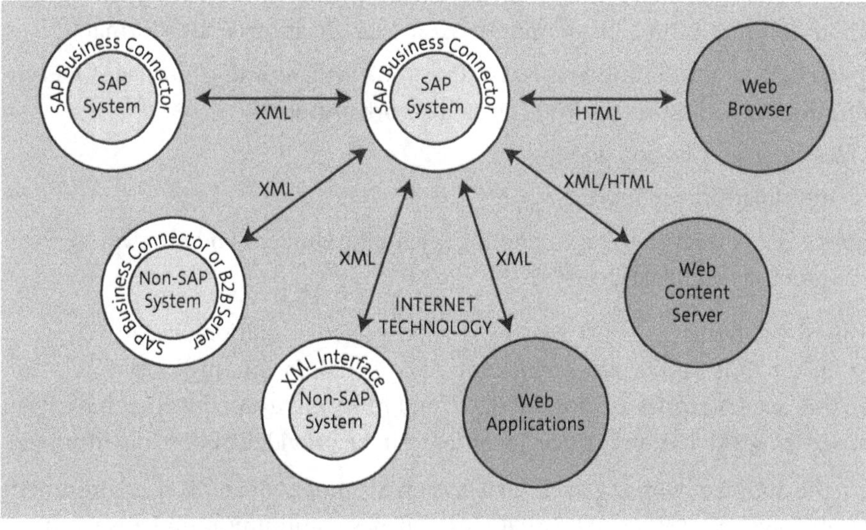

Figure 10.3 Overview of SAP Business Connector

All SAP functionality accessible through Business Application Programming Interfaces (BAPIs) and IDocs may be made available to business partners as secure XML-based services over the internet.

The internal data structures and BAPIs aren't necessary for the other apps. This XML service layer makes SAP functionality accessible to other applications, either inside the company or with outside partners over the internet.

A two-way connection to and from SAP systems is made possible through SAP Business Connector. Real-time integration between supplier stocks and your SAP system; multivendor product, pricing, and availability information and your purchasing application; or fulfillment and order tracking software and your shippers' internal systems are examples of applications.

SAP Java Connector

A Java application can connect with SAP systems using the RFC protocol via SAP Java Connector (SAP JCo) middleware, which is a development component. Both inbound RFCs (Java calls ABAP) and outbound RFCs (ABAP calls Java) are supported by SAP JCo.

Advantages of SAP JCo for creating Java applications that use SAP include the following:

- High performance Java Native Interface (JNI)-based RFC middleware
- Supports R/3 3.1I and higher releases (and other SAP components that have BAPIs or Remote-Enabled Function Modules [RFMs])
- Supports inbound (Java client calls BAPI or RFM) and outbound (ABAP calls Java server) RFC communication
- Supports synchronous RFC (sRFC), transactional RFC (tRFC), queued RFC (qRFC), and background RFC (bgRFC) protocols
- Supports classic RFC data serialization as well as XML RFC (xRFC), Binary ABAP Serialization XML (basXML), and column based/fast serialization (cbRFC) formats
- Supports RFC communication via classic Common Programming Interface for Communication (CPIC) and via the WebSocket network protocol
- Provides client connection pooling
- Offers multiplatform support
- Complete and correct code page handling (including Unicode and non-Unicode single- and multibyte code pages)

SAP Business Connector uses SAP JCo as a key component. Additionally, SAP JCo is put into use in the SAP NetWeaver Application Server (AS), facilitating communication between the Java Enterprise Edition 2.0 (J2EE) and ABAP environments that have been integrated. Using SAP JCo, an invoice document can be provided online to customers.

You have the option of working in either a Java environment or an ABAP environment in SAP NetWeaver AS (SAP_BASIS). SAP JCo establishes communication between these parts. An overview of the SAP NetWeaver AS's architecture is shown in Figure 10.4 with SAP JCo serving as the bridge between the Java and ABAP environments.

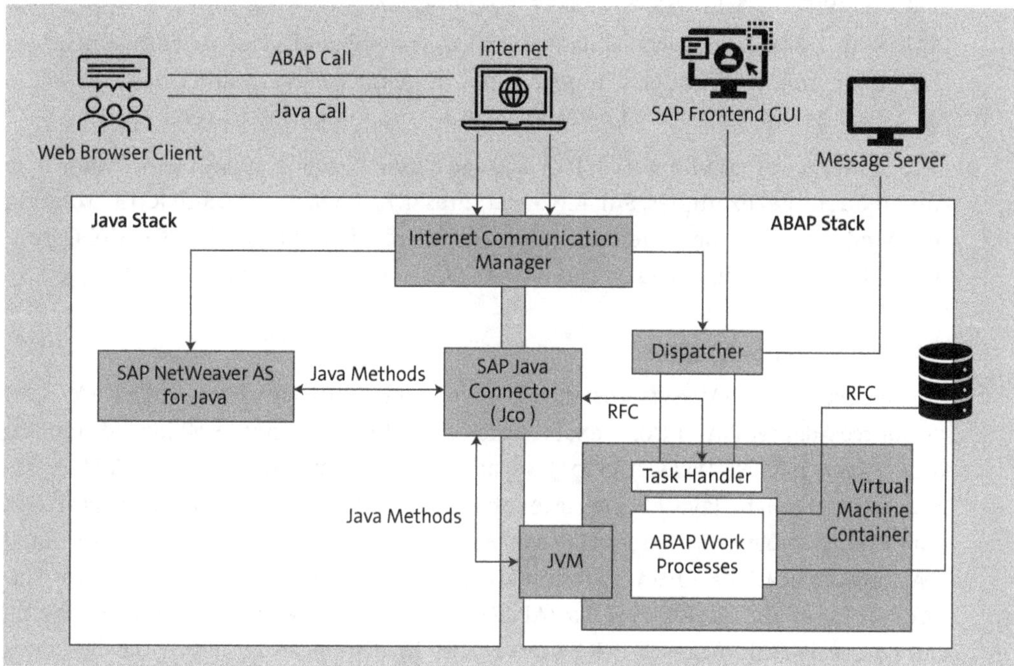

Figure 10.4 Java/ABAP Coupling in SAP NetWeaver Application Server Using SAP JCo

Data transfer between SAP JCo and the ABAP environment is performed by means of an RFC (with gateway communication by means of CPIC), as shown in Figure 10.5.

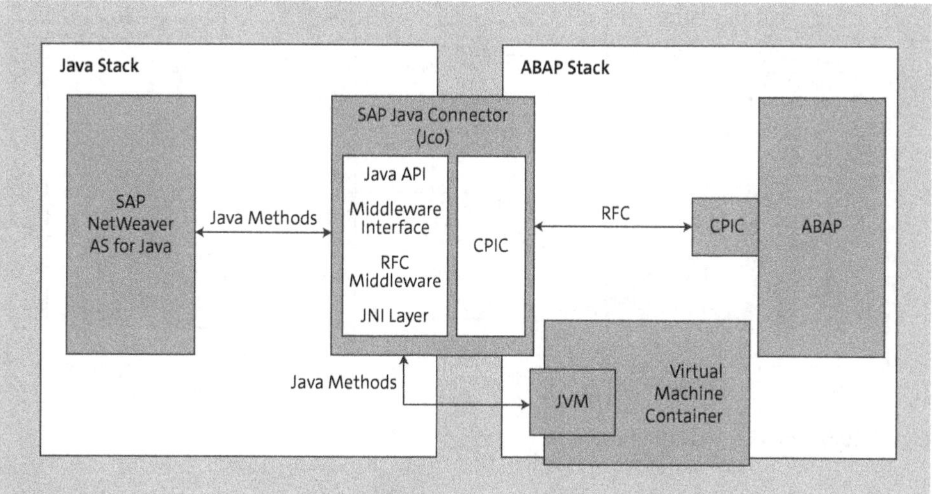

Figure 10.5 Communication between ABAP and Java Using CPIC (Application Server)

Release 3.1.7 of SAP JCo includes fixes and new functionality as described in SAP Note 3276799. It also contains all fixes and enhancements of previous patch releases, including SAP Notes 3204567, 3115707, 3036882, and 2947370.

SAP Note 2786882 describes which operating systems and versions with which Java Development Kits (JDKs)/Java Runtime Environments (JREs) are supported by SAP JCo 3.1, as well as when SAP JCo 3.1 maintenance and support period will end.

The Software Development Kit (SDK) package of SAP JCo 3.1 includes the software, documentation, and examples. SAP JCo 3.1 requires JDK/JRE 8, 11, or 17. SAP JCo 3.1 running on Windows operating systems requires the Microsoft Visual Studio 2013 C/C++ runtime libraries to be installed on the system. (See SAP Note 2786882.)

SAP Java Resource Adapter

A connection for SAP systems that is compatible with Java EE is called SAP Java Resource Adapter (SAP JRA). It enables the connection of remote SAP Java EE application servers (SAP NetWeaver AS for Java) and SAP NetWeaver AS for ABAP. SAP JCo provides the standard Java EE interfaces thanks to SAP JRA, which makes it easier to communicate between SAP NetWeaver AS for ABAP and heterogeneous, distributed SAP Java applications. An SAP Java application may call functions on a remote SAP NetWeaver for AS for ABAP server via SAP JRA, and RFCs can be used to handle incoming function calls. This makes SAP JRA a server and client application. Figure 10.6 illustrates the architecture of SAP JRA, which is an add-on for SAP JCo. It comes with SAP NetWeaver AS for Java as an integrated component.

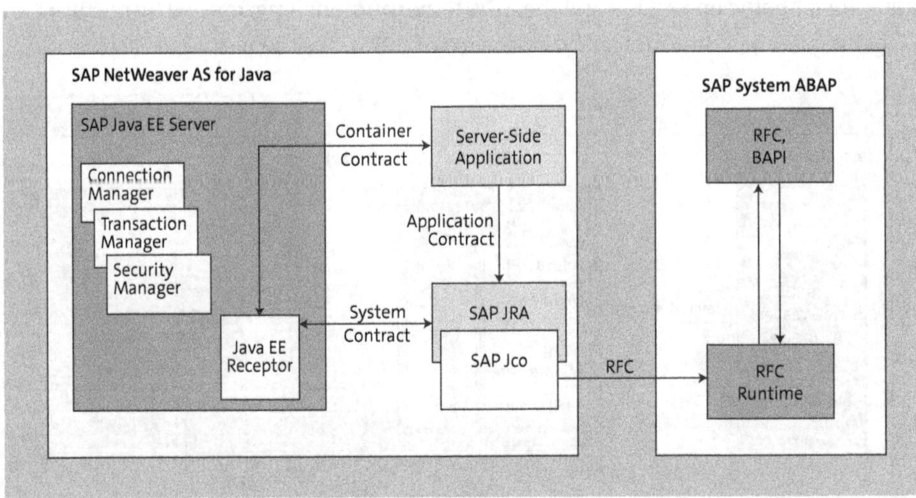

Figure 10.6 SAP JRA Architecture

SAP JRA can be used with Enterprise Java Beans to develop apps in your SAP system. SAP JRA also reduces the implementation effort while developing apps for an SAP system using various J2EE application servers.

Note

The J2EE Connector Architecture 1.5 requirements are currently being implemented. Go to *http://s-prs.co/v571705* for additional information about J2EE and the J2EE Connector Architecture.

SAP provides different resource adapters for use with the SAP Business Intelligence Java Software Development Kit (BI Java SDK) in business intelligence (BI). These allow applications built with the BI Java SDK to interface with disparate data sources. One example is BI Java Connectors; Java Connectors are software components or libraries that enable Java applications to connect to and interact with various BI systems and tools. These connections allow developers to collect, manipulate, and visualize data from BI sources by facilitating communication between Java-based applications and BI systems.

BI Java Connectors provide a standardized mechanism for developers to connect to BI systems, conduct queries, retrieve data, and do other activities without having to create complicated code for low-level communication protocols. Instead, they offer a higher-level interface that isolates the underlying technological aspects, making working with BI data easier.

BI Java Connectors can be used to connect to major BI systems such as Microsoft Power BI, SAP BusinessObjects, and others.

Consider compatibility with your selected BI system, simplicity of integration with your Java application, performance characteristics, and the degree of community support or documentation available for the connection when choosing a BI Java connection for your project.

SAP Connector for Microsoft .NET

Communication between SAP systems and the Microsoft .NET platform is made possible through the SAP Connector for Microsoft .NET (NCo), which is a development environment. With the help of this connector, you can create a variety of applications in Microsoft Visual Studio .NET, including web forms, Windows forms, and console programs. It also supports RFCs and web services. You can use any popular programming language, including Visual Basic .NET, C#, or Managed C++, with the NCo. You can use .NET programs as a client and as a server using the NCo.

You can use the NCo and SAP .NET Proxy Wizard to create the following:

- .NET Windows and web form apps that have access to SAP business objects (BAPIs).
- Client apps for the SAP server for which you can use RFCs or HTTP/Simple Object Access Protocol (SOAP)/XML (outside-in)
- RFC server apps that operate in a.NET environment and may be installed from the inside out (from the SAP system)

You can use Microsoft Visual Studio .NET to do the following:

- Create proxy objects via the integrated Proxy Wizard.
- Use any common programming language that is fully compatible with Microsoft Visual Studio .NET.
- Connect SAP tables and structures to Windows and web form controls (data binding).
- Use secure authentication mechanisms such as Single Sign-On (SSO), Kerberos, and Microsoft Passport (see *http://s-prs.co/v571706* for more information).

For the connection between the NCo and the SAP system, you can use both RFC and SOAP. The protocol used is determined by the release state of your SAP system and whether it's available inside or outside of the firewall. Figure 10.7 is a graphical representation of NCo that shows how the RFC layer and SOAP layer are communicated.

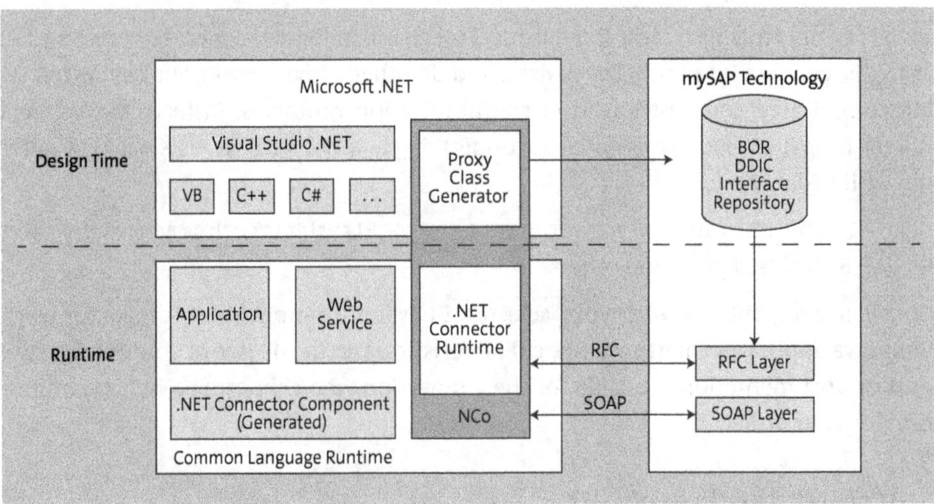

Figure 10.7 SAP Connector for Microsoft .NET: Architecture Overview

You can make function calls to your SAP system and receive the results if you use a .NET program as a client. Follow these steps to use the .NET connector:

1. Create Active Server Pages (ASPs) that invoke an SAP system function.
2. Develop a Windows form application that allows for an interactive user guide.
3. Using NT batch processing, create a console application to accept information from the SAP system.
4. Create a web service to give SOAP access to your SAP system (prior to SAP Version 6.20).

If you employ a .NET program as an RFC server, your SAP system will treat it the same way it would another SAP system. It's capable of sending a function call to the .NET server and receiving the response. Follow these steps to use this .NET connector:

1. Retrieve information (e.g., supplementary client data) needed to complete a customer sales order from an external server.
2. Retrieve information (prices, delivery data, travel bookings, etc.) from an external server that is necessary for an evaluation in the SAP system.
3. Send emails from your SAP system.
4. Send IDocs from your SAP system to an external system.

There are a couple SAP connector options that you can choose from:

- **SAP Connector for Microsoft .NET 3.1**
  SAP Connector for Microsoft .NET 3.1 (NCo 3.1) allows developers to use BAPIs and remote-enabled function modules in any .NET application (inside-out). You can also access .NET components from any ABAP application by implementing an RFC server in .NET (outside-in). It's the successor of SAP Connector for Microsoft .NET 3.0 (NCo 3.0). See SAP Note 3152653 for details about availability and platform requirements.

  Prerequisites: NCo 3.1 supports .NET framework versions 4.6.2 up to 4.8.1 as runtime. It requires the Microsoft C++ Runtime Dynamic Link libraries (DLLs) version 14.x (this version is contained in the Visual C++ Redistributable Package for Visual Studio 2015, 2017, 2019, and 2022). You can get the installer from the Microsoft download page at *http://s-prs.co/v571701*. Setting up direct communication between two business applications requires the usage of a direct connection.

- **SAP Connector for Microsoft .NET 3.0**
  SAP Connector for Microsoft .NET 3.0 (NCo 3.0) allows developers to use BAPIs and RFMs in any .NET application (inside-out). You can also access .NET components from any ABAP application by implementing an RFC server in .NET (outside-in). See SAP Note 856863 for details about availability and platform requirements.

  Prerequisites: The .NET 2.0 variant of NCo 3.0 requires the Microsoft C++ Runtime DLLs version 8.x (this version is contained in Microsoft Visual C++ 2005 SP1 Redistributable Package). See SAP Note 684106 for details about downloading and installing this runtime. The .NET 4.0 variant of NCo 3.0 requires the Microsoft C++ Runtime DLLs version 10.x (this version is contained in Microsoft Visual C++ 2010 Redistributable Package). You can get the installer from the Microsoft download page at *http://s-prs.co/v571702*.

SAP NetWeaver Remote Function Call Software Development Kit

From version R/3 4.6C up to the current SAP S/4HANA systems, the SAP NetWeaver RFC Software Development Kit (SDK) provides a C/C++ interface for connecting to SAP systems. It enables the creation of both programs that can be called from ABAP (RFC servers) and programs that call ABAP functionality (RFC clients).

The native RFC protocol is used. The SAP NetWeaver RFC SDK also provides broad protocol functionality, such as all essential RFC features in SAP systems, exceptional performance, and extreme versatility.

It's thread-safe to use the SAP NetWeaver RFC SDK. This makes it possible for users to create effective multithreaded RFC applications. The "classic" RFC Library and the SAP NetWeaver RFC SDK are incompatible. If you want to switch over to the SAP NetWeaver RFC SDK, you'll need to make adjustments to any current applications that use the old RFC Library.

10.1.3 Remote Function Call/Web Service

We communicate with SAP systems while working with them using tools such as the SAP GUI, Web GUI, and SAP Fiori apps, among others. An SAP system may need to communicate with non-SAP systems or other SAP systems for the purpose of exchanging information, depending on the nature of this activity. The common SAP interface for transferring data between SAP systems is RFC. To accomplish the needed action, RFC invokes a function that is to be run in the system. Depending on the anticipated work, the SAP system offers a variety of RFCs.

This section will examine and explain the various RFC offerings in SAP. These RFC types provide versatility and dependability in connecting SAP systems with one another and with external systems, facilitating frictionless data interchange and process coordination.

Remote Function Calls Overview

The standard SAP interface for communication across SAP and non-SAP systems is the RFC. It's normally a function call in a remote system, which is why it's called a Remote Function Call. There are currently several distinct RFC variations available, each of which has unique characteristics and serves a particular function, so let's take a closer look:

- **Synchronous RFC (sRFC)**
  In sRFC, the calling program waits until the requested processing steps on the remote server is completed before continuing to operate locally. In other words, both systems must be operational at the time of the call.
- **Asynchronous RFC (aRFC)**
  The calling program sends the request to the remote system and then returns to work locally. The desired processing steps is run in isolation on the remote system. If the remote system can't be reachable at the time of the call, the RFC client's asynchronous calls are dropped.

The main characteristic of aRFC is that as soon as the call is made, function control goes back to the caller program. aRFCs provide their parameter information straight to the server rather than to the database. The user can interface with the remote system via aRFCs. The called server must be prepared to respond when the caller initiates an aRFC. The aRFC can return results to the caller program.

When real-time communication with a distant system is established, it's always a good idea to use an aRFC because execution in the calling program shouldn't be stopped until the results of the called function module have been achieved.

- **Transactional RFC (tRFC)**
  Like aRFC, the tRFC communication ensures that if a request is received several times due to network issues, it will only be processed once by allocating a transaction ID (TID). In contrast to aRFC, tRFC doesn't require the distant system to be up when the RFC client program starts the call. Up until the destination system is ready, the data is stored in the source system.

  The failed requests, identified by their TID, are attempted to be sent again when the report program RSARFCSE is called in the background on a regular interval. Implementation of tRFC is recommended if you want to maintain the transactional sequence of the calls.
- **Queued RFC (qRFC)**
  This RFC communication extends the tRFC. In this technique, all requests are queued (inbound and outbound queues) and handled in succession only if all previous calls are completed satisfactorily. This approach ensures that all requests are handled in the order in which they are received.

  This type of RFC will be used in SAP Supply Chain Management (SAP SCM; SAP Advanced Planner and Optimization [SAP APO]) systems for core interface (CIF) queues and other applications where requests must be handled in a certain sequence. Implementation of qRFC is recommended if you want to guarantee that several transactions are processed in a predefined order.
- **Background RFC (bgRFC)**
  In terms of performance and functionality, bgRFC is the replacement for tRFC and qRFC. Therefore, it's strongly advised to use bgRFC as opposed to tRFC and qRFC.

Local Data Queue

In RFC communication, the Local Data Queue (LDQ) is an exceptional instance. The system doesn't actively send any data in this case. Instead, the pull principle dictates that it be kept locally until it's invoked by an outside program (e.g., mobile device).

> **Note**
>
> Background communication encompasses all aRFC methods (tRFC, qRFC, bgRFC, and LDQ).

Depending on the type of partner system, you can configure several types of connections in the RFC interface, as shown in Figure 10.8.

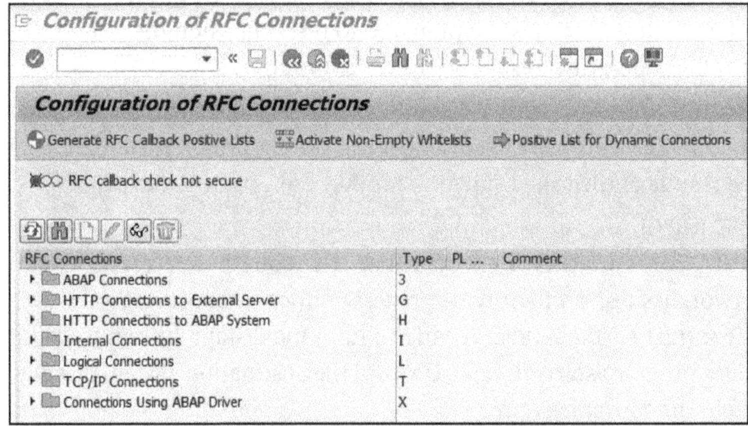

Figure 10.8 Overview Screen: Configuring RFC Connections

There are four possible scenarios:

- **ABAP - ABAP**
  Two SAP systems—technically two SAP NetWeaver AS for ABAP systems—communicate with each other in this case. The RFC interface is used to communicate data between the two partner systems (Figure 10.9). It's not essential to convert the data format again.

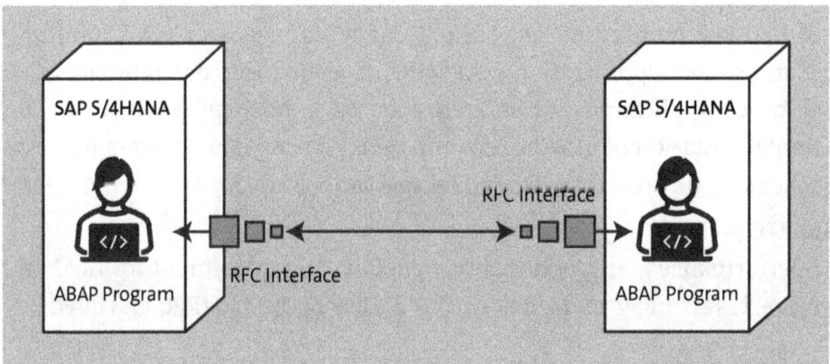

Figure 10.9 SAP NetWeaver AS for ABAP to SAP NetWeaver AS for ABAP

- **ABAP - Other system**
  In this case, the partner system is not an SAP system. The RFC API is used to build an interface that translates the ABAP date format, as shown in Figure 10.10.

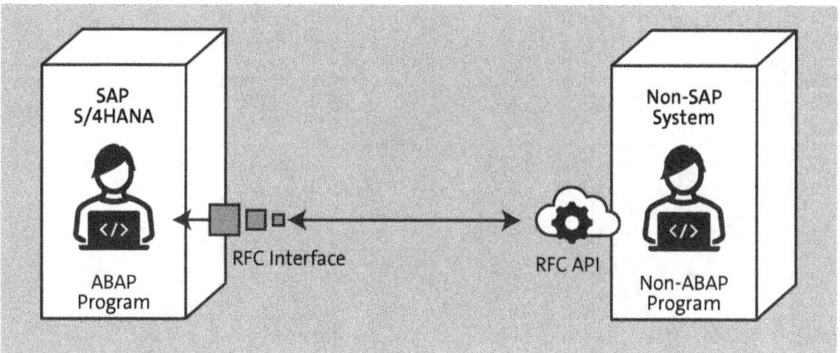

Figure 10.10 SAP NetWeaver AS for ABAP to Non-SAP Server

- **ABAP - Java (SAP)**
  The communication in this case doesn't have to take place between two different systems. But inside SAP NetWeaver AS, communication between SAP NetWeaver AS for ABAP and SAP NetWeaver AS for Java also needs the RFC interface, as shown in Figure 10.11.

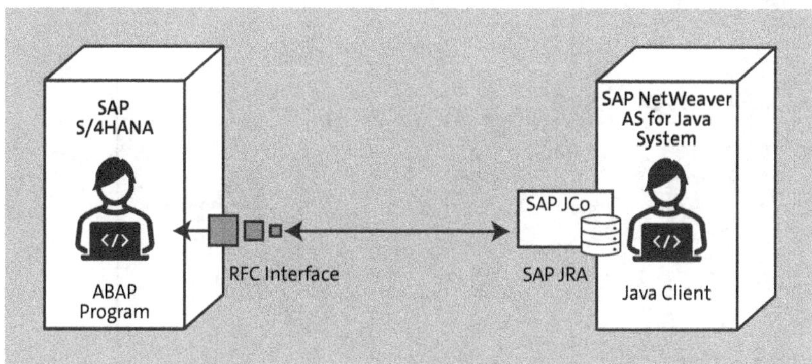

Figure 10.11 SAP NetWeaver AS for ABAP to SAP NetWeaver AS for Java

The interface function and data type mapping are handled by SAP JCo at the SAP NetWeaver AS for Java end. The SAP JRA, which is based on the Java EE standard design, lags behind SAP JCo. Additionally, this case study may be used to communicate with an outside SAP NetWeaver AS for Java instance.

- **ABAP - Java (external system)**
  In this instance, SAP NetWeaver AS for ABAP interacts with a Java platform that is not based on SAP NetWeaver AS for Java. In this instance as well, SAP JCo handles the data mapping and interface function for the non-SAP Java system, as shown in Figure 10.12.

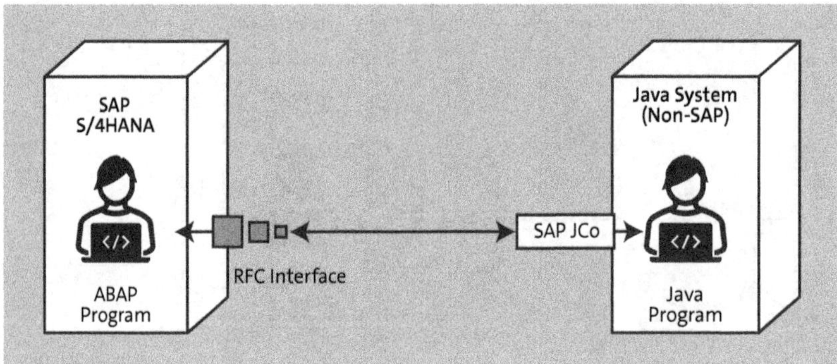

Figure 10.12 SAP NetWeaver AS for ABAP to Non-SAP Java System

RFC destination is one of the finest ways to link an ABAP system to an external system. Let's look at how to create and maintain HTTP RFC destinations for external systems:

1. Go to Transaction SM59, and select **HTTP Connection to External Server**. This brings you to the screen shown in Figure 10.13.

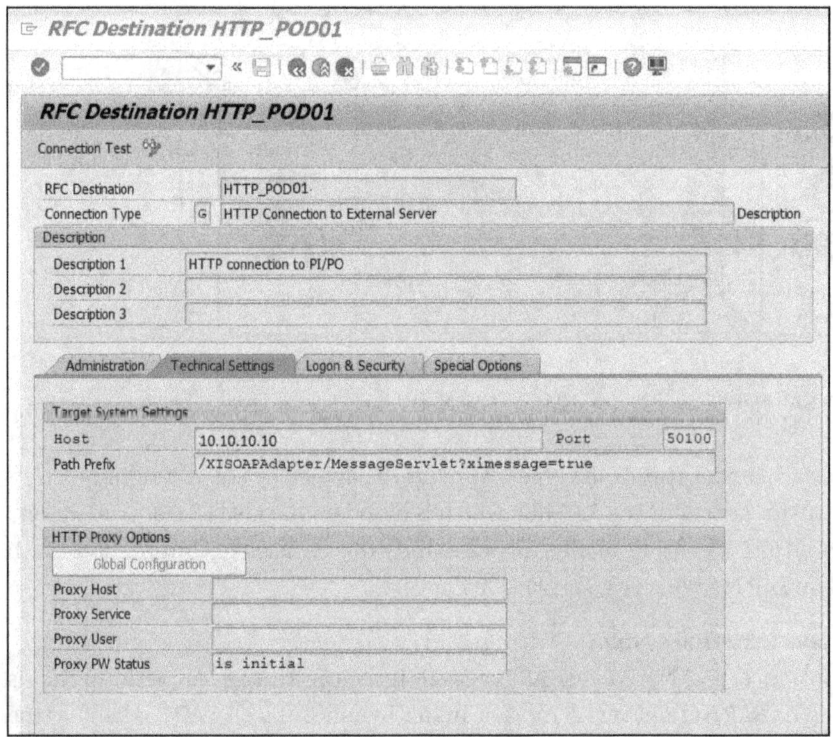

Figure 10.13 Create RFC Destination for Connection Type G

2. Maintain the parameters as shown in Table 10.1.

Parameter	Values
RFC Destination	Enter a unique RFC name.
Connection Type	Select **G**.
Description 1	Give a proper description.
Host	Enter the IP address/server name of your target system you want to connect.
Port	Enter the port number of the target system you want to connect.
Path Prefix	At the time when the connection to this destination is initiated, the system inserts this subpath before .
HTTP Proxy Options	Configure a proxy for HTTP connections. You can determine the proxy host and service, as well as users and passwords for the HTTP connection.

Table 10.1 RFC Creation: Possible Field Entries to Fill

3. If you click on the **Logon & Security** tab shown earlier in Figure 10.13, you can make various entries depending on the chosen connection type. Figure 10.14 shows what possible entries can be maintained.

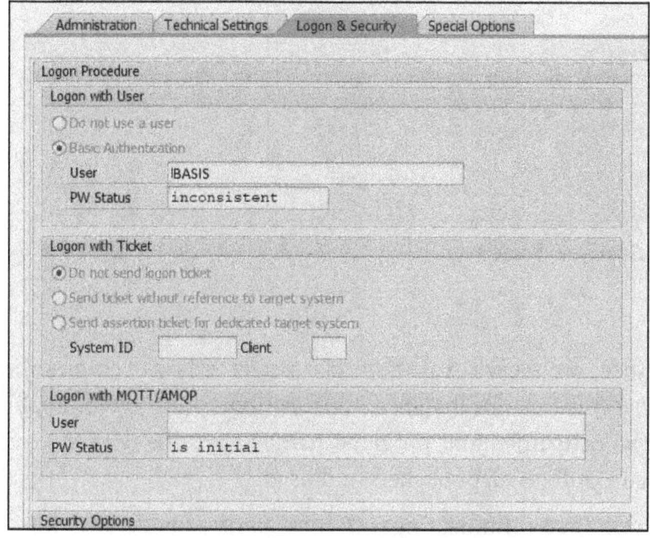

Figure 10.14 RFC Destination Logon & Security Tab Details

4. Select the connection you require. You can choose from different logon procedures as outlined in Table 10.2.

Parameter	Values
Logon Procedure	Choose the radio button to either use a user or don't use a user.
Basic Authentication	Basic Authentication is a standard HTTP authentication method. When the user logs on to the target system, a user ID and a password are provided for authentication.
Description 1	Give a proper description.
Host	Enter the IP address/server name of the target system you want to connect (refer to Figure 10.13).

Table 10.2 RFC Creation: Possible Entries for the Logon & Security Tab

Once you save the RFC connection, your connection is ready, but you need to test whether it's truly ready or not by clicking the **Connection Test** button in the application toolbar. It should have the status for **Status HTTP Response 200** for successful connection establishment, as shown in Figure 10.15.

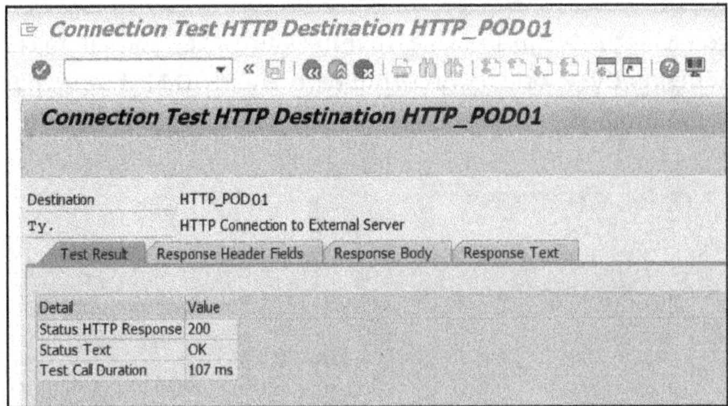

Figure 10.15 Connection Test Overview Screen

Web Service

In service-oriented architectures, web services are a standardized method of communication between client applications that, for instance, consume data from server applications. Web services may be consumed and provided using SAP NetWeaver AS for ABAP. Services can be created, configured, published, and used.

It's a very effective approach while communicating with third-party providers. In third-party logistics scenarios, there are a number of vendors who are publishing their web service, and the provider of a web service is generally called a *service provider*. The service provider will have an XML-based Web Service Description Language (WSDL) document corresponding to the web service. When a web service is published, the service provider submits information about itself as well as a description of the service to the services registry.

The users using this service are known as *service consumers*. In simple terms, any application that accesses the web service is also known as a service consumer. A *services registry* it just like a directory which contains all the information about web services. The Universal Description, Discovery, and Integration (UDDI) services registry is the standard register that is used. The application gets the information it needs from the service description, which is typically stored in the services registry.

You can build ABAP web services using Transaction SE80 (ABAP Workbench), you can use existing BAPIs or RFMs to set up web services, or you can develop a new web services in Object Navigator. ABAP web services can be used for communication between SAP systems and between SAP and non-SAP systems.

A web service provider can be built using existing RFC modules or WSDL documents. It can also be modeled in the Enterprise Services Repository (ES Repository) or in the ABAP backend directly. You can create a service consumer for your web service provider.

When business processes running on multiple distributed systems and applications need to send or receive information from third-party logistics vendors, you can think of the integration scenario concept as well. There are certain third-party logistics vendor systems where you might not have proper gateways installed or configured to create services there; in that case, you can create a Transaction SICF service (REST). Transaction SICF is normally used to maintain services for HTTP communication using the Internet Communication Manger (ICM) and Internet Communication Framework (ICF).

Let's see an example where we need to send some details regarding warehouse, material, plant, batch, and transfer order. We need to send the relevant data/information from SAP HANA to a third-party logistics service provider vendor.

To fulfill this requirement, you need to follow certain steps in the SAP S/4HANA system. First, you'll create a handler class, using these steps:

1. Use Transaction SE24, and create a new class called **ZCL_LOG_WM_GET_HANDLER**.
2. Select the **Interfaces** tab, and add interface **IF_HTTP_EXTENSION**, as shown in Figure 10.16.

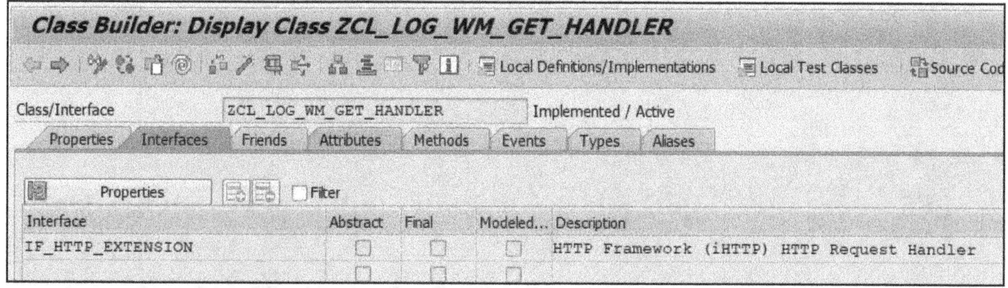

Figure 10.16 Handler Class

3. You can write your code in method HANDLE_REQUEST and send the desired response in JavaScript Object Notation (JSON), as shown in Figure 10.17.

Class Builder Class ZCL_LOG_WM_GET_HANDLER Display

Ty.	Parameter	Typing	Description
	SERVER	TYPE REF TO IF_HTTP_SERVER	HTTP Server

Method IF_HTTP_EXTENSION~HANDLE_REQUEST active

```
* Set the converted data to Browser
  server->response->set_data( data = lv_json ).

ENDMETHOD.
```

Figure 10.17 Method in the Handler Class

4. Create a new node in Transaction SICF (see Figure 10.18), and activate the service by right-clicking on the newly created service and selecting **Activate Service** (see Figure 10.19). By default, the service will be in inactive state after you initially save it.

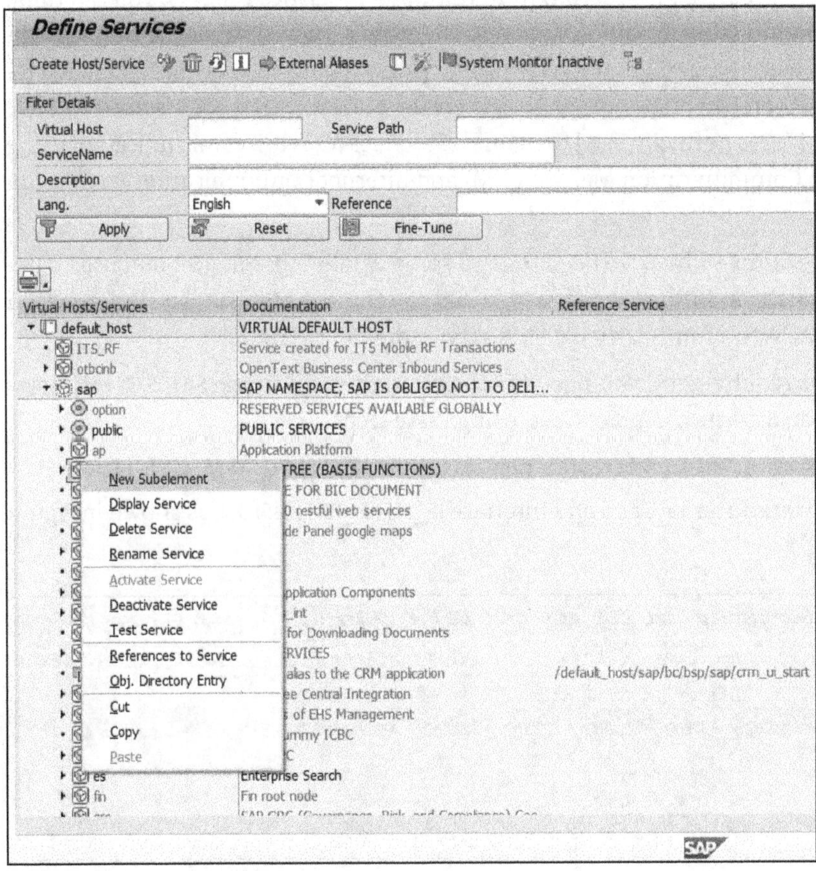

Figure 10.18 Create the Node in Transaction SICF

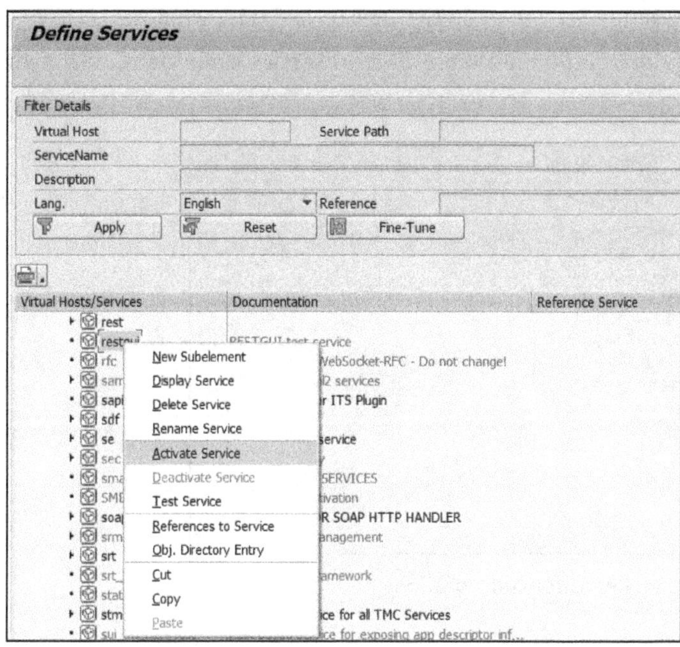

Figure 10.19 Activate Service

5. Once a service is activated, you can double-click in the created service and then click on the **Change** button (or press Ctrl+F1).
6. Go to the **Handler List** tab shown in Figure 10.20, assign the handler class created earlier (refer to Figure 10.16), and click on the Save button.

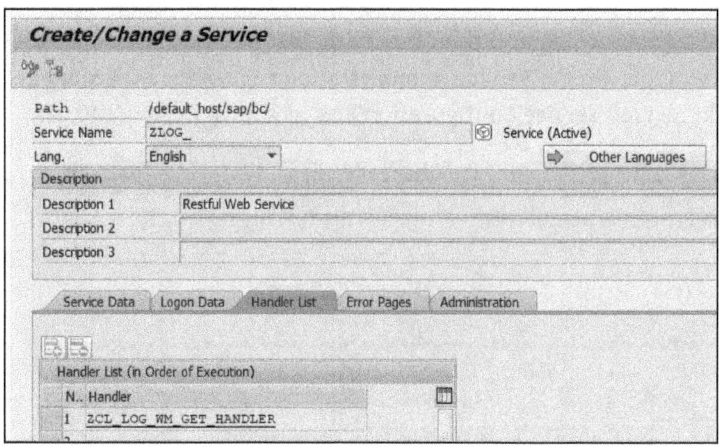

Figure 10.20 Adding a Created Handler in the Created Node

Next, we'll generate a SOAP-based WSDL from the Service-Oriented Architecture (SOA) Manager. WSDL is an XML-based interface description language used to define the functionality provided by a web service. Follow these steps:

1. In SAP GUI, enter Transaction SE84 to open the **Object Navigator** window.
2. Expand **Enterprise Services**, and then double-click on **Service Definitions**, as shown in Figure 10.21.

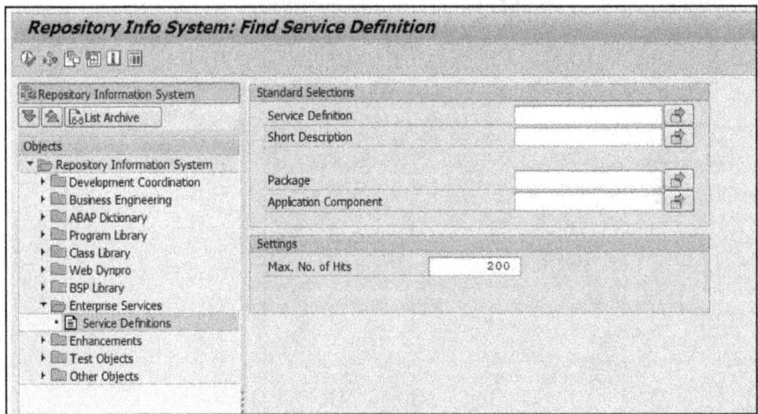

Figure 10.21 Transaction SE84: Repository Information

3. The **Repository Info System: Find Service Definition** window opens.
4. In the **Service Definition** field under **Standard Selections**, enter "Z_MATERIAL_STOCK_READ_DEF".
5. Click **Execute** (F8).
6. Double-click **Z_MATERIAL_STOCK_READ_DEF**. The **Display Service Definition** window shows the details for your web service. Check whether your web service has the state **Active**.
7. Enter Transaction SOAMANAGER, and then log on to SAP NetWeaver to open the **SOA Management** window. On the **Service Administration** page of the **SOA Management** window, click the **Web Service Configuration** link, as shown in Figure 10.22.

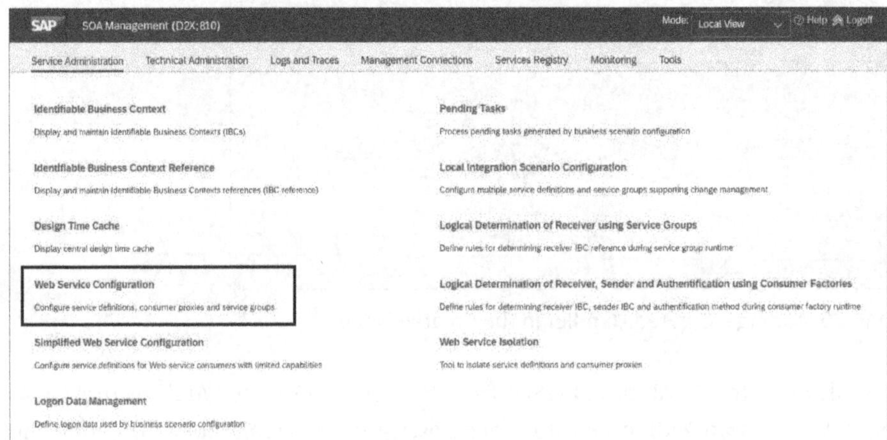

Figure 10.22 SOA Management

8. Enter the service definition in the **Object Name** field, and **Search** for the service definition in Figure 10.23.

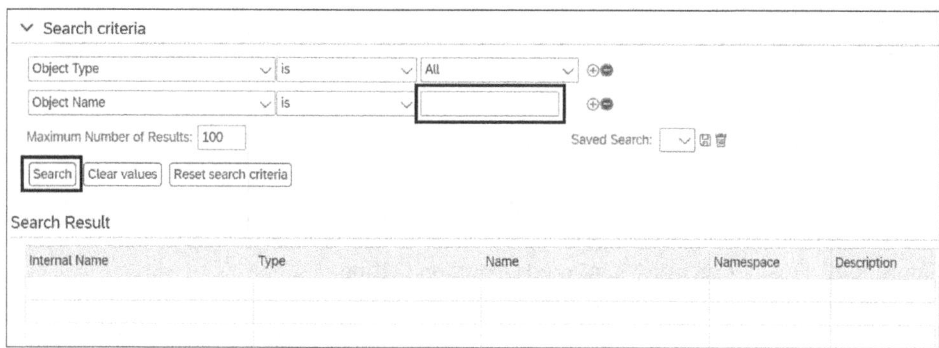

Figure 10.23 SOA Management Search Object

9. You'll find your search result under **Search Result**. Double-click it to see the **Define Services and Binding** Screen.
10. Click on the **Create Service** button (Figure 10.24), and a new Screen for **Configuration of New Bindings for Service Definition** opens.
11. Fill in the **Service Name** and **New Binding Name**, and click on **Next**.

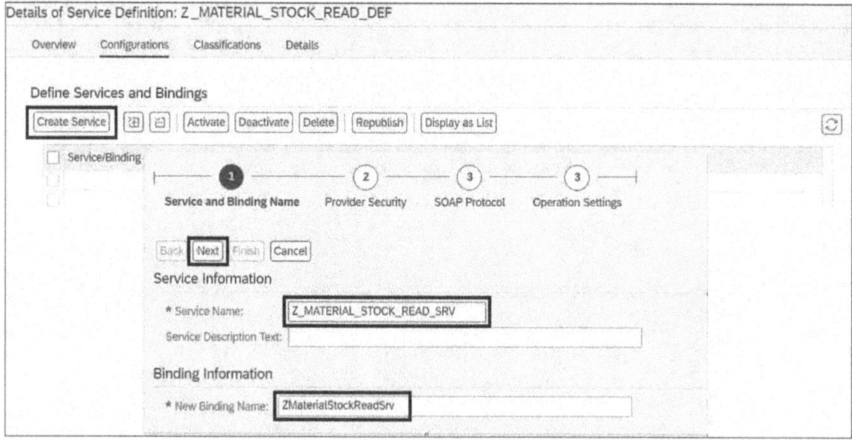

Figure 10.24 SOA Create Service

12. In the **Provider Security** screen, select the option based on the business requirements, and click on **Next**.
13. In the **SOAP Protocol** page, you don't need to update anything, so just click on the **Next** button.
14. In the **Operation Settings** page, click on the **Finish** button (Figure 10.25).
15. The screen shown in Figure 10.26 opens. Click on the **Open Service Binding** icon. The **WSDL Generation for Binding** screen opens.

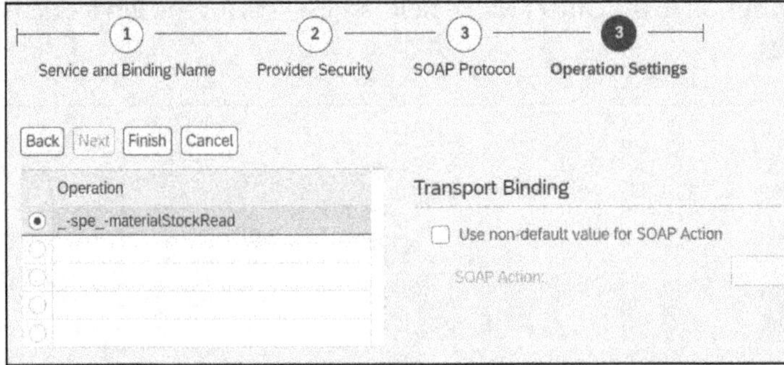

Figure 10.25 Provider Security, SOAP, and Operation Settings

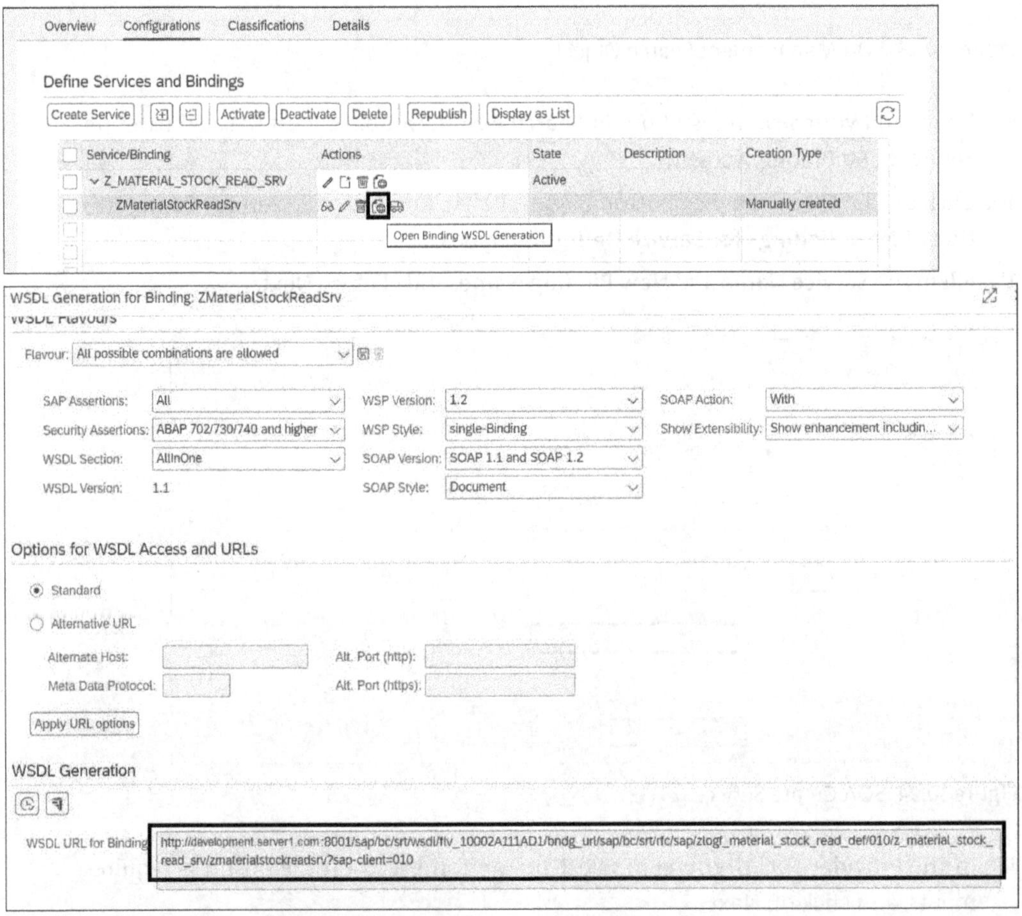

Figure 10.26 WSDL Generation for Binding

16. Scroll down to **WSDL Generation**, and click on the execute icon near the WSDL URL for Binding.

Now, this generated WSDL can be easily consumed by third-party vendors.

10.1.4 Cloud Integration

Diverse demands of complex enterprises are frequently satisfied by goods and services from outside sources. You require middleware to facilitate data communication between your SAP S/4HANA system and external systems. Through middleware, SAP S/4HANA can access data gathered from your third-party products, and the reverse is also possible. Without this connectivity, your data will be kept in isolated databases.

It's clear that SAP has considered this, as seen by the middleware solutions it provides to clients. One of these choices is Cloud Integration, the cloud-based middleware that allows customers to link their SAP S/4HANA systems with third-party logistics services. These services may be on-premises, cloud-based, SAP, or non-SAP. Real-time data transfer between these systems is made possible by the service. Processing of messages might take place inside a department, across an organization, or across businesses.

As shown in Figure 10.27, Cloud Integration is an open integration platform as a service(iPaaS). We mean "open" in the sense that the platform is open source. Developers have a variety of options for their integration and development needs, including Java, JavaScript, Node.js, and Cloud Foundry.

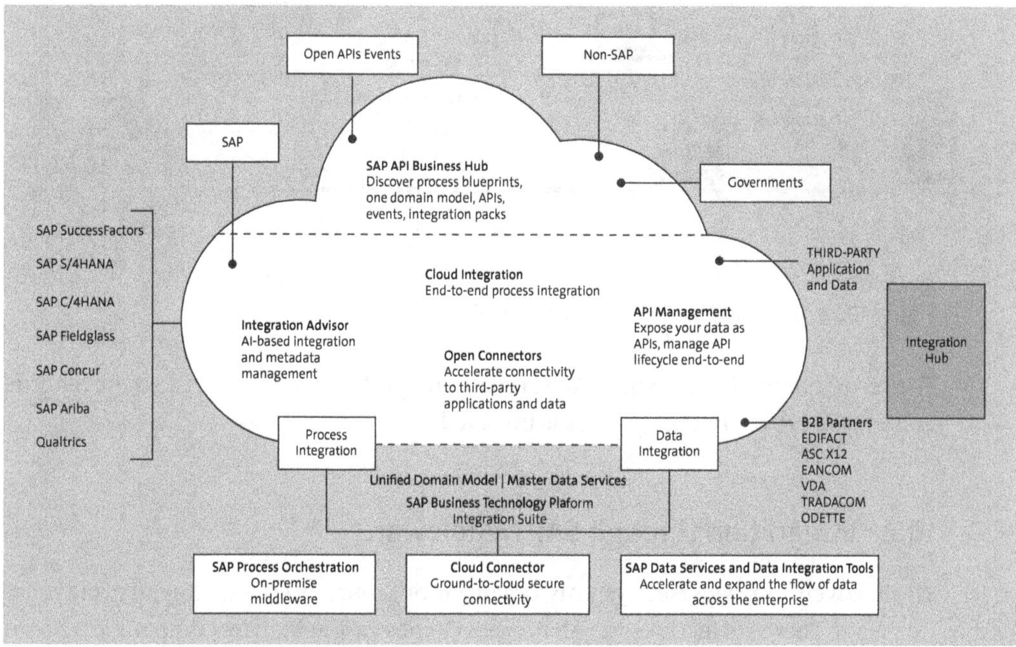

Figure 10.27 Cloud Integration: An Open Integration Platform as a Service

> **Note**
>
> There are currently certain limitations when working in the Cloud Foundry environment. For more information on the limitations, see SAP Note 2752867.

As Cloud Integration is cloud based and offered as a service, organizations don't need to invest in in-house infrastructure and hardware. Customers may use the cloud to create web and mobile apps by using the SAP HANA database management technology.

The following features are available for Cloud Integration:

- Integration content
- Connectivity
- Message processing
- Quality of service
- Adapter development
- API-based integration development
- Monitoring and operations
- B2B integration
- APIs
- Data storage

Many countries today require that companies register eDocuments to government tax authorities such as delivery and invoice notes (business-government [B2G] communication). Using SAP Cloud Platform Integration (with SAP-delivered integration content), SAP S/4HANA can send eDocuments to government agencies and get confirmation using the API eDocument Framework, as illustrated in Figure 10.28.

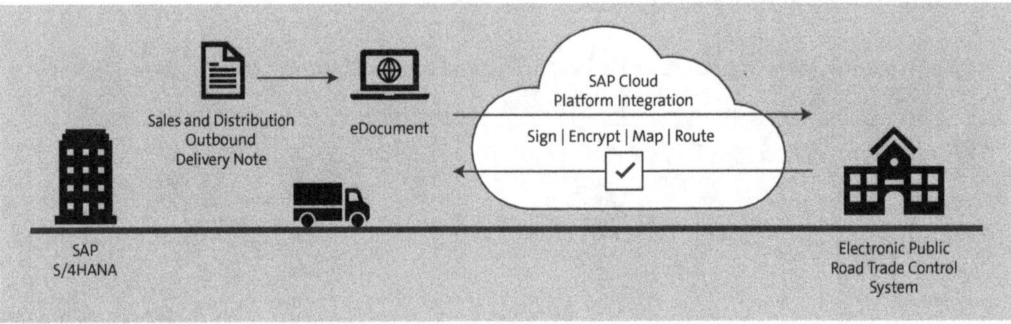

Figure 10.28 External Partner Integration: B2G

The recommendation is to use SAP standard integration content (APIs) wherever integration with government agencies is involved.

10.2 Integration through SAP Middleware

To effectively use dispersed systems, organizations must interface, integrate, and monitor them. There are features in SAP Process Orchestration that are compatible with organizational IT architectures. Conversely, SAP Process Integration enables businesses to provide external services in accordance with SOA.

To provide synchronous and asynchronous communication, SAP Process Orchestration is an application integration middleware program created by SAP that allows customers to interface their SAP systems with SAP or non-SAP systems. The most recent version following the launch of SAP Process Integration is SAP Process Orchestration.

In this section, we'll look at several integration techniques and processes which use middleware or direct interfacing solutions that SAP S/4HANA provides to interface with other systems or third-party logistics providers.

10.2.1 Overview of SAP Process Integration and SAP Process Orchestration

SAP Process Integration is an enterprise integration platform that enables smooth integration between SAP and non-SAP systems, as well as between business-to-business (B2B), business-to-customer (B2C), and application-to-application (A2A) applications. The goal of SAP Process Integration is to make it possible for businesses to effortlessly interchange data between their internal applications and external platforms. System Landscape Directory (SLD), integration building, and architectural monitoring are all done using SAP Process Integration servers during design, configuration, and runtime.

To automate and improve business operations, SAP Process Orchestration makes it simple to synchronize data between various platforms. SAP Process Orchestration includes all of PI's functionalities, including Business Rule Management (BRM), Business Process Management (BPM), ES Repository, B2B Collaboration, Cloud Integration, and Application Integration, in a single Java stack. All required tools are provided by SAP Process Orchestration.

Table 10.3 shows some of the key features of SAP Process Integration and SAP Process Orchestration.

SAP Process Integration		SAP Process Orchestration	
Feature	**Description**	**Feature**	**Description**
Centralized system	A centralized system is provided by SAP Process Integration that can be used to link various applications and systems. This aids organizations in having a unified perspective of their business processes and managing them effectively.	Centralized process modeling	Users can centrally model and manage business processes with the help of SAP Process Orchestration. Users can now monitor, improve, and control operations from a single source.

Table 10.3 Key Features of SAP Process Integration and SAP Process Orchestration

SAP Process Integration		SAP Process Orchestration	
Data transfer	Data exchange across various systems, applications, and databases is made possible by SAP Process Integration. Data may be sent across different systems in a safe and dependable manner.	Process automation	Users may automate procedures with SAP Process Orchestration. This implies that manual jobs are swapped out with automated ones, which lowers costs, increases efficiency, and ensures that operations are carried out properly.
Integration	SAP Process Integration enables organizations to seamlessly connect various applications and systems. Data may be readily shared and synchronized because to the universal platform, it offers to connect various applications and systems.	SOA	Users can easily design, deploy, and manage services thanks to SAP Process Orchestration's SOA. This guarantees that services are accessible when needed and helps shorten the development time.
Automation	Data synchronization and transmission across many programs and systems are automated using SAP Process Integration. It enables businesses to speed up operations, decrease human error, and save time.	Integration capabilities	Integration capabilities for linking various systems are provided by SAP Process Orchestration. As a result, processes run more smoothly, and mistakes are less likely to occur as a result of improved data flow across systems.
Security	Data synchronization and transmission across various applications and systems are made possible by the safe platform offered by SAP Process Integration. Additionally, it makes sure that all data transfers are safe and uncompromised.	Real-time monitoring	Real-time process monitoring is possible for users thanks to SAP Process Orchestration. This facilitates the rapid detection and resolution of any issues that may develop while a procedure is being carried out.

Table 10.3 Key Features of SAP Process Integration and SAP Process Orchestration (Cont.)

One of the main benefits of using SAP Process Integration/SAP Process Orchestration is the smooth construction of interfaces using SAP integration technologies such as the Application Link Enabling (ALE)/IDoc Framework, BAPIs, RFCs, and the ABAP proxy framework. However, SAP also provides a set of tools for adding SAP backend systems to systems horizontally.

10.2.2 Integration Approach and Methodology

The goal of SAP's integration strategy is to create an integrated, intelligent suite of SAP business applications; therefore, integrating end-to-end processes that include SAP systems, partners, or third-party solutions is a major focus.

The integration strategy used by SAP extends well beyond simple application technical integration: integrating data, technology, and business processes completely yields major commercial benefits, such as improved invention speed.

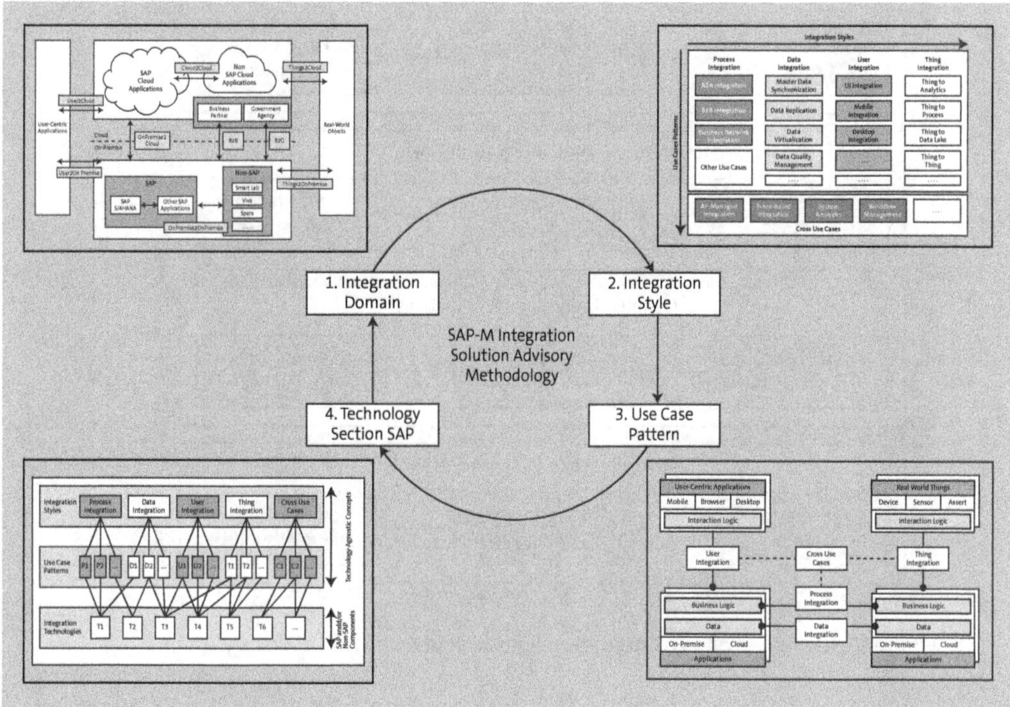

Figure 10.29 Integration Approach and Methodologies

The approach provides a framework with extra best practices for cloud and hybrid IT landscapes, as well as integration use-case patterns and architectural blueprints. Its scope goes beyond just technological considerations and includes organizational elements of enterprise integration, such as creating an Integration Center of Excellence and defining integration roles and their responsibilities. It's a flexible framework whose ideas may be used with SAP integration technologies or those made by other suppliers. Additionally, you can modify and improve its ideas and information to suit your requirements. You can gradually implement the technique in your organization with the help of a clearly defined adoption route.

With the four-step process (see Figure 10.29) that SAP customers and partners are implementing, the integration approach enables the methodical selection of integration solutions.

From establishing to implementing an enterprise integration strategy, the SAP Integration Solution Advisory Methodology provides a clear adoption route through four stages. Every stage is made up of a number of substages that detail the situation, the process, and any relevant data. This methodology's current phases and processes are depicted in Figure 10.30.

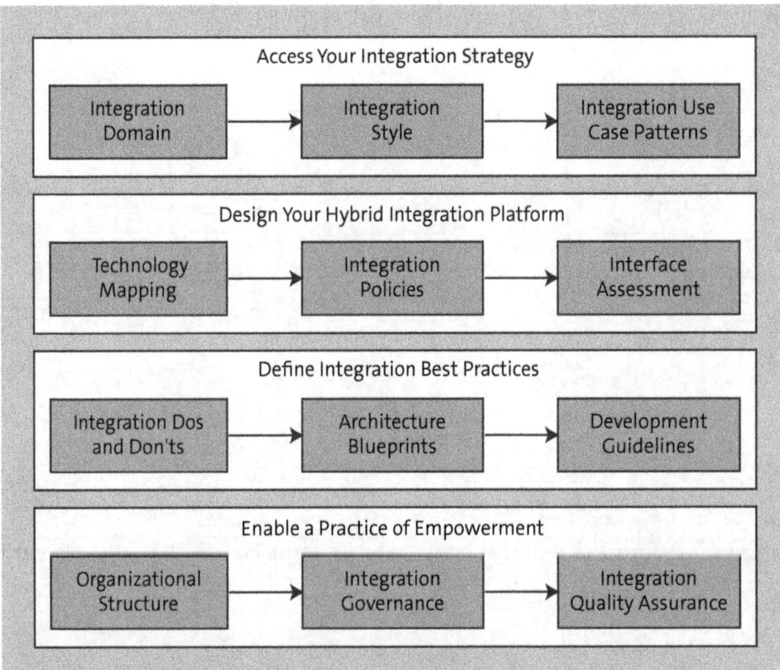

Figure 10.30 Phases of SAP Integration Solution Advisory Methodology

Assess Your Integration Strategy

The SAP Integration Solution Advisory Methodology provides a step-by-step process for customers and partners to evaluate an organization's existing integration strategy: the scoping of integration styles and related use-case patterns is done after a high-level scoping on the level of integration domains. Enterprise architects can choose the templates from SAP that are appropriate for their organization now and/or in the medium future. The following provides a definition of integration domains, integration styles, and use-case patterns.

The process is as follows:

1. **Determine integration domains.**
   Customers and partners review their current integration architecture as part of the first stage by identifying present and foreseeable integration domains. The integration domains in a hybrid system landscape include on-premises to cloud, user to cloud, and more (see Figure 10.31).

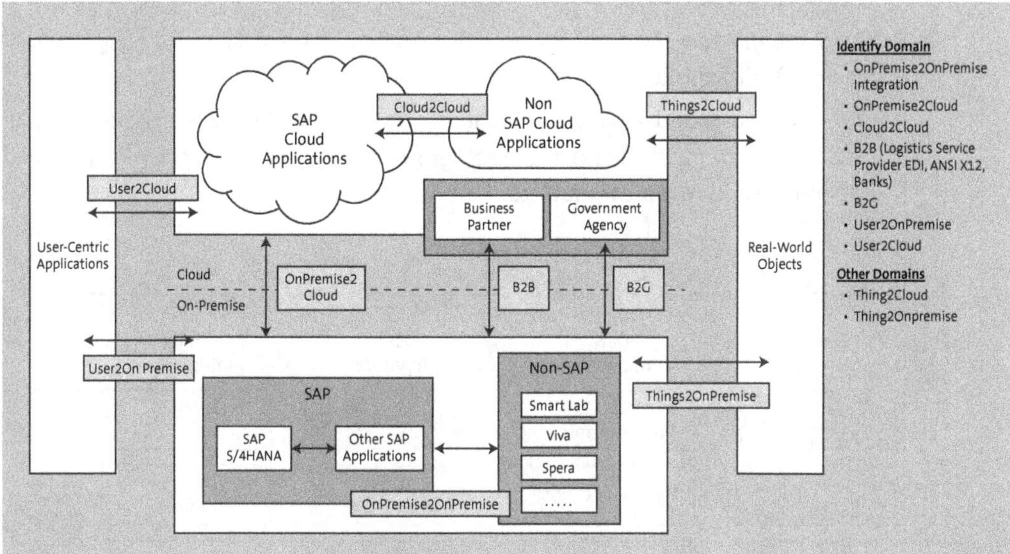

Figure 10.31 Overview of Integration Domains

> **Note**
>
> Integration domains also cover the opposite direction. For example, the integration domain Cloud2OnPremise is also applicable for integration scenarios sending data from business applications that are deployed on-premises to cloud-based business applications.
>
> Typically, an enterprise architect conducts the scoping of integration domains and oversees the entire IT landscape in an organization.

2. **Determine your integration style.**
   The next step is to discover pertinent integration styles after you've chosen the appropriate integration domains. Integration scenarios between a business application and its target are realized according to major integration archetypes, which are represented by five integration styles (see Figure 10.32):

- *Process integration* style connects business processes across applications, for example, invoice posting from third-party logistics vendor to SAP S/4HANA.
- *Data integration* style allows you to synchronize or access data across applications, for example, high volume of data replication and one-time initial load.
- *User integration* style allows you to connect user-centric applications, for example, user access SAP Fiori launchpad, which integrates SAP S/4HANA using OData. Tiles are provided for executing specific transactions.
- *Thing integration* style allows you to connect real-world object (e.g., sensor machine) with applications.
- *Cross use cases* build a category for all integration-related use cases that complement one or more of the four core integration styles, for example, API-managed integration and workflow management.

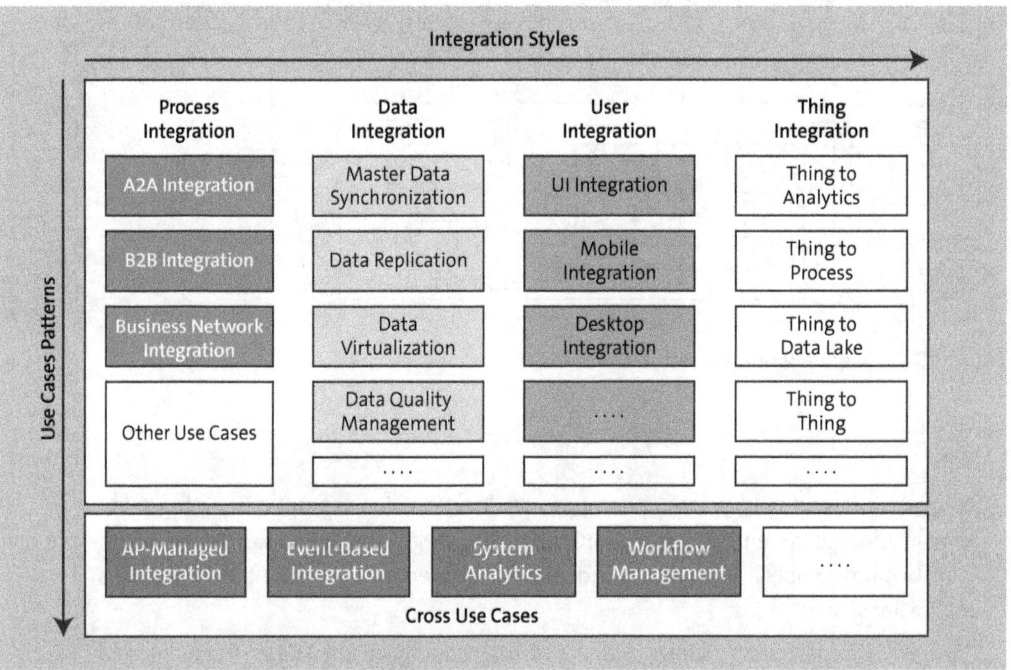

Figure 10.32 Overview of Integration Styles

Each integration style may be used in a variety of integration domains and is independent of technology. An integration style can be complemented by a cross-use pattern. All integration-related use cases that may be integrated with one or more of the five primary integration types fall under this category. For instance, API-managed integration offers comprehensive lifecycle management for APIs that you can use in integration scenarios focused on users or processes.

3. **Determine the integration use-case pattern.**
   The final stage in evaluating your integration strategy is to determine the integration use-case patterns that are pertinent to your company. Use-case patterns, which outline commonly occurring integration use cases in corporate landscapes, may be applied to each integration style to make it more polished. Cross-use-case patterns, such as API-managed integration or event-based integration, can enhance one or more of the five fundamental integration styles because they are founded on the same concepts. Integration use-case patterns are specified in a manner that is independent of technology, just like integration domains and integration styles.

A list of integration use-case patterns is provided in the SAP Integration Solution Advisory Methodology that you can use as a jumping-off point for your business. If necessary, you can nimbly modify the collection of integration use-case patterns, for example, by adding customer-specific examples or removing those that aren't applicable to your business.

The scoping of integration use-case patterns is carried out by an integration architect who is well-versed in integration requirements across the company (see Figure 10.33):

- **A2A integration**
  Business process integration between two applications, for example, inspection data/receipt order process integration between SAP S/4HANA and third-party logistics vendors.
- **B2B integration**
  B2B with partners, customers, or vendors, for example, purchase order/delivery integration with logistics service providers.
- **Business network integration**
  This collection of tools and services enables SAP applications like SAP Business Suite and SAP S/4HANA, among others, to communicate and exchange business documents in cross-application processes with other business networks such as SAP Business Network, SAP Fieldglass, and non SAP-solutions (both on-premise and cloud) using HTTP and web services.

Use-case patterns for data integration include the following:

- **Master data synchronization**
  Synchronizes master data objects within an SAP system that are similar from a business perspective but technically different and thus enables the seamless integration of different SAP applications into your business processes—for example, customer/vendor data synchronization from master data management (MDM) to SAP S/4HANA.
- **Data replication**
  Provides large volume data replication (general ledger summary) from SAP S/4HANA to warehouse management (WM).

- **Data virtualization**
  Provisioning of a virtual data layer by federation of queries to remote data sources without physical data movement.
- **Data quality management**
  Integration of data from multiple sources for cleansing, matching, or consolidation.

Use-case patterns for user integration include the following:

- **UI integration**
  Users accessing SAP Fiori launchpad for tiles.
- **Mobile integration**
  Using a travel/expense app.
- **Desktop integration**
  Outlook integration with a CRM system.

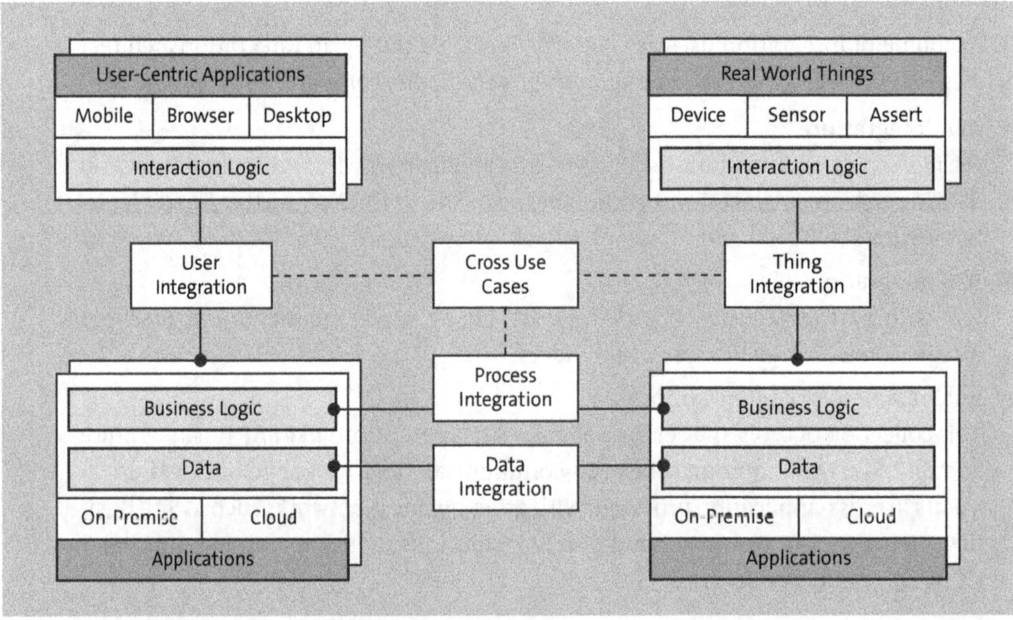

Figure 10.33 Catalog of Integration Use-Case Patterns

Design Your Hybrid Integration Platform

You create the future hybrid integration platform during this phase. The word "hybrid" refers to the SAP Integration Solution Advisory Methodology's capacity to accommodate all pertinent integration domains and styles, not only hybrid deployment methods. A hybrid integration platform also has to take into account the requirements of many personas, including integration professionals, application developers, and line of business users.

To define the extent of your future hybrid integration platform, you must undertake a technological mapping. You then create integration policies that specify when to use particular platform technologies. Conducting an interface evaluation is a continual task that integration developers often perform as the third step: when a new integration scenario is required, interface assessments assist integration developers in selecting the best integration technology or technologies (see Figure 10.34).

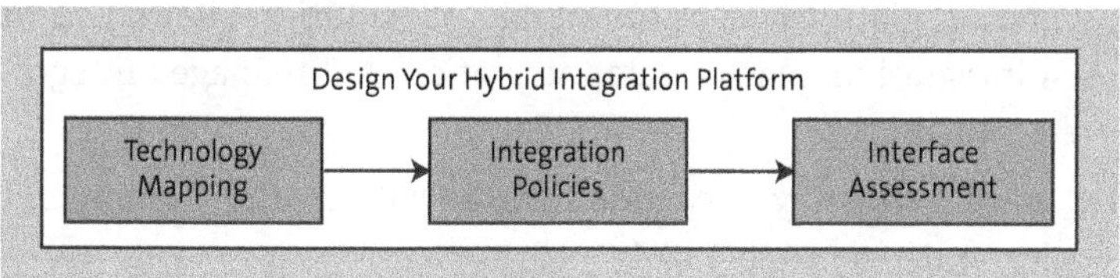

Figure 10.34 Steps of the "Design Your Hybrid Integration Platform" Phase

You develop the following assets as part of the SAP Integration Solution Advisory Methodology's second phase:

- Technology mapping that explains which integration technologies to use for each in-scope integration style and use-case pattern (see Figure 10.35). Most of the time, an n:m mapping is possible instead of a 1:1 mapping between integration use-case patterns. Therefore, more than one integration technology is needed for an integration use-case pattern, and many integration technologies offer several integration use-case patterns.

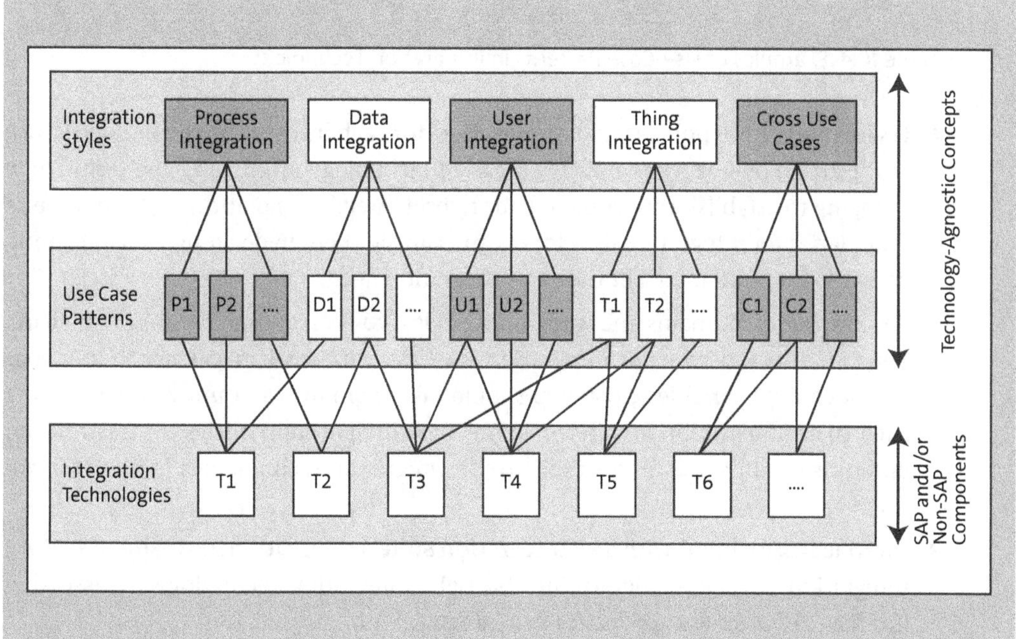

Figure 10.35 Technology Mapping

Examples include the following:

- **Integration use case requiring more than one integration technology**
  For the B2B integration use case, some customers leverage a B2B gateway solution to support the exchange of EDIs with trading partners, together with an Enterprise Service Bus (ESB) to integrate with internal business applications.
- **Integration technology supporting more than one integration use-case pattern**
  An enterprise integration platform as a service (EIPaaS) solution allows you to implement the A2A integration, master data integration, API-managed integration, and more use-case patterns.

Based on the business requirement, you can identify the pattern and recommend integration technologies, as listed in Table 10.4.

Use-Case Patterns	Integration Technologies
■ A2A integration ■ SAP2SAP ■ NonSAP2NonSAP ■ NonSAP2SAP or SAP2NonSAP*	■ SAP Process Orchestration/Cloud Integration ■ External provider such as Dell Boom
B2B integration (e.g., integration with logistics service providers)	SAP Process Orchestration
Business network integration (e.g., Ariba)	Cloud Integration
Master data synchronization	SAP Process Orchestration/Cloud Integration

Table 10.4 Examples of Use-Case Patterns in Integration Technology

- **List of integration policies that are specified for each integration technology in use**
  You provide one or more policies for each technology after using the technology mapping to establish the scope of your hybrid integration platform. These serve as boundaries for integration developers and can aid them in deciding which technologies are best suited for a brand-new integration need.

  For integration domains and integration styles, you can define integration policies. You can also add relevant integration use-case patterns if necessary. If any consumer context variables exist, you should take them into account when designing such rules. Integration architects define such integration policies. In this role, an enterprise architect oversees the current integration technologies in an organization.
- **Interface assessment with SAP Integration Suite's Integration Assessment feature**
  Integration developers determine the right integration technology by assessing interfaces of business applications to integrate with.

Define Integration Best Practices

After the hybrid integration platform's architecture is finished, best practices for integration are established to protect the execution of integration scenarios. This makes it possible to properly transfer integration requirements to internal or external integration developers and can establish the parameters for the quality control of existing integration scenarios. Enterprise architects can produce architectural blueprints for the hybrid integration platform that address the pertinent integration use-case patterns (including integration domains) for this purpose. The scope and interactions of integration technologies with business applications are described and documented in these designs (see Figure 10.36).

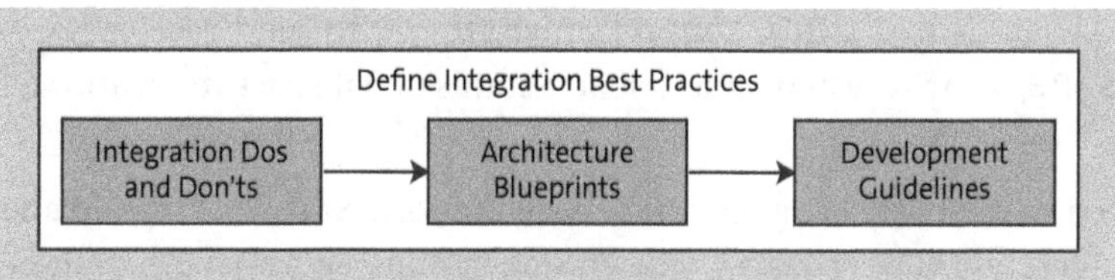

Figure 10.36 Steps of the "Define Integration Best Practices" Phase

Enable a Practice of Empowerment

Once the best practices for integration have been established, they may be implemented internally and outside (toward integration developers). On the basis of the established integration approach, project teams may then be given the freedom to design interfaces in an agile manner. This stage offers a feedback mechanism for adopting the integration strategy's lessons learned.

This use case covers organizational factors and procedures, such as the creation of an integration competency center. Additionally, this contributes to the recognition of integration as a profession inside an organization, supporting and facilitating projects related to digital transformation. Additionally, this technique allows for the scaling up of integration knowledge across the organization: introducing spaces for self-service integration that are run by business personas (also known as citizen integrators), for instance (see Figure 10.37).

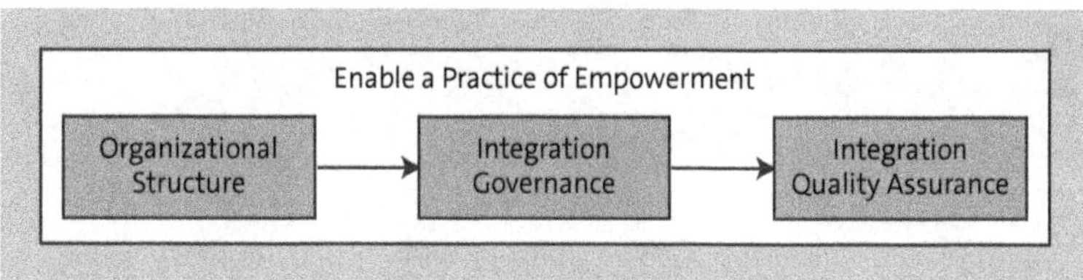

Figure 10.37 Steps of the "Enable a Practice of Empowerment" Phase

10.2.3 IDoc Interface/Application Link Enabling

The IDoc interface is used for exchanging data between the SAP S/4HANA system with various external trading partner systems, and the ALE technology is used for exchanging data between SAP S/4HANA systems and non-SAP systems. The IDoc format is used for exchanging data between the systems that have a common data structure format.

As you know, there are numerous routes available to connect with third-party systems, so here we'll discuss a simpler design where any third-party logistics vendors can exchange data with the SAP S/4HANA system though SAP Process Integration/SAP Process Orchestration as middleware.

Because we're dealing with external system via SAP Process Integration/SAP Process Orchestration, we've configured a tRFC connection via TCP/IP to send/receive IDocs from the sending SAP S/4HANA system and to send a status update back in the form of an IDoc interface.

Technically, we need to establish a connection between the SAP S/4HANA system and the SAP Process Integration/SAP Process Orchestration system. Because both of these systems are under the same platform (i.e., SAP), we can easily connect them by setting up an RFC connection.

To create RFC connections in SAP S/4HANA, use Transaction SPRO, and follow IMG navigation path **Application Server • IDoc Interface/Application Link Enabling (ALE) • Logical Systems • Create RFC Connections**; alternatively, you can use Transaction SALE, as shown in Figure 10.38.

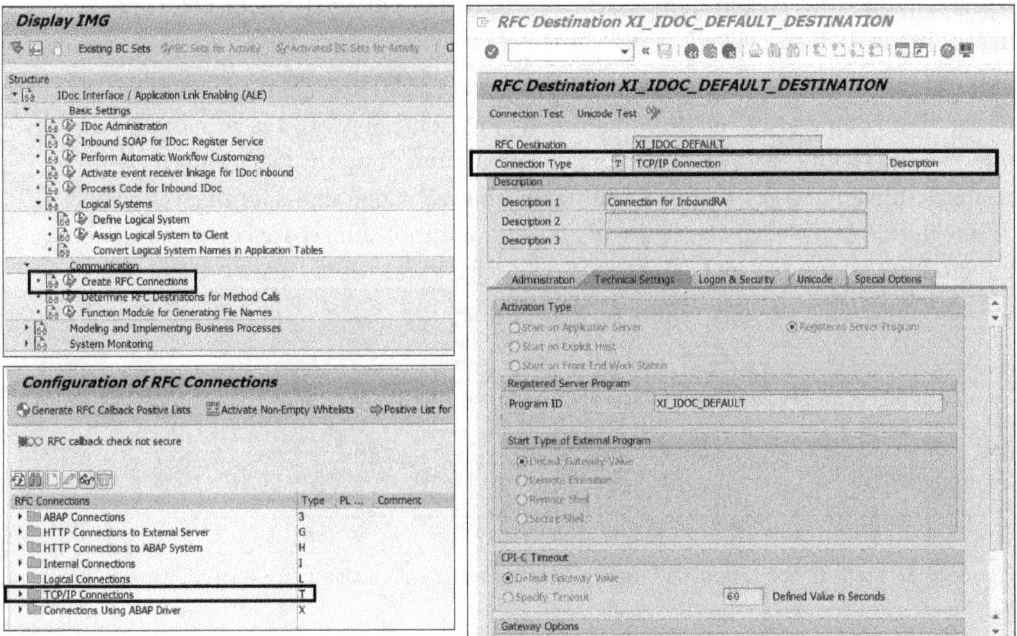

Figure 10.38 Overview of RFC Connection

In addition, you can directly create the RFC connection by executing Transaction SM59. Refer to Chapter 9 for details on how to set up RFC connections.

Mapping of external third-party logistics vendors will be done at the SAP Process Integration/SAP Process Orchestration server using third-party providers (e.g., EDICOM, Seeburger, etc.).

10.2.4 Batch Processing Jobs

Per implementing client requirement(s), there may be multiple systems involved where interface jobs are needed to be scheduled in different systems (SAP S/4HANA, SAP Business Warehouse [SAP BW], other SAP systems, and non-SAP systems). It's recommended to use common platforms/tools for batch scheduling. This will allow you to define dependencies for batch programs between different applications systems, for example, bank interfaces in which SAP S/4HANA generates the file using a batch program. On successful completion for the SAP S/4HANA job, the SAP Process Orchestration job must be triggered to drop files into the bank's SFTP folder. A common automation scheduler will also help in managing all errors from a central place instead of from separate systems.

In the absence of a batch tool, you'll need to schedule jobs in respective systems; for example, SAP S/4HANA, other SAP systems, SAP BW systems, and/or non-SAP system jobs will be scheduled in their respective systems natively. You can manage dependencies on different jobs by defining time intervals manually. You'll manage batch job errors in respective environments separately.

Following is a short list of software that are phenomenal SAP job schedulers:

- ActiveBatch
- Redwood
- Stonebranch
- JAMS Scheduler
- Tidal Software
- SAP Job Scheduling Service

10.3 Connection to a Business-to-Business/Electronic Data Interchange Solution via Application Programming Interfaces

SAP Process Orchestration supports integration with APIs and web services-based communication for SAP and/or any non-SAP system integration. APIs can be integrated with SOAP, REST, OData, and other protocols, as illustrated in Figure 10.39.

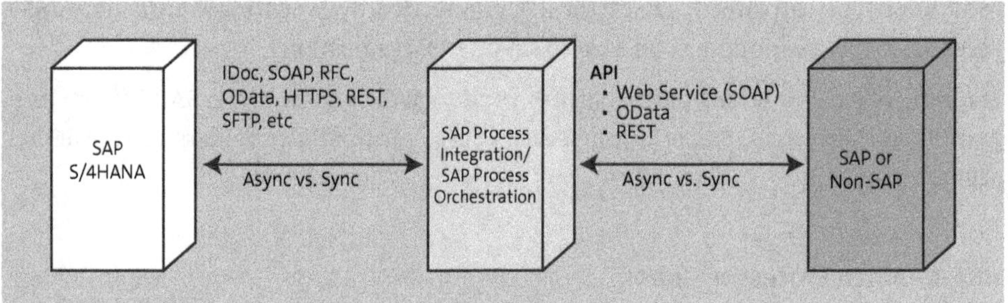

Figure 10.39 Communication Based on Web Services and APIs

Table 10.5 lists the standard adapters that SAP Process Orchestration supports.

Adapters	Description
AS2	Applicability Statement 2 (AS2) protocol based on HTTPS
EDISeparator	B2B for EDI separator
OFTP	Odette File Transfer Protocol (OFTP) for EDI
X400	X400 for EDI B2B add-on
OData	Open Data Protocol (OData)
BC	Business Connector
CIDX	Chemical Industry Data Exchange
FILE	File supports FTP or FTPS
HTTP_AAE	Hypertext Transfer Protocol (HTTP) supported using the Advanced Adapter Engine within SAP Process Orchestration
IDOC_AAE	Intermediate Document (IDoc) supported using Advanced Adapter Engine within SAP Process Orchestration
JDBC	Java Database Connectivity (JDBC) – database connection
JMS	Java Messaging Service (JMS)
Mail	Email Support
MarketPlace	Marketplace Adapter MarketSet Markup Language (MML) messages from marketplaces
REST	Representational State Transfer (REST) adapter
RFC	RFC (SAP protocol)
RNIF	RosettaNet Implementation Framework (RNIF)

Table 10.5 Standard Adapters Supported in SAP Process Orchestration

Adapters	Description
RNIF11	RosettaNet Implementation Framework (RNIF) version 11
SOAP	Web service: Simple Object Access Protocol (SOAP)
WS_AAE	Webservice Reliable messaging (WSRM) for web services
SFTP	Secure (or SSH) File Transfer Protocol

Table 10.5 Standard Adapters Supported in SAP Process Orchestration (Cont.)

In the following sections, we'll discuss the technical architecture of SAP Process Orchestration and the B2B adapter that operates on the SAP NetWeaver Process Integration Adapter Framework, as well as several use cases of logistics service provider business integration situations.

10.3.1 Logistics Service Provider-EDI Integration

The SAP NetWeaver Process Integration Adapter Framework, which is based on the Java Connector Architecture (JCA), serves as the foundation for the SAP Process Integration B2B add-on. The typical message delivery choices (best-effort, exactly-once, exactly-once-in-order), an automated retry mechanism, and information logging are thus enabled thanks to its features. B2B adapters for SAP Process Integration are configured, monitored, and operated in the same manner as other adapters. The B2B add-on supports all legal SAP Process Integration deployment options (ABAP + Java, Java alone) and runs entirely within the Java stack.

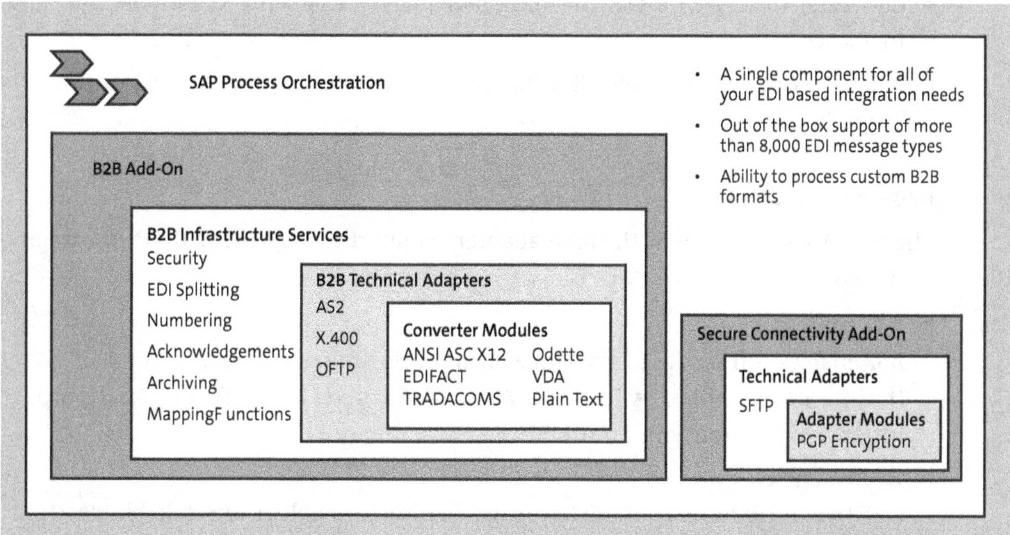

Figure 10.40 SAP Process Orchestration B2B Add-On Overview

As shown in Figure 10.40, SAP Process Orchestration provides B2B add-ons for EDI-based B2B communication called *B2B add-on* and *secure connectivity add-on*. This is a complimentary component with the SAP Process Orchestration solution. It offers AS2, X.400, OFTP, and SFTP adapters, infrastructure services, and converter modules for EDI translations (see Figure 10.41).

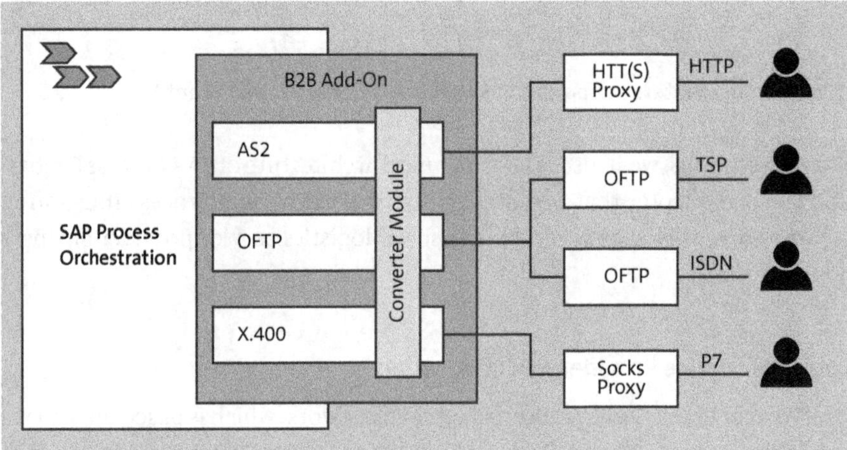

Figure 10.41 B2B Add-On Connectivity Options with SAP Process Orchestration

In the following sections, we'll look more closely at the B2B Add-On for SAP Process Integration scope. Three construction pieces make up the B2B Add-On:

- B2B protocol adapters enabling a standards-based exchange of B2B documents with business partners
- Converter modules for transforming native EDI and plain text formats into XML messages
- B2B infrastructure services allowing you to manage and process EDI documents properly

B2B Protocol Adapters

The B2B Add-On comes with three adapters supporting specific B2B/EDI messaging protocols:

- **AS2 adapter**
  AS2 enables companies to exchange business documents securely and reliably over the internet as defined by RFC 4130. AS2 is based on HTTP and Secure/Multipurpose Internet Mail Extensions (S/MIME).
- **OFTP adapter**
  OFTP was introduced by the Odette organization to enable secure, reliable electronic message exchange (point-to-point or indirectly via a value-added network [VAN]).

OFTP2 is a specification for secure document exchange over the internet, Integrated Services Digital Network (ISDN), and X.25 networks.

- **X.400 adapter**
  The X.400 adapter is used for VAN-based communication using the P7 protocol to connect to X.400 mailboxes.

Converter Modules

The B2B Add-On includes a collection of adapter modules that enable conversion of native EDI messages to EDI XML messages and vice versa. Additionally, a plain text module is offered to handle unique B2B communication types similarly. Due to PI's enhanced processing capabilities for XML communications, this content conversion is advantageous.

For inbound and outbound communication, the converter modules can be used with any transport protocol adaptor. The following B2B/EDI standards are supported in all common message types and versions by them:

- ANSI X.12
- EDIFACT
- TRADACOM
- Odette
- VDA
- Plain text

B2B Infrastructure Services

In addition to the technical connectivity, infrastructure services linked to B2B/EDI are offered that ensure safe and expert B2B/EDI message exchange with your business partners:

- **Security through key management**
  The SAP NetWeaver AS for Java key store views are the core location for managing digital certificates and keys.
- **EDI numbering**
  This is the creation of transmission numbers used to automatically add continuous counters to EDI messages that are being sent out.
- **Handling acknowledgements**
  Acknowledgements are produced and managed in accordance with the relevant B2B/EDI protocol standard.
- **B2B message archiving**
  The AS2, X.400, and OFTP adapter modules' basic archiving capabilities are extended by the adapter module. On the file system, messages can be saved with additional information (e.g., an IDoc number in the filename, business-partner-specific archiving, etc.).

- **B2B mapping functions**
  For EDI/B2B communication, you require a library of particular mapping operations. Additionally, a few simple EDI to IDoc mapping templates are included for testing.
- **EDI splitting of inbound messages**
  More explanation is needed for this service than for the others. For dividing incoming EDIs made up of numerous messages, an adaptor called an EDI separator is needed. The following segments, according to the EDI standard, are separated:
  - EDIFACT is split per service string advice (UNA), interchange header (UNB), and message header (UNH) segment.
  - Interchange control header (ISA), functional group header (GS), and transaction set header (ST) segments are divided according to ANSI X.12 standards. They are also divided by the start section for VDA.

  The messages may also be filtered according to criteria such as message type, version, sender, and receiver. Additionally, character set conversion is supported.

In an integration scenario, the client has a requirement to send or receive confirmation notification (inbound and outbound) and stock status information, inventory adjustment and reconciliation to and from logistics service providers. Businesses use third-party organizations to manage warehousing and distribution activity. These third-party organizations are referred to as logistics service providers. In the following list, we'll walk through five different logistics service provider use cases:

- **Use case 1: Inbound delivery integration from a logistics service provider**
  The inbound delivery confirmation interface will be used to send an electronic confirmation from a logistics service provider to the business for scenarios when product is received at the logistics service provider warehouse on behalf of a business. Once the receipt has been completed into an agreed upon status in the logistics service provider's system, the logistics service provider will send a confirmation message to the business containing the goods receipt information pertaining to a specific inbound delivery. This inbound confirmation message will be routed through middleware SAP Process Integration and into SAP S/4HANA to allow for goods receipt to be posted for the inbound delivery, as shown in Figure 10.42.

  This interface will be used to support the following different scenarios after a logistics service provider receives a product into its warehouse:
  - Receiving from external vendor
  - Receiving during intracompany/intercompany stock transfer order (STO) processes
- **Use case 2: Outbound delivery integration to a logistics service provider**
  The business needs to communicate to logistics service providers about any incoming shipment of goods. The inbound delivery notification interface will be used to send an electronic notification to the logistics service provider for all circumstances when product needs to be received at their site.

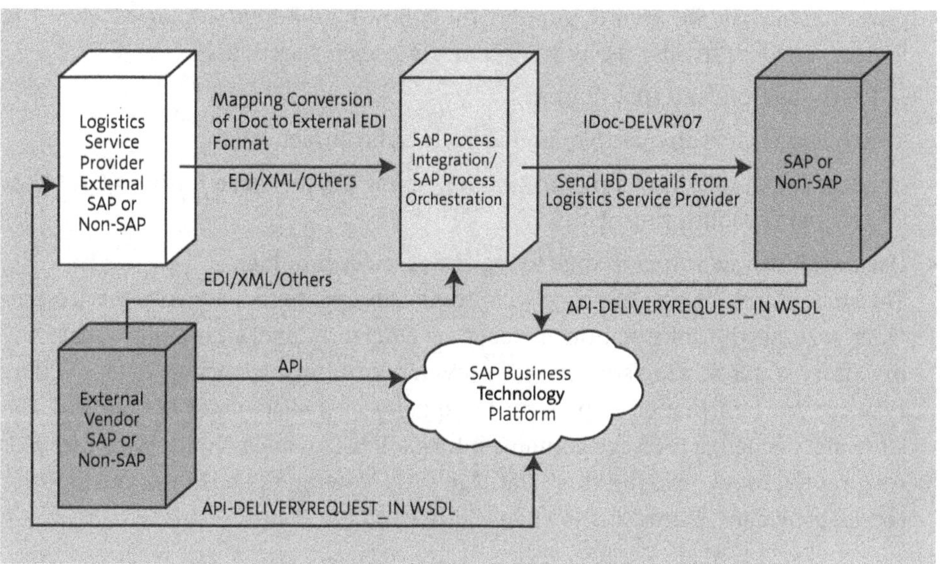

Figure 10.42 Inbound Delivery Integration with a Logistics Service Provider

As shown in Figure 10.43, an inbound delivery will be generated in the business's SAP S/4HANA system to facilitate the receipt of product in the logistics service provider location. An output from the inbound delivery triggers the interface to logistics service provider containing details of the product that the logistics service provider will receive.

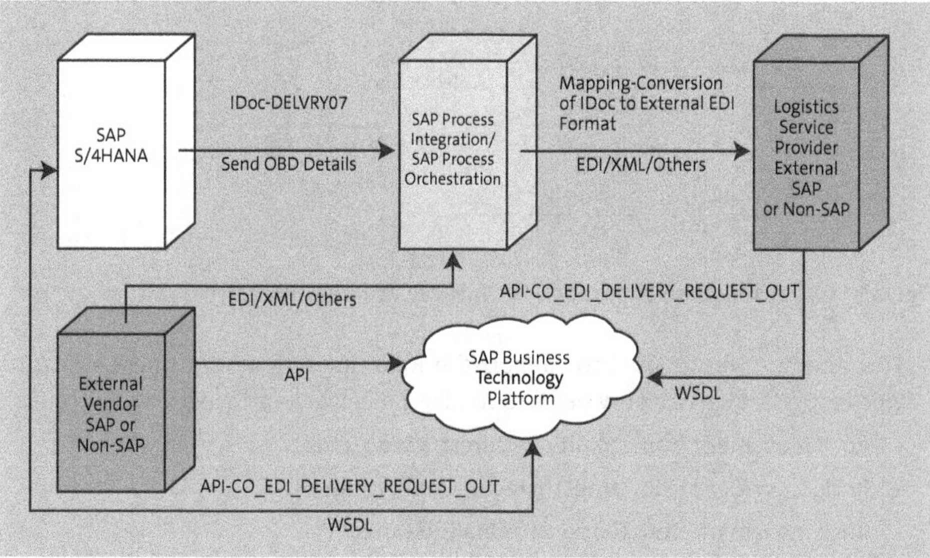

Figure 10.43 Outbound Delivery Interface: Business to Logistics Service Provider

This interface will be used to support the following different scenarios when the logistics service provider receives a product into their warehouse:

- Receiving from external vendor
- Receiving from third party manufacturers/subcontractor
- Receiving during intracompany/intercompany STO between business/SAP User Experience Monitoring plants

- **Use Case 3: Stocks status change to logistics service provider**
  The stock status change message will provide an update to a logistics service provider with all the information it needs to efficiently and accurately update the inventory status at a logistics service provider location. Logistics service providers may use this information to plan further quality-related activities at their site. The stock status change message contains information that a logistics service provider requires to always have the stock status in sync between the business and logistics service provider systems, as shown in Figure 10.44.

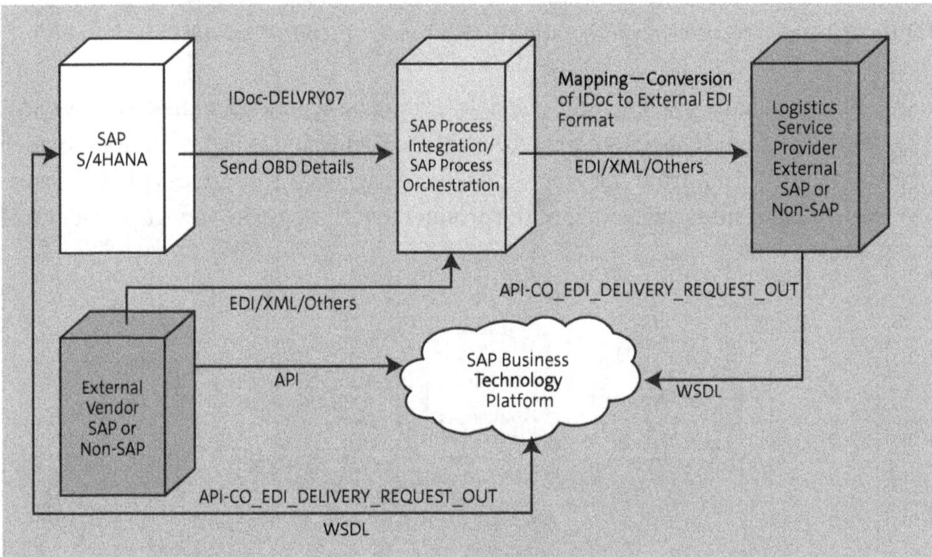

Figure 10.44 Stock Status Notification SAP S/4HANA and Logistics Service Provider

This interface will be used to support the following different business scenarios when a logistics service provider is to receive a message of a change in stock status:

- Stock movement from **Quality** to **Unrestricted** status
- Stock movement from **Unrestricted** to **Quality** status
- Stock movement from **Blocked** to **Quality** status
- Stock movement from **Quality** to **Blocked** status
- Stock movement from **Blocked** to **Unrestricted** status
- Stock movement from **Unrestricted** to **Blocked** status

- Stock movement from **Blocked** to **Scrap** status
- Stock movement from **Return** to **Blocked** status
- Stock movement from **Return** to **Blocked Reversal**

At the SAP Process Orchestration level, you must configure the following:

- The stock status change message will provide an update to a logistics service provider with all the information that provider requires to efficiently and accurately update the inventory status at the logistics service provider location. This message will be routed through middleware SAP Process Integration from SAP S/4HANA to the logistics service provider.
- We're using the AS2 adapter as the receiver channel to send the message to the logistics service provider.
- Message transformation occurs at the SAP Process Orchestration level.
- We're using the IDOC_AAE adapter as a sender channel to receive the message from SAP S/4HANA.

- **Use Case 4: Inventory adjustments from the logistics service provider**
  Inventory adjustments from the logistics service provider is the interface aiming to receive certain inventory-related changes, including quantity, status, location, and so on from a logistics service provider. This is an inbound interface which receives the inventory adjustments that can potentially be processed in a business's SAP S/4HANA system.

  The capability of the solution is to take an inventory adjustment message from a logistics service provider and have it processed in the SAP S/4HANA system. The processing of an adjustment message only occurs after all the described requirements have been addressed, including filtering, validation, augmentation, and workflow approval (see Figure 10.45).
- **Use Case 5: Inventory reconciliation from the logistics service provider**
  The business requires a reporting functionality to compare stock levels in SAP S/4HANA at any given time, with that of a logistics service provider warehouse. The scope of this interface is to capture stock changes that have taken place within the logistics service provider warehouses (sent by the logistics service provider through an interface with a date and time stamp) and store them in an SAP S/4HANA custom database table. This will facilitate the BI/SAP BW system to extract inventory data for logistics service provider–specific stock data along with SAP S/4HANA stock figures for the same set of locations and stock types from SAP S/4HANA on a real-time basis, as shown in Figure 10.46.

  Based on the comparison of the stock-relevant data, as obtained from two different sources (one from the logistics service provider and the other from SAP S/4HANA), the system will display a user-friendly report to help business users with requisite information for manual inventory reconciliation afterward.

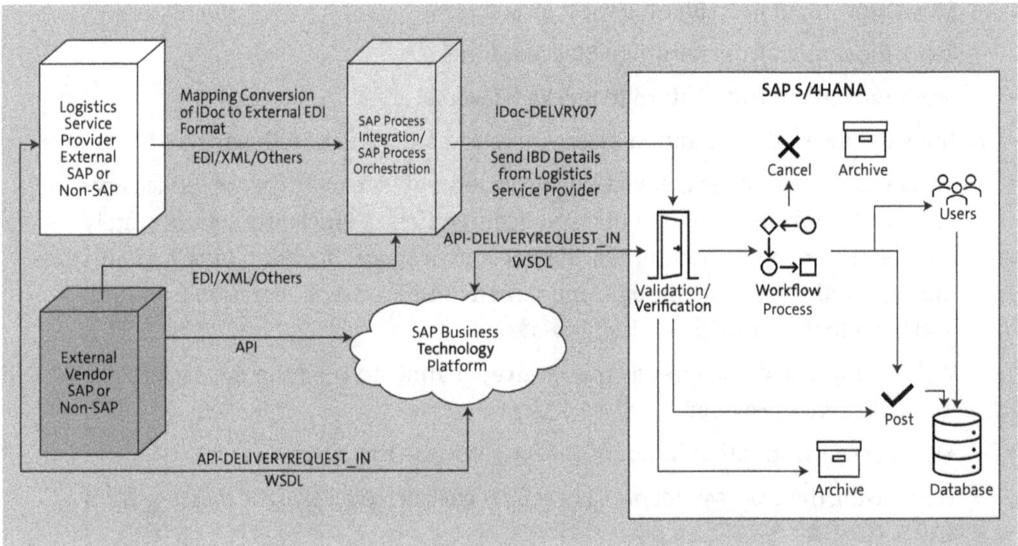

Figure 10.45 Inventory Adjustments: External System and SAP S/4HANA

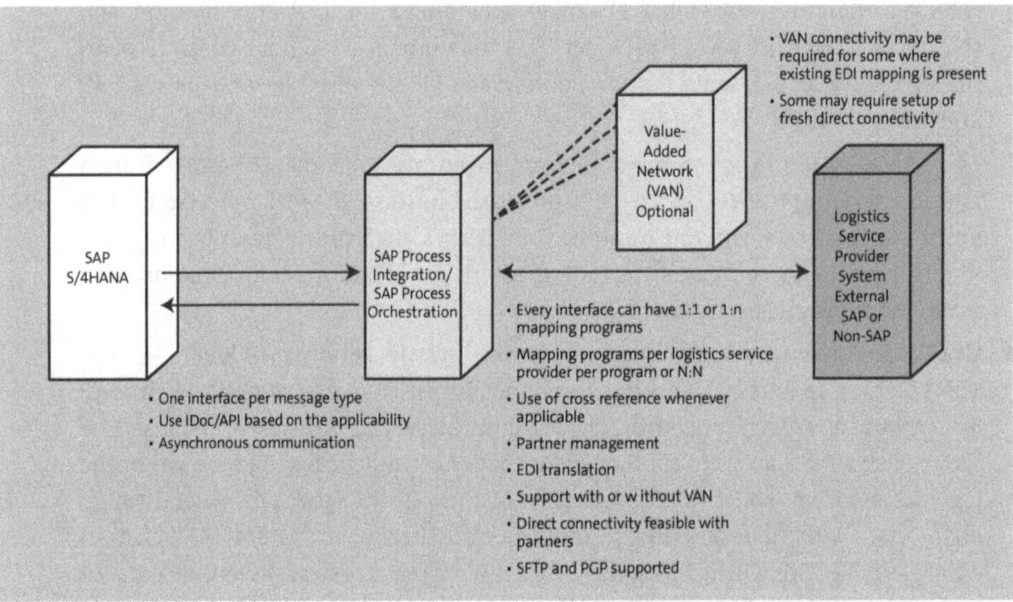

Figure 10.46 Use Case Logistics Service Provider Integration

10.3.2 Application Programming Interface Management Using SAP Business Technology Platform

You can find SAP APIs on the SAP Business Accelerator Hub at *https://api.sap.com* (see Figure 10.47). This is a central public repository to discover SAP's available APIs by customer, partner, and so on. You can also test APIs using the SAP Business Accelerator Hub.

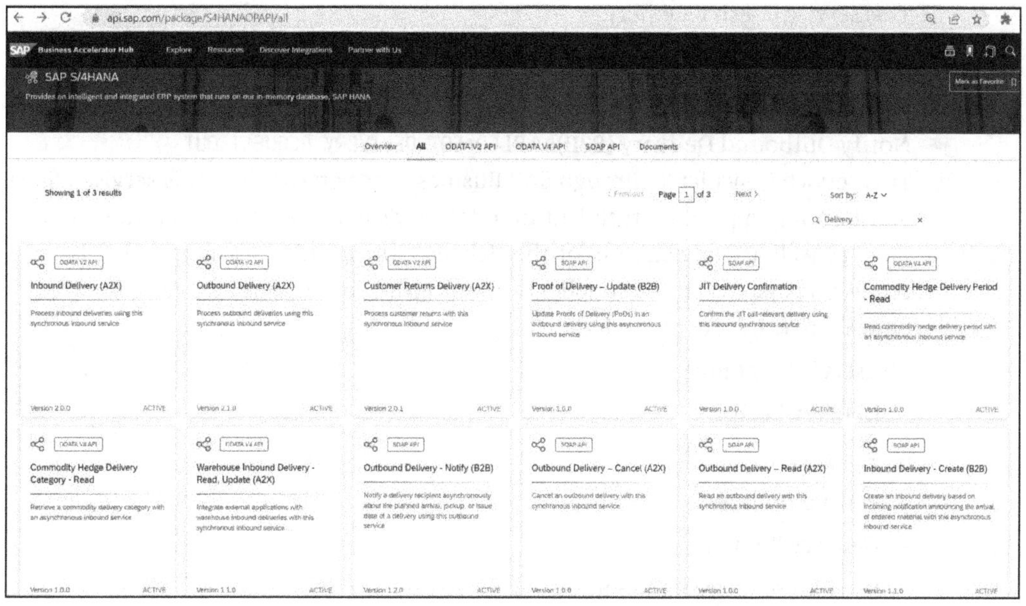

Figure 10.47 SAP Business Accelerator Hub Overview

Across your system landscape, business operations may be connected via the following APIs:

- **Create Inbound Delivery (B2B): API `DELIVERYREQUEST_IN`**
  This service is available through SAP Business Accelerator Hub. This service allows you to generate an inbound delivery based on an incoming notification of a product's expected arrival. Buyers want to be notified in advance whether the provider has sent the things they have bought. This helps them better organize the product receiving process and prepare for subsequent operations in advance.

  The service gets communication from the supplier about the scheduled arrival, issue date of a particular outbound delivery, and specifics about the contents (delivery amount, assigned serial numbers, batch numbers of the items). It then uses this information (together with the information from the purchase order for the products in issue) to generate an incoming delivery.

  The service contains the following elements:

  - Delivery header
  - Partner
  - Partner address
  - Delivery item
  - Serial number assignments
  - Batch split hierarchy details

This service doesn't support:

- Dangerous goods processing
- Material substitution

- **Notify Outbound Delivery (B2B): API** `CO_EDI_DELIVERY_REQUEST_OUT`

  This service is available through SAP Business Accelerator Hub. This service allows you to notify a product recipient of a delivery's expected arrival, pickup, or issue date, as well as the contents of the delivery. Using this service and the delivery request message type, the supplier can send an advanced shipment notice (ASN) of a delivery to the product receiver.

  The service contains:

  - Delivery header
  - Partner
  - Partner Address
  - Delivery item
  - Serial number assignments
  - Batch split hierarchy details

  This service doesn't support:

  - Material substitution

- **Update Proof of Delivery (B2B): API** `PROOFOFDELIVERYREQUEST_IN`

  This service is available through SAP Business Accelerator Hub. It has fields that allow you to provide proof of delivery of the items in a specific outbound delivery. You can enter multiple quantities for a given item. This service compares the proof of delivery quantities with the delivery quantities and calculates any differences. Alternatively, you can enter differences manually. You can also enter a reason for any differences, or if the field is empty, the service automatically determines a reason. You can monitor messages transferred by this service using the SAP Application Interface Framework monitoring.

  You can edit the following information for each item in an outgoing delivery:

  - Actual quantities received.
  - Differences between delivery and receiving quantity:
    - You can provide a justification for any discrepancies. SAP advises delivering a clear justification for the amount discrepancy. If no explanation is given, the system will establish the default explanation on its own.
    - The service message header includes details on the message header data, including the sender and recipient who are involved, as well as the date and time.
    - This service doesn't allow you to cancel PoD.

10.4 Summary

Whenever types of integration are discussed in the SAP world, two famous kinds come to mind: point-to-point and middleware integration. We discussed the technical intricacies and communication protocols involved in these two options, as well as covered how each of these options have advantages and disadvantages.

On one hand, a point-to-point connection is a tightly coupled and robust integration enabling accelerated development in the short term, allowing for simpler data mapping, hassle-free error handling, and easy extensibility.

Middleware hides the complexities of the underlying operating system and network to facilitate the easy integration of new and legacy systems. It's a means for connecting clients to servers, clients to clients, and servers to servers without having to navigate through many operating systems, networks, or resource server layers. For bigger and more complex connections and dynamic or frequently changing processes, implementing a full-scale middleware tool makes sense.

Chapter 11
Business Rule Framework Plus

Business rules can benefit firms in several ways, including streamlining operations and lowering overhead. Programming business rules into workflows and applications can result in increased efficiency of the processes, consistency in execution, and resultant output and scalability when needed. SAP S/4HANA presents a robust architectural platform to enable both functional and technical consultants to design with flexibility, simulate the output prior to deployment, and reduce dependency on technical coding.

To help organizations remain responsive and agile, decision process automation applications help manage business rules independently from other business computing processes. Business Rule Framework plus (BRFplus) provides a common platform to business stakeholders, functional consultants, and developers to maintain flexible decision logic with minimal coding and enable complex rulesets to be enforced consistently across large environments.

BRFplus and its features provide the following benefits:

- **Higher degree of transparency**
  Explicit documentation of business rules and decisions instead of technical coding known only to technical developers.
- **Decreased maintenance and testing costs**
  BRFplus applications are decoupled from applications and can be used across application areas as global or local applications. Business or functional users can modify the data parameters, simulate, and deploy the application quickly. This results in a drastic reduction in change cycle time, quick response to changes requirements, reduced downtime, and minimal or no code changes.
- **Improved administration**
  BRFplus enables change history maintenance, ownership of application and objects, and role-based authorization, thereby, fixing clear responsibility and ownership of rules and business practices.
- **Convenient graphical representation**
  In BRFplus, rules can be displayed in a graphical and practical way to help users maintain it easily. Rules can be maintained as decision tables, decision trees, flow models, binary rules, scorecard and text-based rules, and so on.

The following sections delve deeper into BRFplus advantages, user interfaces (UIs), components of BRFplus, the logical construct of a BRFplus applications, and details of each object type within the BRFplus application.

11.1 Overview

BRFplus offers an extensive user UI and application programming interface (API) for designing and processing business rules without scripting. You can reuse these rules in many applications and model them in an understandable fashion.

Here are some instances when BRFplus is used by applications:

- **Decision services**
  Procedures are automated such as claims processing, customer service management, and credit approval.
- **Validation/error detection**
  Diagnosis is performed with monitoring/alerting and detection of invalid data and states.
- **Classification/matching**
  This includes derivation/classification such as customers, products, or risk, as well as matching responsibilities, suitable products, or locations.
- **Calculation**
  Costs, overhead, risk, and surcharges are calculated.
- **BRFplus as a technical configuration engine**
  BRFplus is a powerful ABAP-based rules engine, which is used in many SAP applications (including SAP ERP, SAP S/4HANA, SAP Customer Relationship Management [SAP CRM]), SAP Master Data Governance, and other custom applications.
- **BRFplus for output management now**
  This is a new functionality in SAP S/4HANA that allows the BRFplus functionality in SAP S/4HANA output management.

As shown in Figure 11.1, major BRFplus components include application, data objects, expressions, actions, functions, rules, and catalogs. A BRFplus function provides the interaction between a business rule modeled with BRFplus and an application using that rule. No matter how complicated the business logic of a rule may be, it's contained within the function.

Rules are implemented as expressions that are assigned to a function. The rule input is known as the context, and the rule output is called the result. Context and result consist of data objects of one of the following types: element, structure, and table.

BRFplus also supports features such as simulation, trace, transport, and XML export and import.

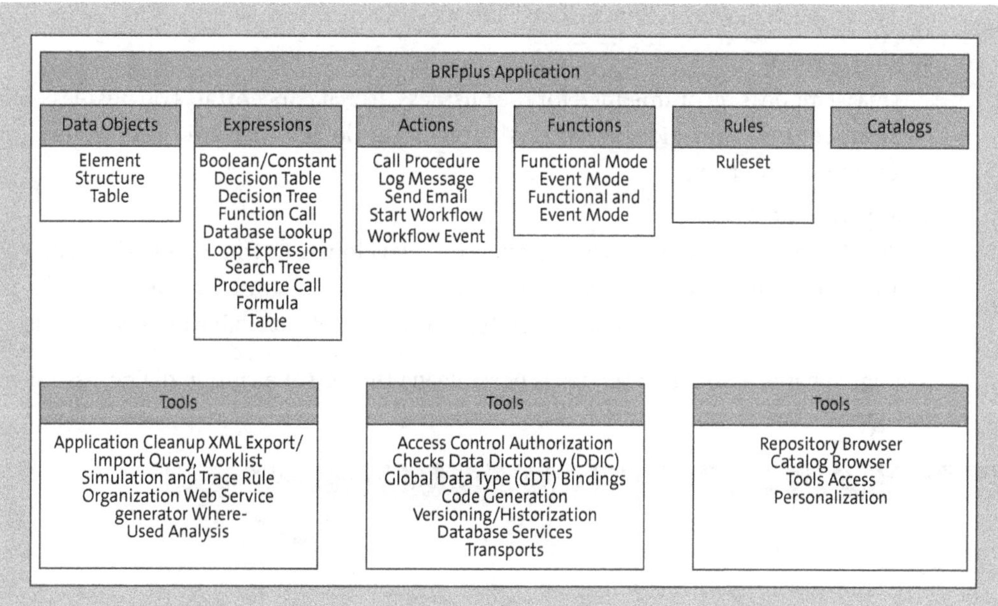

Figure 11.1 BRFplus Components and Tools

11.2 Business Rule Framework Plus User Interface

The BRFplus UI has a workbench that enables users to maintain their business rules. It provides the options to create, modify, and delete BRFplus objects and provides access to all the BRFplus objects available in a system. The workbench enables users to model all the BRFplus-specific rules content, and it provides tools to integrate a BRFplus application into the host application from which the BRFplus function needs to be called. This can be accomplished by implementing the respective calls in the source code of the host application. The workbench greatly facilitates that task by generating web services, function modules, and code snippets that can be used to connect the host program ABAP codes to a particular BRFplus function.

The BRFplus workbench can be opened in one of the following ways:

- In the SAP GUI screen with Transaction BRFPLUS
- In the web browser with the BRFplus workbench URL that is specific for your system landscape

Prerequisite

The role of BRFplus administrator (SAP_BC_FDT_ADMINISTRATOR) has been added to the user profile. Alternatively, a custom role with a subset of the authorizations contained in SAP_BC_FDT_ADMINISTRATOR can be used.

The workbench has the following two UI modes:

- **Simple mode**
  This UI mode is recommended for the business user. It concentrates on business-relevant settings and hides most of the technical details that are only needed for administrative tasks.
- **Expert mode**
  This UI mode is for the business expert or developer. In this mode, the user has access to a number of additional tools needed for administrative or cleanup tasks. In addition, some of the more technical functions are only available in expert mode.

Switching between the two modes is possible via the **Personalization** dialog, as shown in Figure 11.2.

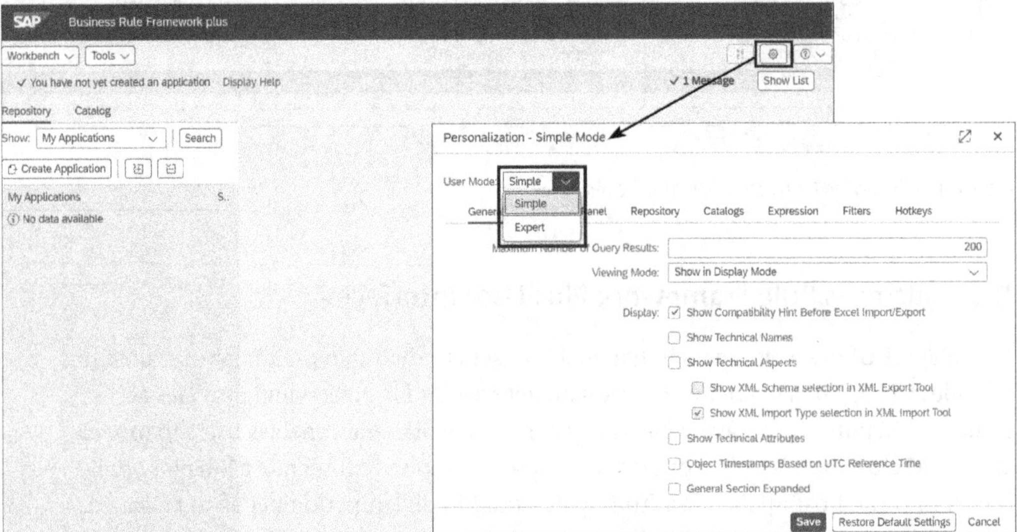

Figure 11.2 BRFplus Workbench: UI Modes

11.3 Business Rule Framework Plus Applications

In this section, we'll refrain from getting in-depth into the BRFplus framework and architecture as this may cause a diversion from the original topic. However, we'll take a quick look at the other integrated features that come with SAP S/4HANA, such as output management.

Various BRFplus objects are contained within a BRFplus application object. The BRFplus application can be viewed as a representation of a certain system functionality that you want to improve with some BRFplus-modeled business rules. Therefore, it's recommended to combine all the BRFplus objects that are involved in a particular system functionality into a single BRFplus application.

The application object type is at the top of the hierarchy of classification for objects. An application must be given ownership of every object.

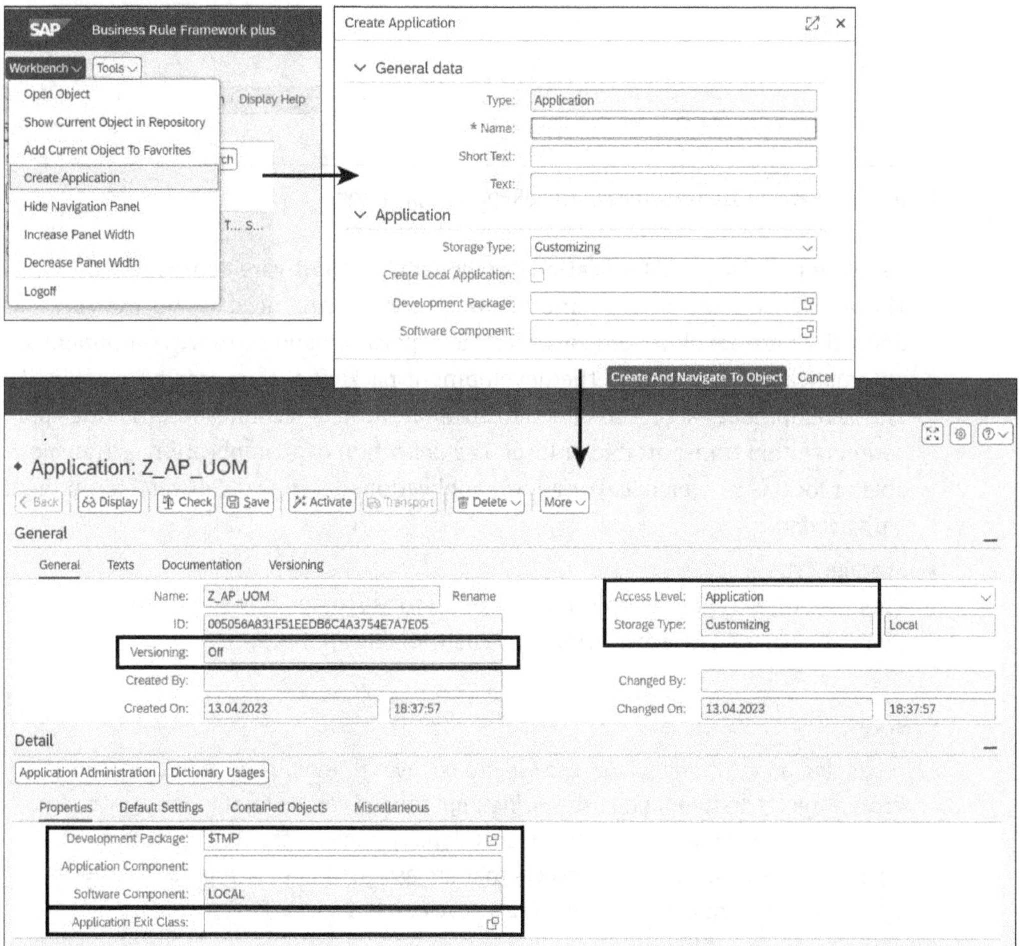

Figure 11.3 BRFplus Application Fields

As shown in Figure 11.3, the BRFplus application can be created from the workbench by clicking the **Workbench** dropdown menu on the left side of the screen and choosing **Create Application**.

The following fields in the BRFplus application carry special significance:

- **Versioning**
  When enabled, the system will retain previous versions of an application in the system database after the application has been modified (Figure 11.4). The newly created objects (within the application) inherit the versioning setting that has been defined by the application.

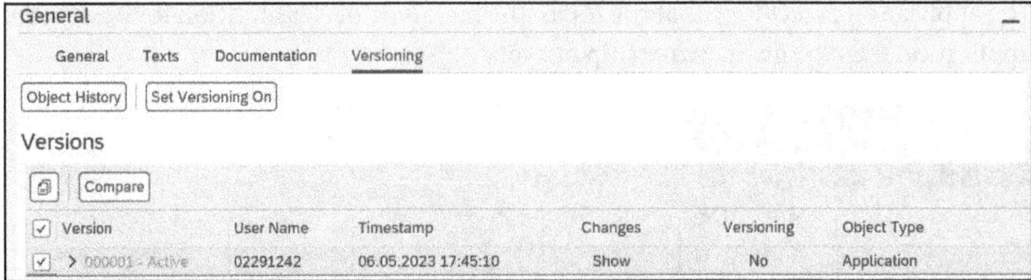

Figure 11.4 Version Maintenance in the BRFplus Application

- **Development Package, Application Component, and Software Component**
  The application component and software component need to be the same as defined for the development package. The application and software component are automatically derived from the development package if they aren't set explicitly. The development package assigned to an application determines whether the application is either transportable or local. The definition of an application as transportable or local is irreversible. For a local application, defining a software component isn't needed.
- **Storage Type**
  There are three types of application referred to as storage types: system application (default), master data application, and customizing application.

> **Note**
>
> Depending on the storage type, the system behaves differently when it comes to system transports. System and customizing applications can be local while master data applications are always local. The storage type as well as the locality setting of an application are inherited by all objects that are assigned to that application. All three types of applications can be assigned to a development package.

The storage type must be defined at the time of creating the application. Once the new application has been created, this setting can't be changed anymore.

Applications for customization that SAP delivers are always installed in client 000. If you need to copy such an application into a system client that is used for development, you can accomplish this with the help of a Customizing type transport request. (See SAP Note 1908227.)

- **Application Exit Class**
  If the BRFplus application requires special processing routines (e.g., for auxiliary calculations, additional validations steps, etc.), such additional functionality can be implemented with methods of an ABAP class. The exit class must be registered for

the application to which it belongs. Only one exit class can be can be maintained for a BRFplus application.

- **Access Level**
  The access level for an object can be defined based on the scope of objects that are allowed to access the current object. Access levels are supported for all object types in BRFplus. By default, each object is assigned the most restrictive access level (**Application**).

Note

It's always recommended to use the most restrictive access level and only extend the scope if needed. This is because with an extended scope, the risk of unwanted impacts may increase. Conversely, restricting the access level of an object (e.g., from **Top Component** to **Application**) can lead to unexpected results if the object is already used by an external application that depends on the extended access rights.

11.4 Business Rule Framework Plus Data Objects

Data objects are the most important factor when designing a BRFplus application. A data object describes data and serves as a data carrier in the context or input to the BRFplus application and/or the result of an expression or an action. There are three main data objects—elements, structures, and tables—which we'll discuss in the following sections.

11.4.1 Element

An elementary data object is helpful when used in a comparison operation such as against a range expression. An elementary data object must have the subtypes described in Table 11.1.

Element Type	Description	Data Binding
Text	Free text or character strings with a maximum allowable length of 255 characters. The maximum length of text data objects refers to the maximum string length that can be stored in the database.	No
Number	Number with decimals.	No
Boolean	Boolean value.	No

Table 11.1 Elementary Data Objects

Element Type	Description	Data Binding
Amount	Number and currency.	No
Quantity	Number and unit of measure (UoM). In addition to the mandatory definition of the UoM at runtime, you can optionally specify the quantity dimension at design time (e.g., length, mass, time). This additional specification has the following effects: At runtime, the system checks if the UoM provided is valid for the specified dimension. If data objects with a specified quantity dimension are used in calculations of a formula expression, the system performs additional formula consistency checks to detect semantic errors that could lead to an error at runtime.	No
Timepoint	▪ Local date ▪ Local time ▪ Local date and time ▪ Universal time (UTC) time stamp ▪ Local date and time with offset to universal time (UTC)	No
ABAP Data Dictionary (DDIC)/Global Data Type (GDT) objects	Data objects can be bound to elements in the DDIC or to the GDT in the Enterprise Services Repository Browser (ES Repository Browser). *Example*: To access a list of ISO country codes, you can just look up a data element in the DDIC that is used for the same purpose and has the country codes already assigned as a list of domain values. Upon binding the BRFplus data object to that data element, the data object assumes the same data type and field length as the bound data element, and all the domain values are made available to the data object.	Yes

Table 11.1 Elementary Data Objects (Cont.)

Figure 11.5 shows a screenshot of elementary data object creation. This can be a free-text element or can be tagged to a DDIC data object, if required.

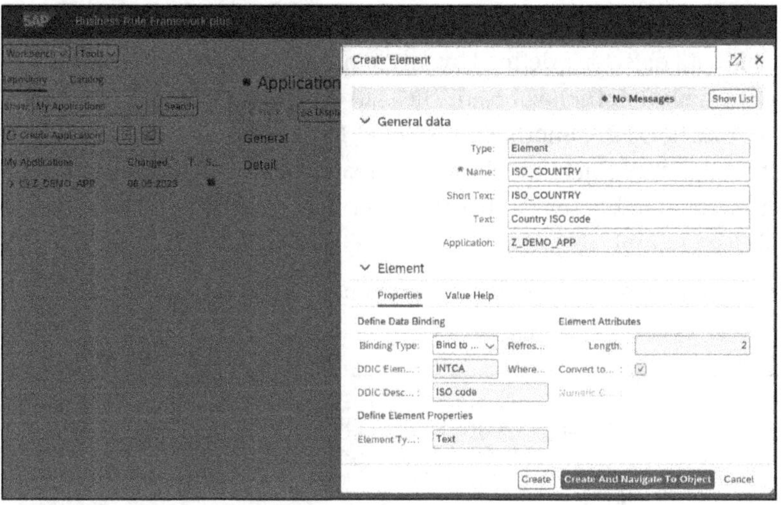

Figure 11.5 Data Dictionary Binding for Elementary Data Types

11.4.2 Structure

A structure is a combination of an arbitrary number of elements, a combination of structures, or a combination of both. The fact that a structure can consist of other structures (which again may consist of structures, etc.) enables you to model data structures of almost any desired degree of hierarchical complexity.

Figure 11.6 provides a BRFplus data structure created using a DDIC binding of table T001W.

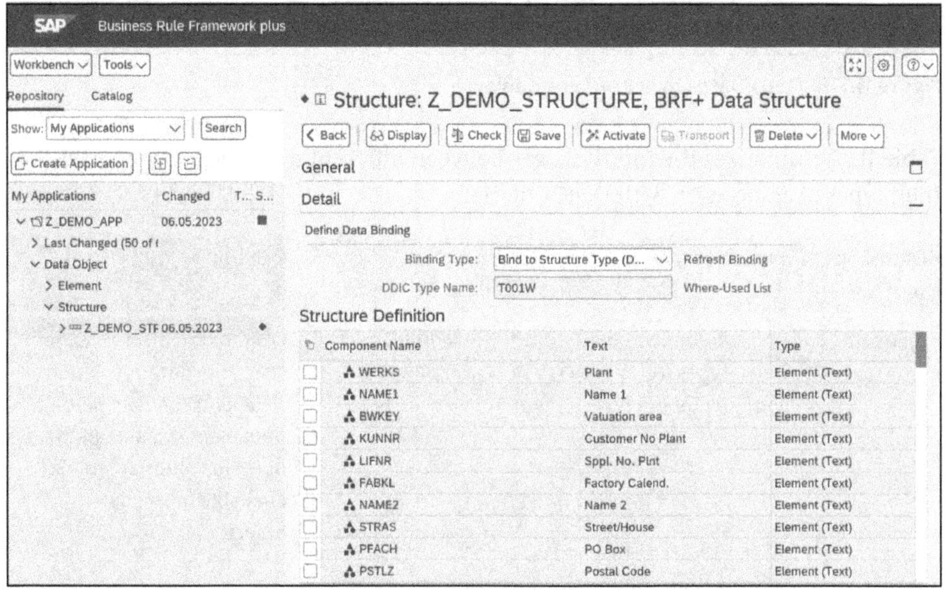

Figure 11.6 BRFplus Data Structure with DDIC Binding

11.4.3 Table

The fields in a table data object are defined by assigning either exactly one element or one structure to the table. In other words, the field structure of a table isn't implicitly built up by adding one field after another to the table. Rather, the table can only adopt the fields of a structure data object that has been created before. If you choose, however, to assign an element rather than a structure data object to a table, this results in a one-column table (i.e., a simple list).

Figure 11.7 shows a BRFplus data object table created by referencing transparent table A816.

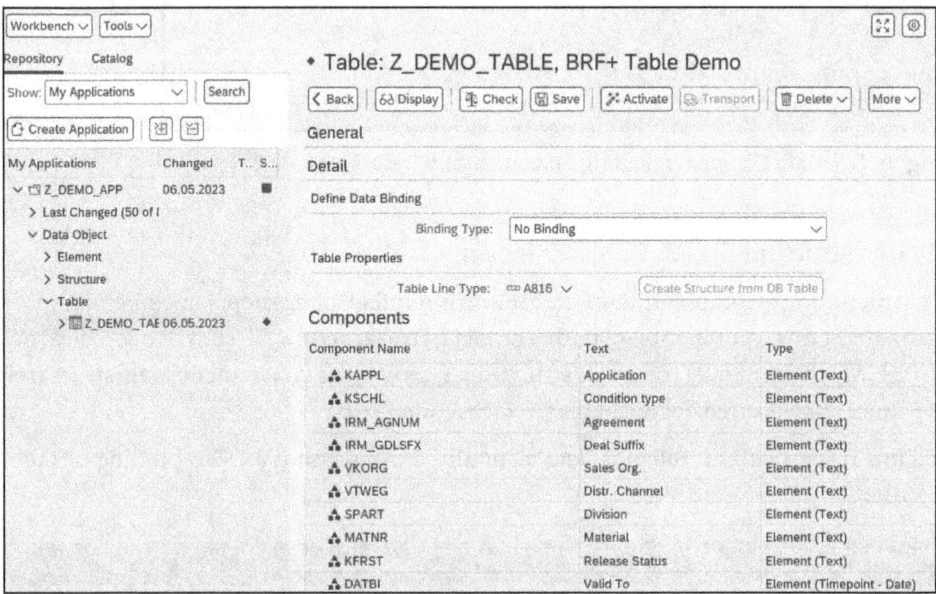

Figure 11.7 BRFplus Data Object of Type Table

Table 11.2 summarizes the interlinkages between different data object types as well as their types of association with DDIC objects.

Data Object Type	Consisting of	Supported Binding Types	Remarks
Element	Predefined data types (refer back to Table 11.1)	▪ No binding ▪ DDIC object	DDIC data object assumes the same data type and field length as the bound data element, and all the domain values are made available to the data object.

Table 11.2 Data Object Types in BRFplus Application

Data Object Type	Consisting of	Supported Binding Types	Remarks
Structure	■ Element (0..n) ■ Structure (0..n) ■ Table (0..n)	■ No binding ■ DDIC structure ■ Core data services (CDS) view	DDIC structure with all contained fields is imported automatically into BRFplus.
Table	■ Element (0..1) ■ Structure (0..1)	■ No binding ■ DDIC table type	Table type structure with all contained fields is imported automatically into BRFplus.

Table 11.2 Data Object Types in BRFplus Application (Cont.)

11.5 Expressions

An expression is a fundamental BRFplus element that allows modeling of business rules. Each expression type defines a self-contained computational unit with a well-defined logic. Expressions use a context or nest other expressions to calculate, determine, or derive a result. An expression can be considered to be an instance of an expression type behaving according to the expression type's logic.

An expression can be reused as a nested building block for a rule, starting with a function as its assigned top expression. Custom expression types can also be created for use in rules. BRFplus comes with a number of predefined powerful expression types as listed here:

- Boolean expression
- Business rule management system (BRMS) connector expression
- Case expression
- Constant expression
- Database lookup expression
- Decision table expression
- Decision tree expression
- Dynamic expression
- Formula expression
- Function call expression
- Loop expression
- Procedure call expression
- Random number expression
- Search tree expression

- Table operation expression
- Value range expression

While these predefined expression types cover a broad range of business rule usage scenarios, it's still possible for a scenario to arise in which the predefined expression types aren't sufficient or aren't easy to use. To overcome such potential shortcomings, BRFplus provides the option of creating your own custom expression types.

The conceptual approach of custom or user-defined expression types is exactly the same as for the built-in expression types. In other words, the built-in expression types are set up by SAP using exactly the same technical infrastructure that can be used for your own custom expression types. Therefore, custom expression types behave just like the built-in ones in terms of performance, access control, and so on because they are treated as a horizontal scope extension of the BRFplus concept rather than as a vertical add-on.

These built-in expression types are frequently used in BRFplus applications. However, keeping the third-party logistics perspective in mind, a few of these expression types can be commonly used for such integrated business scenarios. Let's take a look at those specific expression types.

11.5.1 Boolean Expression

The Boolean expression is used to test a logical condition with the help of a number of predefined expression templates with two or three operands that cover many use cases. When defining a Boolean expression, you replace the formal operands (<1>, <2>, etc.) by BRFplus objects that can be evaluated to a Boolean value (context data objects of type Boolean or expressions with a Boolean result data object).

Following are the predefined templates for frequently used logical conditions:

- Any operand is true.
- All operands are true.
- <1> and <2>
- <1> or <2> or <3>
- (<1> and <2>) or <3>

11.5.2 Constant Expression

This is the simplest of all expression types. It allows you to create an expression and set a constant value that is returned after processing. A constant expression can be used for various purposes:

- Define result values for other expressions.
- Define test parameters for comparison operations.
- Define a value list for elementary data objects.

There are six types of constant expressions: Text, Number, Boolean, Amount, Quantity, and Timepoint.

11.5.3 Formula Expression

This expression type allows you to set up formulas to perform a wide range of calculations. The formula expression offers the basic mathematical operations (addition, subtraction, multiplication, division), more advanced operations (e.g., exponentiation or binary AND), comparisons (greater than, less than, etc.), and a wide variety of formula functions for different purposes and data types (date and time calculation, string handling and manipulation, conversion functions, system functions, etc.). Additionally, comments can also be maintained in a formula to help others better understand the purpose of a particular calculation step.

11.5.4 Decision Table

Business rules can be maintained by business users in the form of decision tables, either locally or globally. A decision table expression is used to compare a given input against a list of predefined combinations of conditions. Each combination has an associated individual or combination of outputs as a result, which is returned as the expression result when the input data matches any particular combination.

The decision table is built up as a table with the following elements and settings:

- One or more columns representing the different conditions against which the input data is compared
- One or more columns used to hold the result that is returned by the expression if all the conditions in a row are satisfied
- One or more rows that are unique in terms of their combined conditions

A decision table expression sequentially processes business rules based on a set of inputs. During runtime, the input data that is supplied to the decision table is compared against the defined conditions, starting with the first row. All cells are evaluated by checking the column value against a number of range comparisons. For each cell, a Boolean result is returned. If all cells in a row are evaluated as true, the evaluation stops, and the expression returns the result associated with the current conditions. Otherwise, processing continues with the next table row until either a matching set of conditions is found or the end of the table is reached.

Following are a few key things to consider while maintaining a decision table:

- **Result columns and result data objects**
  A decision table can also have multiple result columns. This can be achieved either by assigning a structure data object as the result data object or simply by providing several elementary data objects as result columns. In the latter case, the system

offers to create a new structure data object automatically, containing the selected elements. This structure is then automatically assigned as the decision table's result data object.

- **Single versus multiple match mode**
  A decision table can be run in two different modes of operation, depending on the setting of the **Return all matches found** indicator:
  - *Single match mode* (**Return all matches found** is set to **False**): In single match mode, the input data is tested against the conditions defined in a table row. If the input satisfies all the conditions, the associated result is returned and processing of the decision table stops. In other words, the system returns the result associated with the first matching row. This is the default mode.
  - *Multiple match mode* (**Return all matches found** is set to **True**): In multiple match mode, the input data is tested against the conditions defined in all the table rows. All matching results are collected in the result data object and returned as the result of the evaluation.
- **Initial value**
  If no matching condition is found for a given input, the decision table raises an exception instead of returning a result. You can override this default behavior by setting the **Return initial value if no match is found** indicator. If this indicator is set, the decision table returns the initial value that is defined for the type of the result data object. With this setting, it's possible for you to consider the unsuccessful processing as a defined result rather than an error and continue with processing.
- **Mandatory input**
  With this indicator set, the system forces you to provide a condition value at design time for the respective column in each row that you enter in the table definition. Such a condition can therefore not be used for providing an unspecific empty condition that would be evaluated as true regardless of the input data.
- **Column protection**
  One of the big advantages of BRFplus is that it makes the criteria for the business decisions in your company transparent and easy to change. While such transparency can be counted as an asset in itself, it can still be inappropriate for certain parts of your business. Here, you may want to decide that the decision criteria for a particular business case will be protected against unauthorized change or even be completely invisible. You can accomplish this with the help of the **Column Accessibility** setting, both for condition columns and result columns of a decision table:
  - **Full Access (Changes Allowed)**: With this setting, a user can deliberately change the values of the column cells. This is the default.
  - **Display Only (No Changes)**: With this setting, a user can see the values in the column cells without being able to make any changes.

- **Hidden**: With this setting, the column is hidden completely in the BRFplus workbench at design time as well as at runtime. Without checking the table settings, a user doesn't receive any visual indication that the column exists.

- **Table data export and import**
  You can exchange the condition and result data of a decision table with Microsoft Excel. This enables you to share your table data with people who may not have access to BRFplus but might give input from a business perspective. The decision table data can be exported to Microsoft Excel by navigating to the decision table and clicking the **Export to Excel** button to download the data to your local machine. This will also download the structure of the decision table in Microsoft Excel that helps users maintain large volume of data in an offline mode. Then, you can copy and paste the downloaded data in the decision table directly. For optimal performance in a BRFplus decision table, the maximum recommended number of rows is 10,000. (Refer to SAP Note 2585774 for more information on this.)

- **Table data maintenance**
  Due to the sequential processing of the rules defined in the table from top to bottom, we recommend that you define the rules starting with the most specific condition in the first row, followed by rules with a decreasingly specific set of conditions assigned. Otherwise, situations could occur where a given input that satisfies an unspecific condition would lead to the expression result defined for that condition, without giving the system a chance to test whether there is a more specific condition that would match the input data even better.

  Figure 11.8 shows a typical BRFplus decision table where the input data combination is shown on the left-hand side and output column is shown on the right.

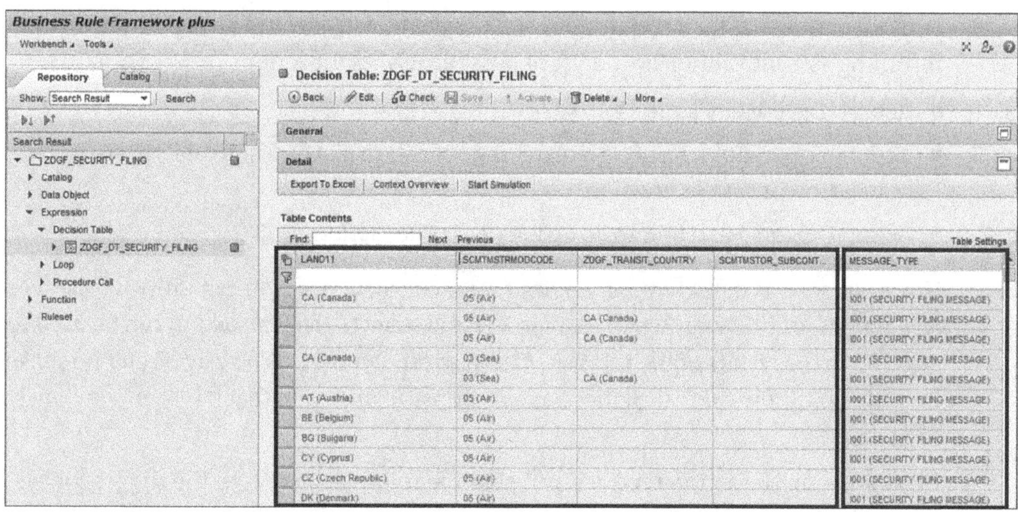

Figure 11.8 BRFplus Decision Table Evaluation Security Filing Requirements

11.5.5 Function Call Expression

A function call expression calls a BRFplus function, maps the expression context to the function context, and returns a result. A function call expression can call different functions based on the input.

For example, a function call expression can be used to map the third-party logistics UoM with your own system-specific UoM. The function call expression calls the respective functions based on the country and third-party logistics provider and passes the context data.

11.5.6 Procedure Call Expression

A procedure call expression enables the execution of function modules as well as of methods of an ABAP object-oriented class in the backend system. This expression type can be used directly to execute procedures that are stored in a database.

The following types of procedures can be assigned to a procedure call expression:

- Function modules
- Methods of ABAP classes
- Database procedures

11.6 Actions

Actions are used to define the interactive part of BRFplus. Under certain conditions, actions can be used as an alternative to expressions. For example, in a decision table, you can define that for a condition row, the system will perform an action rather than returning a result value. Here's a list of possible usages for actions:

- Action triggered by a rule
- Action triggered as a result of a matching condition in a decision table, decision tree, search tree, or case expression
- Action triggered as a follow-up action of another action

However, actions don't have any output. As a consequence, you can only use them as an element for the final step in each processing branch. Here, an action can be used to respond to the results of the rule calculation so far. But there is no way of, and no sense in, performing any rule calculations as a response to an action that has been previously triggered.

Depending on the action type, it's still possible to use an action as the trigger for system activities of considerable complexity. This is true, for example, for Workflow or

Call Procedure action types. But still, once an action has been carried out, it's not possible to define any further processing as a specific response to any kind of changes that may have been made by that action.

Some of the action types correspond to a related expression type. To avoid confusion over whether a given object is an action or an expression, the action type names are all marked by a trailing (Act) in the system. In addition, the action type names follow the linguistic pattern "verb - object" (e.g., Call Procedure (Act), as opposed to the Procedure Call expression type), where possible.

BRFplus supports the following action types:

- Call procedure action
- Log message action
- Send email action
- Start workflow action
- Workflow event action

Figure 11.9 shows the **Action** types available within a BRFplus application.

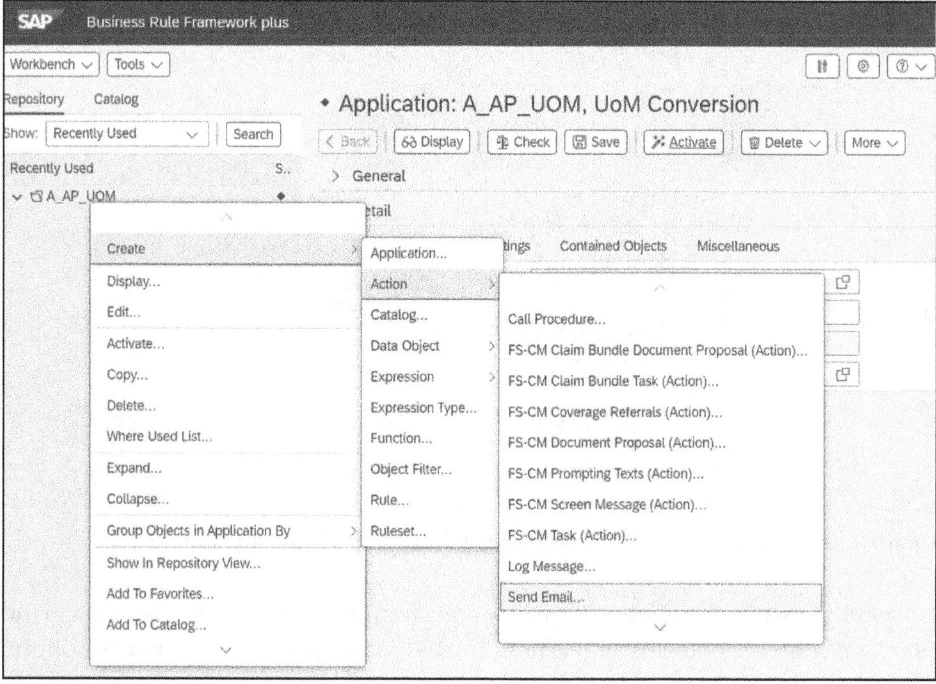

Figure 11.9 BRFplus Action Types

11.7 Function

A function is the rule interface in BRFplus and acts as a link between the application code and the BRFplus code. Once the function is properly maintained and checked for correctness, it provides the option to enable generation of an ABAP code snippet by the system, and this code can be used for easy integration of a BRFplus function into the desired application.

A function carries a *context* and a *result*. It imports the context from the calling application and passes the context data to the assigned top expression or ruleset for further processing. After processing, the function returns the calculated result to the calling application.

Figure 11.10 shows a schematic representation of how the entire BRFplus application works and function is the central or key object that controls the functioning of the BRFplus application.

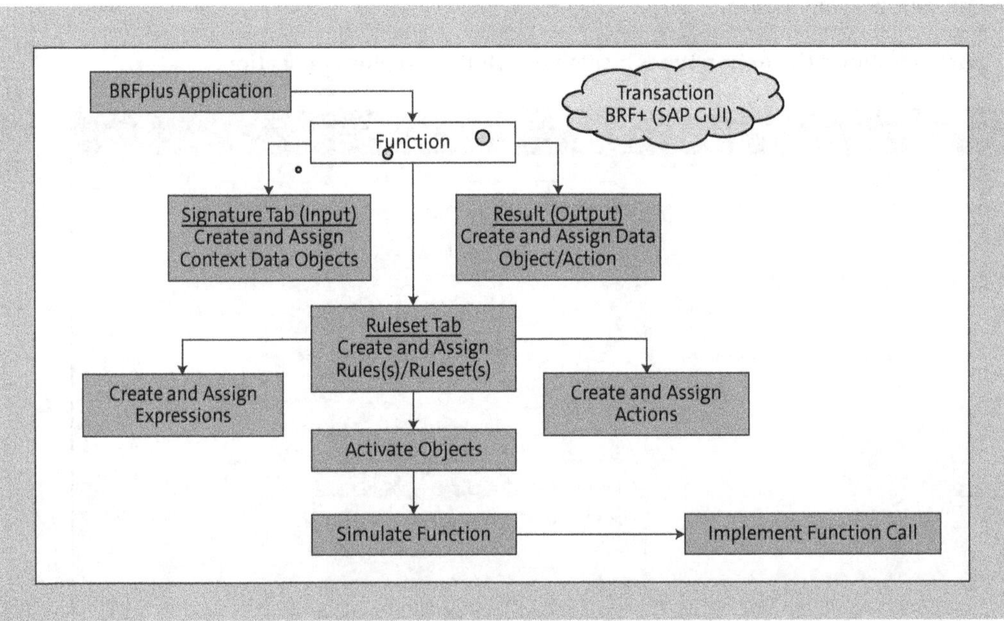

Figure 11.10 BRFplus Function Controls the Operation of the Entire Application

Let's look at mandatory fields in functions and how they contribute to operation of the function within the BRFplus application. The BRFplus application acts as a repository, and it's essentially the function object that controls the application of intended business rule logic. Figure 11.11 shows various features and fields of a BRFplus function, which we'll cover in more detail in the following sections.

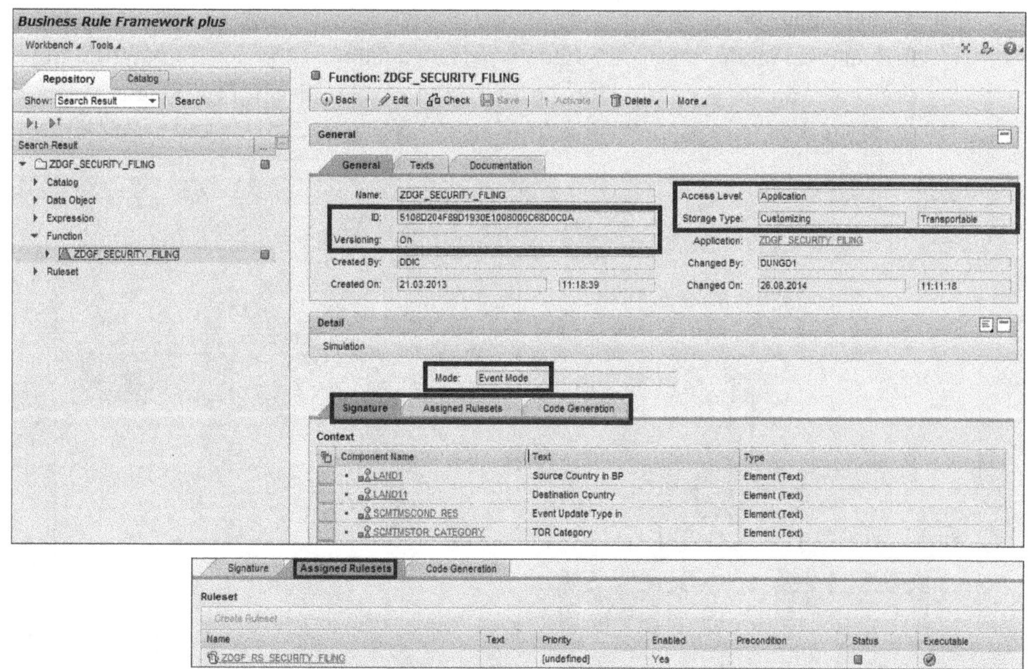

Figure 11.11 Important Characteristics of the BRFplus FUNCTION

11.7.1 Mode of Operation

A function can be designed for one of the following modes of operation:

- **Functional mode**
  In functional mode, function execution starts with the assigned top expression. From this top expression, the processing may run through any number of nested subexpressions until a result is returned.
- **Event mode**
  In event mode, the function is associated with a list of rulesets that are executed according to their execution priority and their position in the list. Figure 11.12 shows comparison between function and event modes of operation.
- **Functional and event mode**
  This mode is a combination of the two modes mentioned previously. At runtime, the function starts processing the assigned top expression. Once the expression evaluation is finished, function execution is continuous with the associated rulesets.

  Figure 11.13 shows how function and event modes work in combination.

- **Analytical mode**
  In analytical mode, rules are based on persisted data in CDS views. Analytical functions use a database context for the query and a table object for the results. Analytical mode is supported in BRFplus from SAP NetWeaver 7.51 SP00 and SAP S/4HANA.

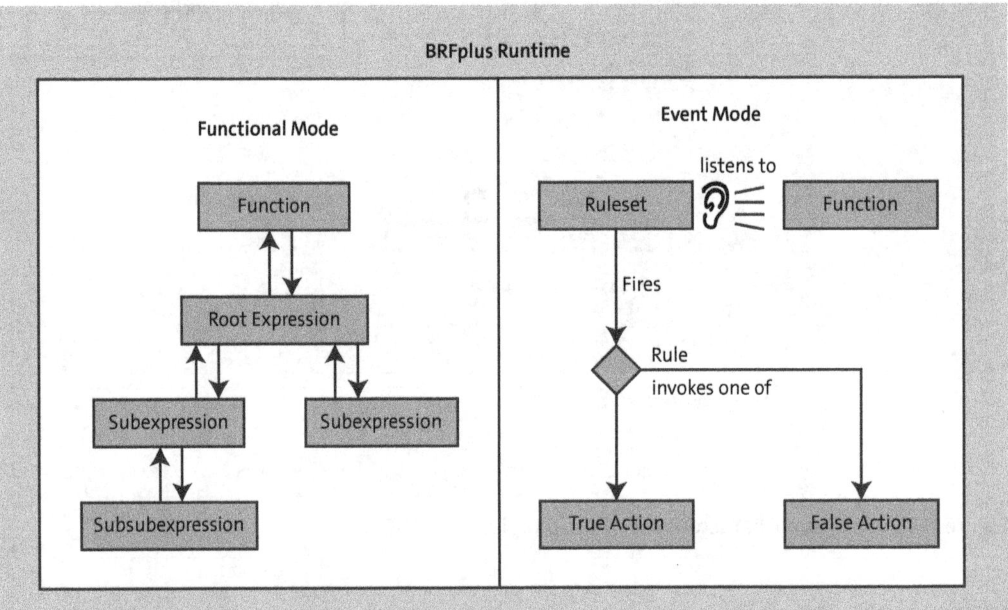

Figure 11.12 Comparison of Functional with Event Modes

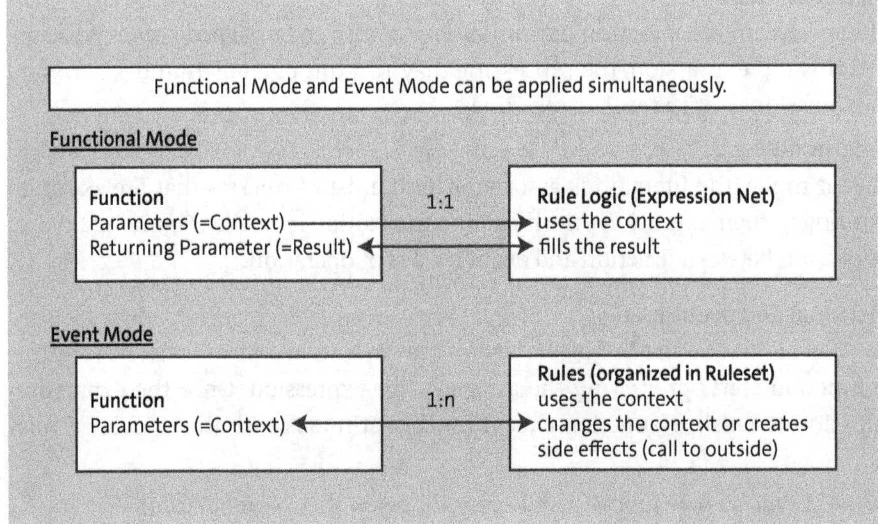

Figure 11.13 Functional and Event Mode Together

11.7.2 Signature

The signature function consists of two parts:

- **Context**
  The context is a container for data objects that you can assign as import parameters for the function. You choose the context data objects of a function according to the requirements of the calling application by which the BRFplus function is invoked. In addition, the context data objects of a function define the scope of objects that can be accessed by the expressions that are evaluated during function execution.
- **Result data object**
  The result data object returns a result value that has been calculated by the expressions that the function consists of.

  Figure 11.14 shows the BRFplus function and its various features, including the BRFplus function name ❶, BRFplus tools ❷, operation mode ❸, context in BRFplus function ❹, and the result data object ❺. This is an example of outbound messages to third-party logistics partners where the output IDoc will contain the logistics service provider warehouse ID based on input data field values appearing under context. You may also notice a top expression representing a decision table along with functional mode (1:1). So, the input field values will be passed to decision table, and, in turn, the resulting column output (WH_ID) will be returned as the result data object. The figure also shows some standard tools such as simulation, function code generation, traces, and so on available in the BRFplus function.

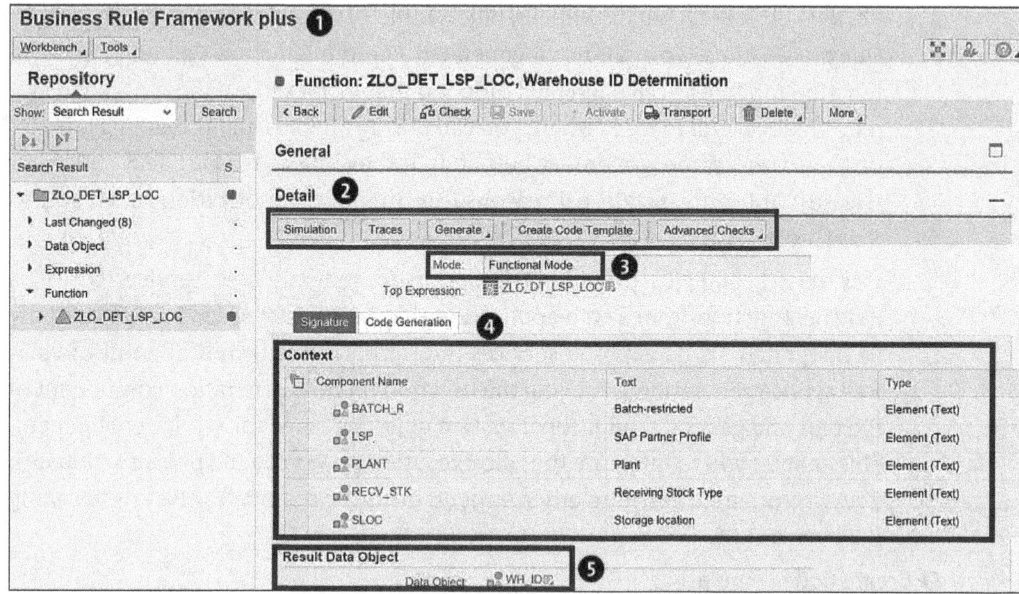

Figure 11.14 BRFplus Function with Operation mode, Signature, Result Object, and Tools

11.7.3 BRFplus Tools

Now we'll look at the tools and features available with the BRFplus function. These tools help is analyzing the BRFplus structure and its functioning, as well as provide options to generate code for integration with the host application.

Figure 11.15 shows such tools and features within the BRFplus function:

❶ **Simulation**
Option to simulate the function processing to test how the application will behave with given sets of inputs. This is done prior to code generation of the function. It provides an in-depth insight into the system status for every single step the system takes during function processing.

❷ **Traces**
BRFplus gives the option to create a processing log to keep track of all processing steps during function execution. The trace information is stored in the system and can be reviewed at any point in time.

❸ **Generate**
For a BRFplus function, you can take advantage of different services that automatically generate code and other objects that are needed to bridge the gap between the ABAP backend system where a BRFplus function is hosted and other systems or environments that want to access BRFplus. These services are needed to address different usage scenarios. These can be categorized as follows:

- *Code templates*: A predefined snippet of ABAP code that can be used to call a BRFplus function from the application or program in the same system.
- *Function module*: You can use a generated Remote Function Call (RFC)-enabled function module to call a BRFplus function that resides in a remote backend system. This is useful in cases where you can't bring BRFplus and the calling application into the same system—whether it for technical reasons (e.g., different releases) or for organizational reasons (e.g., missing authorization to develop in the remote system).
- *Web service*: BRFplus provides the option to generate web services to call a BRFplus function from a web application that isn't ABAP-based or that can only be integrated via its exposed services interface. From a technical point of view, web service generation is reusing the function module generation component of BRFplus and generates additional system objects needed for service enablement. This enables you to integrate the rule execution power of a BRFplus function into a service-oriented software environment, or into a remote ABAP system calling BRFplus via RFC.

❹ **Create Code Template**
BRFplus comes with predefined code templates that you can use to let the system generate an ABAP code snippet. After having made some minor adaptations to this snippet, you can copy and paste it into the code section of your application from

where you want to invoke BRFplus rule execution. You can choose whether the function is called with or without trace information, and you can decide whether the explanatory in-line comments are visible in the code or not.

❺ **Advanced Checks**

In addition to the standard consistency checks that are available for each BRFplus object via the **Check** button, the system offers additional checks for the following functions:

- **Assigned Rulesets**: With this check, the system analyzes the correctness and consistency of all the rulesets that are assigned to a function. This check is only available for functions running in event mode or in functional and event mode.
- **Un-versioned Objects**: With this check, the system drills down through the hierarchy of all objects assigned to a function and determines whether there are any objects for which versioning hasn't been switched on.
- **Inactive Objects**: With this check, the system drills down through the hierarchy of all objects assigned to an activated function and determines whether there are any objects that are currently inactive. For each inactive object, the system sends a corresponding message to help you analyze the object.

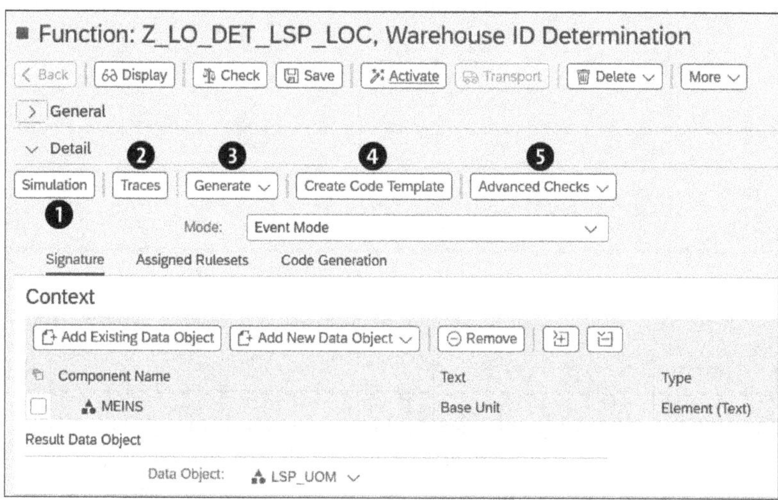

Figure 11.15 BRFplus Tools and Features

11.7.4 Assigned Ruleset

A ruleset is a collection of rules to be processed in a particular business case. It serves as an entry point for rule processing and can also be maintained with a pre-condition that serves to determine if the ruleset or logic is to be applicable in a particular scenario.

When **Mode of Operation** in a BRFplus function is set to either **Event Mode** or **Functional and Event Mode** on the Properties tab, it's mandatory to assign one or more rulesets. There are two ways to assign ruleset to the function:

- A new ruleset can be created when the function is in edit mode.
- An existing ruleset can be assigned to the function. However, in this case, you have to navigate to the ruleset and define the desired function as the ruleset's trigger. Once this assignment has been made, the ruleset is listed on the **Assigned Rulesets** tab of the function.

If there is more than one ruleset assigned to a function, the rulesets are triggered according to the order in which they appear in the function's list of assigned rulesets. You can influence the execution order for the rulesets by defining a priority for each involved ruleset. The priority is a numeric value between 0 and 99 that can be defined at the header section of the ruleset. The default priority for a newly created ruleset is 0, meaning that there is no priority defined for that ruleset. The list of assigned rulesets is sorted by that priority in ascending order so that the ruleset with the lowest priority value is executed first, followed by the ruleset with the second-lowest priority, and so on.

Figure 11.16 provides a snapshot of assigned rulesets to a function along with its priority and status.

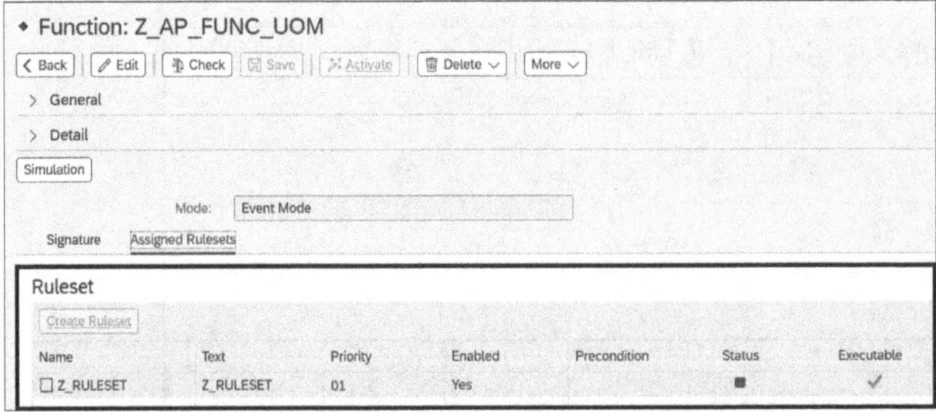

Figure 11.16 BRFplus Function with Assigned Ruleset (Ruleset Execution Priority Manually Set as 01)

11.7.5 Rules

A rule object is used as the technical representation of a simple business rule to be applied to a particular business case. Rules can be adapted to the increasing complexity of business cases by adding additional actions or expressions to the rule and by combining any number of rules to a ruleset. Actually, rules can only be triggered in the context of a ruleset to which a rule is assigned. They can't be triggered as stand-alone objects.

A rule consists of an embracing If condition tested by the system once the rule is triggered. Based on the result of the condition, the rule performs a set of operations. The condition must have a True branch, whereas the False branch is optional.

Figure 11.17 shows the BRFplus rule construct for acceptance/rejection of credit amounts based on a credit score.

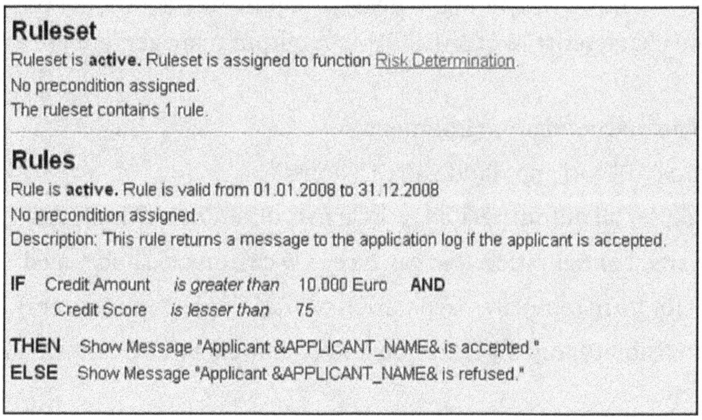

Ruleset

Ruleset is **active.** Ruleset is assigned to function Risk Determination.
No precondition assigned.
The ruleset contains 1 rule.

Rules

Rule is **active.** Rule is valid from 01.01.2008 to 31.12.2008
No precondition assigned.
Description: This rule returns a message to the application log if the applicant is accepted.

IF Credit Amount *is greater than* 10.000 Euro **AND**
Credit Score *is lesser than* 75

THEN Show Message "Applicant &APPLICANT_NAME& is accepted."
ELSE Show Message "Applicant &APPLICANT_NAME& is refused."

Figure 11.17 Example of BRFplus Rule Logic Based on Credit Score

In a rule, you can access and manipulate context data. This seems surprising at first glance because the rule itself doesn't have a context of its own. Rather, accessing context data from within a rule means accessing the context data that is available to the rule through the ruleset to which the rule is assigned.

The following operation types are supported by the rule:

- Triggering an action
- Processing an expression (result of the expression can be used to update the context)
- Initializing or updating context data objects

11.8 Business Rule Framework Plus in Output Management

In this section, you'll see how SAP S/4HANA is designed and what is the precise scope of coverage for output types. It's essential to also look at the flexibility regarding its integration with complex output requirements.

11.8.1 SAP S/4HANA Output Control

SAP S/4HANA output management includes cloud qualities such as extensibility enablement, multitenancy enablement, and modification-free configuration. Configuration is

BRFplus based and accessible to customers. Therefore, the complete configuration differs from the configuration that is used when output management is based on table NACE.

In addition, in SAP S/4HANA, the target architecture is based on Adobe Document Server and Adobe Forms, and the form determination rules (along with other output parameters) use BRFplus functionality for message determination.

It's integrated in the SAP S/4HANA standard and doesn't require any additional purchase or license. The main characteristics of SAP S/4HANA output management are as follows:

- Available in a cloud or an on-premise environment
- Natively integrated into SAP Fiori apps (and SAP GUI apps)
- Offers a unified solution for all output-related tasks across different SAP modules
- Comes with a predelivered configuration that can be easily customized and scaled
- Provides a new option for form templates (separation of static layout and content)
- Supports standard extensibility for configuration and documents (form and email templates)

BRFplus base provides option flexibility for output parameter determination in the following ways;

- Determination can be based on application-specific fields that are extendable.
- Predelivered content can be applied.
- CDS views ensure extensibility.

Output management in SAP S/4HANA is designed for a specific scope and may not exactly compare with existing frameworks such as table NAST. It caters to functions for sending outbound messages (e.g., printouts, emails, EDI) from transactional applications and to business receivers, as well as documenting the output result. However, this doesn't include system integration scenarios or data replication scenarios. It has a business-document-centric design, meaning it's not a standalone functionality but is always embedded into the processing of business documents, such as invoices, purchase orders, dunning letters, and so on. Figure 11.18 shows the output options available in SAP S/4HANA on-premise and cloud editions.

However, there may be a concern as to which of the options customer should opt for when implementing SAP S/4HANA. While doing this, you may also need to look at the following implementation scenarios:

- SAP older versions were implemented before SAP S/4HANA (i.e., an output solution already existed).
- SAP S/4HANA is newly implemented with no existing output solution.

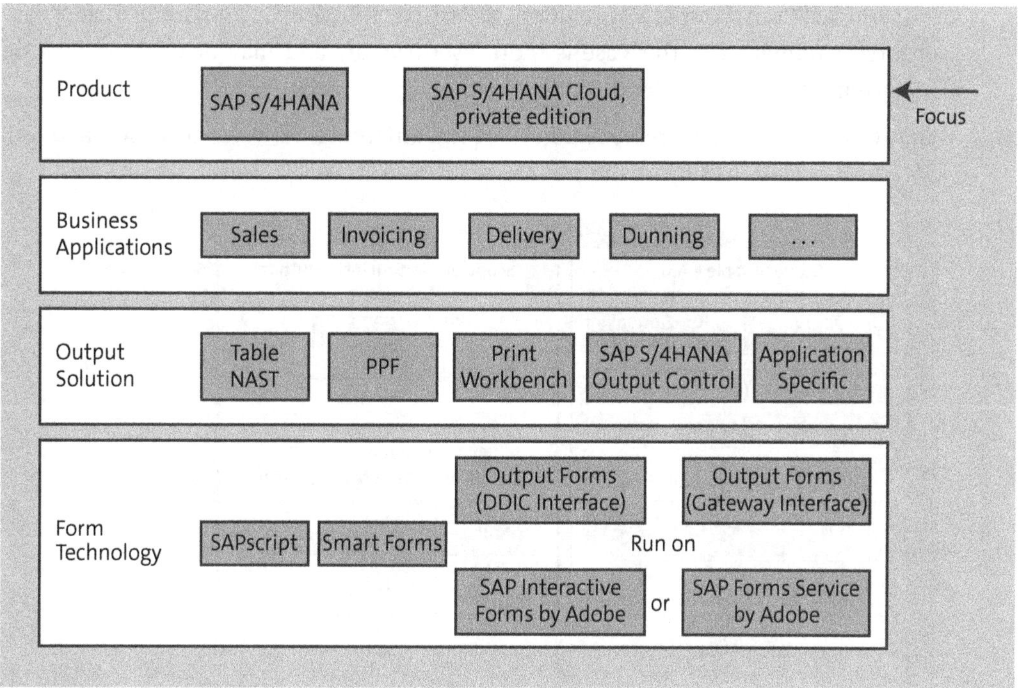

Figure 11.18 Different Output Options Available in SAP S/4HANA

Table 11.3 presents a comparison of all output management features available for the previously mentioned scenarios.

Edition	Available Output Solutions	Default	Switchable?
SAP S/4HANA	Existing output solution (e.g., table NAST) SAP S/4HANA output control	Table NAST	Yes
SAP S/4HANA Cloud, private edition	Existing output solution (e.g., table NAST) SAP S/4HANA output control	Table NAST	Yes
SAP S/4HANA Cloud	SAP S/4HANA output control	SAP S/4HANA output control	No

Table 11.3 Flexibility in Output Control during Implementation of SAP S/4HANA

11.8.2 Limitations of SAP S/4HANA Output Management

Table NAST is a generic framework that can be used not only for output but also for integration, workflow, special functions, and for literally anything that can be implemented. This flexibility in table NAST comes with the price of very high complexity and most of the time requires consulting intervention.

SAP S/4HANA output control provides only output functionality. Maintenance of conditional parameters in this case is more user-friendly and may or may not require intervention from consultants.

Figure 11.19 is a pictorial representation of the difference between table NAST and SAP S/4HANA output management options.

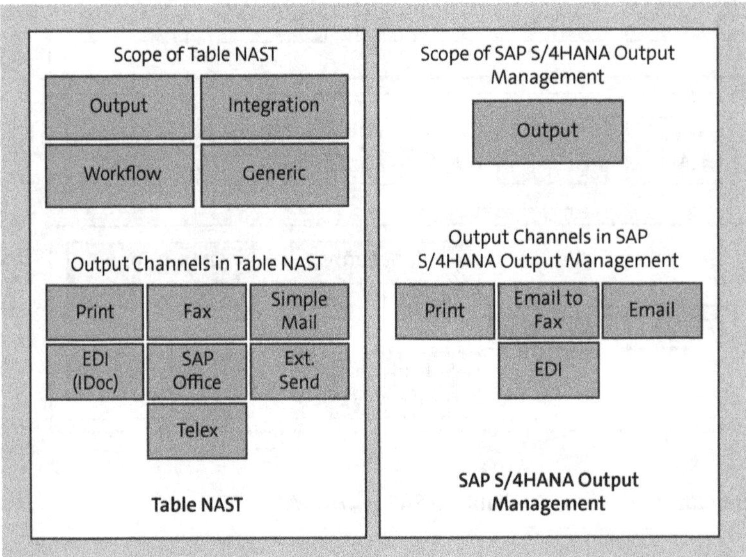

Figure 11.19 Table NAST versus SAP S/4HANA Output Management

SAP S/4HANA output control isn't designed to completely replace table NAST, but only to cover the output part (print, email, partly EDI).

System integration, workflow, and special functions aren't in the scope of SAP S/4HANA output control. Of course, there is a possibility to implement a mixed approach to both running side-by-side for different sets of requirements. This means using SAP S/4HANA output control for applications previously using table NAST only makes sense in exceptional cases. The functionality of SAP S/4HANA output control should be enabled only when it's found to match the business requirements exactly. Replacing table NAST with this in a haste may jeopardize the design.

Note

We purposely didn't touch upon how to configure the BRFplus output in SAP S/4HANA in this section, as it's available in multiple blogs and publications, such as the one at the following URL: *http://s-prs.co/v571704*. The idea here was to show which method is more suitable for third-party logistics integration, considering the fact that output requirements (for third-party logistics partners/logistics service providers) may be more complex in nature. Typically for third-party logistics integration scenarios, table NAST is a preferred option.

11.9 Managing Custom Rules with Business Rule Framework Plus

So far, we've had a look at various features of BRFplus and how the SAP S/4HANA output management benefits from it. How can BRFplus applications be used in logistics service provider scenarios?

Traditionally, third-party logistics integrations used many technical developments to handle custom rules for business logic, managing custom tables for one-to-many, many-to-one, or even one-to-one conversions. This required involvement of consultants, and managing frequent changes involved higher total cost of ownership (TCO).

Using BRFplus to implement rules, logic, and conversions helps avoid technical complexity, puts more flexibility in the hands of business users, requires less involvement of consultants, and reduces the overall TCO. In addition, BRFplus applications are fully extendable and provides authority-based access to users to make it secure.

When deciding which system you need to maintain BRFplus applications, consider the following options:

- Because BRFplus is an ABAP-based framework, it's ideally suited for integration into ABAP-based system environments, such as an expansion to an already-existing SAP Business Suite component. Use of the Java-based SAP Business Rules Management (SAP BRM) is an option if your system environment is mostly Java-based or if you intend to design business rules for a system landscape based on service-oriented architecture (SOA). SAP NetWeaver 7.1 EhP 1 or later is required to use SAP BRM.
- Another possible scenario is one in which both parts are employed simultaneously: Both engines may be used in their respective areas in a mixed environment that includes both ABAP and Java components. Calling between BRFplus and SAP BRM is possible. This implies that, for instance, rules can be maintained in SAP BRM and called remotely into BRFplus, and vice versa.
- Finally, using BRFplus's web services may be a good option even in a SOA setting. This is accurate in situations when an ABAP backend system houses the bulk of the data that needs to be handled. Using BRFplus in this situation offers benefits in terms of performance, size, and the accessibility of the integrated components.

Some examples of requirements during third-party logistics integrations are presented here:

- The UoM at the logistics service provider's end (in their system) may differ from your system. It may also be called something different in their system. Another variation of this is that if the logistics service provider is global, they may have a different UoM based on the location: country or region.
- For logistics service providers with a global setup, their warehouses may be located in the region catering to local units of customers. Top logistics companies such as DHL, DB Shenker, Bollore, and so on may have regional warehouses catering to shipper's

distribution centers and plants. Such a linkage can be maintained easily using a BRFplus decision table.

- Another example of rule logic is in the determination of stock types for incoming information from the logistics service provider's end. Many of them follow their own specific terminology for identifying stock types (e.g., damaged, new priced, short dated, scrapped, quarantined, etc.). When they send stock quantity and type information for inventory reconciliation, the relevant stock type can easily be determined by application of the rule logic or decision table.
- Inventory information can be routed to a selected number of recipients based on location, stock status, and so on.

11.10 Summary

BRFplus has many more usages other than simple application of rules and matching. It can use formula calculation, perform workflow triggering, and employ complex nested rule logic to derive the outcome using certain specific inputs from business transactions. Once designed and put in use, it may hardly require any change in its basic design criteria. However, only a portion of the application, such as decision table entries, can be made available to a specific business user group for frequent maintenance. The versatility, scalability, and extensibility (further enhancement through ABAP) makes it a suitable choice for implementing business logic, and this helps in getting rid of redundant custom tables and coding logics.

This chapter explained the details of the BRFplus application and its components. It covered in detail the various data object types, expressions, actions, functions, ruleset, and rules within a BRFplus application. These together are held in a BRFplus application acting as a container. The BRFplus function acts as the central feature of the BRFplus application and binds all the other objects together in a logical construct. Function has tools such as simulation and trace that help users check for correctness of the BRFplus logic. Other features include automatic code generation or creation of function modules and web services that help in binding the BRFplus application host program, which can be either ABAP-based or non-ABAP systems. Finally, we focused on comparing the standard output management using BRFplus with that available in the traditional use of table NAST. Due to the complex output requirements of a typical third-party logistics integration scenario, the decision seemed to favor the traditional approach.

Chapter 12
Monitoring and Control

Interface monitoring and error handling involves monitoring and tracking of messages exchanged through the interfaces. It also empowers business users to monitor and troubleshoot business-related processing issues promptly and without technical intervention. SAP S/4HANA provides multiple options such as IDoc and electronic data interchange (EDI) monitors, along with the Application Interface Framework (AIF).

Monitoring and resolution of interface message issues always has been a challenge and consumes substantial time and effort. The struggle for timely resolution of tickets and effectively managing escalations is real. This chapter describes some of the latest monitoring features available in SAP S/4HANA and how they facilitate user interaction in tracking and error resolution.

12.1 IDoc Monitoring and Dashboard

Usually, business users face multiple issues while monitoring IDocs, which can be classified into multiple segments:

- **Business process areas**
  There may be multiple interfaces specific to process areas, for example, order-to-cash, procure-to-pay, and logistics. Grouping and monitoring the IDocs based on process area can help businesses with the assignment of failed IDocs to respective teams leading to faster resolution.
- **IDoc statistics**
  Businesses must analyze how many IDocs have hung, failed, and been successfully processed, but doing so manually can be quite taxing.
- **IDoc classification by error code**
  Sometimes, the interface design necessitates classifying the failed IDocs based on their status. They may need to route the IDocs further based on status codes for the next possible action. Other possibilities are to check on the details of failed IDocs based on error types, error description, status, and value. This can give a clear picture to the IT user monitoring the IDocs.

- **Monitor critical or recurring IDocs**
  Another area of interest is to keep an eye on repeatedly failing IDocs and their criticality or priority.

Standard SAP ERP provides multiple options to monitor and analyze IDocs through Transaction WEDI with a menu option for all IDoc-relevant transactions. All these have been retained in SAP S/4HANA as well. Figure 12.1 shows the standard menu options available using Transaction WEDI.

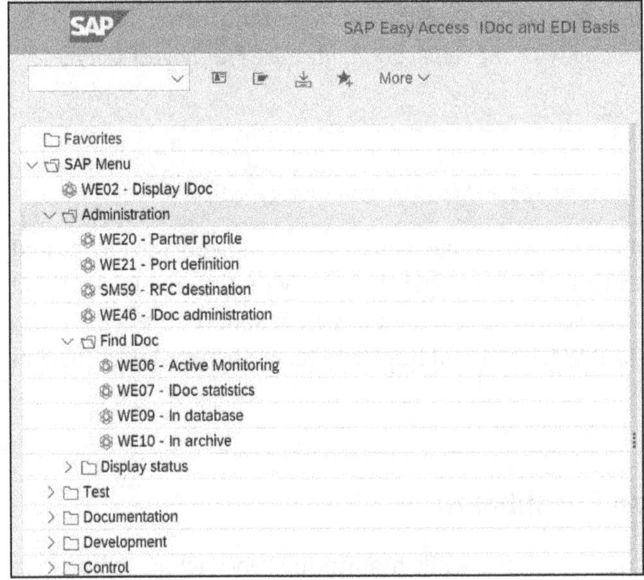

Figure 12.1 Standard IDoc Monitoring Tools in SAP ERP and SAP S/4HANA (Transaction WEDI)

Table 12.1 lists a few standard monitoring and testing tools used by functional consultants.

Transaction	Description
WE02	Display IDoc
WE05	IDoc Lists
WE07	IDoc Statistics
WE11	Delete IDocs
WE19	Test Tool

Table 12.1 Standard SAP Transactions Used for IDoc Monitoring

In addition to these transactions, SAP S/4HANA also offers improvised Transaction WLF_IDoc (SAP Fiori app called IDoc Processing) that is more user-friendly and provides a cockpit view of IDocs. As highlighted in Figure 12.2, several selection screen options are available under various tabs. The **Criteria for data record** option can even be used to search for specific data within segments (as is possible in the traditional Transaction WE09 approach). After the selection is executed, the resulting screen presents options to view and process data per the requirements.

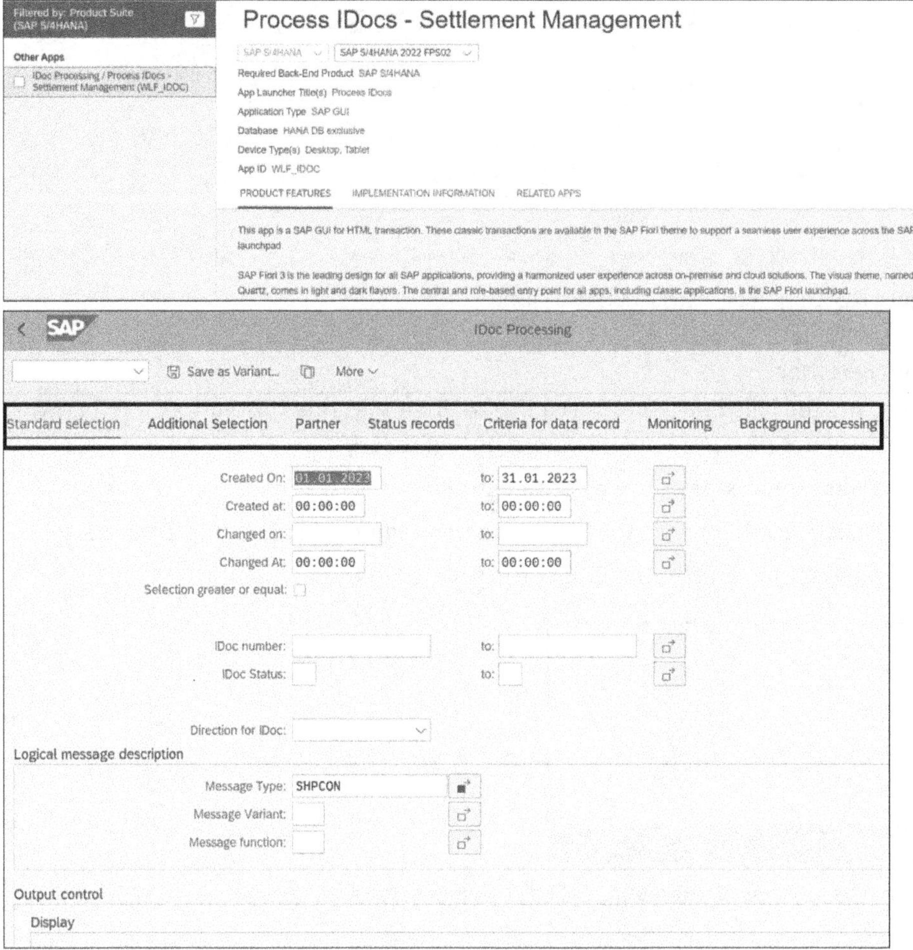

Figure 12.2 SAP Fiori App and Selection Screen View of Transaction WLF_IDoc

Figure 12.3 shows a view of the output from Transaction WLF_IDoc.

As highlighted, there are multiple options to display and process IDocs further:

❶ **Display IDoc**
Display IDoc details.

❷ **Edit IDoc**
Edit IDoc segment-specific data.

IDoc number	Number	Status	Stat. Gr.	Links	Log	Te Date created	Time creat..	Changed On	Changed At	Description	Direct..	Message Type	Msg. Var.	Message function
15471666	218	03				02.01.2023	00:00:17	02.01.2023	00:01:02	Data passed to port OK		SHPORD		
15471668	278	03				02.01.2023	00:00:24	02.01.2023	00:01:02	Data passed to port OK		SHPORD		
15471669	160	03				02.01.2023	00:00:28	02.01.2023	00:01:02	Data passed to port OK		SHPORD		
15477136	60	03				02.01.2023	12:00:25	02.01.2023	12:01:02	Data passed to port OK		SHPORD		
15477415	88	03				03.01.2023	00:00:09	03.01.2023	00:01:03	Data passed to port OK		SHPORD		
15477416	88	03				03.01.2023	00:00:11	03.01.2023	00:01:03	Data passed to port OK		SHPORD		
15477417	116	03				03.01.2023	00:00:13	03.01.2023	00:01:03	Data passed to port OK		SHPORD		
15483126	31	16				03.01.2023	00:00:17	03.01.2023	00:01:52	Functional acknowledgment positive		SHPORD		
15483127	63	16				03.01.2023	00:00:23	04.01.2023	01:16:11	Functional acknowledgment positive		SHPORD		
15483128	28	16				03.01.2023	00:00:24	03.01.2023	00:01:52	Functional acknowledgment positive		SHPORD		
15487149	294	03				03.01.2023	09:00:11	03.01.2023	09:01:02	Data passed to port OK		SHPORD		
15487150	214	03				03.01.2023	09:00:14	03.01.2023	09:01:02	Data passed to port OK		SHPORD		
15489327	31	16				03.01.2023	15:00:22	03.01.2023	15:09:44	Functional acknowledgment positive		SHPORD		
15489328	31	16				03.01.2023	15:00:23	03.01.2023	15:09:44	Functional acknowledgment positive		SHPORD		
15489329	31	16				03.01.2023	15:00:25	03.01.2023	15:09:44	Functional acknowledgment positive		SHPORD		
15489330	31	16				03.01.2023	15:00:27	03.01.2023	15:09:44	Functional acknowledgment positive		SHPORD		
15489944	85	03				04.01.2023	00:00:13	04.01.2023	00:01:02	Data passed to port OK		SHPORD		
15489951	85	03				04.01.2023	00:00:16	04.01.2023	00:01:02	Data passed to port OK		SHPORD		
15489955	82	03				04.01.2023	00:00:19	04.01.2023	00:01:02	Data passed to port OK		SHPORD		
15489957	120	03				04.01.2023	00:00:22	04.01.2023	00:01:02	Data passed to port OK		SHPORD		
15489989	90	03				04.01.2023	00:00:25	04.01.2023	00:01:02	Data passed to port OK		SHPORD		

Figure 12.3 Output of Transaction WLF_IDoc

❸ **Process IDoc**

Process the failed or unprocessed IDocs. Upon selecting this button, the following three options are presented, as shown in Figure 12.4:

- **Online** (process the IDoc in the foreground),
- **Dialog (BD87)** (process the IDoc using the traditional Transaction BD87 route),
- **Background** (process the IDoc in the background).

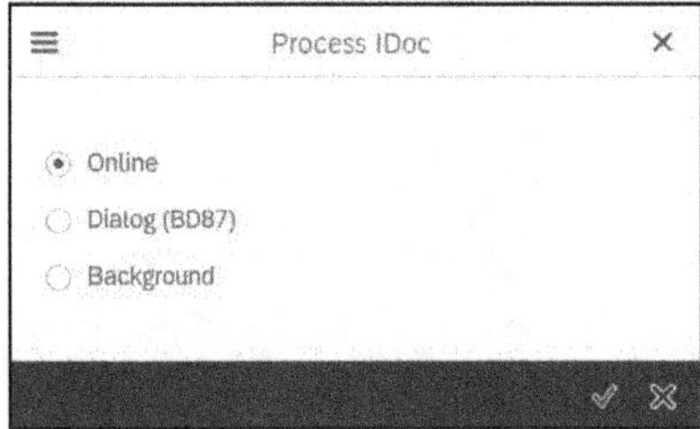

Figure 12.4 Options for Processing Failed or Unprocessed IDocs

❹ **Send IDoc via RFC**

Send the IDoc via Remote Function Call (RFC) to a remote SAP/ABAP-based system.

❺ **Compare IDoc**

Compare two IDocs that appear the same, but one has failed. If the status message for the failed IDoc isn't enough to identify the error, you can list the differences between a successful IDoc and the unsuccessful one and compare each segment to find the issue.

There are other features such as displaying the count of IDocs in a summary or tree view, sorting and filtering, downloading, and mailing.

Regarding third-party logistics integration, the message types can be grouped by process area with business process–specific parameters such as sales area for order-to-cash, plant and storage location for logistics, and purchase organization and/or purchasing groups for procure-to-pay. Standard tools such as Transactions WE02 or WE09 can be copied and modified to custom transactions with process-based selection options and screen variants.

In addition, SAP has constantly endeavored to build a monitoring tool based on business content and from on-premise SAP S/4HANA 1511 onward, additional monitoring tools have been introduced, as follows:

- Transaction EMASN: IDoc Monitor for Inbound Ship. Notification
- Transaction EMFOR: IDoc Monitor for Forecast/JIT Delivery Schedule
- Transaction EMINV: IDoc Monitor for Incoming Invoices
- Transaction EMORD: IDoc Monitor for SD Orders
- Transaction EMTRA: IDoc Monitor for Inbound Transports

These monitoring transactions are specific to process areas. Figure 12.5 ❶ shows selection screen options for Transaction EMINV. You may observe that it provides selection options to input business data ❷, and you can view the document status ❸.

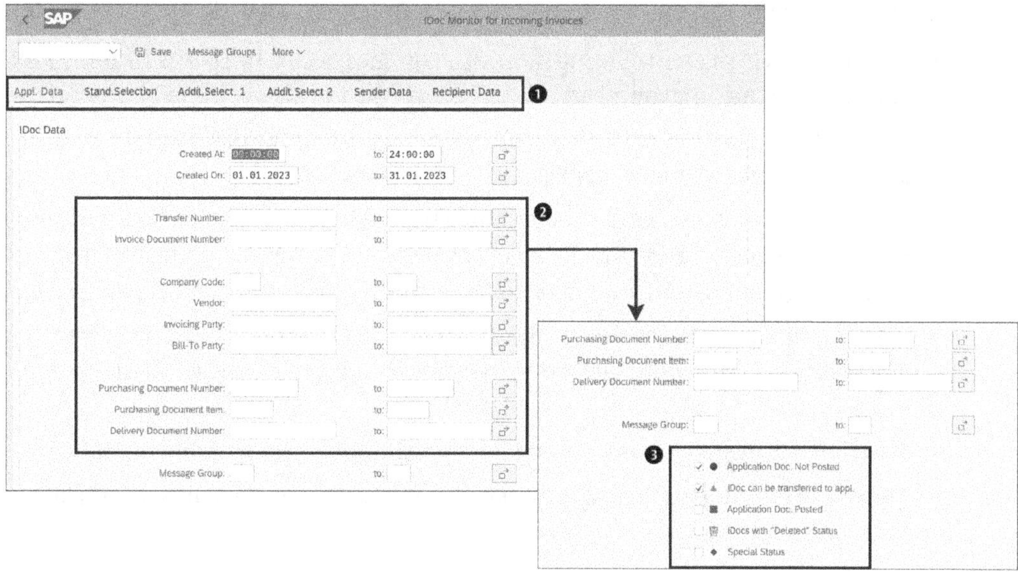

Figure 12.5 Selection Screen for Transaction EMINV

Figure 12.6 shows the output of Transaction EMINV. It provides options to view the details of IDocs, process IDocs, and so on.

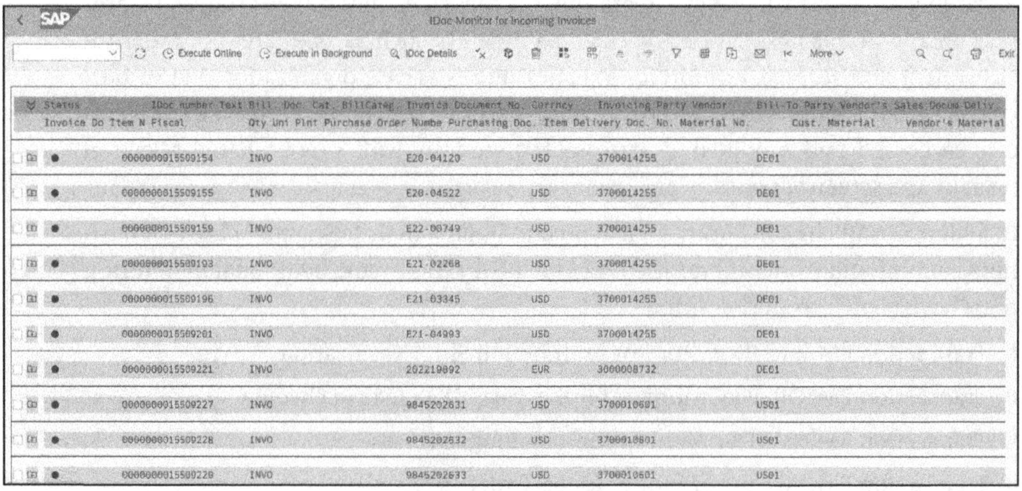

Figure 12.6 Output Screen of Transaction EMINV

Similar to Transaction EMINV, the IDoc monitor for sales orders, Transaction EMORD, provides selection screen options to allow inputs by functional area that enable segregation of responsibilities between sales and shipping teams.

As shown in Figure 12.7, the new monitor for sales orders has a few significant upgrades:

- Messages can be searched by business contents such as sold-to party, ship-to party, and customer material number.
- Standard organization entities such as sales organizations, distribution channels, and sales offices are now available as part of the selection.
- Multiple types of messages can be searched together. As an example, for a customer, the orders, requests, and order changes can all be pulled up with a single search, just by inputting the customer number and selecting the type of messages to retrieve.
- It's not mandatory to remember the basic types and message types. However, the **Std. selection** tab provides the option to search based on message and basic types for IDocs.
- Selections can be made based on the status of the document (**Document Status**).
- Additional selection options let you specify message-specific details such as partner function and so on.

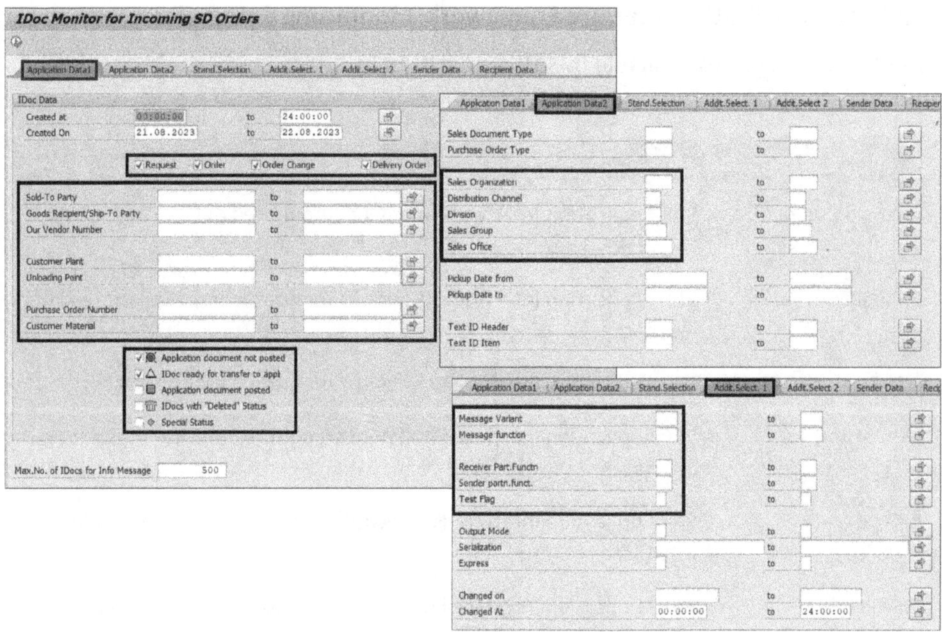

Figure 12.7 Selection Screen Options for Transaction EMORD

12.2 SAP Application Interface Framework

SAP Application Interface Framework enables you to easily develop, run, and monitor all the application interfaces your business may need. AIF is an integral part of SAP S/4HANA and SAP S/4HANA Cloud Edition for interface development, error and monitoring handling. Additionally, it's also offered as an add-on for the SAP Business Suite.

While SAP Application Interface Framework has a local monitoring tool integrated, the messages sent through SAP Application Interface Framework can be monitored also with SAP Focused Run to provide a centralized monitoring for all interface types and all systems in the SAP landscape.

SAP Application Interface Framework is the standard monitoring and interface hub for SAP S/4HANA and the SAP S/4HANA Cloud applications. SAP Application Interface Framework is now part of the SAP S/4HANA core component and no longer needs to be installed as an add-on to the system.

Technical Prerequisite

For monitoring of SAP Application Interface Framework messages, the following technical installations are needed:

- FRUN 2.0 Feature Pack 01
- SAP S/4HANA running on SAP Basis 7.54 and SAP_ABA 75E or higher

12.2.1 SAP Application Interface Framework: Monitoring Categories

Monitoring categories available in SAP Application Interface Framework are shown in Table 12.2.

Category	Description
AIF-bgRFC(queue)	SAP Application Interface Framework monitored queued background RFC (bgRFC) interfaces
AIF-bgRFC(trans)	SAP Application Interface Framework monitored transactions bgRFC interfaces
AIF-IDoc	SAP Application Interface Framework monitored IDoc interfaces
AIF-PI-runtime	SAP Application Interface Framework monitored process integration interfaces
AIF-qRFC	SAP Application Interface Framework monitored queued RFC (qRFC) interfaces
AIF-tRFC	SAP Application Interface Framework monitored transactional RFC (tRFC) interfaces
AIF-SYNC_LOG	SAP Application Interface Framework monitored Sync Log interfaces
AIF-XML	SAP Application Interface Framework monitored XML interfaces
AIF-WS-runtime	SAP Application Interface Framework monitored web service interfaces

Table 12.2 Monitoring Categories Available in SAP Application Interface Framework

Note

Transaction /N/AIF/ERR is used to monitor currently running interfaces and messages in the SAP S/4HANA system. The SAP Application Interface Framework setup is done in Transaction /N/AIF/IFMON.

Figure 12.8 shows a bird's-eye view of the system architecture involving SAP S/4HANA, the SAP Application Interface Framework add-on, Cloud Integration, and SAP Process Orchestration integration.

SAP Application Interface Framework menus can be accessed using Transaction /N/AIF/. Alternatively, In the **SAP Easy Access** menu, choose **Cross Application Components • SAP Application Interface Framework**. Figure 12.9 shows the **SAP Easy Access** menu available under **SAP Application Interface Framework**.

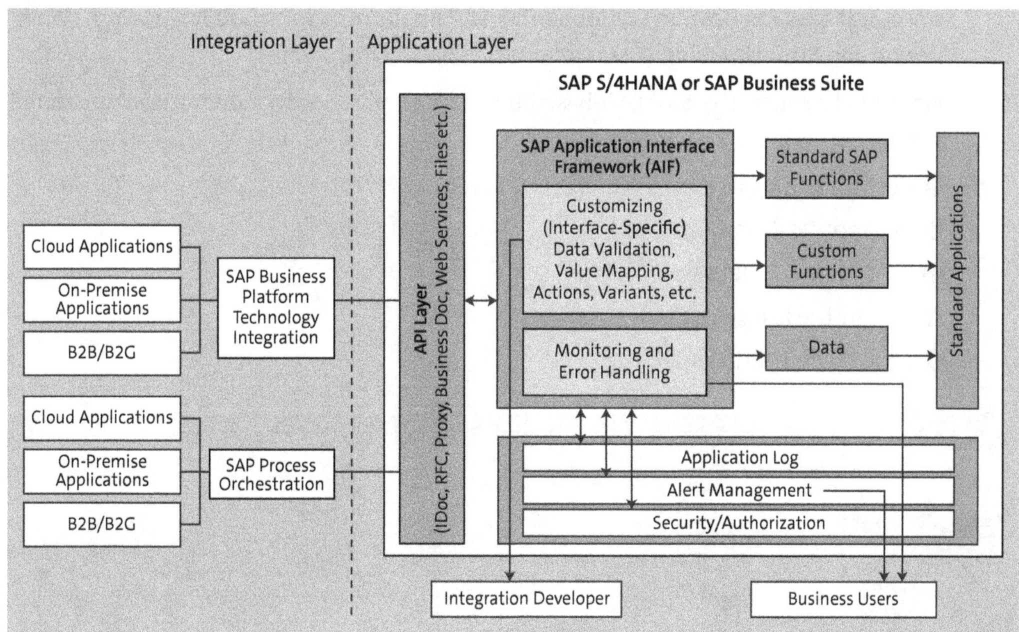

Figure 12.8 System Architecture Involving AIF

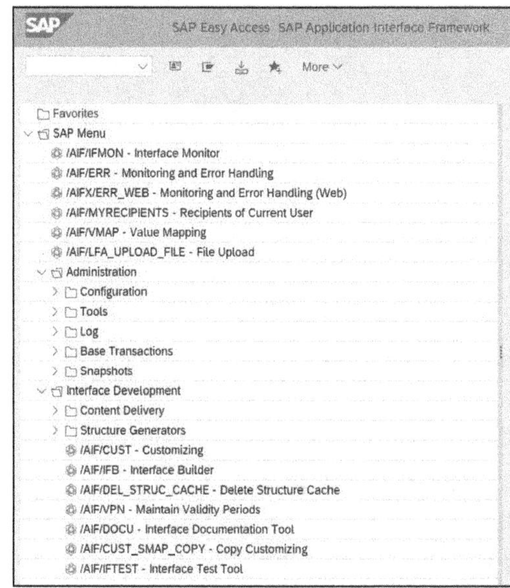

Figure 12.9 SAP Easy Access Menu for SAP Application Interface Framework

12.2.2 Monitoring and Error Handling

Both monitors based on SAP GUI and Web UI are available in SAP Application Interface Framework. SAP GUI-based UIs can be opened by using the **SAP Easy Access** menu and selecting **Cross Application Components • SAP Application Interface Framework • Monitoring and Error Handling** (Transaction /n/AIF/IFMON).

Figure 12.10 shows the SAP GUI-based interface monitor with summarized and detail views:

❶ Custom namespace for IDoc (e.g., DESADEV, DELVRY07)

❷ Mail notification sent

❸ IDoc message (date-wise)

❹ Restart the IDoc (if failed)

❺ Message indicating status

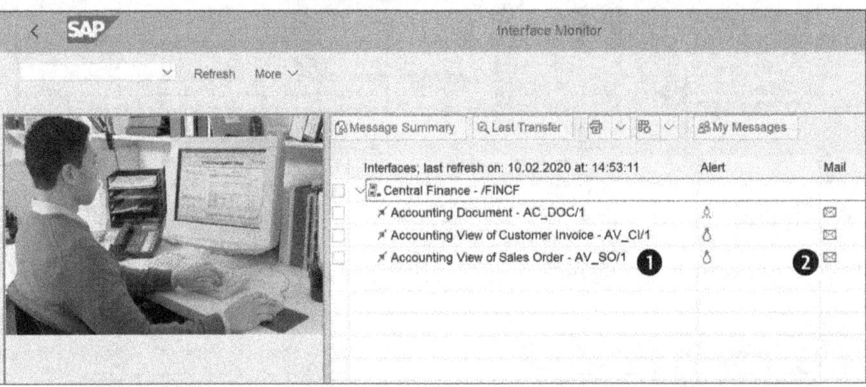

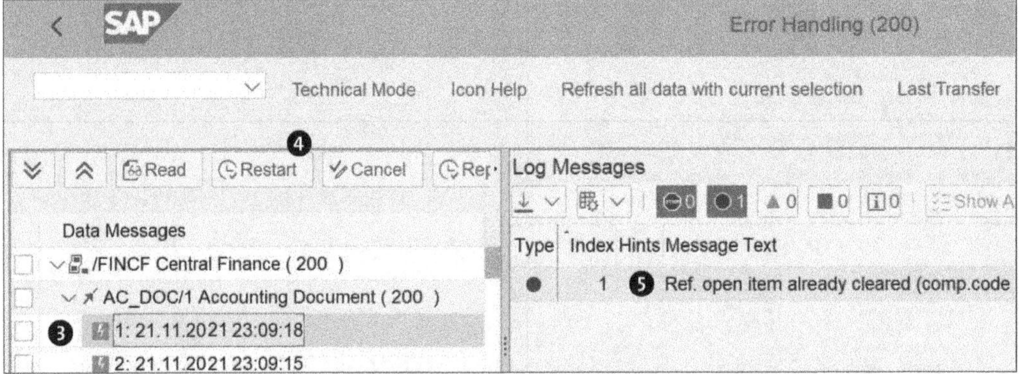

Figure 12.10 SAP GUI Interface Monitor in SAP Application Interface Framework

The web UI of the interface monitor can be accessed by using the **SAP Easy Access** menu and selecting **Cross Application Components • SAP Application Interface Framework • Monitoring and Error Handling (Web)** (Transaction /N/AIFX/ERR_WEB).

Figure 12.11 shows the web-based UI of the interface monitor with summarized and detail views.

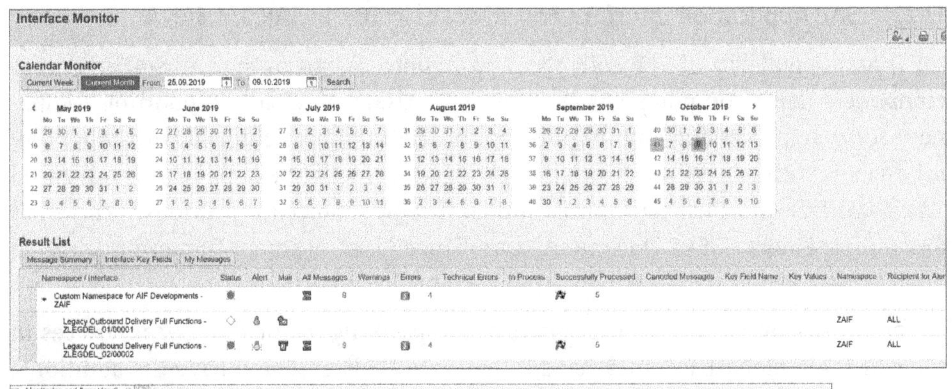

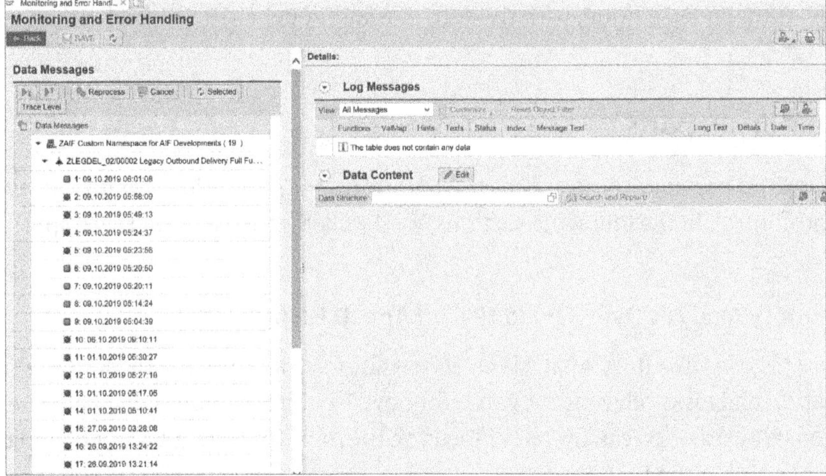

Figure 12.11 Web-Based UI of Interface Monitor in AIF

12.2.3 Customizing SAP Application Interface Framework

Use Transaction /N/AIF/CUST to customize the application programming interface (API) framework. Only Customizing for AIF is displayed. In the **SAP Easy Access** menu, choose **Cross Application Components • SAP Application Interface Framework • Administration • Configuration • Customizing.**

Figure 12.12 shows the Customizing menu options available in AIF.

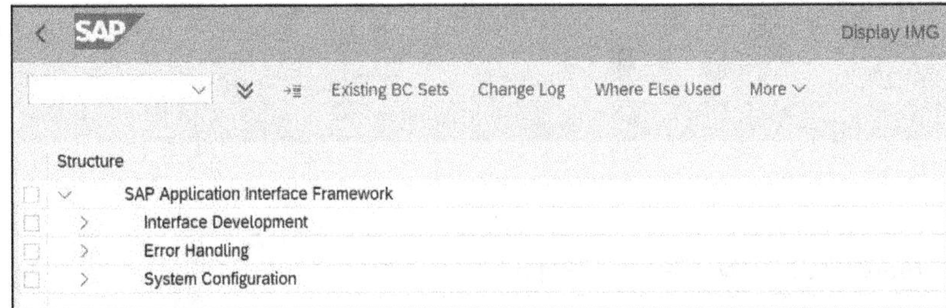

Figure 12.12 Customizing Menu in SAP Application Interface Framework

12.2.4 SAP Application Interface Framework License in SAP S/4HANA

AIF is an integral part of SAP S/4HANA and shipped as part of the SAP S/4HANA core component (Basis component S4CORE/SAPSCORE). Usage of AIF as a supporting component for only processing and monitoring the AIF interfaces shipped with other SAP products is included in the respective product license, for example, SAP Advanced Track and Trace for Pharmaceuticals, SAP S/4HANA Finance (or previously known as SAP Simple Finance), SAP Document Compliance (or eDocument framework), and so on.

Beyond this scope, that is, if customers develop, configure, or monitor interfaces in their own AIF namespace, they need to have an AIF license and have the additional product AIF 1.0, extended edition, with the software component AIFGEN installed.

12.3 Inventory Reports

SAP S/4HANA 2022 provides some very useful SAP Fiori apps as solutions for inventory analytics and control. In the following sections, we'll walk through each of these apps.

12.3.1 Inventory Analysis Overview (SAP Fiori App ID F3366)

The overview page in Inventory Analysis Overview shows at a glance the most important information and tasks relevant for you right now. The information is displayed on a set of actionable cards. You can therefore focus on the most important tasks, enabling faster decisions and immediate action.

Figure 12.13 provides a snapshot of an SAP Fiori report on inventory analysis based on a group of actionable metrics.

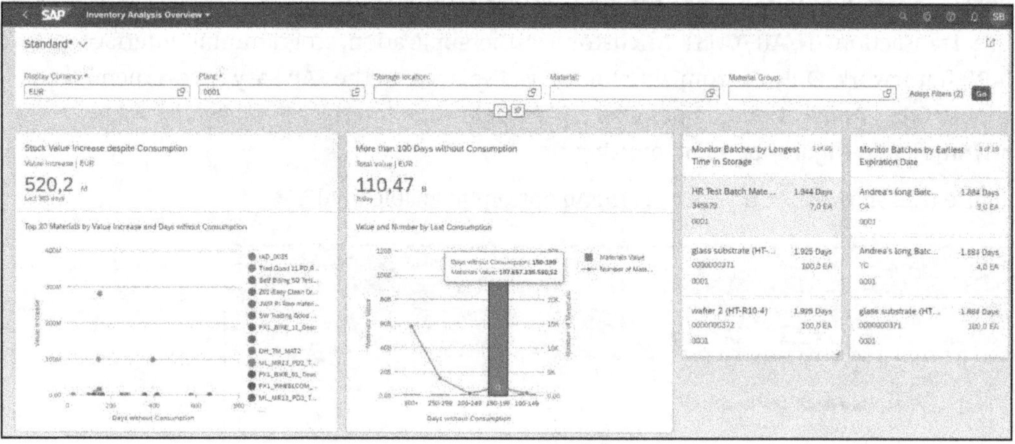

Figure 12.13 Inventory Analysis Overview App Report

The cards or blocks show you the most important information ranked according to relevance. For example, on the **Stock Value Increase despite Consumption** card, you can monitor, at a glance, the top 20 materials that had the highest stock value increase despite consumption in the past 365 days. You can get more information about individual materials by clicking on the corresponding dots in the chart.

Depending on the specified filter criteria, you can see all unexpected stock value increases, monitor batches over time, or recognize slow-moving materials. For graphical or analytical cards, selecting the card takes you to more detailed analytical information.

12.3.2 Inventory KPI Analysis (SAP Fiori App ID F3749)

This app can be used to monitor inventory key performance indicators (KPIs) effectively to ensure forecast and inventory accuracy in a timely manner and visualize this in a meaningful way. This enables identification of critical KPIs to monitor the inventory flow or investigate potential problems. 12

As an example, stock values for a time period can be compared with the previous period to identify potential outliers that could impact the fixed capital situation. This also provides the option for filtering on these outliers and displaying only these particular stocks. The app offers various KPI filters that display potential outliers in a frequency distribution chart (see Figure 12.14).

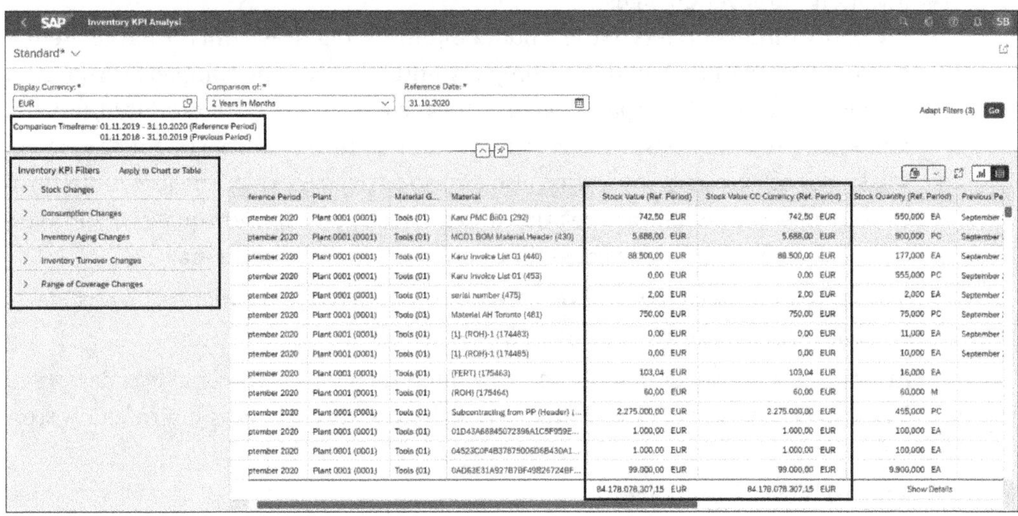

Figure 12.14 Inventory KPI Analysis App Report

The app supports five KPIs and calculates them based on each material stock (stock-identifying fields) in the selected time period:

- **Stock Changes**
  Displays the stock value changes of the filtered materials with the current value in the company code currency for time periods.
- **Consumption Changes**
  Displays the consumption changes of the materials in the company code currency for time periods. The consumption value is the result of the consumption multiplied by the current price.
- **Inventory Aging Changes**
  Inventory aging measures the average number of days inventory sits unsold in the warehouse. An aging inventory report is used to identify slower-moving SKUs and build strategies to increase the inventory turnover ratio and reduce dead stock. The inventory aging changes are defined as a percentage relationship between average stock quantity and consumption quantity in the selected time period:

  Inventory aging changes (%) = (Average stock quantity in period ÷ Consumption quantity in period) × 100

  Average stock over time period = (Sum of stock quantities ÷ Number of days)

  For this KPI, note that inventory aging includes only goods issues that are marked as consumption. Any reversal postings aren't considered.
- **Inventory Turnover Changes**
  Inventory turnover refers to the amount of time that passes from the day an item is purchased or manufactured by a company until it's sold. One complete turnover of inventory means the company sold the stock that it purchased or produced, less any items lost to damage or shrinkage. A low inventory turnover ratio might be a sign of weak sales or excessive inventory, also known as overstocking. The inventory turnover changes are defined as a relationship between consumption quantity and average stock quantity in the selected time period (quarters, months, days):

  Inventory turnover changes (no. of time units) = (Consumption quantity in period ÷ Average stock quantity in period)

  An inventory turnover determines how often the average material was destocked and restocked in a specific time frame. A high inventory turnover indicates low storage costs.
- **Range of Coverage Changes**
  Stock coverage is a metric that shows the period—usually expressed in days—during which a company can meet customer demand with the available inventory in its warehouse. For measuring this KPI, you divide the amount of stock stored in your facility by your average sales for a given period. The range of coverage changes

is defined as a relationship between the current stock quantity and the average consumption in the selected time period:

Range of coverage changes (days) = (Current stock quantity ÷ Average consumption)

Average consumption = (Consumption in period ÷ Number of days in period)

The range of coverage provides information at the stock level in relation to demand. It indicates how long a stock quantity will last given a specific average daily requirement.

12.3.3 Analyze Stock in Date Range (SAP Fiori App ID F6185)

This app supports analysis and monitoring of goods movement postings for one or more materials in a defined date range. The app displays stock quantities and values on the defined analytical start and end date. Figure 12.15 shows analysis of stock movements and differences within a start and finish date. The detail page (not shown in this figure) provides a comprehensive analysis of all goods movements that happened within the defined date range.

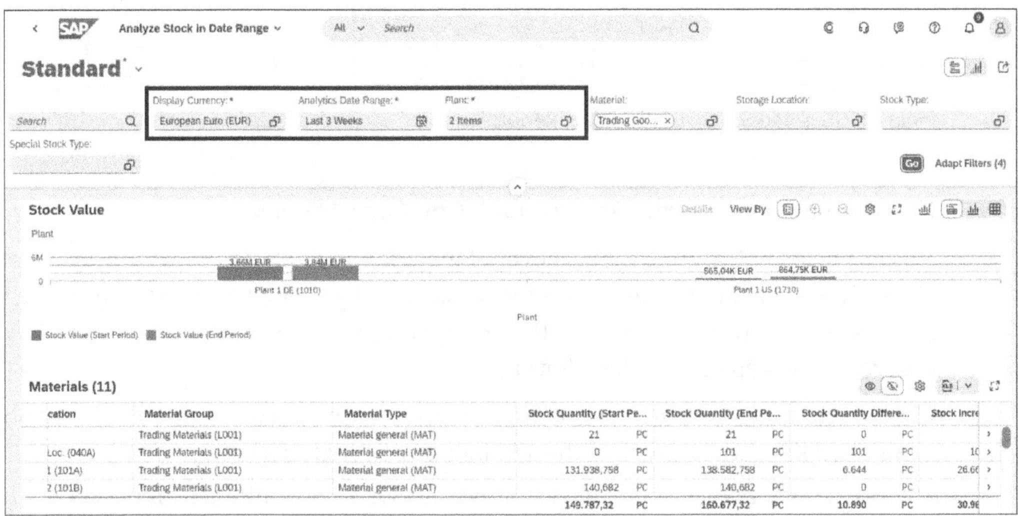

Figure 12.15 Analyze Stock in Date Range App Report

12.3.4 Dead Stock Analysis (SAP Fiori App ID F2899)

This app supports monitoring and enabling time-dependent investigations of dead stock. Dead stock represents a combination of locations and materials where the stock value has increased over a defined analysis period despite existing consumption postings within that period.

The analysis of the dead stock identifies materials that could potentially have a negative impact on the company's cash flow and profitability. Based on these results, follow-on activities can be decided such as discontinuance, write-off, or transferring to a different location. Figure 12.16 provides a snapshot of the Dead Stock Analysis app for a selected time frame.

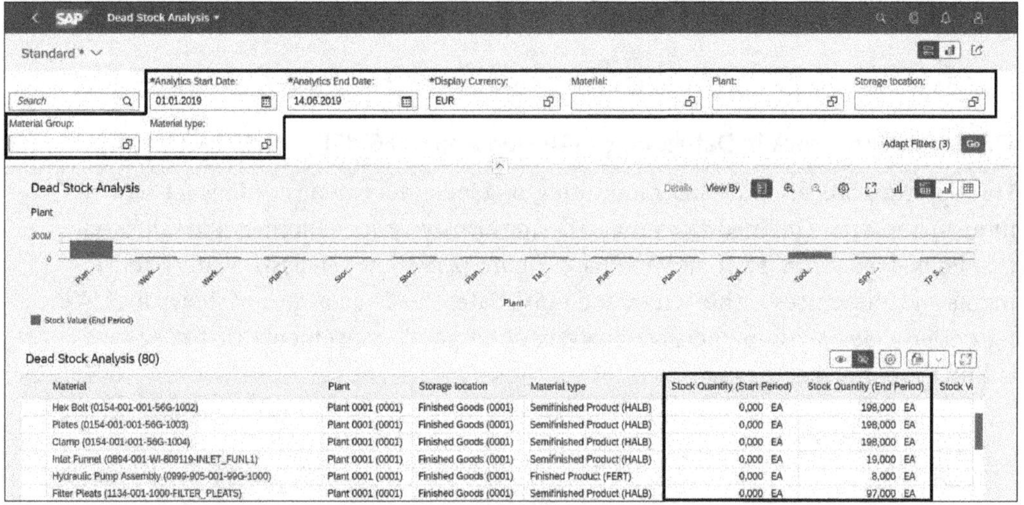

Figure 12.16 Dead Stock Analysis App Report

12.3.5 Inventory Turnover Analysis (SAP Fiori App ID F1956)

This app helps is analyzing the turnover of materials in a selected plant. The app provides you with detailed information about materials that could have turnover-related issues according to turnover definition, depending on the time period selected. Key features of this app include the following:

- Research turnover-related problems down to a single material of interest, based on different filter criteria such as the ABC indicator or material type.
- Identify materials for a detailed turnover analysis using different representation forms, for example, scatter or bubble charts.
- Make use of personalized settings to aid the analysis. Default filters can be defined for the following fields:
  - **Plant**
  - **Material**
  - **Material Group**
  - **Material Type**

Figure 12.17 shows the Inventory Turnover Analysis app report for a selected time frame.

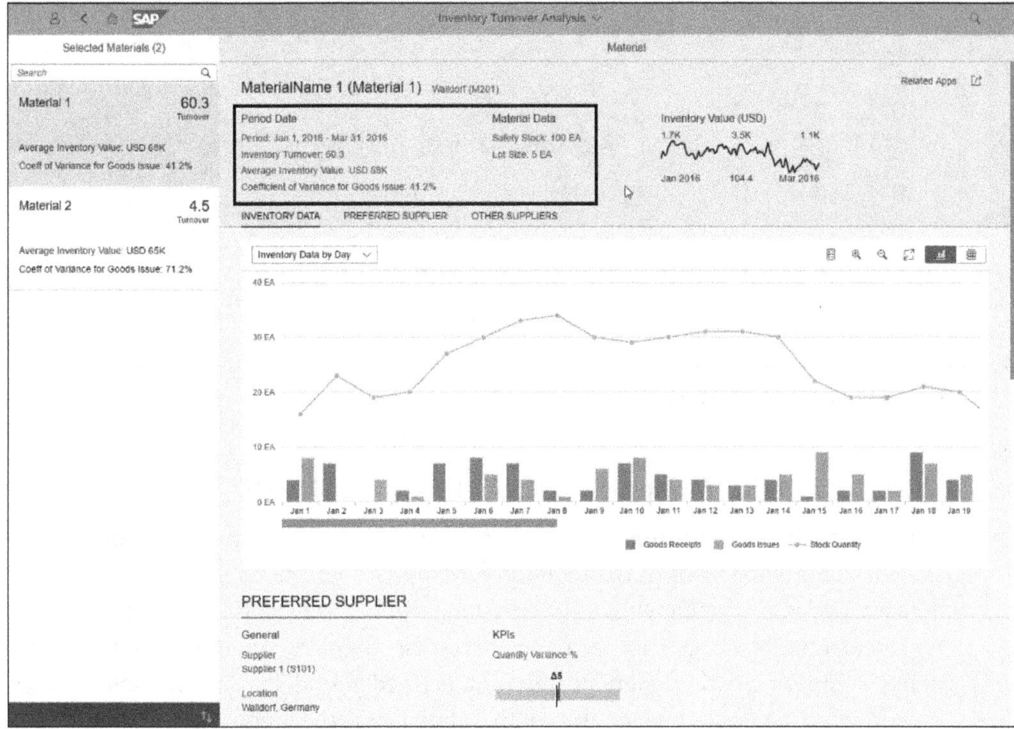

Figure 12.17 Inventory Turnover Analysis App Report

12.3.6 Goods Movement Analysis (SAP Fiori App ID W0055)

With this app, you can analyze the goods movements in your company. The Goods Movement Analysis app has the following features:

- Drill down to the individual material document item.
- Define filters to narrow down the scope of search per the need.
- Choose from many different available dimensions and key figures.

The app supports technical features and options such as **Sort**, **Totals and Subtotals**, and **Export to Excel**.

As shown in Figure 12.18, the Goods Movement Analysis app report captures goods movements such as receipts and issues, and the same can be seen in a graphical format and can be drilled down to reach the individual material level.

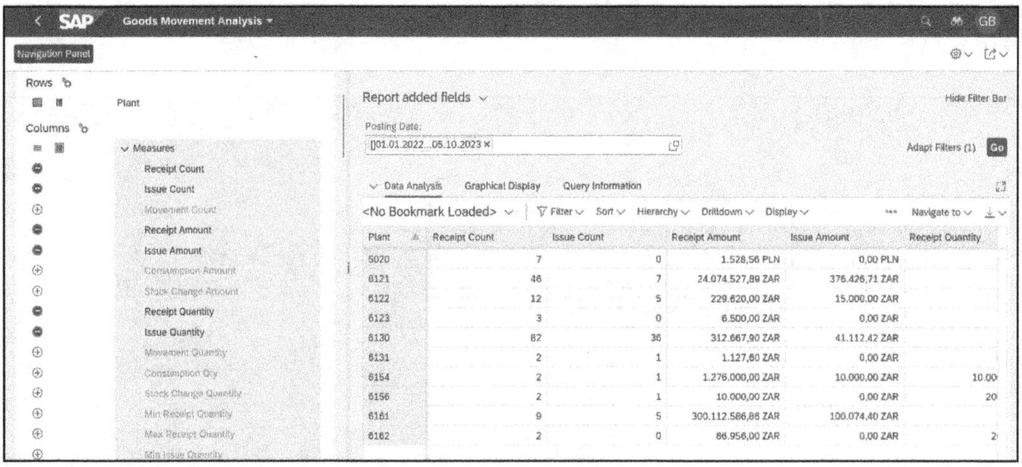

Figure 12.18 Goods Movement Analysis App Report

12.3.7 Slow or Non-Moving Materials (SAP Fiori App ID F2137)

This app enables monitoring and performing time-dependent investigations of the slow or nonmoving SKUs in the inventory. Based on these results, immediate follow-on activities (e.g., scrapping or stock transfers) can be decided. Nonmoving materials represent a combination of locations and materials where the stock level isn't zero, and no consumption postings were posted for a defined period. It's also possible to track slow-moving materials. Key features of this app include the following:

- Monitor slow or nonmoving materials.
- Monitor how much capital is tied up by the slow-moving items, for example, by checking the total value in the chosen display currency.
- Monitor and compare materials with low consumption based on the consumption-to-stock ratio with the **Slow-Moving Indicator**. Therefore, identification and comparison of critical materials based on individual criteria can be done.
- Monitor materials without consumption (**Slow-Moving Indicator** = **0**) based on analysis for an individual time period.
- Define follow-up actions such as scrapping or stock transfers, or inform concerned teams about the stock situation.

Figure 12.19 provides a snapshot of the Slow or Non-Moving Materials app report.

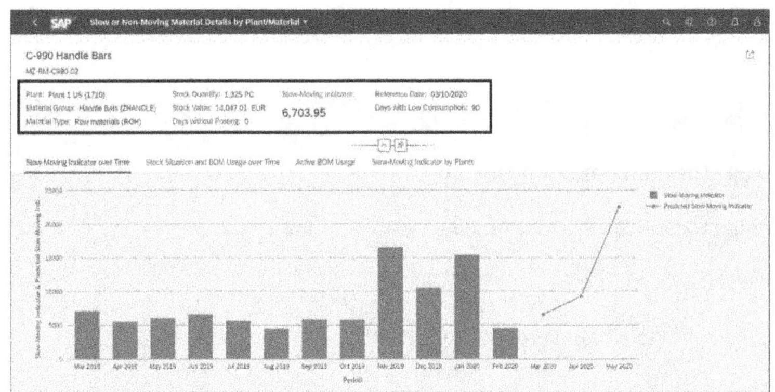

Figure 12.19 Slow or Non-Moving Materials App Report

12.3.8 Custom Inventory Reports

In a third-party logistics integration scenario, there could be requirements to capture inventory snapshots or compare inventory changes over time. Such requirements may require developing a custom report with the following options:

- GUI-based reports in SAP S/4HANA
- Business intelligence (BI) reports that extract data from the SAP HANA database

Figure 12.20 shows a sample SAP GUI report for inventory comparison of third-party logistics stocks.

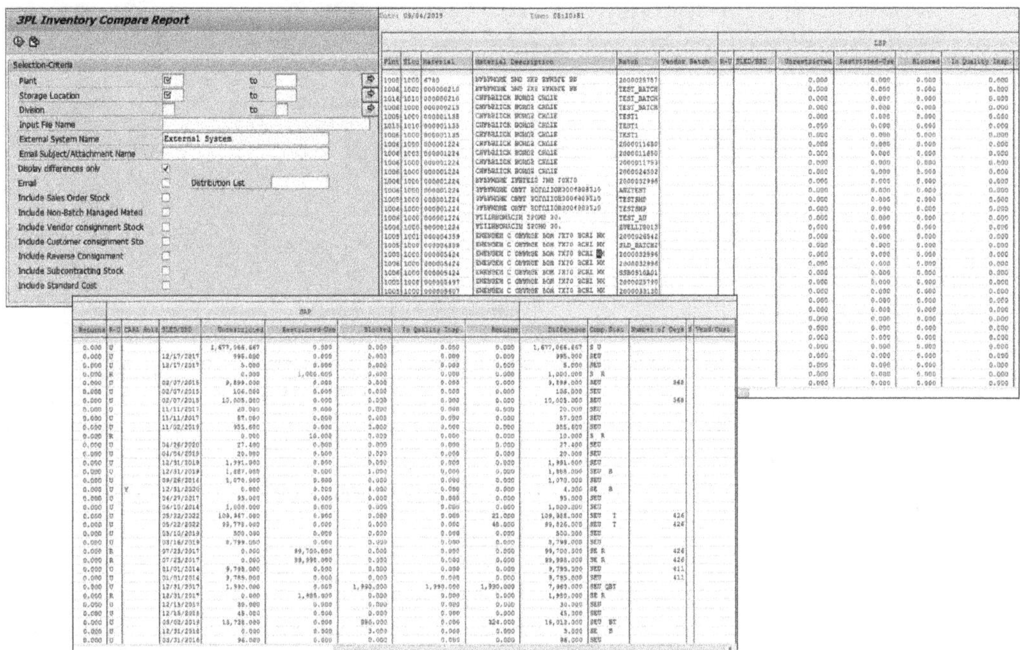

Figure 12.20 SAP GUI Inventory Comparison Report for Third-Party Logistics Inventory

Relevant standard SAP tables for normal and batch-specific inventory data are listed in Table 12.3.

SAP Table	Description
MARD	Storage Location Data for Material
MARD_LGORT	Storage Locations with Maintenance Status for Input Help
MARC	Plant Data for Material
MCH1	Batches (if Batch Management Cross-Plant)
MCHA	Batches
MCHB	Batch Stocks
MCHB_CHARG	Batch View
MCHBH	Batch Stocks: History

Table 12.3 Standard SAP Tables Containing Details of Normal and Batch-Specific Inventory

Figure 12.21 shows a sample BI report prepared using SAP S/4HANA standard extractors for stock data along with custom extractors to fetch data from custom tables containing logistics service provider data.

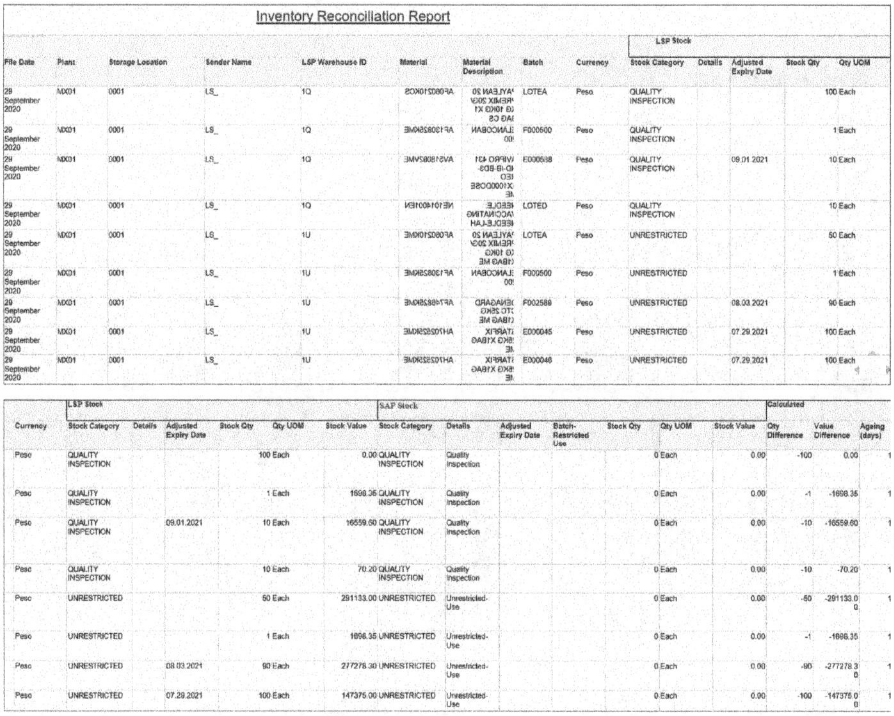

Inventory Reconciliation Report

									LSP Stock				
File Date	Plant	Storage Location	Sender Name	LSP Warehouse ID	Material	Material Description	Batch	Currency	Stock Category	Details	Adjusted Expiry Date	Stock Qty	Qty UOM
29 September 2020	MX01	0001	LS_	1Q	[illegible]	[illegible]	LOTEA	Peso	QUALITY INSPECTION			100	Each
29 September 2020	MX01	0001	LS_	1Q	[illegible]	[illegible]	F000500	Peso	QUALITY INSPECTION			1	Each
29 September 2020	MX01	0001	LS_	1Q	[illegible]	[illegible]	E000588	Peso	QUALITY INSPECTION		09.01.2021	10	Each
29 September 2020	MX01	0001	LS_	1Q	[illegible]	[illegible]	LOTED	Peso	QUALITY INSPECTION			10	Each
29 September 2020	MX01	0001	LS_	1U	[illegible]	[illegible]	LOTEA	Peso	UNRESTRICTED			50	Each
29 September 2020	MX01	0001	LS_	1U	[illegible]	[illegible]	F000500	Peso	UNRESTRICTED			1	Each
29 September 2020	MX01	0001	LS_	1U	[illegible]	[illegible]	F002588	Peso	UNRESTRICTED		08.03.2021	90	Each
29 September 2020	MX01	0001	LS_	1U	[illegible]	[illegible]	E000045	Peso	UNRESTRICTED		07.29.2021	100	Each
29 September 2020	MX01	0001	LS_	1U	[illegible]	[illegible]	E000046	Peso	UNRESTRICTED		07.29.2021	100	Each

	LSP Stock						SAP Stock							Calculated		
Currency	Stock Category	Details	Adjusted Expiry Date	Stock Qty	Qty UOM	Stock Value	Stock Category	Details	Adjusted Expiry Date	Batch-Restricted Use	Stock Qty	Qty UOM	Stock Value	Qty Difference	Value Difference	Ageing (days)
Peso	QUALITY INSPECTION			100	Each	0.00	QUALITY INSPECTION	Quality Inspection			0	Each	0.00	-100	0.00	1
Peso	QUALITY INSPECTION			1	Each	1698.35	QUALITY INSPECTION	Quality Inspection			0	Each	0.00	-1	-1698.35	1
Peso	QUALITY INSPECTION		09.01.2021	10	Each	16559.60	QUALITY INSPECTION	Quality Inspection			0	Each	0.00	-10	-16559.60	1
Peso	QUALITY INSPECTION			10	Each	70.20	QUALITY INSPECTION	Quality Inspection			0	Each	0.00	-10	-70.20	1
Peso	UNRESTRICTED			50	Each	291133.00	UNRESTRICTED	Unrestricted-Use			0	Each	0.00	-50	-291133.00	1
Peso	UNRESTRICTED			1	Each	1698.35	UNRESTRICTED	Unrestricted-Use			0	Each	0.00	-1	-1698.35	1
Peso	UNRESTRICTED		08.03.2021	90	Each	277278.30	UNRESTRICTED	Unrestricted-Use			0	Each	0.00	-90	-277278.30	1
Peso	UNRESTRICTED		07.29.2021	100	Each	147375.00	UNRESTRICTED	Unrestricted-Use			0	Each	0.00	-100	-147375.00	1

Figure 12.21 BI-Based Inventory Comparison Report for Third-Party Logistics Inventory

12.4 Summary

The first part of this chapter covered the importance and means of monitoring and controlling data flow across interfaces with third-party logistics. This is of utmost importance for the success of any third-party logistics integration scenario to have seamless data flow to and from third-party logistics and other relevant legacy systems to an SAP S/4HANA system. Then, we focused on relevant SAP Fiori-based inventory monitoring and reporting tools and also looked at a few custom inventory comparison reports for third-party logistics stock comparison with the company's book inventory. The goals of inventory analysis include lowering costs, managing cash flow, and ensuring goods availability per demand of the customers.

The Authors

Gautam Bhattacharya is an SAP solution lead in the supply chain management space with specializations in sourcing and procurement, logistics, and Business Rule Framework plus. He has more than 21 years of experience related to the design, development, and implementation of materials management, quality management, supplier relationship management, business rule automation, and SAP S/4HANA legacy integration. His recent engagements include overseeing design and implementation of end-to-end global intercompany processes and 3PL integration. As a postgraduate in engineering, Gautam started his career in domain functions of metals and welding industries. He holds certifications in materials management, SAP Supplier Relationship Management, SAP S/4HANA Sourcing and Procurement, and SAP Ariba Procurement.

Mehfuze Ali Molla is a senior consultant at IBM. He has more than 15 years of SAP-related experience with ABAP and Basis, covering all project stages including requirements determination, gap analysis, business process reengineering, issue resolution, configuration, custom code specifications, testing, training, go-live assistance, and post implementation support. Mehfuze has hands-on experience in full lifecycle implementations (a number of which have been for SAP S/4HANA), support projects, and technical upgrades.

Index

U

V

W

X

Interested in reading more?

Please visit our website for all new book and e-book releases from SAP PRESS.

www.sap-press.com